STEEL DESIGNERS HANDBOOK

BRANKO E. GORENC *(1928–2011)* was a Fellow of The Institution of Engineers, Australia, and held a degree in Civil Engineering from the University of Zagreb, Croatia. He practised in the field of structural steel design for over five decades, specialising in the areas of conceptual framing design and analysis, member and connection design. He led teams of designers on projects around the world, including on such notable structures as sports facilities and stadiums, wide-bodied aircraft hangars, airport terminals and steel-framed buildings for commerce and industry.

RON TINYOU holds the Degree of Bachelor in Engineering from The University of Sydney. He is a member of The Institution of Engineers, Australia and a life member of the American Society of Civil Engineers (ASCE). Ron practised mainly in structural engineering over a wide range of industrial and hydraulic structures. Subsequently he was appointed Senior Head Teacher at the Sydney Institute of Technology, teaching structural engineering and lecturer at the University of Technology, Sydney, specialising in steel structures.

ARUN A. SYAM holds Bachelor and Masters degrees in engineering from The University of Sydney, is a Corporate Member of The Institution of Engineers, Australia, and has a Certificate in Arc Welding. Following his studies he was employed as a Structural Design Engineer with a major fabricator, several consulting engineering firms and has held all senior technical positions with the Australian Institute of Steel Construction (now Australian Steel Institute). He has had significant involvement with steel design and fabrication, Standards Australia, steel industry associations, national steel issues, industry publications and software, welder certification. He lectures on steelwork around the world. Arun has authored and edited numerous well-known steelwork publications and journals and is currently the Tubular Development Manager at OneSteel.

8th EDITION

STEEL DESIGNERS' HANDBOOK

BRANKO E. GORENC, RON TINYOU & ARUN A. SYAM

A UNSW Press book

Published by
NewSouth Publishing
University of New South Wales Press Ltd
University of New South Wales
Sydney NSW 2052
AUSTRALIA
newsouthpublishing.com

First published 1970
Second edition 1973
Third edition 1976
Fourth edition 1981
Fifth edition 1984, reprinted with minor revisions 1989
Sixth edition 1996, reprinted 2001, 2004
Seventh edition 2005
Eighth edition 2012

National Library of Australia Cataloguing-in-Publication entry
Author: Gorenc, B. E. (Branko Edward)
Title: Steel designers' handbook/Branko E. Gorenc, Ron Tinyou, Arun A. Syam.
Edition: 8th ed.
ISBN: 9781742233413 (pbk.)
9781742245942 (ePDF)
Notes: Previous ed.: 2005.
Subjects: Steel, Structural – Handbooks, manuals, etc.
Civil engineering – Handbooks, manuals, etc.
Structural design – Handbooks, manuals, etc.
Other Authors/Contributors: Tinyou, R. (Ronald). Syam, Arun.
Dewey Number: 624.1821

Design, typesetting and diagrams DiZign Pty Ltd

Cover photographs Di Quick; full details on p 464

Disclaimer

All reasonable care was taken to ensure the accuracy and correct interpretation of the provisions of the relevant standards and the material presented in this publication. To the extent permitted by law, the authors, editors and publishers of this publication:

(a) will not be held liable in any way, and

(b) expressly disclaim any liability or responsibility

for any loss, damage, costs or expenses incurred in connection with this publication by any person, whether that person is the purchaser of this publication or not. Without limitations this includes loss, damage, costs and expenses incurred if any person wholly or partially relies on any part of this publication, and loss, damage, costs and expenses incurred as a result of negligence of the authors, editors and publishers.

Warning

This publication is not intended to be used without reference to, or a working knowledge of, the appropriate current Australian and Australian/New Zealand Standards, and should not be used by persons without thorough professional training in the specialised fields covered herein or persons under supervisors lacking this training.

Contents

Preface

Since the release of the seventh edition of the *Steel Designers' Handbook* in 2005, there have been significant changes to Australian and other national Standards, key design references and aids as well as noteworthy advances in the research and development of steel structures. These changes precipitated the need for this eighth edition of the Handbook.

One of the more significant changes to Standards included a major amendment to the 1998 version of AS 4100 *Steel Structures*. This amendment was released on 29 February 2012. Amongst other aspects, the key changes due to the amendment to AS 4100 are summarised in Section 1.14 and other related parts of the Handbook and, with grateful acknowledgement of Standards Australia, are listed in Appendix D of the Handbook. Additionally, there have been significant changes to design aids and models for structural steel connections.

The use of this Handbook is not intended to be a stand-alone steel design code. The authors recommend that readers take this text as a map and guide to AS 4100 and related publications. Readers should also refer to AS 4100, its Commentary, related Standards and other relevant publications, to gain a suitable appreciation of current structural steel design requirements.

This text is intended to cover enough material to enable the design of everyday structural steel frames, members and connections. Appendix A includes a thoroughly revised and expanded bibliography, and a substantial updated list of related Standards and references. The Handbook, together with the information included in the Appendices, should provide the reader with a solid background to a variety of structural steel design tasks.

As noted in previous editions, the following key points should be considered when using the Handbook:

- Significant reference is made to other key design aids and publications (e.g. Australian Steel Institute (ASI) Design Capacity Tables, etc.) for quick design calculations.
- Tips, shortcuts and design/fabrication economics are presented where possible.
- Useful links and references are provided to other Standards, websites, manufacturers and suppliers in the steel construction and related industries (no other similar hard-bound publication provides this consolidated information).
- As is normal practice, and in line with the typical precision of data used in structural design, all calculations and worked examples are generally done to *three* (3) significant figures—hence there may be some very minor numerical rounding when comparing calculated or listed values with those in other references.
- Linear interpolation of tables may generally be undertaken.

- The worked examples are for illustrative purposes and consequently some may depart from actual detail practice (e.g. bolt threads excluded from the shear plane, etc.).

It is with deep regret that we note the passing of the Handbook's first author Branko Gorenc in June 2011. Branko's legacy through his visionary structural steel designs and publications are well known and he will be truly missed.

Lastly, the authors gratefully acknowledge the support and assistance provided by the Handbook's publisher (UNSW Press), together with the valued inputs provided by Paul Uno (Cement and Concrete Services), Peter Russell (Sitzler), Tim Hogan (steel consultant), Australian Steel Institute (Margrit Colenbrander, Dr Peter Key), OneSteel and others who were kind enough to offer feedback. As always, the authors welcome further comments, observations and questions from readers in the interests of ongoing improvement.

May your steel designs be fruitful ones!

R. Tinyou and Arun A. Syam

Vale
Branko Edward Gorenc

(20 July 1928–4 June 2011)

Branko Gorenc was arguably one of Australia's most pre-eminent structural engineers. His background included working as a young engineer in Croatia on structures such as bridges, buildings and various infrastructure projects. He arrived in Australia with his wife Pauline in 1958, then worked for the Latrobe Valley Water & Sewerage Board (1958-62 in Traralgon, Victoria), then in Sydney at Gutteridge Haskins and Davey (GHD) (1962-65) and Crooks Michell & Stewart (1965–68). This was followed with work in the mining industry with Nabalco, at Gove (now Nhulunbuy) in the Northern Territory. Branko's time there was most fulfilling as he was involved in developing the region from scratch over many years (1968–73).

Branko then broadened his background in industrial and utility structures to include significant architect-designed building projects. In 1973 he joined the prominent engineering consultancy, Macdonald Wagner & Priddle (which later became Macdonald Wagner, Connell Wagner and, currently, Aurecon). Branko retired from Connell Wagner several times, his last 'retirement' occurring in the early 2000s. Branko worked on various engineering undertakings for Connell Wagner during his 'retirement' and he thought the lead-up to Sydney 2000 Olympics and thereafter was the most creative part of his life.

Branko's extensive knowledge could be applied to many different types of projects. Many of his peers had the view that he was a creative designer in the European mould—that is, he understood engineering, architecture and art. It has been said his style was reminiscent of that of great engineering designers such as Luigi Nervi of Italy and Santiago Calatrava of Spain.

Branko also had the ability to communicate through images. He was a self-taught perspective artist and this endeared him to architects. He was also a keen model maker and used models extensively to develop conceptual and detail designs as well as possible failure mechanisms.

Many considered Branko an 'engineer's engineer'—the engineer to whom other engineers go when they have problems they cannot solve. He was the optimal hybrid of academic and practitioner. These attributes made Branko an effective communicator as he made complex engineering issues readily understandable. The ongoing popularity of this Handbook since 1970 is a tribute to his skills.

The projects Branko was involved with in over five decades are too numerous to mention here. Some of the more iconic (and highly awarded) ones include:

- Sydney Olympic Park, Homebush: Athletic Stadium, Aquatic Centre, Hockey Stadium
- Sydney Airport: Qantas Domestic Terminal (T3), Virgin Domestic Terminal (T2—formerly the Ansett terminal)
- Wembley Stadium, England.

Branko was also very well known for his work on various committees and professional bodies including the Standards Australia code committees for steelwork, cranes and silos. In addition to committee work, Branko was well known for his technical publications (including the *Steel Designers' Handbook* which he initiated himself in 1970 and *Design of Crane Runway Girders*). He also produced numerous technical papers for engineering journals and conferences.

Besides engineering, Branko's interests included art, languages, science, travel and the spiritual life. He was also a humanitarian involved in overseas aid.

Branko passed away on 4 June 2011 and leaves behind his wife Pauline, children Francis and Belinda and grand-children Stephen, Shaun and Nicholas.

At Branko's funeral on 9 June 2011, respected Principal Structural Engineer at Aurecon and close colleague Boyne Schmidt poignantly and accurately reflected:

> *Branko: Structural Engineering was your world—the rest of us were just visiting.*
> *You are highly respected and much loved. May you rest in peace!*

Arun A. Syam

Co-Author, *Steel Designers' Handbook*, 6th, 7th and 8th Editions

chapter 1

Introduction

1.1 Developments in steel structures

Early steel structures in bridges, industrial buildings, sports stadia and exhibition buildings were fully exposed. At the time no special consideration had been given to aesthetics. The form of a structure was driven by its function. Riveted connections had a certain appeal without any further treatment. As the use of steel spread into commercial, institutional and residential buildings with their traditional masonry facades, the steel structure as such was no longer a principal modelling element and became utilitarian, merely a framework of beams and columns.

The role of steel started to change with the trend towards lighter envelopes, larger spans, and the growing number of sports and civic facilities in which structural steel had an undisputed advantage. Outstanding lightweight structures have been constructed in the past four decades. Structural framing exposed to full view has taken many forms, including space frames, barrel vaults, cable stayed and cable net roofs. The trend continues unabated with increasing boldness and innovation by designers.

The high visibility of structural framing has brought about a need for more aesthetically pleasing connections, where the architect might outline a family of connection types. In this instance, standardisation on a project-to-project basis is preferred to universally applied standard connections. Structural designers and drafters have been under pressure to re-examine their connection design. Pin joints often replace bolted connections, simply to avoid association with industrial-type joints. Increasingly, 3D computer modelling and scale models are used for better visualisation. Well-designed connections need not be more expensive because fabrication tools have become more versatile. Even so, it is necessary to keep costs down through simplicity of detailing and the maximum possible repetition.

In many other situations, structural steelwork is also used in 'non-visible' (e.g. behind finishes), industrial and resource applications. In these instances, where possible, general standardisation of connections across all projects is worthwhile. This makes structural steel framing more attractive in terms of costs, reduced fabrication and erection effort, without any reduction in quality and engineering efficiency.

Therefore, constructing in steel provides the designer with a panoply of solutions from which to innovate. In Australia and throughout the world there are fine examples of structural steel being used in many outstanding commercial buildings as well as in large span structures, reticulated domes and barrel vaults, space truss roofs, cable nets and other lightweight structures. A way for the designer to partake in this exciting development is to visit a good library of architecture and engineering technology or to contact resource centres within the relevant industry associations (e.g. Australian Steel Institute (ASI), Welding Technology Institute of Australia (WTIA), Galvanizers Association of Australia (GAA), Heavy Engineering Research Association (HERA) in New Zealand and Steel Construction New Zealand (SCNZ).

To be successful in the current creative environment, structural steel designers need to shed many of the old precepts and acquire new skills. One essential element is a basic understanding of the behaviour of structural steel and the use of a design or modelling methodology that adequately reflects this behaviour while emphasising efficiency and economy. Such a methodology is embodied in the limit states design philosophy incorporated in key design Standards, such as AS 4100 (Steel Structures) and NZS 3404 (Steel Structures Standard). The mastery of such methods is an ongoing task, which constantly expands as one delves deeper into the subject.

1.2 Engineering design process

The structural engineer's ('designer') involvement with a project starts with the design brief, setting out the basic project criteria. The designer's core task is to conceive the structure in accordance with the design brief, relevant Standards, statutory requirements and other constraints. Finally, the designer must verify that the structure will perform adequately during its design life.

It has been said that the purpose of structural design is to build a building, bridge or load-bearing structure. In this context the designer will inevitably become involved in the project management of the overall design and construction process. From a structural engineering perspective, the overall design and construction process can be categorised sequentially as follows.

(a) Investigation phase:
- site inspection
- geotechnical investigation
- study of functional layout
- research of requirements of the statutory authorities
- determination of loads/actions arising from building function and environment
- study of similar building designs.

(b) Conceptual design phase:
- generation of structural form and layout
- selecting materials of construction
- constructability studies
- budget costing of the structural options
- evaluation of options and final selection.

(c) Preliminary design phase:

- estimation of design actions and combinations of actions
- identification of all solution constraints
- generation of several framing systems
- preliminary analysis of structural framework
- preliminary sizing of members and connections
- preliminary cost estimate
- quality assessment of the design solution
- client's review of the preliminary design
- reworking of the design in line with the review.

(d) Final design phase:
- refining the load/action estimates
- final structural analysis
- determination of member types and sizes
- detail design of connections
- study of the sequence of construction
- quality review of the final design (QA)
- cost estimate
- client's review of the design and costing
- modification of the design to meet client's requirements.

(e) Documentation phase:
- preparation of drawings for tendering
- writing the specifications
- preparing bills of quantities
- final structural cost estimate
- preparing a technical description of the structure
- quality review of the tender documentation (QA)
- client's approval of the tender documentation
- calling tenders.

(f) Tendering phase:
- preparing the construction issue of drawings
- assisting the client with queries during tendering
- assisting in tender evaluation and award of contract.

(g) Construction phase, when included in the design commission (optional):
- approval of contractor's shop drawings
- carrying out periodical inspections
- reviewing/issuing of test certificates and inspection
- final inspection and certification of the structure
- final report.

The process of development and selection of the structural framing scheme can be assisted by studying solutions and cost data of similar existing structures. To arrive at new and imaginative solutions, the designer will often study other existing building structures and then generate new solutions for the particular project being designed.

Much has been written on design philosophy, innovation and project management, and readers should consult the literature on the subject. This Handbook's main emphasis is on determination of action (i.e. load) effects and the design of frames, members and connection details for low-rise steel structures. The theory of structural mechanics does not form part of the Handbook's scope and the reader should consult other texts on the topic.

1.3 Standards and codes of practice

The designer has only limited freedom in determining nominal imposed actions, setting load factors and serviceability limits. This information is normally sourced from the appropriate statutory or regulatory authority e.g. National Construction Code series (NCC[2011]), previously the Building Code of Australia (BCA), by the Australian Building Codes Board (ABCB), which is gazetted into State legislation and may in turn refer to relevant 'deemed to comply' Standards (AS 4100 etc.).

Design Standards have a regulatory aspect, and set down the minimum criteria of structural adequacy. This can be viewed as the public safety aspect of the Standards. Additionally, Standards provide acceptable methods of determining actions (e.g. forces), methods of carrying out structural analyses, and sizing of members and connections. This gives the design community a means of achieving uniformity and the ability to carry out effective quality-assurance procedures. Standards also cover the materials and workmanship requirements of the structure (quality, testing and tolerances), which also impact on the design provisions. The degree of safety required is a matter of statutory policy of the relevant building authorities and is closely related to public attitudes about the risk of failure.

A list of some of the relevant Standards and their 'fitness' aspects is given in Table 1.1.

Table 1.1 List of relevant steelwork Standards

Standard	Fitness aspect—design
(a) AS Loading Standards	
AS 1170, Part 1	Dead and live loads and load combinations (not referenced in NCC[2011])
AS 1170, Part 2	Wind loads (not referenced in NCC[2011])
AS 1170, Part 3	Snow loads (not referenced in NCC[2011])
AS 1170, Part 4	Earthquake loads [2007]
(b) AS/NZS structural design actions replaces AS 1170 referred to in:	
AS/NZS 1170, Part 0	General principles
AS/NZS 1170, Part 1	Permanent, imposed and other actions
AS/NZS 1170, Part 2	Wind actions
AS/NZS 1170, Part 3	Snow and ice actions
(c) Other standards:	
AS 2327	Composite construction
AS 4100	Steel Structures. Includes resistance factors, materials, methods of analysis, strength of members and connections, deflection control, fatigue, durability, fire resistance.
AS/NZS 4600	Cold-formed steel structures

Note: At the time of writing this 8th edition, the National Construction Code series (NCC[2011]) replaced the Building Code of Australia (BCA). Unless noted otherwise, the above Standards are specifically referred to in the NCC[2011].

continued

Table 1.1 List of relevant steelwork Standards (continued)

Standard	Fitness aspect—design/material quality
AS 1111	ISO metric hexagon bolts (Commercial bolts)
AS/NZS 1163	Cold-formed structural steel hollow sections
AS/NZS 1252	High-strength bolts, nuts and washers
AS/NZS 1554, Parts 1–7	Welding code
AS/NZS 3678	Hot-rolled plates
AS/NZS 3679, Part 1	Hot-rolled bars and sections
AS/NZS 3679, Part 2	Welded I sections

A more exhaustive listing of Australian and other standards of direct interest to the steel designer is given in Appendix A.

This edition of the Handbook is generally intended to be used with AS 4100:1998 Steel Structures and its subsequent amendment, which is in limit states format. Commentary is also given on related loading/action Standards.

1.4 General structural design principles

For the purposes of this text, the term 'structure' includes structural members, connections, fasteners and frames that act together in resisting imposed actions (loads, pressures, displacements, strains, etc.). The essential objective of structural design is to define a structure capable of remaining fit for the intended use throughout its design life without the need for costly maintenance. To be fit for its intended use the structure must remain stable, safe and serviceable under all actions and/or combinations of actions that can reasonably be expected during its service life, or more precisely its *intended* design life.

Often the use or function of a structure will change. When this occurs it is the duty of the owner of the building to arrange for the structure to be checked for adequacy under the new imposed actions and/or structural alterations.

Besides the essential objectives of adequate strength and stability, the designer must consider the various requirements of adequacy in the design of the structure. Of particular importance is serviceability: that is, its ability to fulfil the function for which that structure was intended. These additional criteria of adequacy include deflection limits, sway limits as well as vibration criteria.

1.5 Limit states design method

The 'limit state of a structure' is a term that describes the state of a loaded structure on the verge of becoming unfit for use. This may occur as a result of failure of one or more members, overturning instability, excessive deformations, or the structure in any way ceasing to fulfil the purpose for which it was intended. In practice it is rarely possible to determine the exact point at which a limit state would occur. In a research laboratory the chance of determining the limit state would be very good. The designer can deal only with the notion of *nominal* limit states, as determined by the application of the relevant limit states Standards.

The first step in verifying the limit state capacity of a structure is to determine the most adverse combination of actions that may occur in the lifetime of the structure. The usual way of determining the actions is to comply with the requirements of the relevant loading Standard (e.g. AS 1170.X or AS/NZS 1170.X) and/or other relevant specifications. In special situations the designer could arrange for a statistical/probabilistic analysis of actions to be carried out by an accredited research organisation. This would entail determining actions and their combinations, such that the structure will have an acceptably low risk of failure or unserviceability. An example of such a special situation might be a large, complex roof structure for which the wind actions are not given in the wind loading code.

In addition to loads, the structure may be subjected to such actions as strains due to differential temperature, shrinkage strains from reinforced concrete elements if incorporated, weld shrinkage strains, and deformations induced by differential settlement of foundations.

With actions determined, the next stage in the design procedure is to determine the internal action effects in the structure. In the vocabulary of the limit states design method, the term 'design action effect' means internal forces determined by analysis: axial forces, bending moments or shears. It is up to the designer to select the most appropriate method of structural analysis (see Chapter 4).

With regard to the strength limit state, the following inequality must be satisfied:

(Design action effect) ≤ (Design capacity or resistance)

or, symbolically,

$$E_d \leqslant \phi R$$

where the design action effect, E_d, represents an internal action (axial force, shear force, bending moment) which is obtained by analysis using factored combinations of actions *G*, *Q* and *W*. In other words, the design action effect E_d is a function of the applied design actions and the structural framing characteristics (geometry, stiffness, linkage).

In calculating design action effects, actions are factored and combined in accordance with the loading code. Action combination factors vary with the type of action, combination of actions and the relevant limit state, with the typical values ranging between 0.4 and 1.5 as detailed in Section 1.6 below.

The capacity reduction factors, ϕ, are intended to take account of variability in strength of material and constructional uncertainties. Different capacity reduction factors are used with different structural element types. Typical values are between 0.60 and 0.90 for the strength limit state.

The statistical/probabilistic relationship between action effects and capacity are illustrated in Figure 1.1. The interplay between the design action effect and design capacity is illustrated by the separation (or gap) between the probability curves for design action effects and design capacity.

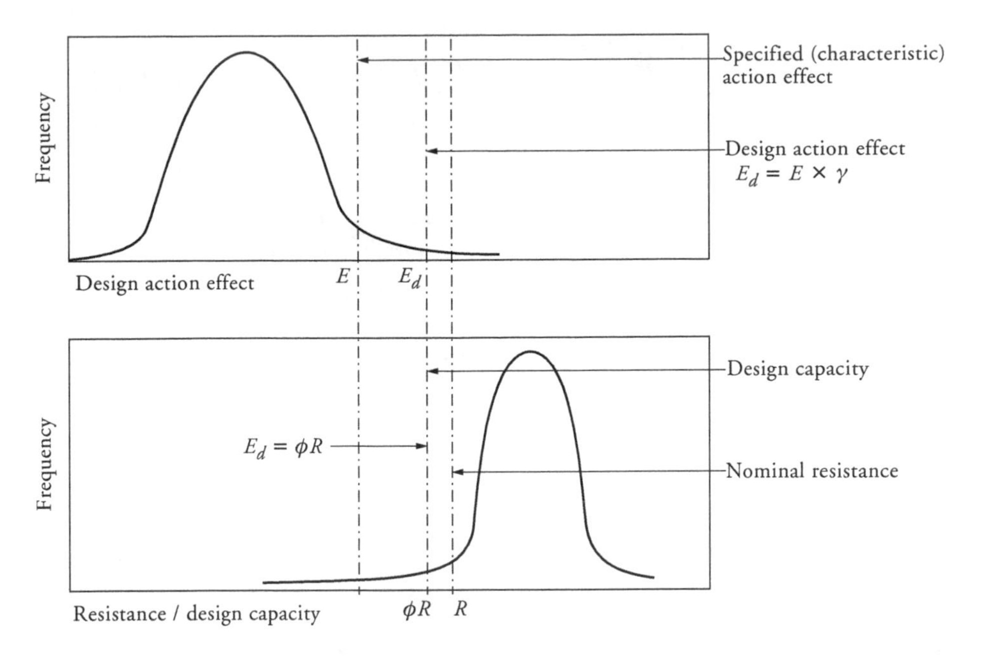

Figure 1.1 Relationship between action effects and resistance/capacity

The limit states method entails several limit states. This is illustrated by Table 1.2. The procedure described above applies to all limit states.

Table 1.2 Limit states for steel structures

Limit state	Design aspect addressed
Strength	Resistance against yielding, fracture or collapse under predominantly 'static' actions
Stability	Resistance against overturning, sliding
Serviceability	Limit of satisfactory service performance (deflections, sway, vibration, etc)
Fatigue	Resistance against premature fatigue-induced fracture
Fire	Resistance against premature collapse in a fire event
Earthquake	Endurance against low-cycle, high-strain seismic loads
Brittle fracture	Resistence against fracture at temperatures below notch ductile transition
Durability	Resistance against loss of material by corrosion or abrasion

The term 'static actions', in Table 1.2 applies to actions that, although variable in time/space, do not repeat more than, say, 20 000 times during the design life of the structure. Wind action on a building structure is regarded as quasi-static. However, wind action on a slender mast, chimney or other wind-sensitive structures is treated as a dynamic-type action. Dynamic action is often induced by machines having rotating or translating parts. A glossary of the terms used in the limit states method is given in Section 1.13.

1.6 Combination of actions

As noted in Section 1.5, in order to determine the relevant design action effects (e.g. the maximum moment on a beam), the critical design actions (e.g. externally imposed loads) must initially be assessed.

When acting in a singular manner, design actions such as permanent (G), imposed (Q) and wind (W) loads are generally variable in magnitude. The variability is more pronounced for the Q and W loads, and they will consequently have a higher load factor (as noted in Section 1.5).

Realistically, design actions generally do not act in a singular manner but in combination with each other. This combination of actions brings on another dimension of variability. The variability is allowed for in a combination of actions (i.e. load combination) with different and/or additional factors being applied to the relevant individual nominal action. A case in point is, say, peak wind and earthquake actions, which may have high load factors when acting individually but, as such events are rare in combination, the combination factors would be very low—if considered at all. In many instances, a combination of actions is considered specifically in AS/NZS 1170, Part 0 which supersedes AS 1170 Part 1. Table 1.3 illustrates some examples of action factors and their general combinations.

Table 1.3 Examples of some typical action factors and combinations (AS/NZS 1170.0 provisions)

Combination no.	Action combination factors for strength limit state
1	$1.35G$
2	$1.2G + 1.5Q$
3	$1.2G + 1.5\psi_l Q$
4	$1.2G + \psi_c Q + W_u$
Other	See AS/NZS 1170.0

where

G = permanent actions (self-weight or 'dead' action)

Q = imposed actions ('live' action due to occupancy and use)

W_u = wind actions (ultimate)

ψ_l = load factor for determining quasi-permanent values of long-term actions (varies between 0.4 and 1.0)

ψ_c = combination factor for imposed actions (varies between 0.4 and 1.2).

Readers of previous editions of this Handbook and those knowledgeable of changes published by Standards Australia in 2002 will note that there has been a change in terminology due to the significant revision of the AS/NZS 1170 suite of 'loading' Standards. Such changes have seen the general term 'load' replaced by 'action', 'dead' by 'permanent', 'live' by 'imposed', to name a few. Specific changes to load factors are also noted—e.g. the change in the load factor for permanent actions acting either singly or in combination.

Further aspects of design actions and their combinations are considered in Chapter 3.

1.7 Strength limit state

The object of design for strength is to ensure that the structure as a whole, including all of its members and connections, have design capacities in excess of their respective design action effects. The basic strength design criterion is that the structure must be designed so that its probability of failure in service is as near to zero as practicable. In AS/NZS 1170.0 this is taken as five percentile values in a probability distribution curve. It should be noted that zero probability of failure is an ideal that could not be achieved in real life. Engineering design aims to reduce the failure probability to a figure less than that generally regarded as acceptable to the public at large (often about 1 in 100 000 per year, per structure).

The basic inequality for Strength Limit State design to AS 4100 is:

(Design action effect) $\leqslant \phi$ (Nominal capacity)

For example:

(Design bending moment) $\leqslant \phi$ (Nominal bending capacity)

(Design axial compression force) $\leqslant \phi$ (Nominal compression capacity)

The main features of Strength Limit State design to AS 4100 are as follows:

1. The structure is deemed to be of adequate strength if it can be shown that it can resist the least favourable design action combination without exceeding the limit state of strength.
2. Load factors are applied to the specified actions sometimes termed 'characteristic' actions. The load factors range from 0.40 to 1.50 for the strength limit state (refer Chapter 3).
3. The design action effects (bending moments, axial, and shear forces) are computed from 'factored' loads and their combinations.
4. The computed member and section capacities (ultimate resistances) are factored down using capacity reduction factors.
5. The capacity reduction factors for steel structures range from 0.6 to 0.9, depending on the type of the member or connection and the nature of forces.

Table 1.4 gives the values of the capacity reduction factor ϕ in AS 4100.

Table 1.4 Values of capacity reduction factor ϕ in AS 4100

Element	ϕ
Steel member as a whole	0.90
Connection component (excluding bolts, pins or welds)	0.90
Bolted or pin connection	0.80
Ply in bearing	0.90
Complete penetration butt weld	0.90 (0.60)
Longitudinal fillet weld in RHS, t <3 mm	0.70
Other welds	0.80 (0.60)

Note: Figures in brackets apply to category GP welds.

1.8 Serviceability limit state

The term 'serviceability' applies to the fitness of the structure to serve the purpose for which it has been designed. The actions used in verifying the serviceability limit state are combined using load factors of 1.0 (e.g. 1.0 G + 1.0 $\psi_s Q$). Some of the serviceability limit states include:

- deflections, sways and slopes
- vibration affecting human comfort or mechanical plant performance
- loss of material due to corrosion or abrasion
- bolt slip limit state.

Deflections, sway and slopes need to be limited to maintain the proper functioning of the building and to avoid public concern about its appearance, safety or comfort. AS 4100 gives only the most essential limits on deflections, leaving it to the designer to investigate whether the serviceability requirements are satisfied (Clause 3.5.3). Appendix B of AS 4100 gives a short list of vertical deflection limits, reproduced below in Table 1.5(a) and (b).

Table 1.5(a) Deflection limit factor C_d in $\Delta \leq L/C_d$ (from AS 4100)

Beam type	Loading	Coefficient C_d	
		Beams	Cantilevers
Beams supporting masonry			
(i) No pre-camber	$G_1 + Q$	1000	500
(ii) With pre-camber	$G_1 + Q$	500	250
All beams	$G + C$	250	125

Notes: $(G_1 + Q)$ means actions applied by the wall or partition and subsequently applied imposed actions.
$(G + C)_{LF}$ means the least favourable combination of permanent actions (G) and non-permanent actions (C—e.g. imposed, wind, etc.) which may or may not be in the direction of permanent actions.
Δ = beam/cantilever deflection.
L = span.

Table 1.5(b) Limits of horizontal deflections

Description of building	Limit
Clad in metal sheeting, no internal partitions, no gantry cranes	$H/150$
Masonry walls supported by structure	$H/240$

Note: For buildings with gantry cranes, the sway and deflection limits of AS 1418.18 apply. H = column height. The above horizontal deflection limits are applicable to the eaves level of adjacent frames in industrial buildings.

A comprehensive tabulation of deflection and sway limits for building elements (structural and non-structural) can be found in Appendix C of AS/NZS 1170.0. Woolcock et al. [2011] also provides some authoritative advice on the topic.

1.9 Other limit states

1.9.1 Stability limit state

This limit state safeguards against loss of equilibrium of the structure or its parts due to sliding, uplift or overturning. This is covered in detail in AS/NZS 1170.0.

1.9.2 Fatigue limit state

Design against premature failure from fatigue damage is required where the number and severity of fluctuating loads is above the threshold of fatigue damage.

1.9.3 Fire limit state

The behaviour of the structure in the event of fire is an important design consideration. AS 4100 sets the principles of fire engineering for the common building element types and covers bare steel and most passive fire protection systems, except concrete encasement and filling.

1.9.4 Earthquake limit state

A separate section has been included in AS 4100 to cover special provisions for structures subject to earthquake forces. In particular, the Standard specifies the design features necessary to achieve ductile behaviour. Further useful guidance can also be found in AS 1170.4 and NZS 3404.

1.9.5 Brittle fracture

Although the risk of this type of failure is low, design against brittle fracture under certain conditions must be considered. Section 10 of AS 4100 and Section 2.7 of this Handbook give guidance on design against brittle fracture.

1.10 Other features of AS 4100

The requirements for high-strength bolting are included in AS 4100. Design of welded joints is also fully specified in AS 4100 leaving only the clauses on workmanship, materials, qualification procedures and weld defect tolerances in the welding Standard AS/NZS 1554. AS 4100 also incorporates requirements for Fabrication, Erection, Modification of Existing Structures and Testing of Structures.

1.11 Criteria for economical design and detailing

The owner's 'bias' towards minimal initial cost for the structure, as well as low ongoing maintenance cost, must be tempered by the edicts of public safety, utility and durability. The designer is constrained to work within the industry norms and limits imposed by the statutory regulations and requirements of design and material Standards.

The choice of an appropriate structural system is based on experience. While it is possible to carry out optimisation analyses to arrive at the least-weight structural framing, it is rarely possible to arrive, purely by analysis, at the most appropriate solution for a particular design situation.

In the design of a steel structure the achievement of a minimum weight has always been one of the means of achieving economy. Designing for minimum weight should produce the minimum cost in material but it does not necessarily guarantee the lowest total cost, because it does not take into account the cost of labour and other cost sources. A more comprehensive method of achieving economical design is the optimum cost analysis. Additionally, costs associated with coating systems (e.g. for corrosion or fire protection) can dramatically increase the first cost of structural steelwork. The minimum weight design does have its virtues in structures that are sensitive to self-load, e.g. long-span roofs. The dead weight of such a roof structure needs to be minimised, as it contributes the significant part in the load equation.

Design for optimum cost, while not covered by AS 4100, is part of good engineering design. The term 'optimal cost' applies to the total cost of materials, labour for fabricating the structural elements (members, details, end connections), coating and erection. Erection cost is an important consideration, and advice should be sought from a suitably experienced contractor whenever novel frame solutions are being considered.

A rational approach to assessing the costing of steelwork has been developed (Watson et al. [1996]). This costing method does not assess total fabricated steelwork costs on a 'smeared' $/tonne basis but develops accurate costs from various relevant parameters, which include material supply ($/m), fabrication ($/hour), application of coatings ($/sq.m) and erection ($/lift). Though quite detailed, the rational costing method requires current pricing information (which may also vary on a regional basis), and would seem more suitable for larger projects. To use the method for every small to medium-sized project may not be justifiable on time and fees considerations.

An alternative course of action would be to use the rational costing method on an initial basis to determine the relative economics of joints and other systems and to utilise these outcomes over many projects—much like the practice of standardising connections within a design office. The method is also very useful for quantifying costs in variation assessments. Watson et al. [1996] should be consulted for a detailed understanding of the rational costing method.

Since design costs are part of the total project cost, economy of design effort is also important. There are several ways in which the design process can be reduced in cost, and these are the use of computers for analysis and documentation, use of shorthand methods for sizing the members and connections, and use of standard job specifications. As the design process is almost always a step-by-step process, it is helpful if the initial steps are carried out using approximate analyses and shorthand routines whilst reserving the full treatment for the final design phase.

As always, specifications and drawings are the documents most directly responsible for achieving the planned result, and should be prepared with the utmost care. AS 4100 adds a few requirements on the contents of these documents. Clause 1.6 of AS 4100 stipulates the additional data to be shown on drawings and/or in specifications. These requirements should be studied, as their implementation may not be easy.

1.12 Design aids

Various design aids and computer software packages are available for rapid sizing and assessing the suitability of steel members, connections and other components. Additional publications that provide connection design models, background information and worked examples for all facets of the loading and design Standards are also available—e.g. ASI [2009a, 2004, 2009b], Bennetts et al. [1987, 1990], Bradford et al. [1997], Hancock [2007], Hogan [2011], Hogan & Munter [2007a-h], Hogan & Van der Kreek [2009a-e], Hogan & Syam [1997], OneSteel [2011, 2012a,b], Rakic [2008], Syam [1992], Syam & Chapman [1996], Thomas et al. [1992], Trahair et al. [1993c,d]. A useful summary of the more readily available design aids is given by Hogan [2009]. An excellent reference on the background to AS 4100 is provided by Trahair and Bradford [1998], with Woolcock et al. [2011] and ASI [2010] providing some very good practical guidance on the design of portal framed buildings.

1.13 Glossary of limit states design terms

Action A new term to represent external loads and other imposed effects. The word 'load' is used interchangeably with 'action' in this Handbook—see 'Nominal action'.

Action combination factor The factor applied to specified nominal actions within a combined actions equation.

Capacity A term to describe structural resistance—see 'Nominal capacity'.

Capacity reduction factor, ϕ The factor applied to the nominal computed member or connection capacity to safeguard against variations of material strength, behaviour and dimensions.

Design action (load) The product of (nominal action) × (load/combination factor).

Design action effect (load effect) Internal action such as axial load, bending moment or shear in a member, arising from the application of external actions. Design action effects are calculated from the design actions and appropriate analysis.

Design capacity The capacity obtained by multiplying the nominal capacity by the capacity reduction factor.

Load (action) factor The factor applied to a specified action to safeguard against load variations.

Nominal action Defined as the following acting on a structure: direct action (e.g. concentrated/distributed forces and moments) or indirect action (e.g. imposed or constrained structural deformations).

Nominal capacity The capacity of a member, section or connection at the strength limit state, e.g. axial load at the onset of yielding in a stub column.

Specified action The action of the intensity specified in a loading Standard.

1.14 Recent code changes for this 8th edition

Since the release of the seventh (2005) edition of this Handbook, there have been numerous changes in various Standards, codes and other supporting technical/R&D literature on steel construction. The more substantial changes have included:

(a) Significant amendment to AS 4100

During 2011, an amendment to AS 4100 was drafted by Standards Australia Committee BD-001. This document was subsequently issued for Public Comment (DR AS 4100 AMD 1 and then further considered and balloted by Committee BD-001. The ballot process, via all positive votes, then confirmed the amendment should be enacted. Amendment No.1 to AS 4100-1998 ("AS 4100 AMD 1" as listed in Appendix A.2) was published on 29 February 2012 and, through the courtesy of Standards Australia with grateful acknowledgement by the Authors & Publishers, is also reprinted in Appendix D.

The major amendments to AS 4100 include:

- The inclusion of higher strength steel up to a design yield stress 690 MPa. This provision permits the use of quenched and tempered type structural plate steels (compliant with AS 3597) and replaces the previous limit of 450 MPa which are typically available for structural steel hollow sections and mild steel plates.
- Changes to the Structural design actions Standards—the AS/NZS 1170 series of Standards—i.e. general principles (including the new provisions on structural robustness—i.e. Section 6 of AS/NZS 1170.0); design permanent, imposed, other actions; Wind actions; Snow and ice actions; and Earthquake actions in Australia—see comments below.
- Changes to how AS 4100 defines and handles seismic design due to the splitting of the joint Australia/New Zealand design actions (e.g. loads) Standard. AS 4100 specifically recognises AS 1170.4 [2007] which was also revised—see comments below.
- Changes to "acceptance of steels" (Clause 2.2.2 of AS 4100) from a design perspective. These provisions were tightened up in line with changes to the steel material Standards (AS/NZS 1163, AS/NZS 3678, AS/NZS 3679.1 and AS/NZS 3679.2) subsequent to industry concern on global sourcing of (plain and fabricated) steel materials and steel quality.
- Further recognition of steel bridge design being done to AS 5100.
- A better reflection on steel product and grade availability.
- A change to the bearing buckling capacity of I- and C-section webs (Clause 5.13.4 of AS 4100) where only one flange is effectively restrained against lateral movement.
- For members subject to axial compression, the differentiation of members which are subject to torsional or flexural-torsional buckling and those members not subject to those buckling modes. Prior to the amendment, AS 4100 did not consider such instabilities as it assumed that all members within its scope are subject to only flexural (Euler-type) buckling when subject to axial compression.
- Connection "Components" (i.e. plate connection elements such as cleats, gusset plates, brackets, etc.) are to be designed to specific design capacity provisions within

AS 4100 as well for "Block Shear". Block shear was not considered specifically in AS 4100 (nor its predecessor Standards—unlike in North America etc.) and is to be assessed when the Connection Component is subject to a design shear force and/or design tension force.

- Clarification on Filler Plates used in connections (Clause 9.3.2.5 of AS 4100) whereby the bolt nominal shear capacity is reduced in a linear manner for Filler Plates which are above 6 mm and below 20 mm in thickness.
- In line with changes to AS/NZS 1554, new diagrams to calculate design throat thickness and weld size for fillet welds at angled connections.
- In line with changes to the welding consumable Standards and AS/NZS 1554, new weld metal consumable designations (now based on process), slight (conservative) changes to nominal tensile strength of weld metals (f_{uw}), the introduction of a new weld metal with f_{uw} = 550 MPa and the introduction of weld metals for joining quenched and tempered plate.
- Clarification and updates to some aspects of the Brittle Fracture provisions (Section 10 of AS 4100) and the inclusion of a new provision (Clause 10.4.3.4) on non-complying conditions.
- Minor update and clarification on the Fatigue provisions (Section 11 of AS 4100).
- Minor update on the Fire provisions (Section 12 of AS 4100).
- Complete revision to the Earthquake provisions (Section 13 of AS 4100) with respect to the Australian perspective of changes in the Structural Design Actions series of Standards—i.e. AS/NZS 1170 Parts 0, 1, 2, 3 and AS 1170.4. In 2007, the latter Standard had been significantly revised and AS 4100 was totally misaligned with it prior to AS 4100 AMD 1—see below for further comments.
- Minor update and clarification of bolt hole sizes (Clause 14.3.5.2 of AS 4100).
- Complete revision to the Suggested Horizontal Deflection Limits in Paragraph B2 of AS 4100 with due reference given to Appendix C of AS/NZS 1170.0.
- Updates on other revised reference Standards (e.g. for steel materials, bolting, welding, etc.)
- Some new notation and corrections where required.

The key items noted above are further considered in their relevant section of this Handbook.

(b) Changes to the Structural Design Actions suite of Standards

The "harmonisation" of Standards between Australia and New Zealand began during the 1990s and this was manifested in many "AS/NZS" documents which were embraced by both countries. Part of this included the harmonisation of the "Structural Design Actions" suite of Standards. In Australia they were previously known as the following:

"Before" harmonsiation—Australian Standards

- AS 1170 Minimum design loads on structures (known as the SAA Loading Code)
- AS 1170.1 Part 1: Dead and live loads and load combinations
- AS 1170.2 Part 2: Wind loads

- AS 1170.3 Part 3: Snow loads
- AS 1170.4 Part 4: Earthquake Loads (released in 1993).

Since 2002, the above Standards were withdrawn, then significantly revised and published as the following (with the previous versions being withdrawn):

"After" harmonisation—Australian (only) Standard

- AS 1170 Structural design actions
- AS 1170.4 Part 4: Earthquake actions in Australia (released in 2007).

"After" harmonisation—Joint Australian/New Zealand Standards

- AS/NZS 1170 Structural design actions
- AS/NZS 1170.0 Part 0: General principles
- AS/NZS 1170.1 Part 1: Permanent, imposed and other actions
- AS/NZS 1170.2 Part 2: Wind actions
- AS/NZS 1170.3 Part 3: Snow and ice actions.

"After" harmonisation—New Zealand (only) Standard

- NZS 1170 Structural design actions
- NZS 1170.5 Part 5: Earthquake actions—New Zealand.

The changes "before" and "after" harmonisation of the above Standards are somewhat self-evident with firstly the creation of Part 0 after harmonisation. Part 0 sets out the general principles for these suite of Standards and includes:

- Structural design procedures and general confirmation methods for both the Ultimate and Serviceability Limit States.
- Information on annual probability of exceedance. This is in Section 3 for NZ only due to their Standards framework and regulatory requirements and Appendix F (normative) for Australia for design events not given elsewhere.
- Combination of actions (i.e. "load"/action factors and their "load"/action combinations—though other combinations may be required which are not noted in the Standard) for ultimate limit states (stability, strength, peculiar actions such as snow, liquid pressure, rainwater ponding, ground water and earth pressure). This is considered to be the most important and commonly referred part (Section 4) of the Standard.
- A new section on Structural Robustness.
- Further guidelines for the Serviceability Limit States.
- Plus information on Special Studies and use of Test Data for design.

The remaining parts of the AS, AS/NZS and NZS 1170 Structural Design Actions suite of Standards set out to identify and quantify the necessary Nominal Actions prior to being factored and combined with other Design Actions in AS/NZS 1170.0.

Prior to 2005, parts of the Structural Design Actions changes as noted above were covered in the seventh edition of this Handbook. Since the publication of the seventh edition of this Handbook the following changes have occurred in the "1170" series of Standards:

- AS/NZS 1170.0 was issued with Amendment numbers 4 (April 2011) and 5 (September 2011).
- AS/NZS 1170.1 was issued with Amendment number 2 (January 2009).
- AS/NZS 1170.2 was revised as a second edition (March 2011).
- AS/NZS 1170.3 was issued with Amendment number 1 (April 2007).
- AS 1170.4 was revised as a second edition (October 2007). It preceded the original Standard AS 2121-1979 which was subsequently revised and redesignated as AS 1170.4-1993. As is noted by the prefix "AS" instead of "AS/NZS" for this Standard, a decision was made by Standards Australia and Standards New Zealand and their stakeholders not to harmonise this as a joint Standard. The reasons for this are many and varied and include differing loading and design philosophies, plus the significantly differing severity of earthquakes in the two countries. Consequently, there are separate Earthquake Actions Standards between Australia and New Zealand, whereas the other structural design actions Standards are common to both.

Commentaries from the Standards bodies are also available on structural design actions though, at the time of writing this Handbook, there are was no Commentary published to AS 1170.4 [2007].

(c) Updates and/or changes to supporting documents

Where possible, this edition also notes updates/changes to supporting documents which have been subject to recent changes in philosophies and/or research, testing, etc. in their respective areas. These have been included in the relevant Handbook section and/or noted in the Bibliography (Appendix A) of the Handbook. In particular, there has been a significant update in the connection design models/aids area.

1.15 Further reading

- Steel and the architect's perspective: Ogg [1987], though slightly dated this is an excellent high quality Australian publication on the topic. Additionally, AISC(US) [2002], Bruno et al. [2009] and Meyer-Boake [2011] are also very good references in this area.
- Use of Design Capacity Tables (DCTs): Syam & Hogan [1993].
- Minimum requirements for information on structural engineering drawings: Clause 1.6 of AS 4100, Syam [1995], Tilley [1998] and ASI [2012a].
- Information pertaining to steel detail (workshop) drawings: ASI [2003].
- Information technology in the steel industry: Burns [1999], Colombo [2006] and Munter [2006].

chapter 2

Material & Design Requirements

2.1 Steel products

The main elements in steel building construction consist of steel plates, standard sections and compound sections. An infinite variety of structural forms can be derived from these simple elements. Plates and standard sections are regarded as the *fundamental* elements (see Figure 2.1): that is, compound sections can be made from plates and sections. The designer has the freedom to compose special sections subject to dictates of economy. Some commonly used compound sections are shown in Figures 5.1 and 6.13.

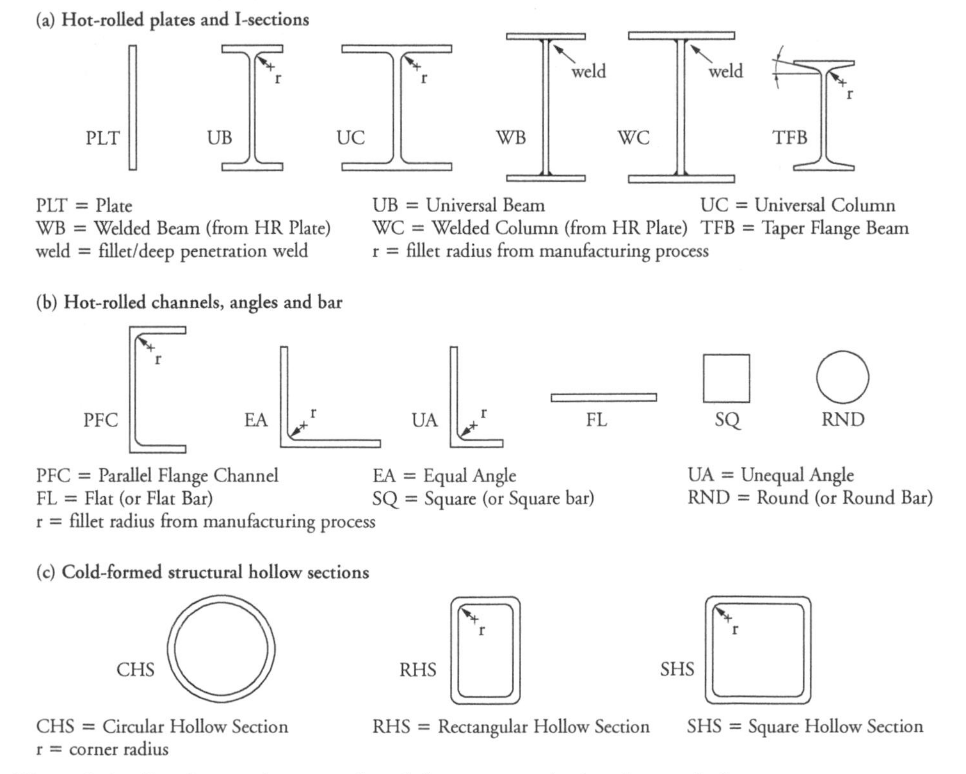

Figure 2.1 Fundamental structural steel elements: standard sections and plate

For reasons of economy, unless there are other specific criteria to be observed (e.g. minimum mass, headroom restrictions, etc.), the designer's best strategy is to choose standard sections in preference to compound sections. Typically, the more readily available standard sections are the following:

Hot-rolled sections: (OneSteel [2011]):
- universal beams (UB)
- universal columns (UC)
- parallel/taper flange channels (PFC/TFC)
- taper flange beams (TFB)
- equal/unequal angles (EA/UA)
- flat bar (with rolled edges).

Standard welded products—three-plate girders (OneSteel [2011]):
- welded beams (WB)
- welded columns (WC).

Structural steel hollow sections—cold-formed (OneSteel [2012a]):
- circular hollow sections (CHS)
- square hollow sections (SHS)
- rectangular hollow sections (RHS).

Plate product information and technical data can be found in Bluescope Steel [2010].

The above product classification is not exhaustive. Generally, a division is made between hot-rolled and open cold-formed products for the design and fabrication of steel structures. This Handbook's scope is primarily to consider the provisions of AS 4100 Steel Structures (which could be regarded as a hot-rolled product design code). The scope of AS 4100 applies to members and connections consisting of hot-rolled plates and sections, though it does also consider cold-formed hollow section members that were traditionally manufactured by hot-forming operations. The inclusion of cold-formed hollow sections within a design Standard as AS 4100 is due to the fact that such sections behave in a similar manner to hot-rolled open sections—specifically, member buckling modes. Further restrictions to the scope of AS 4100 are discussed in Section 2.4.

The design and fabrication of cold-formed steel structures is treated in AS/NZS 4600 and its related material standards. The treatment of AS/NZS 4600 and other aspects of (open-type) cold-formed steel structures is outside the scope of this Handbook, though some mention is made of the material aspects of this form of construction. The reader is directed to the AS/NZS 4600 Commentary and Hancock [2007] for an authoritative treatment of this subject. Very useful software, Coldes (see Appendix A.4) is also available for designing to AS/NZS 4600.

2.2 Physical properties of steel

Plotting the stress versus strain diagram from data obtained during tensile tests permits a better appreciation of the characteristic properties of various steel types. Figure 2.2 depicts the typical stress–strain diagrams for mild steel and low-alloy steel.

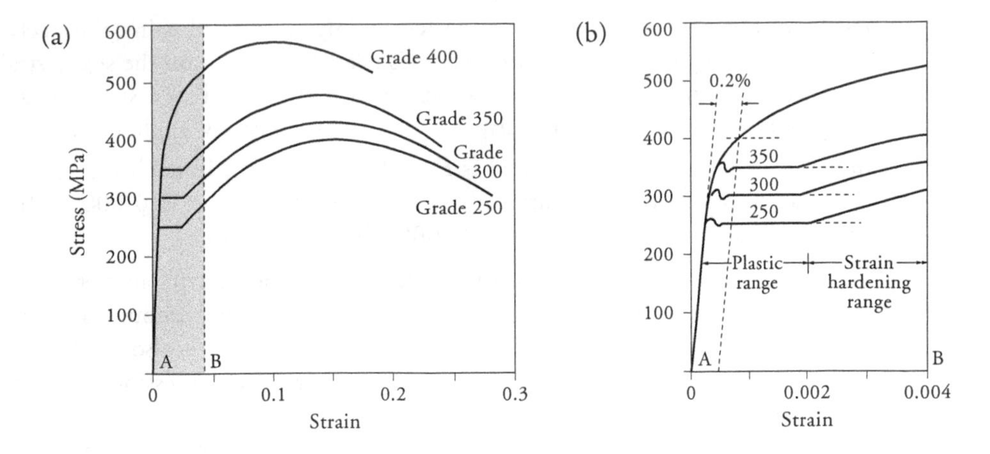

Figure 2.2 *Typical stress–strain diagrams for hot-rolled finished steel Grades 250, 300, 350 and 400*
(a) Complete diagram
(b) Enlarged portion of the diagram (a) in region A–B

The following definitions are provided to explain the various terms.

Elastic limit The greatest stress that the test piece can sustain without permanent set.

Elastic range That portion of the stress–strain diagram in which Hookes' law of proportionality holds valid.

Permanent set The strain remaining in the test piece after unloading to zero stress; also termed plastic strain, plastic elongation or permanent elongation.

Plastic range That portion of the stress–strain curve over which the stress remains approximately constant for a certain range of strains.

Proof stress See definition in yield stress.

Proportional limit The greatest stress which the tensile piece can sustain without deviating from Hookes' law of proportionality.

Reduction in area The difference between the cross-sectional areas at rupture and that at the beginning of the test, expressed as a percentage of the latter.

Strain Any forced change in the dimensions of a body; usually unit strain is meant: that is, change in dimension per unit length.

Strain hardening range The portion of the stress–strain curve immediately after the plastic range.

Stress–strain diagram The curve obtained by plotting unit stress as ordinate against corresponding unit strain as abscissa (using the initial cross-sectional area).

Ultimate elongation Maximum elongation of a test piece at rupture expressed as a percentage increase of the original gauge length.

Ultimate tensile strength (denoted as f_u or UTS) The maximum stress that a tensile piece can sustain, calculated as a quotient of the ultimate force on the original area.

Yield point The lowest stress at which the strains are detected to grow without a further increase of stress.

Yield stress (denoted as f_y) The stress corresponding to the yield point. For steels without a clearly defined yield point, the yield stress is related to a point on the stress–strain curve at which, after unloading to zero, the permanent set is equal to a strain of 0.002 (i.e. elongation of 0.2%) (see Figure 2.2(b))—the term 'proof stress' applies to this case.

Young's modulus of elasticity (denoted as E) The slope of the initial linear elastic portion of the stress–strain curve. E varies in the range of 190 000 to 210 000 MPa, and for design purposes is approximated as 200 000 MPa in AS 4100.

The yield stress, f_y, is regarded as one of the most important parameters in design. It varies with chemical composition, method of manufacture and amount of working, and though this value is determined from uniaxial tension tests it is also used to determine maximum 'stresses' for flexure, transverse shear, uniaxial compression, bearing etc., or combinations of these design action effects.

Table 2.1 lists the physical properties that are practically constant (at ambient conditions) for all the steels considered in this Handbook. These properties apply at room temperature. At elevated temperatures the properties are subject to variation, as indicated in Table 2.2. As can be seen at temperatures above 200°C the steel properties start being markedly lower than at room temperature, and the coefficient of thermal expansion starts rising significantly. This is of particular importance for structures required to operate at elevated temperatures (some industrial structures) and structures subjected to fire.

Table 2.1 Physical properties of steel for design to AS 4100

Property	Value
Young's modulus of elasticity, E	200 000 MPa
Shear modulus, G	80 000 MPa
Coefficient of thermal expansion, α_T at 20°C	11.7×10^{-6} per °C
Poisson's ratio, ν	0.25
Density, ρ	7850 kg/m^3

Table 2.2 Properties of steel at elevated temperatures (degrees Celsius) for design

Temperature °C	20	100	200	300	400	500
Reduction factor for E	1.0	0.97	0.93	0.87	0.78	0.65
Reduction factor for f_y	1.0	1.0	1.0	0.84	0.70	0.55
Multiplier for the coefficient of thermal expansion, α_T	1.0	1.03	1.09	1.15	1.21	1.29

Other mechanical properties of interest to the designer of special structures are:

Fatigue strength The stress range of a steel specimen subjected to a cyclic reversal of stresses of constant magnitude, which will cause failure at a certain specified number of cycles (usually 1 million cycles). The method of assessment for fatigue loading of the AS/NZS 1163, AS/NZS 1594, AS/NZS 3678 and AS/NZS 3769 steels is given in Section 11 of AS 4100.

Creep strength Long-term exposure to temperatures above 300°C can severely reduce the strength of steel because of the effect of creep. For example, Grade 250 mild steel exposed to 400°C for a period of 10 000 hours can fracture by creep at a stress equal to one-half of the UTS specified at room temperature. Steels for high-temperature applications have been specially developed, and advice from the steel makers should be sought in each case.

Bend radius The minimum radius to which a plate can be bent at room temperature without cracking. This is important for both plates and sections that undergo forming by presses, though plates are more critical, as they generally possess a higher bend radius to 'section' depth ratio. AS/NZS 3678 presents information on minimum bend radii for plates, which is dependent on the direction of rolling. Riviezzi [1984] presents bend radius limitations for sections, which are dependent on whether the bend radius is in the major or minor principal bending axis plane.

Hardness For special applications, where abrasion resistance or indentation resistance is a design factor. Special steels are available for this particular design application.

Impact properties at specified temperature Fracture toughness is an energy term that describes the ductile/brittle behaviour of steels at various service temperatures. Due to the presence of cracks or other types of discontinuities in regions of high local stress, brittle fracture may be possible when the steel has a low fracture toughness. In this instance these cracks may no longer allow the steel to behave in a ductile manner and subsequently propagate at high speed to cause (catastrophic) brittle failure.

Steels generally possess a characteristic temperature property, the transition temperature; above it the steel is predominantly (notch) ductile and below it the steel is predominantly brittle. Low fracture toughness and subsequent brittle fracture may then arise if the service temperature of a steel is below its transition temperature. The toughness of a particular steel is dependent on its grade, manufacture and thickness.

Impact property is also an important parameter for cold-formed hollow sections. Specific impact properties are important to guard against brittle behaviour when such sections are subject to dynamic or impact loads. This parameter becomes more important for thicker wall (i.e. >6 mm) cold-formed hollow sections.

Steels can be supplied with minimal absorbed energy (i.e. fracture toughness) requirements for test pieces at temperatures of 0°C and –15°C. These grades are referred to as notch-ductile and generally have the L0 and L15 subgrade designation. Section 10 of AS 4100 offers guidance on designing against brittle fracture.

2.3 Steel types and grades

Steel is an extremely versatile material available in a very wide range of properties and chemical compositions to suit every field of technology. Not all steels produced by steel makers are suitable for structural applications where the following properties are of paramount importance.

2.3.1 Weldability

Weldability is a relative term; all steels can be welded with due care, but steels for structural purposes should be able to be welded with relative ease and without complicated procedures. Structural steels must be tolerant to small imperfections in welding—at least up to certain specified limits.

2.3.2 Ductility

Ductility is an essential property, as the whole concept of structural steel design is based on ductile behaviour of all parts of the structure. For steel, there is a fundamental relationship between the elongation of a tensile test piece at fracture and the degree of ductility, but the designer should not rely too heavily on this; it is all too easy to reduce the ductility in the real structure by improper detailing and poor workmanship. The majority of fractures in service have occurred in the immediate vicinity of joints and abrupt changes in ductility brought about by a triaxial stress condition in these areas.

2.3.3 Low cost-to-strength ratio

The high strength of steel is naturally important when considering the range of strengths available. The ratio of cost to strength may be of real interest in the selection of a steel type for a particular structure. Many high-strength steels offer an economical solution for tensile and flexural members, although, understandably, availability has a direct bearing on the cost of these items. ASI [2009b] provides some general cost indices for the varying strengths of steel grades.

2.3.4 Availability of sections and plates

The availability of some steels in hot-rolled sections (universal, channels, etc.) in very high-strength grades is not as good as for the mild steel grades, although it is advisable to make enquiries about the availability of the higher-strength grades. Large quantities can always be produced, but it takes more time to place them on the steel maker's rolling program. The same is applicable for plates, though a much larger variety of this product is available. Conversely, higher-strength grade hollow sections (i.e. AS/NZS 1163 Grade C450L0 for RHS/SHS and AS/NZS 1163 Grade C350L0 for CHS) are more readily available. Bluescope Steel [2010] and OneSteel [2011, 2012a] provide further information on the availability of various manufactured steel sections and plates.

Structural steels may be grouped as follows:

(a) *Carbon and carbon/manganese steels* (typically 230–350 MPa yield stress)
These steels derive their strength from alloying with carbon and manganese. They are generally known as *structural steels* and are produced in relatively high tonnages. Because of their widespread use, they are readily available in all standard sections and plates. These steels are generally supplied in the fully killed (fully deoxidised) condition. AS/NZS 3678 and AS/NZS 3679 cover the material specifications, chemistry, mechanical properties, methods of manufacture, tolerances on dimensions and supply requirements. For general structural purposes, the most applicable grades are Grade 300 for hot-rolled sections (OneSteel 300PLUS specification) and Grade 250 or 350 for structural plates. Where slightly enhanced strength is required, Grade 350 can be supplied for hot-rolled sections. Steel plates of enhanced notch ductility and tensile strength are manufactured to AS 1548 (steel plates for boilers and unfired pressure vessels).

(b) *High yield strength, low-alloy steels* (typically 320–450 MPa yield stress)
These steels are similar to those in (a) above, except for the addition of small quantities of alloying elements useful in producing a fine-grain steel. Because the grain-refining elements seldom exceed 0.15%, these steels are known as low-alloy steels. Generally these steels are fully killed. Grade 400 is used for production of welded sections (WB and WC). Grade 350 structural plates are available. Cold-formed hollow sections are generally available in this steel type—see (d) below.

(c) *Low-alloy weathering steels* (typically 350 MPa yield stress)
By their chemical nature these steels are similar to those in (b) above, except for a further addition of chromium, nickel and copper (up to 2.1%, total) for a greatly enhanced resistance to atmospheric corrosion. These alloying elements cause the steel surface to weather to a uniform patina, after which no further corrosion takes place. This allows the use of unpainted steelwork in regions away from marine environment and heavy pollution. These steels are not commonly produced and are only available, in plate form, direct from the steel mill.

(d) *Structural steel hollow sections* (typically 250–450 MPa yield stress)
The material specifications for hollow sections are covered in AS/NZS 1163. In line with current overseas practice this Standard considers only hollow sections manufactured by cold-forming operations (hence the 'C' prefix before the grade designation, e.g. C250, C350 and C450). Hollow sections for structural purposes produced in Australia are manufactured only by cold-forming and electric resistance welding (ERW). Consequently, stress relieving after the forming and welding operation (at ambient temperatures) is now no longer required. The current range of C250, C350 and C450 grades of steel for hollow sections are readily available to meet the notch-ductile L0 (e.g. C450L0) requirements of AS/NZS 1163. The L0 rating is typically available from Australian tube manufacturers and should be generally specified (see 'Impact properties…' in Section 2.2). RHS/SHS are generally available in Grades C450L0 or C350L0 and CHS in Grade C350L0.

(e) *Heat-treated carbon/manganese steels* (typically 500–600 MPa yield stress)
These steels are manufactured from feed derived from rolled steels, somewhat similar to those listed in (a) and (b) above but having enhanced levels of micro-alloys. The steel is then subjected to a combination of heating and cooling (quenching and tempering). This changes the microstructure of the steel to raise its strength, hardness and toughness. In Australia these steels are manufactured only in plate form and comply with AS 3597.

(f) *Heat-treated alloy steels* (typically 500–690 MPa yield stress)
These steels are the most advanced (and most costly) constructional steels of weldable quality currently available. Except for significant increases of carbon and manganese content, the overall chemistry such as Cr, Ni and Mo and method of manufacture are similar to those in (e) above. Plate products of this type of steel comply with AS 3597, and are manufactured in Australia by Bisalloy Steel.

The steels listed in (e) and (f) are used for structural purposes when the saving of mass is of prime importance—for example, in long-span beams and bridges, high-rise building columns and vehicle building. There is an increased use of these types of steel in the construction industry and the Amendment to AS 4100 (AS 4100 AMD 1 see Appendix D) now incorporates such steels within its provisions—see Section 1.14(a).

The above grouping of steels has been arranged in order of increasing yield stress and increasing unit cost of raw product. Except for hollow sections, the expertise required for welding also increases roughly in the same order.

Other steels complying with relevant Standards (e.g. AS 1397, AS 1548, AS/NZS 1594, AS/NZS 1595) are available for steel flat products. These include steels for cold-formed structural steel (other than tubular), tank and boiler applications; they are mentioned here as their application is outside the scope of this Handbook.

The following (rationalised) product-based Australian Standards cover the steels normally used in building construction:

- *AS/NZS 1163 Cold-formed structural steel hollow sections*
  Cold-formed Grade C250/C350 circular hollow sections (CHS) and Grade C350/C450 rectangular and square hollow sections (RHS and SHS) suitable for welding are considered in this Standard. These sections are manufactured by cold-forming and subsequent electric resistance welding (ERW) operations. See Table 2.3 for additional strength details.

- *AS 3597 Structural and pressure vessel steel: Quenched and tempered plate*
  This Standard covers the production, material, supply and other technical requirements of 500, 600 and 620–690 MPa (depending on thickness) quenched and tempered plate. See Table 2.4 for additional strength details.

- *AS/NZS 3678 Structural steel: Hot-rolled plates, floor plates and slabs*
  This is an 'omnibus' Standard covering the specification of steels of plate products grouped in (a) and (b) above—that is, Grades 200, 250, 300, 350, 400 and 450. Subgrades L0 and L15 cover steels of enhanced notch ductility; a minimum Charpy V-notch value of 27 J is obtainable at temperatures above 0°C for subgrade L0 and –15°C for L15. The Standard also covers the material specification for weather-resisting steels. See Table 2.3 for additional strength details.

- *AS/NZS 3679, Part 1 Structural steel: Hot-rolled structural steel bars and sections*
  This is another 'omnibus' Standard covering the specification of steels of hot-rolled sections (universal sections, taper flange beams, angles, parallel/taper flange channels and flat bars) for structural and engineering purposes in ordinary weldable grades. Grades include the commonly specified Grades 250 and 300 as well as 350 and the subgrades of L0 and L15 as in AS/NZS 3678. See Table 2.3 for additional strength details. Before October 1994, sections of Grade 250 were produced by BHP (as it was known at the time) as the base grade.

- *AS/NZS 3679, Part 2 Structural steel: Welded I sections*
  This Standard provides the production, material, supply and other technical requirements for welded I-type steel sections for structural and engineering purposes in ordinary weldable and weather-resistant weldable grades. Steel grades include Grade 300 and 400 steel and the subgrades of L0 and L15, as noted in AS/NZS 3678. Flange-to-web connections are made by deep-penetration fillet welds using the submerged-arc welding (SAW) process. This Standard covers the range of standard welded products

released in 1990 to extend the range of universal sections. See Table 2.3 for additional strength details.

The mechanical properties of structural steels are given in Table 2.3. As can be seen from the table, the yield stress of all steel grades varies slightly from the base figure. This is unavoidable, as the steel receives various amounts of hot and cold working during the product rolling process. In general, the thinner the plate, the higher the yield strength.

Table 2.3 Specification and strengths of typical structural steel products noted in AS 4100.

Standard/Product	Steel grade	Thickness (mm)	Yield stress (MPa)	Tensile Strength (MPa)
AS/NZS 1163	C450, C450L0	All	450	500
CF Hollow sections	C350, C350L0	All	350	430
	C250, C250L0	All	250	320
AS/NZS 3678	450, 450L15	≤20	450	520
Hot-rolled		21 to 32	420	500
plate and floorplate		33 to 50	400	500
	400, 400L15	≤12	400	480
		13 to 20	380	480
		21 to 80	360	480
	350, 350L15	≤12	360	450
		13 to 20	350	450
		21 to 80	340	450
		81 to 150	330	450
	WR350, WR350L0	≤50	340	450
	300, 300L15	≤8	320	430
		9 to 12	310	430
		13 to 20	300	430
		21 to 150	280	430
	250, 250L15	≤8	280	410
		9 to 12	260	410
		13 to 50	250	410
AS/NZS 3679.1	400, 400L0, 400L15	≤17	400	520
Hot-rolled steel		>17	380	520
bars and sections	350, 350L0, 350L15	≤11	360	480
		12 to 40	340	480
		≥40	330	480
	300, 300L0, 300L15	<11	320	440
		11 to 17	300	440
		>17	280	440
	250, 250L0, 250L15	<11	260	410
		11 to 40	250	410
		≥40	230	410

Note: (1) For full listing of steel strengths refer to Section 2 of AS 4100.
(2) Welded I-sections complying with AS/NZS 3679.2 are manufactured from steel plates complying with AS/NZS 3678.
(3) The 300PLUS range of hot rolled sections (OneSteel [2011]) comply with the above strength requirements for AS/NZS 3679.1 Grade 300.

Table 2.4 Mechanical properties of quenched & tempered plates (Bisalloy Steels [2011]) to AS 3597

Property / thickness	Specification / grade					
	Bisalloy 60 (AS 3597 Grade 500)		Bisalloy 70 (AS 3597 Grade 600)		Bisalloy 80 (AS 3597 Grade 700)	
	f_y^*	f_u	f_y^*	f_u	f_y^*	f_u
Minimum yield strength, MPa, for thickness of:						
5	500	590–730	600	690–830	650	750–900
6	500	590–730	—	—	690	790–930
8 to 25	500	590–730	—	—	690	790–930
26 to 32	500	590–730	—	—	690	790–930
33 to 65	—	—	—	—	690	790–930
70 to 100	—	—	—	—	620	720–900

Note:(1) f_y^* is 0.2% proof stress. (2) As noted in Section 1.14(a), the amendment to AS 4100 (AS 4100 AMD 1—see Appendix D) permits quenched and tempered steels to be within the scope of AS 4100. (3) See Table 2.1 of AS 4100 AMD 1 (Appendix D) for the full thickness-strength combinations of quenched and tempered plate to AS 3597.

2.4 Scope of material and design codes

The original scope of AS 4100 precludes the use of:

- steel elements less than 3 mm thick. One exception is that hollow sections complying with AS/NZS 1163 are included irrespective of thickness;
- steel elements with design yield stresses exceeding 450 MPa;
- cold-formed members (other than hollow sections complying with AS/NZS 1163), which should be designed to AS/NZS 4600;
- composite steel–concrete members (these are to be designed to AS 2327, which, at the time of this Handbook's publication, considers only simply supported beams).

Structural steels within the scope of AS 4100 are those complying with the requirements of AS/NZS 1163, AS/NZS 1594, AS/NZS 3678 and AS/NZS 3679. Clause 2.2.3 of AS 4100 permits the use of 'unidentified' steels under some restrictions, which include limiting the design yield stress, f_y , to 170 MPa and the design tensile strength, f_u , to 300 MPa.

As noted in Section 1.14(a), the amendment to AS 4100 (AS 4100 AMD 1—see Appendix D) permits steels with yield stress used in design (f_y) up to and including 690 MPa. This includes AS/NZS 1594—XF500 (with f_y = 480 MPa) and the quenched and tempered plate steels to AS 3597 (Grades 500, 600 and 700 with f_y = 500, 600, and 620–690 respectively).

2.5 Material properties and characteristics in AS 4100

The nominal strengths of the steels considered within AS 4100 are the same as those listed in Table 2.3. AS 4100 does stipulate the design yield stress and design ultimate tensile strengths of steels, which are dependent on the method of forming and the amount of work done on the steel. Table 2.1 of AS 4100 provides the design yield stress (f_y) and design tensile strength (f_u) of relevant steels for design to AS 4100. It should be noted that apart from the cold-formed hollow sections and some AS 1594 steels, f_y is dependent on both grade and thickness, while f_u is dependent only on grade (and not thickness).

In some parts of AS 4100 the designer must determine the 'residual stress' category of the member or member element (e.g. in Tables 5.2 and 6.2.4 of AS 4100) to assess its local buckling behaviour. As with the evaluation of f_y, the amount of 'residual stress' is dependent on the method of manufacture and amount of work done in forming operations. As a guide to designers, the following 'residual stress' categories can be assumed for the above-mentioned tables in AS 4100:

- hot-rolled sections complying with AS/NZS 3679.1: Category HR
- welded sections complying with AS/NZS 3679.2: Category HW
- hollow sections complying with AS/NZS 1163: Category CF.

The following should be noted:

- The HW residual stress has been assumed for the AS/NZS 3679.2 sections due to the nature of the welding operation involved (deep-penetration fillet welds) and subsequent straightening of flanges.
- The residual stress categories HR, HW and CF reduce to SR designation if stress relieving is undertaken after member fabrication.
- It is often difficult in the design office to determine the exact magnitude of residual stresses for the section being considered, and the above categorisation method is sufficient.

2.6 Strength limit state capacity reduction factor ϕ

A feature of the strength limit state design method adopted by AS 4100 is the use of two 'safety factors': the 'load' factor γ_i and the capacity reduction factor ϕ. The load factor γ_i is determined from the Structural Design Actions Standard AS/NZS 1170 for a particular loading combination and considers the uncertainties in the magnitude, distribution and duration of loads as well as the uncertainties in the structural analysis.

The capacity reduction factor ϕ is considered within Table 3.4 of AS 4100 and accounts for the variability in the strength and defects of steel elements and connections. Further information on capacity reduction factors can be found in Sections 1.5 and 1.7 of this Handbook. Table 2.5 summarises these capacity reduction factors.

Table 2.5 Capacity reduction factors for strength limit state design to AS 4100

Type of component	Capacity reduction factor ϕ
Beam, column, tie:	0.9
Connection plates:	0.9
Bolts and pins:	0.8
Welds:	
Complete penetration butt welds:	0.6 for GP category welds
	0.9 for SP category welds
All other welds:	0.6 for GP category welds
	0.8 for SP category welds
	0.7 for SP category welds to RHS sections, $t < 3$ mm

Note: Weld categories GP (general purpose) and SP (structural purpose) reflect the degree of quality control and are described in AS/NZS 1554.

2.7 Brittle fracture

2.7.1 Introduction

Unlike fatigue, the brittle fracture failure mode is a one-off event. The first time the critical loading event occurs in an element containing a critical flaw, the element is liable to fracture. In contrast, the fatigue cracking accumulates, cycle after cycle. The four conditions that can lead to brittle facture are:
- loading at a temperature below the transition temperature of the steel
- relatively high tensile stress, axial or bending
- presence of cracks, notches or triaxial stress states that lower the ductility of the detail
- use of steels having impaired ductility at the lowest service temperature.

The danger of brittle fracture is increased when the ductility of steel is reduced by:
- suppression of yield deformations, as may be caused by triaxial stressing
- the use of relatively thick plates or sections
- impact loading, i.e. high strain rate
- cold bending, such that a strain greater than 1% is induced as a result of fabrication or field straightening
- detailing that results in severe stress concentrations (notches).

Methods of design against brittle fracture include such measures as:
- choosing a steel that is not susceptible to brittle fracture at the minimum service temperature for which the structure is exposed
- lowering the maximum operating stresses
- using details that do not suppress the ductility of steel and contain no notches
- post-welding heat treatment (normalising of welds).
- consulting a metallurgical engineer to advise on appropriate actions.

2.7.2 The transition temperature

The ductility of steel is normally tested at room temperature, say 20°C. At lower temperatures the ductility of steel diminishes. The temperature at which the reduction of ductility becomes significant depends on the ductility of the steel, normally measured by impact energy absorbed in the Charpy test. Impact energy of 27 joules at 20°C would normally be required for the hot-rolled plates and sections. For cold-formed hollow sections the requirement is 27 joules at 0°C.

A practical method of determining the suitability of steel for a particular service temperature is given in Section 10 of AS 4100. The method requires an evaluation of the minimum service temperature, plate element thickness and steel type, which is roughly dependent on the notch ductility. Table 10.4.1 of AS 4100 is reproduced in Table 2.6 for the commonly available steel types in Australia. It is advisable to consult the steel manufacturer on the selection of a suitable steel grade.

The service temperatures for various locations in Australia are those determined by LODMAT isotherms, deemed to be the 'lowest one-day mean ambient temperature'. There are only two regions in Australia where the temperature falls to or slightly below zero (based on LODMAT), namely parts of the southern and central Great Dividing Range. In New Zealand there are extensive areas where sub-zero service temperatures occur (refer to NZS 3404). The selection of steel with appropriate notch toughness becomes more important at lower service temperatures. It should be noted that

LODMAT gives average 24-hour temperatures but hourly minima can be some 5 degrees lower. The Commentary to AS 4100 notes that this may be allowed for by subtracting 5°C off the structure's permissible service temperature and ensuring that this is above the region's LODMAT isotherm value.

Table 2.6 Steel types required for various service temperatures as noted in AS 4100

Steel type	**Permissible service temperature for thickness ranges, mm**					
See Note	≤6	7–12	13–20	21–32	33–70	>70
1	-20	-10	0	0	0	+5
2	-30	-20	-10	-10	0	0
2S & 5S	0	0	0	0	0	0
3 & 6	-40	-30	-20	-15	-15	-10
4	-10	0	0	0	0	+5
5	-30	-20	-10	0	0	0
7A	-10	0	0	0	0	-
7B	-30	-20	-10	0	0	-
7C	-40	-30	-20	-15	-15	-
8C	-40	-30	-	-	-	-
8Q, 9Q & 10Q	-20	-20	-20	-20	-20	-20

Note: Steel types are listed below.

Steel types vs steel specifications

	Specification and Steel grades				
Steel Type	AS/NZS 1163	AS/NZS 1594	AS/NZS 3678 AS/NZS 3679.2	AS/NZS 3679.1	AS 3597
1	C250	Note 2	200, 250, 300 A1006, XK1016	300	-
2	C250L0	-	-	300L0	-
2S	-	-	250S0, 350S0	300S0	-
3	-	XF300	250L15/L20/Y20 250L40/Y40 300L15/L20/Y20 300L40/Y40	300L15	-
4	C350	HA350/400, HW350	350, WR350, 400	350	-
5	C350L0	-	WR350L0	350L0	-
5S	-	-	350S0	350S0	-
6	-	XF400	350L15/L20/Y20 350L40/Y40 400L15/L20/Y20 400L40/Y40	-	-
7A	C450	-	450	-	-
7B	C450L0	-	-	-	-
7C	-	-	450L15/L20/Y20 350L40/Y40	-	-
8C	-	XF500	-	-	-
8Q	-	-	-	-	500
9Q	-	-	-	-	600
10Q		-	-	-	700

Note: (1) Various Grades listed - see relevant section in Table 10.4.4 of Appendix D. (2) See below on the revision of the above two Tables based on the amendment to AS 4100 (AS 4100 AMD 1).

As noted in Section 1.14(a), the amendment to AS 4100 (AS 4100 AMD 1—see Appendix D) sees Table 2.6 change to include Steel Type 2S and 5S (a "seismic" type grade for AS/NZS 3678 plate products (min. f_y = 250 MPa), AS/NZS 3679.1 UB & UC (min. f_y = 300 MPa) and AS/NZS 3679.2 WB & WC (min. f_y = 300 MPa) with high impact properties — 70 Joules at 0 degrees Celsius), Steel Type 8C (AS/NZS 1594 flat product XF500) and the Quenched & Tempered plate Grades of 500, 600 and 700 — Steel Types 8Q, 9Q and 10Q respectively. The AS/NZS 3678 plate range has also been increased within these Tables. See Appendix D also.

As can be read from Table 2.6, a plate stressed in tension and thicker than 70 mm in steel type 1 and 4 is adequate only for a design service temperature of +5°C or higher. This is important for open-air structures and bridges, particularly in colder climates. The remedy is to use a steel with improved notch ductile properties, say type 2 or higher.

Fabrication of structures in low-temperature zones must be carried out with care. Straining beyond the strain of 1% due to cold forming or straightening could produce an effect equivalent to lowering the service temperature by 20°C or more.

2.7.3 Hydrogen cracking

In the presence of water vapour coming in contact with the surface of the molten metal during welding, any hydrogen is freed and gets absorbed into the weld metal. Steel has a great affinity for hydrogen, and if hydrogen is prevented from exiting during cooling, it promotes the formation of Martensite. Owing to the great hardness of the Martensite the weld metal loses ductility, and this can result in cracking in service. Generally, the thicker the plates to be joined, the more care is needed to prevent the hydrogen embrittlement of the welds.

The remedy is to prevent hydrogen from becoming entrapped, first by reducing the weld-cooling rate so that the hydrogen is given more time to escape from the molten pool and the heat-affected zone. Preheating the steel prior to welding is beneficial. Using a low heat input during welding is also a common practice. The second line of defence is shielding the weld pool area from the entry of hydrogen by using inert gas shielding or 'low-hydrogen' electrodes.

It is sometimes advisable to carry out qualification of welding procedures by testing as a part of the welding procedures approvals. There is a need for frequent and reliable inspection during the welding of components, especially those fabricated from thicker plates (>40 mm). Welding inspection should ascertain that the weld defect tolerances are not exceeded and should also include hardness tests of the weld metal and the heat-affected zone. In critical welds there should not be any areas of excessive hardness (see AS/NZS 1554 and WTIA [2004] for further guidance).

2.8 Further reading

- Background to the metallurgical aspects of steel: Lay [1982a].
- Background to the evolution of steel material Standards: Kotwal [1999a,b].
- Standards and material characteristics for cold-formed open sections: Hancock [2007].
- Availability of steel products: Keays [1999].
- Relative costing between steel materials and products: Watson, et al. [1996].
- Steel castings: AS 2074.
- Websites for steel material manufacturers/suppliers: See Appendix A.5.
- Websites for steel industry associations: See Appendix A.6.

chapter 3

Design Actions

3.1 General

Design actions are divided into permanent and imposed actions. The next category are the environment-generated actions such as wind, waves, vibrations, earthquake and actions caused by volumetric changes (e.g. temperature and concrete shrinkage). Design actions are set in the AS/NZS 1170 series of Standards (and AS 1170.4 or NZ 1170.5 for earthquake actions) on the basis of either statistical/probabilistic analyses or calibration studies of currently available data.

The terms 'loads' and 'actions' are used in this text as follows: 'loads' are used in the traditional sense, and 'actions' denote indirect effects such as temperature, weld shrinkage, concrete shrinkage and inertial effects.

There are situations where the designer must determine the design actions for a number of reasons:

- the building owner's intended special use of the structure
- mechanical equipment and its vibrations.

In such instances the designer has to determine the 95 percentile value of the actions. This may require tests and statistical analyses.

To obtain the design action effects (shears, moments etc.), the nominal (characteristic) actions have to be multiplied by 'load factors' to be followed by a structural analysis (see Chapter 4). The action combination rules of AS/NZS 1170.0 are discussed in Section 1.6, with the most common combinations of permanent, imposed and wind actions noted in Table 1.3. Other load combinations can be found in AS/NZS 1170.0.

3.2 Permanent actions

Permanent actions or 'dead loads' are actions whose variability does not change in the structure's life. AS/NZS 1170.1 specifies the structural design actions due to permanent loads. The main actions are calculated from the self-weight of the:

- structure
- permanently fixed non-structural elements
- partitions, fixed and movable
- fixed mechanical, air conditioning and other plant.

AS/NZS 1170.1 provides for partition walls and the effect of non-structural items that are capable of being removed. The loads due to self-weight of the structure and loads due to non-structural elements should often be reviewed during the design process. Load changes occur constantly as members of the design team look for optimum solutions in their disciplines. Unit weight and densities of common materials and bulk solids are given in AS/NZS 1170.1 with selected extracts provided in Tables 3.1 to 3.4 in this Handbook.

Caution should also be exercised with allowances for the mass of connections and fixtures, which can be as high as 10% of the member mass in rigid frames. The allowance for bolts, gussets, stiffeners and other details should preferably be kept below 10% in simple construction. The added mass of these 'small' items could have a significant effect on longer-span structures and a careful review of the dead loads should thus be undertaken at the end of the preliminary design.

Dimensional errors in construction can also result in the variation of dead loads. Similarly, building alterations can result in changes to the permanent and imposed loads. It is important that dead loads be reassessed after significant design changes and followed by a fresh structural analysis.

Table 3.1 Typical unit weights of materials used in building construction

Material	**Unit weight** kN/m^2	kN/m^3	**Material**	**Unit weight** kN/m^2	kN/m^3
Brick masonry, per 100 mm			Plaster render, per 10 mm thickness		
Engineering, structural	1.90		Cement	0.23	
Calcium silicate	1.80		Lime	0.19	
			Gypsum	0.17	
Ceilings/walls/partitions					
Fibrous plaster, per 10 mm	0.09		Roofing, corrugated steel sheet		
Gypsum plaster, per 10 mm	0.10		Galvanized, 1.00 mm thick	0.12	
Fibre-cement sheet, per 10 mm	0.18		Galvanized, 0.80 mm thick	0.10	
			Galvanized, 0.60 mm thick	0.08	
Concrete block masonry, per 100 mm			Galvanized, 0.50 mm thick	0.05	
Solid blocks	2.40				
Hollow blocks, unfilled	1.31		Roofing, non-metallic		
			Terracotta roof tiles	0.57	
Concrete, reinforced using:			Profiled concrete tiles	0.53	
Blue stone aggregate (dense)		24.0			
Laterite stone aggregate		22.0	Stone masonry, per 100 mm		
For each 1% one-way reinforcement, add		+0.60	Marble, granite	2.70	
For each 1% two-way reinforcement, add		+1.20	Sandstone	2.30	
Floor finishes per 10 mm thickness					
Magnesium oxychloride—heavy	0.21				
Terrazzo paving	0.27				
Ceramic tiles	0.21				

Note: See Appendix A of AS/NZS 1170.1 for further information.

Table 3.2 Properties of bulk materials used in materials handling calculations

Material	Weight kg/m ^{3}	Angle of repose degrees	Material	Weight kg/m ^{3}	Angle of repose degrees
Alum, fine	720–800	30–45	Iron ore, haematite	2600–3700	35–40
lumpy	800–950	30–45	crushed	2200–2500	35–40
Alumina, powdery	800–1200	22–33	Kaolin clay, lumps	1010	35
Aluminium hydrate	290	34	Lead arsenate ore	1150	
Asphalt, paving	2160		Lead ores	3200–4300	30
Ash, dry, compact	700–800	30–40	Lead oxides	960–2400	35
Baryte, ore	2220–2400		Lime, ground, burned	960–1400	35–40
powdered	1920–2040		hydrated	640–700	35–40
Barley	670	27	Limestone, crushed	1360–1440	38–45
Bauxite. dry, ground	1090	35	Manganese ore	2000–2240	39
mine run	1280–1440	31	Magnetite ore	4000	35
crushed	1200–1360	30	Marble, crushed	1280–1500	
Brick, dense	2000		Nickel ore	1280–2400	
light	1600		Paper, pulp stock	640–960	
Cement, at rest	1300–1600	30–33	sheet	800–1500	
aerated	960–1200	20–30	Phosphate, super	800–880	45
Cement clinker	1200–1520	30–40	rock, broken	1200–1350	25–39
Chalk, lumpy	1200–1360	35	Potash salt	1280	
fine, crushed	1040–1200	30–35	Potassium, carbonate	820	
Charcoal, porous	290–400	35	chloride	1920–2080	
Chrome ore	2000–2240		nitrate	1220	
Clay, dry	1000–1100		sulphate	670–770	
damp, plastic	1750–1900		Pumice, screenings	640–720	
Coal, anthracite, crushed, −3 mm	960–1140	25–30	Pyrites, lumps	2150–2320	
bituminous	720–800	35–40	Quartz, screenings	1280–1440	
lignite	700–850	35	Salt, common, dry, fine	1120	25
waste	1400	35	dry, cake	1360	36
Coke, loose	370–560	30	Sand, dry	1580–1750	35
petroleum	600–1000	35–40	wet	1750–2080	45
breeze	400–560	30–45	Sandstone, broken	1350–1450	40
Copper ore	2000–2800	35	Sawdust, dry	150–220	36
sulphate	1200–1360	31	Shale, broken	1450–1600	39
Cork, granulated	190–240		Sinter	1600–2150	
Dolomite, crushed	1440–1600		Slag, furnace, dry	1020–1300	25
Earth, damp, packed	1630–1930	40–45	wet	1450–1600	40–45
Felspar, crushed	1440–1750	34	Slate	2800	
Flour, wheat	560–640	40	Stone rubble	2200	
Fluorspar, crushed	1750–1920		Snow, loose, fresh	200–400	
Fly ash	640–720	42	compact, old	600–800	
Granite, crushed	1360–1440	35–40	Sugar, granulated	800–880	25–30
broken	1500–1600		raw, cane	880–1040	36–40
Graphite	2300		Talc screenings	1250–1450	
Gravel, dry	1440–1600	30–38	Terracotta	2080	
wet	1800–1900	10–38	Vermiculite	800	
Gypsum dust, at rest	1490	42	Wheat	800–850	25–30
lumps	1600	40	Wood chips, softwood	200–480	40–45
Glass, window	2600		hardwood	450–500	40–45
Gneiss	2800		Zinc ore	2600–2900	35
Ilmenite, ore	2240–2550				

Table 3.3 Approximate densities of metals

Metal	kg/m³	Metal	kg/m³
Aluminium, alloy	2700	Manganese	8000
Aluminium bronze	7700	Mercury	13600
Brass	8700	Monel metal	9000
Bronze, 14% Sn	8900	Nickel	9000
Copper	8900	Platinum	21500
Gold	19300	Silver	10600
Iron, pig	7200	Steel, rolled	7850
Lead	11350	Tin	7500
Magnesium alloy	1830	Zinc	7200

Table 3.4 Densities of bulk liquids at 15°C (unless otherwise noted)

Liquid	kg/m³	Liquid	kg/m³
Acid, acetic	1040	Kerosene	800
muriatic, 40%	1200	Linseed oil	880
nitric	1510	Mercury	13600
sulphuric, 87%	1800	Milk	1030
Alcohol, 100%	800	Oil, crude	1000
Ammonia	880	heating	995
Aniline	1000	lubricating	950
Benzine	800	vegetable	960
Benzol	900	Petrol	700
Beer	1030	Water, drinking 4°C	1000
Bitumen	1370	100°C	958
Caustic soda, 50% solids	1520	sea, at 20°C	1030
Glycerine	1250	Tar pitch	1180

3.3 Imposed actions

Imposed actions (or 'live' loads) arise from the intended function of the building or structure. They are actions connected with the basic use of the structure and are highly time-dependent as well as randomly distributed in space. Their magnitudes and distribution vary significantly with occupancy and function. Imposed actions vary from zero to the nominal value specified in AS/NZS 1170.1 for most types of intended use. Occasionally, but not very often, they are determined by the designer or prescribed by the owner of the structure.

It would be impractical to try to determine all the loads in a structure by calculating load intensities at different locations. However, AS/NZS 1170.1 provides a uniform, statistically based approach to determine imposed actions. Table 3.5 lists an extract of imposed floor loads as noted in AS/NZS 1170.1. Two load types are noted: uniformly distributed load (UDL) and concentrated load. The reason for considering concentrated loads is that there are some localised loads (e.g. heavy items of furniture, equipment or vehicles) that may not be adequately represented by a UDL.

Readers should also note that, since the introduction of AS/NZS 1170, Parts 0, 1 and 2, some load reduction equations for floor loads have changed and been incorporated in a factor for reduction of imposed floor loads due to area (ψ_a)—see Clauses 3.4.1 and 3.4.2 of AS/NZS 1170.1. Overall, it is argued, the effect is the same. Additionally, there has been some change in the philosophy of loading and magnitude of load to reflect the New Zealand and ISO (International Organization for Standardization) principles. Reference should be made to the Supplementary Commentary of AS/NZS 1170.1 for further information on the changes and use of the Standard.

Table 3.5 Typical values of imposed floor and roof loads

Specific uses	**UDL** kN/m^2	**Concentrated load** kN
Self-contained dwellings	1.5	1.8
Balconies and accessible roof areas		
in self-contained dwellings	2.0	1.8
Offices for general use, classrooms (with tables)	3.0	2.7
Work rooms, light industrial (no storage)	3.0	3.5
Public assembly areas with fixed seats	4.0	2.7
Terraces and plazas	4.0	4.5
Assembly areas without fixed seating	5.0	3.6
Parking garages restricted to cars	2.5	13
Structural elements and cladding of roofs	0.12+1.8/*A*	1.4
	Min 0.25	
Roof trusses, joists, hangers	-	1.4

Note: (1) For further information and detailed tabulation of specific imposed action requirements see AS/NZS 1170.1.
(2) *A* = plan projection of roof area supported by member, in sq.m.

3.4 Wind actions

Wind load intensities and load determination are specified in AS/NSZ 1170.2.

The *site* wind speed is given by:

$$V_{sit,\beta} = V_R M_d (M_{z,cat} \ M_s M_t)$$

where V_R = 3 second gust speed applicable to the region and for an annual probability of exceedance, $1/R$ (500 return period for normal structures)—see Table 3.6

M_d = 1.0 or smaller wind direction multiplier

$M_{z,cat}$ = multiplier for building height and terrain category—see Table 3.7

M_s = shielding multiplier—upwind buildings effect

M_t = topographical multiplier—effect of ramping, ridges

The reference annual probability of exceedance is linked to the risk of failure levels (importance levels) as specified in AS/NZS 1170.0. The design wind forces are determined from the following expression:

$$F = 0.5\rho_{air} (V_{des,\theta})^2 C_{fig} C_{dyn} A_{ref}$$

with $\rho_{air} = 1.2$ kg/m^3 (density of air), then

$$F = 0.0006(V_{des,\theta})^2 C_{fig} C_{dyn} A_{ref}$$

where F is the design wind force in kN

$V_{des,\theta}$ = maximum value of $V_{sit,\beta}$ (see above)
= design wind speed

C_{fig} = aerodynamic shape factor—internal and external pressures

C_{dyn} = dynamic response factor, use 1.0 unless the structure is wind sensitive

A_{ref} = reference area, at height upon which the wind pressure acts, in sq.m.

Table 3.6 Regional wind speed (V_R) for annual probability of exceedance of 1 in 500 (V_{500}) for normal structures

Region	**Wind velocity** for V_{500}	**Cities** in Australia	in New Zealand
A1 to A7	45 m/s	Hobart, Perth, Sydney, Adelaide, Canberra, Melbourne	Auckland, Dunedin, Christchurch, Westport, Wanganui
B	57 m/s	Norfolk Is., Brisbane	
C	66 m/s	Cairns, Townsville Darwin	
D	80 m/s	Carnarvon, Onslow Pt Hedland	
W	51 m/s		Wellington

Note: Refer to AS/NZS 1170.2 for other locations and probability levels.

Table 3.7 Terrain category and height multiplier $M_{z,cat}$ for ultimate limit state design (not serviceability) in regions A1 to A7, W and B.

Height m	**Terrain category** 1	2	3	4
5	1.05	0.91	0.83	0.75
10	1.12	1.0	0.83	0.75
15	1.16	1.05	0.89	0.75
20	1.19	1.08	0.94	0.75
30	1.22	1.12	1.00	0.80
50	1.25	1.18	1.07	0.90
100	1.29	1.24	1.16	1.03

At the time of publication of this Handbook, AS/NZS 1170.2 was last published in 2011 and superseded the 2002 version which also had a supplementary commentary. There was no supplementary commentary published by Standards Australia/New Zealand for the 2011 version. However, Holmes et al [2011] have produced an excellent supporting document for the 2011 version of AS/NZS 1170.2 which includes the basis of the clauses, background information, and additional information on shape factors and dynamic factors which are not provided in, but are compatible with and enhance, the Standard.

In addition to AS/NZS 1170.2 (2011), Holmes et al [2011] also consider the:

- change in rationale of surface roughness for open water aspects
- introduction of Terrain category 1.5 for near coastal breaking water
- removal of Table 4.1(B) and use of 4.1(A)—i.e. $M_{z,cat}$ —for all wind regions.

These changes have ramifications on design wind speed particularly for low rise structures in cyclonic areas. The philosophy associated with consideration of internal pressures has yet again changed, making it mandatory to consider full internal pressures in cyclonic areas. Holmes et al [2011] consider these aspects and many other changes of note since the March 2011 release of AS/NZS 1170.2. At the time of publication of this Handbook, these changes are being drafted into an amendment to AS/NZS 1170.2.

3.5 Earthquake actions

Intuitively, the structural response to wind actions is somewhat immediate to the application of the actions—that is, a typical building structure responds directly to the pressure forces imposed on its surfaces. In contrast, earthquake actions arise from the structure's response to base (foundation) movements. That is, the building structure does not respond to forces imparted directly to it—it responds to translational movements at the base. This means that inertial forces come into play which, coupled with effects from the distribution of the structure's mass and stiffness, may not be synchronised with the base movement (in terms of time and intensity). Though different in the nature of loading, earthquake loads on structures can be modelled in design by using quasi-static loads, much like the design for wind loads.

At the time of the publication of this Handbook, and as noted in Section 1.14(b), Standards Australia and Standards New Zealand had decided that there was to be no joint Standard for Earthquake Actions. Hence, as NZS 1170.5 was previously released in 2004 (applicable to New Zealand only), Standards Australia then published a substantially revised AS 1170.4 in 2007 (to supersede the 1993 version) for application in Australia. Since 2007 there has been a disparity between the definitions and methodologies for the evaluation of Earthquake actions and design (strength, ductility and detailing) for the earthquake requirements in AS 4100 and that in AS 1170.4 [2007]. It should be noted at this juncture that for earthquake design, not only is strength important but also "ductility". The "ductility of a structure" for earthquake design is different to the traditional meaning of ductility in a "general material" sense (i.e. avoidance of brittle failures) where the former term refers to the "ability of a structure to sustain its load-carrying capacity and dissipate energy when responding to cyclic displacements in the inelastic range during an earthquake" (as noted in AS/NZS 1170.4 [2007]).

The objective of AS 1170.4 is to evaluate earthquake actions and the general detailing requirements for structures subject to earthquakes. Due to the abovementioned changes, the substantial revision from AS 1170.4 [1993] to AS 1170.4 [2007] are noted in the Preface of the latter document.

The method for determining ultimate limit state earthquake actions from AS 1170.4 [2007] requires the following parameters to be evaluated:

- determine the *Importance level* (1, Domestic structure (housing), 2, 3, 4) of the structure from Appendix F of AS/NZS 1170.0 and NCC [2011]. Most non-domestic building structures come under *Importance level* 2.
- determine the *Annual probability of exceedance* (P) from Appendix F of AS/NZS 1170.0 and NCC [2011]. In most cases, the P is 1/500 years.
- determine the *Probability factor* (k_p) which is dependent on P. In most cases $k_p = 1.0$.
- determine the *Earthquake hazard factor* (Z) which can be evaluated for the city/town listed in Table 3.2 or a *Earthquake hazard map* from Geoscience Australia in Figures 3.2(A) to (G) of AS 1170.4. Z is equivalent to an acceleration coefficient with a P of 1/500 (i.e. a 10% probability of exceedance in 50 years). Most Australian cities have a Z between 0.05 to 0.10.
- determine the *Site sub-soil class* (A_e , B_e , C_e , D_e , E_e) which ranges from strong rock (A_e) to very soft soil (E_e). Most sub-soil site class conditions in major Australian cities would come under Class C_e (shallow soil).
- determine the *Earthquake design category* (EDC)—i.e. EDC I (minimum static check), EDC II (static analysis) or EDC III (dynamic analysis) which is evaluated from the *Importance level*, ($k_p Z$) for *Site sub-soil class*, and *Structure height*.
- When applicable, EDC I type structures apply only to structures with a height from the base of the structure to the uppermost seismic weight/mass (h_n) being limited to $h_n \leqslant 12$ m. Clauses 5.2 and 5.3 of AS 1170.4 [2007] can be used as a minimum for such structures, and provides a simplified method of equivalent static forces being applied to the centre of mass (or at a floor level for a multi-storey building). The horizontal equivalent static design force at the i level (F_i) is determined by

$$F_i = 0.1 W_i$$

where

W_i = seismic weight of the structure or component at level i
= $\Sigma G_i + \Sigma \psi_c Q_i$ (see Clause 6.2.2 of AS 1170.4 [2007] for further details on this calculation, allowance for ice on roofs and values of ψ_c).

With the frame stiffness (i.e. frame geometry, member/joint stiffness and boundary conditions such as connection type to foundations/connections to other stiffness) already set, F_i is now inputted with the gravity loads and the structural analysis can proceed.

- For EDC II type structures not exceeding 15 m in height (and using the same overall simple method as for EDC I):

$$F_i = K_s \left[\frac{k_p Z S_p}{\mu} \right] W_i$$

where

K_s = factor to account for height of a level in a structure and Sub-soil class

S_p = Structural performance factor which is based on frame and ductility type as noted in Clause 6.5 of AS 1170.4 [2007] and AS 4100 AMD 1 (see Appendix D)

μ = Structural ductility factor which is based on frame and ductility type as noted in Clause 6.5 of AS 1170.4 [2007] and AS 4100 AMD 1 (see Appendix D).

With the frame stiffness (i.e. frame geometry, member/joint stiffness and boundary conditions such as connection type to foundations/connections to other stiffness) already set, F_i is now inputted with the gravity loads and the structural analysis can proceed.

- For EDC II type structures exceeding 15 m in height, the following equivalent static analysis applies as noted in Section 6 of AS 1170.4 [2007] to calculate the horizontal equivalent static shear force (V) acting at the base of the structure (base shear):

$$V = \left[k_p Z C_h(T_l) \frac{S_p}{\mu} \right] W_t$$

where

$C_h(T_l)$ = value of the spectral shape factor for the T_l and the Site sub-soil class (Clause 6.4 of AS 1170.4 [2007])

T_l = fundamental natural period of the structure
= $1.25 k_t h_n^{0.75}$

k_t = factor for determining building period based on framing type

W_t = (total) seismic weight of the structure taken as the sum of W_i for all levels.

To complete the inputs prior to the analysis with the above established V, the [revised] F_i to be distributed vertically along the structure at each level i is determined by:

$$\text{[revised] } F_i = k_{F,i} V = \frac{W_i h_i^{k} V}{\sum_{j=1}^{n} (W_j h_j^{k})}$$

where

$k_{F,i}$ = seismic distribution factor for the i th level

h_i = height of level i above the base of the structure

k = exponent, dependent on T_l

n = number of levels in the structure.

V is not actually used in the structural analysis, and as the frame stiffness (i.e. frame geometry, member/joint stiffness and boundary conditions such as connection type to foundations/connections to other stiffness) is already set, the [revised] F_i is now inputted with the gravity loads and the structural analysis can proceed. However, torsional effects, inter-storey drifts and P-delta effects must be taken into account (as required by Clauses 6.6 and 6.7 in AS 1170.4). Inter-storey drift limits for the ultimate limit state are also set for this structural type (Clause 5.4.4 of AS 1170.4 [2007]).

- For EDC III type structures, a dynamic analysis in accordance with Section 7 of AS 1170.4 [2007] must to be undertaken. Inter-storey drift limits for the ultimate limit state are also set this structural type (Clause 5.5.4 of AS 1170.4 [2007]).

- It should be noted that for all EDC categories, overall vertical earthquake actions need not be considered (except for vertical actions being applied to parts and components from earthquake loads applied to the overall structure).
- Clause 2.2 of AS 1170.4 [2007] also notes that serviceability limits are considered satisfied under earthquake actions on structures with Importance levels of 1, 2 and 3 when they are designed in accordance with the Standard and incorporate the appropriate material design Standards. However, a special study is required on Importance level 4 structures to ensure they remain serviceable.

To re-align the Australian Earthquake design and actions/loads Standards, and as noted in Section 1.14(a), the amendment to AS 4100 (AS 4100 AMD 1—see Appendix D) totally revises Section 13 of AS 4100 which (like before) provides the additional minimum design and detailing requirements for structures, members and connections.

AS 4100 AMD 1 (Appendix D) notes that all the existing Section 13 provisions of AS 4100 be deleted and replaced with its amended provisions. However, such a directive is mostly a "re-badge" (i.e minor alteration in terminology, format and text placement) with the following noteworthy changes:

- Definition changes to align with AS 1170.4 [2007] in terms of altered terminology, modelling and design philosophy.
- Change in the requirement of non-structural elements to accommodate relative movements between storeys (transferring the provision to AS 1170.4 [2007].).
- The deletion of *Earthquake Design Categories* (A, B, C, D, E) to be replaced by ductility-classified groupings based on the Structural ductility factor (μ)—e.g. *'limited ductile' steel structures* (μ = 2), *'moderately ductile' steel structures* (μ = 3) and *'fully ductile' steel structures* (μ > 3).
- For *Bearing wall and building frame systems*, the deletion of the minimum compressive capacity for Brace Members in structures over two storeys.
- Reference is made to NZS 1170.5 and NZS 3404 for μ > 3 steel structures whereas limited guidance was previously given by AS 4100 in this area.
- Deletion of the Clause on *Design requirements for non-building structures.*

See also Section 1.14(a) and (b).

3.6 Other actions

As specified in AS/NZS 1170.0, various other actions must be considered in the design of buildings and other structures. Where relevant, such actions include:

- snow and ice loads (as noted in AS/NZS 1170.3)
- retaining wall/earth pressures
- liquid pressures
- ground water effects
- rainwater ponding on roofs
- dynamic actions of installed machinery, plant, equipment and crane loads
- vehicle and vehicle impact loads
- temperature effects (changes and gradients)
- construction loads

- silo and containment vessel loads
- differential settlement of foundations
- volumetric changes (e.g. shrinkage, creep)
- axial shortening from imposed actions
- special structure requirements and responses.

Due to space constraints, these highly specialised actions are not considered in this Handbook, and are covered in the relevant supporting publications of the topic. The suite of AS/NZS 1170 and AS 1170 standards and their supplementary commentaries shed further light on these other actions, and may provide a starting point for seeking further information.

3.7 Notional horizontal forces

Within the General Design Requirements (Section 3) of AS 4100, Clause 3.2.4 requires designers to consider notional horizontal forces for (only) multi-storey buildings. The horizontal force is equal to 0.002 (0.2%) times the design vertical loads for each particular floor level and is considered to act in conjunction with the vertical loads. The rationale for this provision is to allow for the minimum horizontal actions arising from the 'out-of-plumb' tolerance limits for erected columns in such structures (i.e. 1/500 from Clause 15.3.3 of AS 4100). It should be noted that these 'notional' horizontal forces are for action combinations involving vertical permanent and imposed loads only, and need not be used in combination with the following:

- other imposed horizontal/lateral actions (e.g. wind, earthquake actions)
- 2.5% restraint forces used for the design of beam and column restraints
- any of the limit states except for strength and serviceability.

However, structural steel designs in Australia and New Zealand must not only comply with AS 4100 and NZS 3404 respectively, but also AS/NZS 1170.0. At the time of writing the previous edition, Clause 6.2.2 of AS/NZS 1170.0 specified a "minimum lateral resistance" of the structure as being 2.5% of the sum of vertical actions—i.e. $(G + \psi_c Q)$ for each level. This is much higher than the notional horizontal force requirement in Clause 3.2.4 of AS 4100 but still had to be observed. AS/NZS 1170.0 was then amended in April 2011 (Amendment No. 3) to note the minimum lateral resistance of the structure to be the following percentage of $(G + \psi_c Q)$ for each level for a given direction:

(a) for structures over 15 m tall 1%

(b) for all other structures.......................1.5%.

As noted in Section 1.14(a), the amendment to AS 4100 (AS 4100 AMD 1—see Appendix D) introduces a new provision (Clause 3.2.5 in AS 4100) on "structural robustness" that specifically directs designers to comply with Section 6 of AS/NZS 1170.0. Even though the 0.2% horizontal force rule is still included in Clause 3.2.4 of AS 4100, it is considered superfluous as the 1–1.5% minimum lateral resistance rule in Clause 6.2.2 of AS/NZS 1170.0 will control when considering minimum lateral design loads.

Clause 6.2.3 of AS/NZS 1170.0 also specifies that all parts of the structure should be interconnected, and members used for that purpose should have connections capable of resisting a force equal to 5% of the relevant sum of $(G + \psi_c Q)$ over the tributary area relying on the connection.

3.8 Temperature actions

Any change of temperature results in deformation of the structure. Uniform temperature variation produces internal action effects only where the structure is constrained at support points. Temperature gradients (i.e. non-uniform temperature distribution) result in internal action effects in addition to deformations (bowing). The main factors to consider are the ambient temperature range, solar radiation, internal heating or cooling, and snow cover.

AS/NZS 1170.0, Supplement 1, Appendix CZ lists the extreme temperatures in Australia.

The action combination factor for temperature actions is taken as 1.25. The thermal coefficient of carbon/manganese steels, α_T, is 11.7×10^{-6} per degree Celsius.

3.9 Silo loads

Material properties, loads and flow load factors for bulk storage structures are given in AS 3774, together with the methods of calculating the load effects for the design of the container walls and the support structure.

3.10 Crane and hoist loads

Loads and dynamic factors for the design of crane structures are specified in AS 1418.1. Methods of calculating load effects for crane runway girders and monorails are covered by AS 1418.18.

3.11 Design action combinations

The combinations of design actions are considered in Section 4 of AS/NZS 1170.0 and in Sections 1.6 and 3.1 of this Handbook.

3.12 Further reading

- All the commentaries to the AS/NZS 1170 series of Standards.
- Specific load/action requirements may be required for platforms, walkways, stairways and ladders—see AS 1657.
- Special loads/actions (e.g. construction loads) may be required for composite steel-concrete systems during construction (e.g. placement of fresh concrete) before the composite structural system is effected (i.e. concrete is cured)—AS 2327.1.
- Loads encountered during the erection of steelwork—see AS 3828.
- For floor vibrations—see Murray [1990], Ng & Yum [2005] and Marks [2010].
- The detailed and rigorous aspects of bridge loading and design is considered in AS 5100.

chapter 4

Structural Analysis

4.1 Calculation of design action effects

The objective of structural analysis is to determine internal forces and moments, or design action effects. The input into structural analysis consists of frame geometry, member and connection properties, and design actions. Design actions are factored loads (e.g. permanent, imposed, wind, etc.) applied to the structure (see Chapter 3). The analysis output includes design action effects (M^*, N^*, V^*) and deformations.

It should be noted that at the time of publication there is a slight mismatch between the notation in AS 4100 and AS/NZS 1170. Consequently, the notation used in this Handbook follows the convention noted below:

- AS 4100 notation for design capacities and design action effects using a superscript asterisk, e.g. M^* stands for design bending moment
- AS/NZS 1170 notation is used for general actions.

Methods of structural analysis range from simple to complex, depending on the structural form. The designer must decide which method will be adequate for the task, having in mind the degree of accuracy required. The time required to assemble data for rigorous analysis is a consideration, hence simple methods are usually employed in preliminary design.

It is a requirement of AS 4100 that second-order effects be considered—that is, the interaction between loads and deformations. In other words, the effects of frame deformations and member curvature must be taken into account either as a part of the frame analysis or separately. Prior to conversion to the limit state design, a 'first-order' analysis was all that was required for most structures, though 'second-order' effects were approximately considered in combined actions. In first-order analysis the deformations of the structure are calculated during the process of solution, but the deformations are assumed to be independent of the design action effects. In reality the deformations of the structure and the design action effects are coupled, and the method of analysis that reflects this is termed 'second-order' analysis.

The second-order effects due to changes in geometry during the analysis are often small enough to be neglected, but can be significant in special types of structures such as unbraced multi-storey frames. The second-order effects can be evaluated either by a second-order analysis method or by post-processing first-order analysis results. The latter method, termed the 'moment amplification' method, approximates the second-order effects within the limitations imposed by Section 4.3 of AS 4100 and onwards. The moment amplification method is particularly useful where manual analysis is carried out, e.g. by moment distribution method. Further information on second-order effects can be found in Bridge [1994], Harrison [1990], Trahair and Bradford [1998], and in the Commentary to AS 4100.

Computer programs such as Microstran, Spacegass, Multiframe and Strand (see Appendix A.4) are capable of carrying out first- and second-order analysis, but the designer should be fully conversant with the underlying theory before attempting to use these programs.

Nowadays, the ready availability of user-friendly computer software packages incorporating second-order analysis sees these software packages providing the most effective way to undertake structural analysis. This would be for various structural configurations including elemental, truss, portal frame (Woolcock et al. [2011]), multi-storey, etc. Only in the simplest situations (e.g. elemental with simple loadings, boundary conditions and design actions) are manual methods employed to determine the structural analysis outcomes required by AS 4100.

4.2 Forms of structure vs analysis method

4.2.1 General

The method of structural analysis should be suited to the structural form and to the degree of accuracy deemed necessary, having regard to the consequence of possible risk of failure. The methods of analysis generally in use are:

- elastic analysis (first- and second-order)
- plastic analysis (first- and second-order)
- advanced analysis.

Structural framing often consists of one or more substructures, which are summarised in Table 4.1.

Table 4.1 Structural framing systems

Type	Description
FS1	Isolated beams or floor systems consisting of a network of beams
FS2	Braced frames and trusses with pin-jointed members
FS3	Braced frames with flexibly jointed members subject to the minimum eccentricity rule
FS4	Braced frames with rigidly jointed members
FS5	Unbraced frames (sway frames) with flexibly jointed members, e.g. beam-and-post type frames with fixed bases
FS6	Unbraced (sway) frames with rigidly jointed members
FS7	Frames with semi-rigidly jointed member connections

For framing systems of type FS1 it is sufficient to carry out simple, first-order elastic analysis as long as the axial compressive forces in the beams are relatively small: say, less than 15% of the nominal axial capacity of the member. For such members there is no need to amplify the bending moments. If a beam is subject to significant axial compressive forces (such as a beam forming part of a horizontal wind-bracing system), the moment amplification method will have to be applied (see Section 4.4).

Framing system FS2 with rotationally free connections using actual pin connections can also be analysed by simple, first-order analysis subject to the same limitations as for system FS1. Should transverse loads be applied to a column or between nodes on a truss member, moment amplification is likely to apply.

The difference between framing systems FS3 and FS2 is that flexible connections transfer small bending moments to the columns, and AS 4100 requires a specific minimum eccentricity to be applied in calculating the column end bending moments. The most appropriate method of analysis is first-order elastic analysis followed by a moment amplification procedure for columns. The beams will normally require no moment amplification unless significant axial compression is induced in them, e.g. beams forming part of the bracing system.

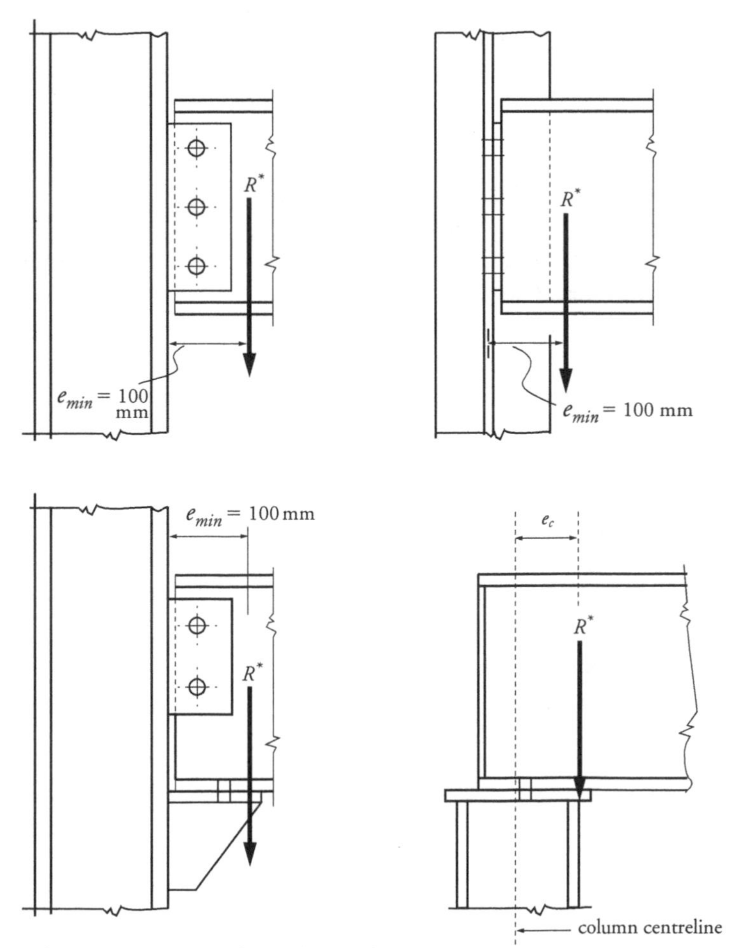

Figure 4.1 Bending moment imposed on columns due to eccentricity

The first-order elastic analysis for framing systems FS1 to FS3 can be quite simple and hand analysis is sufficient in most cases. Such structures can be broken up into individual members for rapid design. A check should be made to ascertain that the connections offer only a minimal rotation resistance. Standard flexible end plates and web cleats are equally suitable.

With framing systems FS4 to FS7, columns attract significant bending moments and therefore the appropriate procedure is to carry out either a second-order analysis or the first-order analysis followed by moment amplification for all members subject to bending and compressive axial force.

It is a requirement of AS 4100 (Clause 4.3.4) that columns carrying simply supported beams (types FS1 and FS2) be designed for a moment arising from the eccentric application of the beam reaction:

$$e_c = d_2 + d_r$$

where d_2 is the distance from the column centre line to the column face adjacent to the beam connection and d_r is taken as the greater of 100 mm or the distance to the centre of bearing, which can be approximated to be equal to one-half of the bracket length in the direction of the beam. Where a beam rests on a column cap plate, the eccentricity e_c is taken as being equal to the distance from the centre line of the column to the column face (see Figure 4.1).

4.2.2 Subdivision into substructures

It is often possible to simplify the analysis by subdividing the total structure into smaller substructures, which are easier to analyse. The subdivision should follow the planes of weak interaction between the substructures.

The best insight into the working of the framing system can be gained by considering the third dimension. Many framing systems are arranged in the form of parallel frames with weak interaction between the parallel frames, which makes it possible to deal with each frame separately. For example, a single-storey warehouse building can be subdivided into portal frames loaded in their plane and infill members (braces, purlins, girts) at right angles to the portal frames.

Where the floor system is relatively rigid in its plane, the frames are forced to deflect laterally by the same amounts but otherwise carry the vertical loads independently. An example of such frame is the multi-storey building frame with a concrete floor over the steel beams. Here the structural model can be simplified for computational purposes by arranging the frames in a single plane with hinged links between them.

Some frames are designed to interact three-dimensionally and should therefore be analysed as one entity. Typical of these are space grid roofs, latticed towers and two-way building frames.

Rigidly joined substructures with no bracing elements rely entirely on the frame rigidity to remain stable. Connections must be designed to be able to transmit all the actions with minimum distortion of the joints. Such frames are usually initially analysed by an elastic or plastic first-order analysis, followed by an assessment of second-order effects to verify their stability against frame buckling.

4.3 Calculation of second-order effects

4.3.1 General

AS 4100 requires that the design action effects due to the displacement of the frame and deformations of the members' action be determined by a second-order analysis or a method that closely approximates the results of a second-order analysis.

All frames undergo some sway deformations under load. Sway deformations of braced frames are often too small to be considered. Unbraced frames rely on the rigidity of the connections, and their sway deformations cannot be neglected. The sway may be caused by lateral or asymmetrical vertical forces but most often by inadequate stability under significant axial forces in the columns.

In performing a simple elastic analysis, the computational process ends with the determination of bending moments, axial and shear forces. Displacements of nodes are determined in the final steps of the analysis, but there is no feedback to include the effects of the changed frame geometry in successive steps of the analysis. Such a method of elastic analysis is known as first-order analysis.

Figure 4.2 shows bending moments and deformations of a single-storey rigid frame subject to vertical and lateral loads. As can be seen, the tops of columns undergo lateral displacement, Δ. Hence they are no longer vertical and straight. Applied vertical forces P_1 and P_2 act on slanted columns and thus tend to displace the nodes further right, with the consequence that the bending moments M_1 and M_2 will increase. This second-order effect is also known as the P-Δ (P-large delta) effect. Additionally, the axial compressive forces in the deformed beam-column members also produce bending moments in the columns, equal to P-δ (that is the P-small delta effect).

It should be noted that the P-Δ effect is primarily due to the relative lateral movement of the member ends from sway frame action. However, the second-order moments from P-δ effects is due to the interaction of individual member curvature (from bending moments) with the axial compression forces present. In this instance there need not be any relative transverse displacement between the member ends (i.e. a braced member) for the second-order moment to occur.

Strictly speaking, member/frame second-order effects can be noticed in the following action effects:

- *bending moments:* from the interaction of axial compression, member curvature from flexure and sway deflections
- *bending moments:* from flexural straining (additional curvature deformations)
- *axial loads:* from axial straining (i.e. shortening or lengthening)
- *bending moments and axial loads:* from shear straining effects.

The last three second-order effects are not significant for typical steelwork applications and are not specifically considered in the body of AS 4100. If required, commonly available structural analysis programs provide non-linear options to consider these second-order effects (see Appendix A.4). Shear straining effects are somewhat rare and can arise from very stubby members with relatively high shear loads. Hence, from an AS 4100 perspective and practically speaking, the only second-order effects to be considered are changes to bending moments from the interaction of axial compression, member curvature and sway deflections.

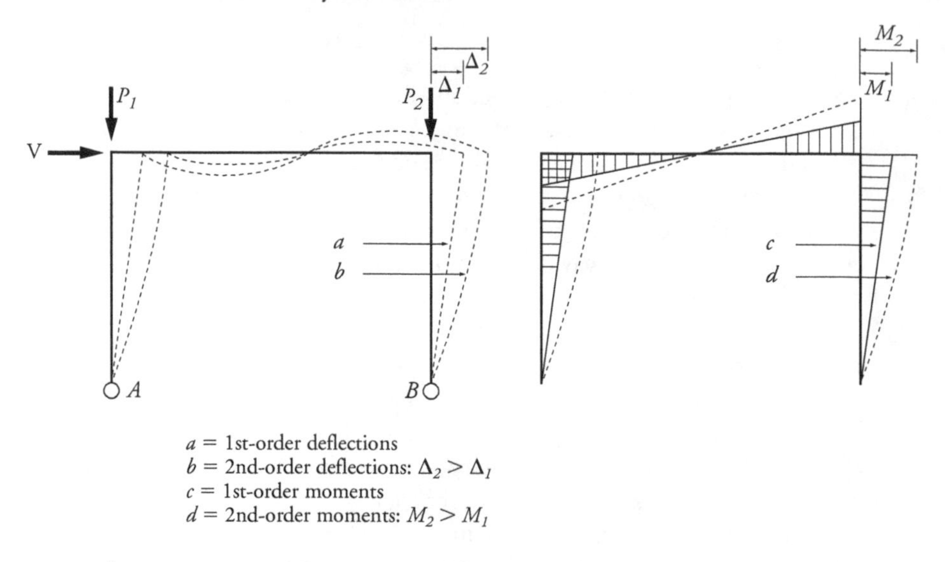

Figure 4.2 *Moment amplification on a single-storey rigid frame*

When a rational second-order elastic analysis is carried out, the design action effects M^*, V^*, N^* are obtained directly from the analysis. The analysis tracks the magnitudes of all displacements as it proceeds to the evaluation of design action effects until all displacements converge. If convergence is not achieved, the structure is regarded as being unstable. No further amplification of bending moments needs to be applied.

There are two strategies for avoiding excessive design effort: one is to use a simplified procedure such as the moment amplification method, and the other is to employ a computer program (e.g. 'Spacegass', 'Microstran', 'Multiframe'—see Appendix A.4) suited to the task. Modelling of the structure for a second-order analysis should be carried out with great care, as the structure must be fully modelled, including the secondary (restraint) members.

AS 4100 allows, as a lower-tier option, replacing the second-order analysis with a simpler manual procedure. Termed the 'moment amplification method' (as described in Clause 4.4 of AS 4100) it can be used for simple structures which can result in an overall saving in time while keeping the process easy to visualise and understand. Section 4.4 describes the method in detail. For further reading on the subject, the reader is directed to Trahair [1992a,b,c,1993a] and Trahair & Bradford [1998].

4.3.2 Escape routes

As discussed earlier, flexural members are normally subject to negligible axial forces and are therefore not subject to second-order effects. Similarly, tension members are not subject to second-order effects. Triangulated frames in which member forces are predominantly axial and no transverse forces are applied between the nodes of the compression chord can also be designed on the basis of first-order analysis alone. This is further elaborated in Sections 4.4.

4.4 Moment amplification method in detail

4.4.1 Basis of the method

As an example, the moment amplification method consists of the following steps:

(a) elastic first-order analysis
(b) calculation of the moment amplification factors
(c) checking that the moment amplification factor δ (or Δ if applicable) does not exceed the value of 1.4
(d) evaluation of design moments in all members subject to the axial compression force, i.e.:

$$M^* \leqslant M\delta$$

An amplification factor greater than 1.4 would indicate that the frame was too flexible and probably would not pass the serviceability check. The options left to the designer in such a case are either to redesign the frame or to try to use an advanced method of analysis and so, one hopes, verify the design.

Any member in the frame subject to axial tension or relatively small axial compressive force (e.g. beams and ties) is assumed to have a moment amplification factor of 1.0.

4.4.2 Moment amplification procedure

A distinction should be made between braced and unbraced (sway) members and frames. In AS 4100 terminology, braced members are those which undergo no sway under load, i.e. no relative transverse displacement between the ends of the members. For example, the members in rectangular frames can be categorised as follows:

- *braced frame* — columns and beams are braced members
- *sway frame* — columns are sway members and beams are braced unless axial compression is significant.

4.4.2.1 Braced members and frames

The procedure for calculating the moment amplification factor is as follows:

(a) Determine elastic flexural buckling load N_{omb} for each braced compressive member:

$$N_{omb} = \frac{\pi^2 EI}{(k_e l)^2}$$

The effective length factor k_e is equal to 1.0 for pin-ended columns and varies from 0.7 to 1.0 for other end conditions, as explained in Section 4.5.

(b) Calculate the factor for unequal moments, c_m. For a constant moment along the member, c_m=1.0, and this is always conservative. For other moment distributions the value of c_m lies between 0.2 and 1.0. The method of calculating c_m is given in Section 4.6.

(c) Calculate the moment amplification factor for a braced member, δ_b, from:

$$\delta_b = \frac{c_m}{\left(1 - \dfrac{N^*}{N_{omb}}\right)} \geqslant 1.0$$

where N^* is the design axial compressive force for the member being considered.

(d) Check that δ_b does not exceed 1.4. Otherwise the frame is probably very sensitive to P-δ effects, and a second-order analysis is then necessary unless the frame stiffness is enhanced.

(e) Multiply bending moments from the first-order analysis by δ_b to obtain the design bending moments:

$$M^* = M_m \delta_b$$

where M_m is the maximum bending moment in the member being considered. This calculation is carried out on a member-by-member basis. See Table 4.2 on the use of the above approximate method for specific structural forms and loading distributions.

4.4.2.2 Sway members and frames

Moment amplification factors for 'sway' members δ_s in regular, rectangular sway frames are calculated as follows:

$$\delta_s = \frac{1}{(1 - c_3)}$$

$$\text{where } c_3 = \frac{(\Delta_s \Sigma N^*)}{(h_s \Sigma V^*)}$$

Δ_s is the translational displacement of the top relative to the bottom in storey height (h_s) from a first-order analysis, ΣN^* is the sum of all design axial forces in the storey under consideration and ΣV^* is the sum of column shears in the storey under consideration. See structural form and loading no. 8 in Table 4.2 for a description of these parameters. The above procedure provides a generally conservative approach and is termed the storey shear-displacement moment amplification method.

Alternatively, δ_s can be calculated from elastic buckling load factor methods such that:

$$\delta_s = \frac{1}{1 - \left(\frac{1}{\lambda_{ms}}\right)} = \frac{1}{1 - \left(\frac{1}{\lambda_c}\right)}$$

where

λ_{ms} = elastic buckling load factor for the *storey* under consideration

$$= \frac{\Sigma\left(\frac{N_{oms}}{l}\right)}{\Sigma\left(\frac{N^*}{l}\right)}$$ for rectangular frames with regular loading and negligible axial forces in the beams

N_{oms} = elastic flexural buckling load (N_{om}) for a sway member (see Section 4.5)

l = member length = storey height

N^* = member design axial force with tension taken as negative and the summation includes all columns within a storey

λ_c = elastic buckling load factor determined from a rational buckling analysis of the *whole* frame

= the lowest of all the λ_{ms} values for a multi-storey rectangular sway frame

If λ_{ms} or λ_c is greater than 3.5, then δ_s will exceed 1.4, which means that a rational second-order analysis is required or the frame stiffness needs to be enhanced. For many situations, and when possible, the storey shear-displacement moment amplification method is used when first-order deflections are known. When these deflections are unknown, the λ_{ms} method is used, but there is the requirement of regular loading on the frame with negligible axial forces in the beams. The λ_c method requires a rational buckling method analysis program that may not be commonly available.

In conjunction with the evaluation of δ_s there is also a requirement to calculate δ_b (see Section 4.4.2.1) for the sway member, i.e.:

$$\delta_b = \frac{c_m}{\left(1 - \frac{N^*}{N_{omb}}\right)} \geqslant 1.0$$

The overall moment amplification factor for a sway member/frame (δ_m) can now be evaluated:

If $\delta_b > \delta_s$ then $\delta_m = \delta_b$, otherwise $\delta_m = \delta_s$.

A value of $\delta_m > 1.4$ indicates that the frame may be very sway sensitive and the above simplified method may have no validity with second-order analysis being necessary.

The final stage is to multiply bending moments from the first-order analysis by δ_m to obtain the design bending moments:

$$M^* = M_m \delta_m$$

This calculation is carried out on a member-by-member basis. See Table 4.2 on the use of the above approximate method for specific structural forms and loading distributions.

4.4.3 Limitations and short cuts

Excluded from the approximate methods in Section 4.4.2 are non-rectangular frames such as pitched portal frames (having a pitch in excess of 15 degrees). Also excluded are frames where beams are subject to relatively high axial compressive loads, highly irregularly distributed loads or complex geometry. These frames must be analysed by a second-order or rational frame buckling analysis, as outlined in AS 4100.

The above methods can be applied successfully to a majority of rectangular frames as long as the following limitations apply:

- live loads are relatively regularly distributed through all bays
- frames are of sufficient rigidity, i.e. the moment amplification factors are less than 1.4
- for members within a frame, the member is not subject to actions from an adjacent member which is a critically loaded compression member and consequently the adjacent member will increase the moments of the member under consideration.

It should also be noted that some references (e.g. AS 4100 Commentary) suggest that second-order effects of less than 10% may be neglected.

Portal frames with sloping rafters may be treated as rectangular frames, provided that the pitch of the rafters does not exceed 15 degrees. Columns of such frames are usually lightly loaded in compression and can be expected to have an amplification factor between 1.0. and 1.15.

Braced rectangular frames will normally fall into the 'unity amplification' category if their columns are bent in double curvature ($\beta_m>0$, $c_m<0.6$), see Section 4.6.

Trusses and other triangulated frames of light to medium construction would normally need no second-order analysis, as the bending moments in members are normally quite small. Where members in the web are relatively stocky in the plane of the truss, say l/r <60, the moment amplification factor should be computed.

In general, it is worth noting that the aspects likely to increase the value of the moment amplification factor are the following:

- a high ratio of design compressive axial load to the elastic flexural buckling load of the member
- bending moments producing single curvature bending combined with relatively high axial compression load.

Table 4.2 gives some examples of members and framing systems with suggested methods of analysis.

Table 4.2 Suggested methods of analysis and use of moment amplification factors

Case	1st or 2nd order (Braced or sway)	Moment amplification factor	Structural form and loading
1	1 Braced	$\delta_b = \dfrac{c_m}{\left(1 - \left(\dfrac{N^*}{N_{omb}}\right)\right)} \geq 1.0$	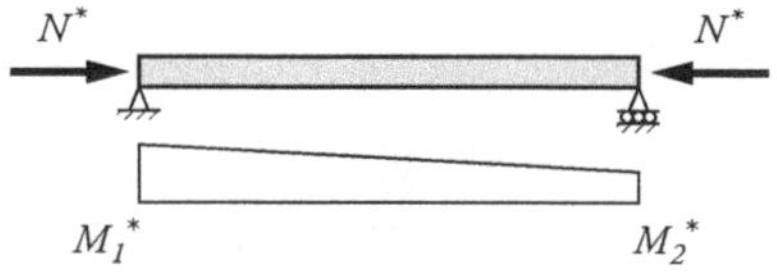 N^* N^* M_1^* M_2^*
2	1 Braced	$\delta_b = 1.0$ (Tension forces)	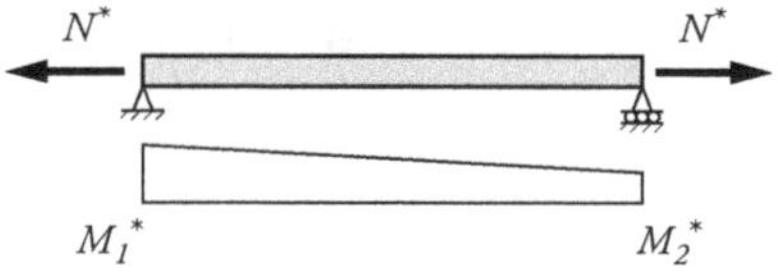 N^* N^* M_1^* M_2^*
3	1 Braced	$\delta_b = 1.0$ (No compressive force)	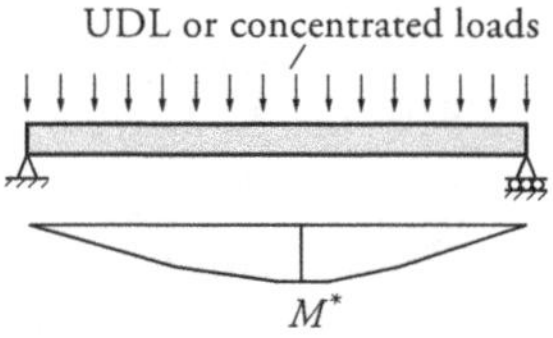 UDL or concentrated loads M^*
4	1 Braced	$\delta_b = \dfrac{c_m}{\left(1 - \left(\dfrac{N^*}{N_{omb}}\right)\right)} \geq 1.0$	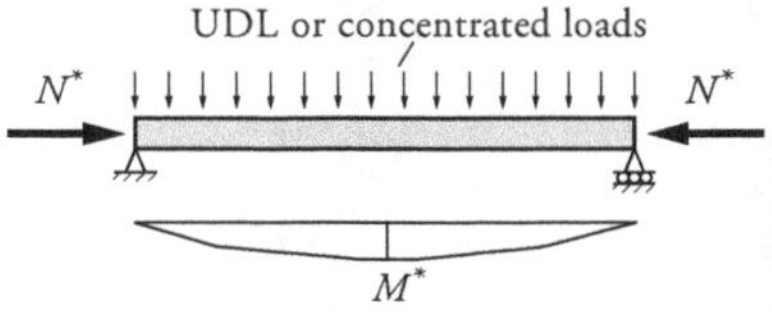 UDL or concentrated loads N^* N^* M^*

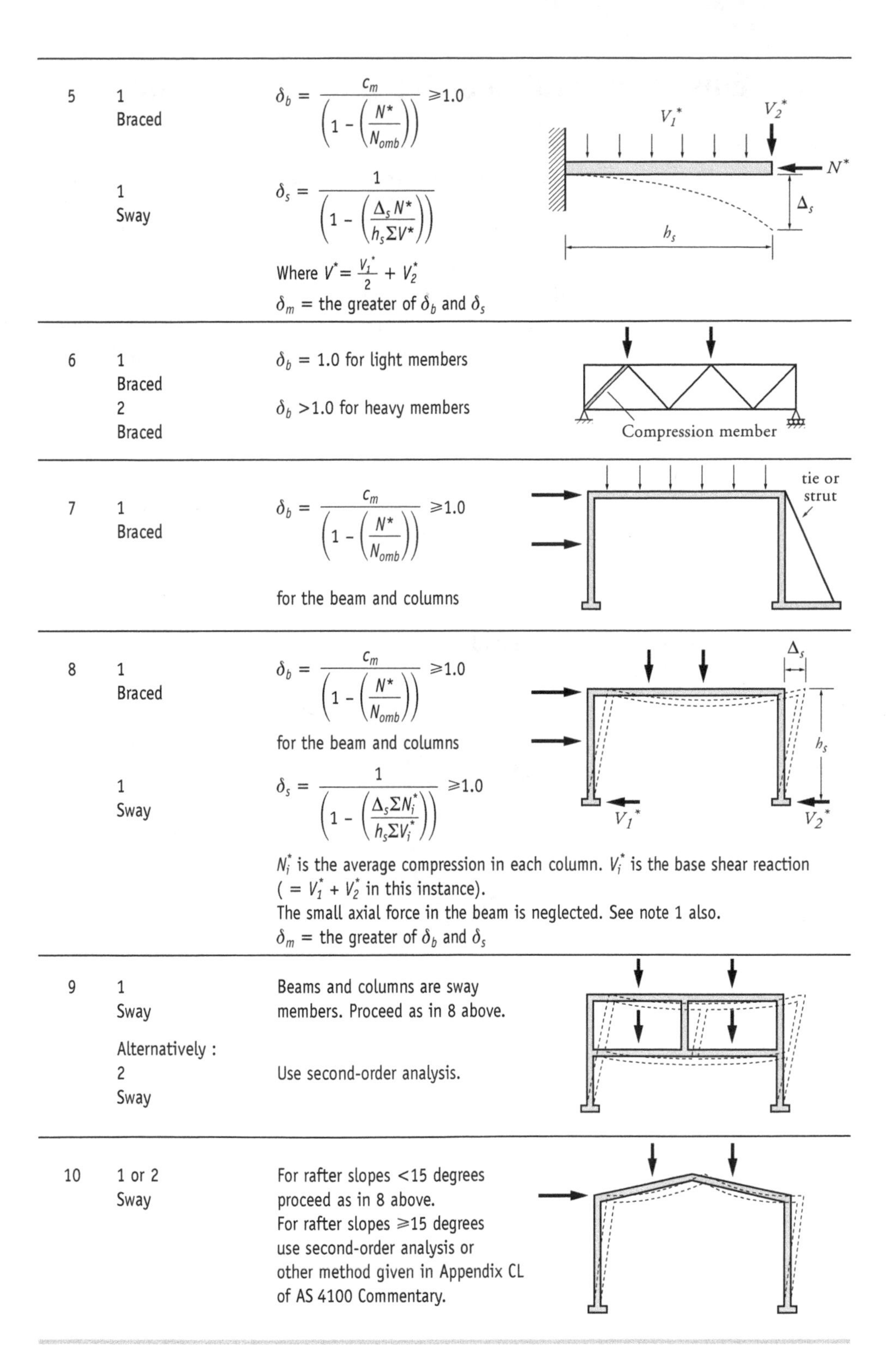

5	1 Braced	$\delta_b = \dfrac{c_m}{\left(1 - \left(\dfrac{N^*}{N_{omb}}\right)\right)} \geqslant 1.0$
	1 Sway	$\delta_s = \dfrac{1}{\left(1 - \left(\dfrac{\Delta_s N^*}{h_s \Sigma V^*}\right)\right)}$ Where $V^* = \dfrac{V_1^*}{2} + V_2^*$ δ_m = the greater of δ_b and δ_s
6	1 Braced 2 Braced	δ_b = 1.0 for light members δ_b >1.0 for heavy members
7	1 Braced	$\delta_b = \dfrac{c_m}{\left(1 - \left(\dfrac{N^*}{N_{omb}}\right)\right)} \geqslant 1.0$ for the beam and columns
8	1 Braced	$\delta_b = \dfrac{c_m}{\left(1 - \left(\dfrac{N^*}{N_{omb}}\right)\right)} \geqslant 1.0$ for the beam and columns
	1 Sway	$\delta_s = \dfrac{1}{\left(1 - \left(\dfrac{\Delta_s \Sigma N_i^*}{h_s \Sigma V_i^*}\right)\right)} \geqslant 1.0$ N_i^* is the average compression in each column. V_i^* is the base shear reaction ($= V_1^* + V_2^*$ in this instance). The small axial force in the beam is neglected. See note 1 also. δ_m = the greater of δ_b and δ_s
9	1 Sway	Beams and columns are sway members. Proceed as in 8 above.
	Alternatively : 2 Sway	Use second-order analysis.
10	1 or 2 Sway	For rafter slopes <15 degrees proceed as in 8 above. For rafter slopes $\geqslant$15 degrees use second-order analysis or other method given in Appendix CL of AS 4100 Commentary.

Note 1: The items 8 and 9 expression for δ_s is based on the storey shear-displacement moment amplification method. The alternative Unbraced Frame Buckling Analysis method (i.e. λ_{ms}) may also be used as an approximate method (see Section 4.4.2.2).

4.5 Elastic flexural buckling load of a member

4.5.1 General

In the previous sections frequent reference is made to the elastic flexural buckling load of a member, N_{om}:

$$N_{om} = \frac{\pi^2 EI}{(k_e l)^2}$$

where I is the second moment area in the relevant buckling mode, k_e is a factor used in effective length calculations and l is the 'system' length of the member—that is, the length between the centres of the intersecting members or footings.

The value of the effective length factor, k_e, depends on the stiffness of the rotational and translational restraints at the member ends. The effective length (l_e) of a column-type member is readily calculated as:

$$l_e = k_e l$$

For a braced member, k_e has a value between 0.7 and 1.0 (see Figure 4.6.3.2 of AS 4100). A sway member will have a k_e larger than 1.0 and has no defined upper limit. Generally, the end restraint condition of a compression member can be divided into two categories: those with idealised end restraints, and those which are part of a frame with rigid connections.

4.5.2 Members with idealised end connections

Clause 4.6.3.2 of AS 4100 lists the applicable effective length factor (k_e) for the combination of idealised connection types at the end of a compression member (e.g. pinned or encased/fixed in rotation and braced or sway in lateral translation). This is summarised in Figure 4.3.

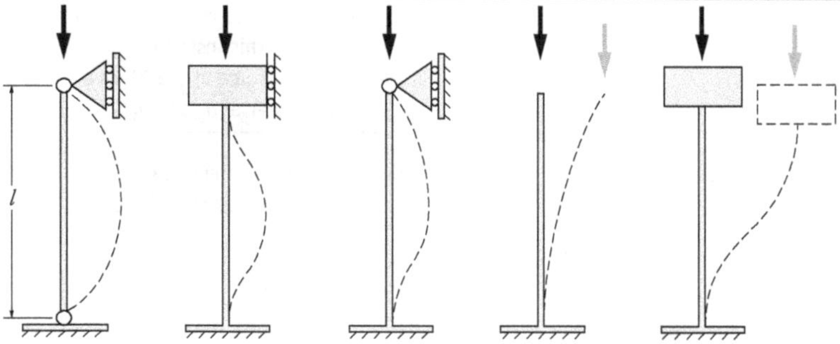

Characteristic	Case				
Rotational end restraint					
Top	PIN	FIX	PIN	NIL	FIX
Bottom	PIN	FIX	FIX	FIX	FIX
Translational restraint					
Top	R	R	R	NIL	NIL
Bottom	R	R	R	R	R
$k_e = \frac{l_e}{l}$	1.0	0.7	0.85	2.2	1.2

Legend: PIN = pinned; FIX = fixed; R = restrained.

Figure 4.3 Effective length factor for members with idealised end constraints

4.5.3 Members in frames

Many end connections of compression members to other steelwork elements cannot be categorised as ideal, as they may fit somewhere between the fixed and pinned type of connection. This is particularly the case for members in frames, as noted in Figure 4.4, where the connection type will influence the buckled shape of the individual members and the overall frame.

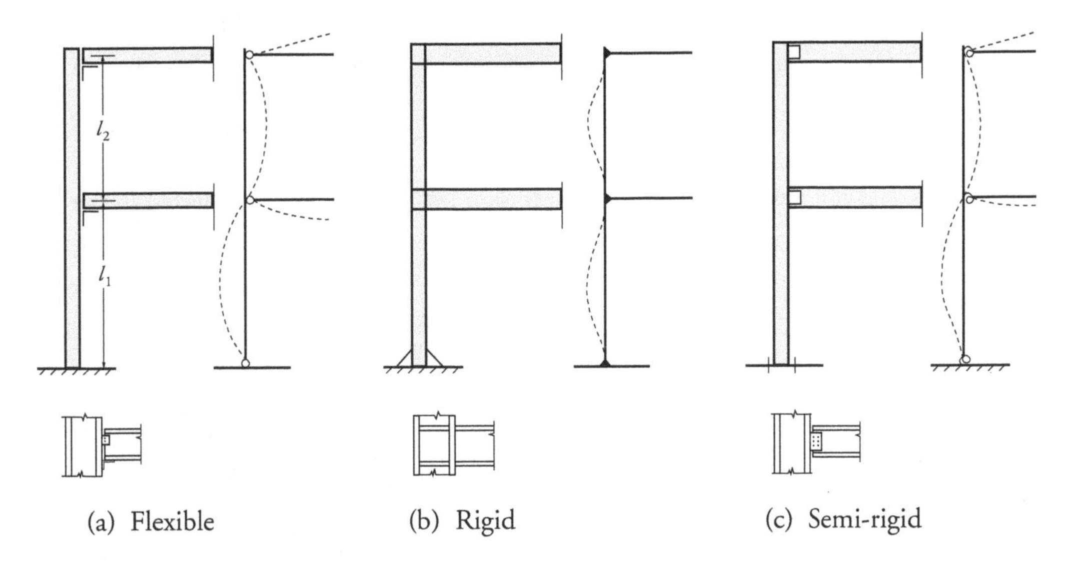

(a) Flexible (b) Rigid (c) Semi-rigid

Figure 4.4 Influence of member connection fixity on buckled member and frame shape

Where a compression member is a part of a rectangular frame with regular loading and negligible axial forces in the beams, the method given in Clauses 4.6.3.3 and 4.6.3.4 of AS 4100 should be used. The method consists of evaluating the restraint stiffness, γ_1 and γ_2, at each column end, that is:

$$\gamma = \frac{S_c}{S_b}$$

where S_c and S_b are the flexural stiffnesses of the columns and beams respectively, meeting at the node being considered, giving the combined stiffnesses:

$$S_c = \Sigma\left(\frac{I_c}{l_c}\right)$$

$$S_b = \Sigma\left(\frac{I_b}{l_b}\right)\beta_e$$

where I_c and I_b are respectively the column and beam second moment of area for the in-plane buckling mode, l_c and l_b are column and beam lengths respectively and β_e is a modifying factor that varies with the end restraint of the beam end opposite the column connection being considered. Table 4.3 lists the values of β_e as noted in Clause 4.6.3.4 of AS 4100.

Table 4.3 Modifying factor, β_e, for connection conditions at the far beam end

Far end fixity	Modifying factor, β_e, for member type being restrained by beam	
	Braced	Sway
Pinned	1.5	0.5
Rigidly connected to the column	1.0	1.0
Rotationally fixed	2.0	0.67

For columns with end restraints to footings, Clause 4.6.3.4 of AS 4100 also stipulates the following:

- column end restraint stiffness, γ, is not less than 10 if the compression member is *not* rigidly connected to the footing, or
- γ is not less than 0.6 if the compression member is rigidly connected to the footing.

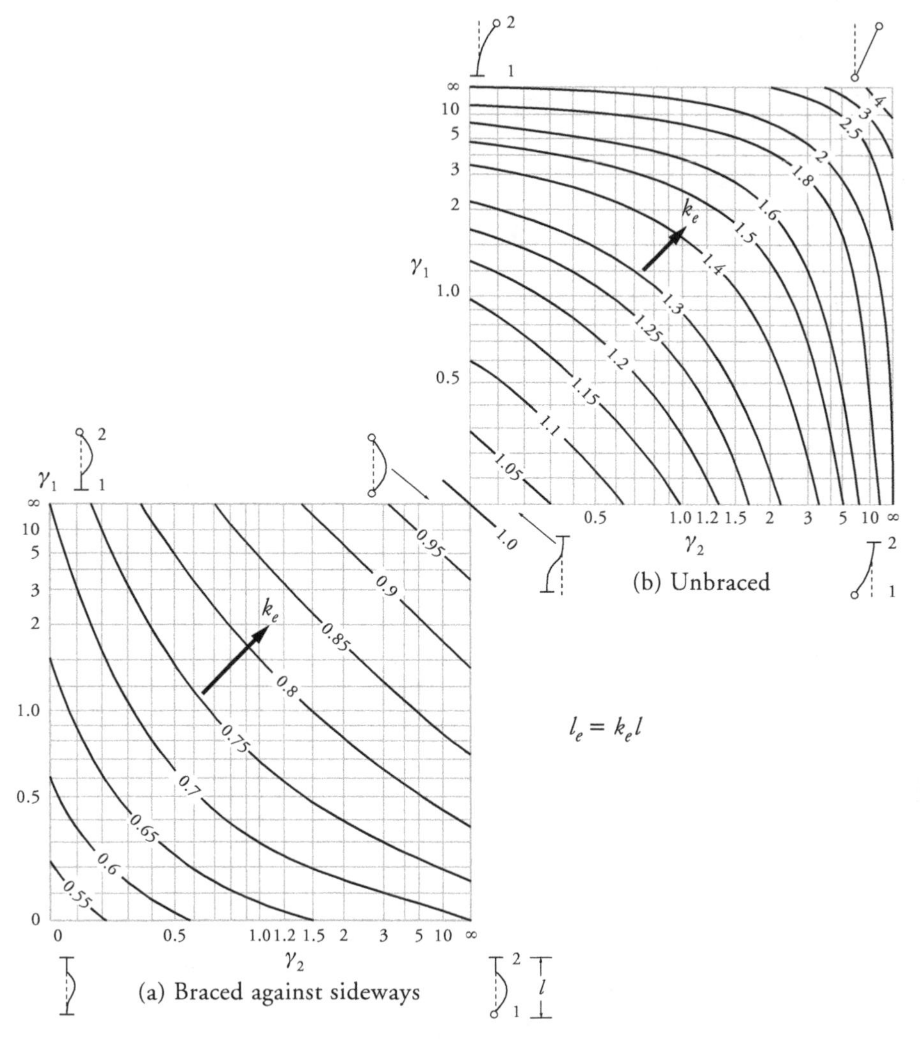

Figure 4.5 Effective length factor k_e in accordance with AS 4100

Based on the above, it is suggested that pinned type footing connections should have $\gamma = 10$ and $\gamma = 0.6$ for rigidly connected footing connections—unless, as AS 4100 points out, a rational analysis can substantiate another value.

Two values of γ are required, one at each end, γ_1 and γ_2. AS 4100 gives graphs for evaluating k_e based on γ_1 and γ_2. One graph is for sway frames and the other for braced frames, and both are reproduced in Figure 4.5. Examples 4.1 and 4.2 in Section 4.7 illustrate the use of these graphs.

For members in triangulated structures, AS 4100 offers two options:

- taking k_e as 1.0 for effective lengths from node to node
- carrying out a rational buckling analysis.

The second option can sometimes be satisfied by using published solutions (e.g. Packer & Henderson [1997]).

4.6 Calculation of factor for unequal end moments c_m

This Section should be read in conjunction with Section 4.4.2 and applies only to the moment amplification method. The highest moment amplification factor occurs when the bending moment is uniform along the member, resulting in $c_m = 1.0$.

Usually, the bending moment varies along the member length and c_m is less than 1.0. The value of c_m is calculated from Clause 4.4.2.2 of AS 4100:

$$c_m = 0.6 - 0.4\beta_m \leqslant 1.0$$

where the coefficient β_m is calculated from the ratio of the smaller to the larger bending moment at member ends:

$$\beta_m = \frac{M_1}{M_2}$$

where M_1 is the numerically smaller moment. The sign of β_m is negative when the member is bent in single curvature, and positive when bent in reverse curvature, thus:

$\beta_m = -1$ for uniform moment distribution

$\beta_m = +1$ for a moment distribution varying along the member from $+M$ to $-M$ (reverse curvature)

The above expressions for β_m are based on bending moment distributions arising from end moments only. However, AS 4100 also permits the above equation for c_m to be used for transverse and moment loads, with β_m being determined as:

(a) $\beta_m = -1.0$ (conservative)

(b) using Figure 4.4.2.2 in AS 4100 with varying bending moment distributions due to uniform distributed loads, concentrated loads, end moments, concentrated mid-span moments, uniform moments, etc. acting either singly or in combination. Figure 4.4.2.2 in AS 4100 is very useful for determining β_m as long as it has a similar bending moment diagram along the member

(c) $\beta_m = 1 - \left(\frac{2\Delta_{ct}}{\Delta_{cw}}\right)$ with $-1.0 \leqslant \beta_m \leqslant 1.0$

where Δ_{ct} is the mid-span deflection of the member loaded by transverse loads together with end moments. The value Δ_{cw} is calculated in the same way but without those end moments, which tend to reduce the deflection.

Table 4.4 summarises the value of β_m as noted in (b) above.

Table 4.4 Summary of β_m values as noted in Figure 4.4.2.2 of AS 4100

Moment distribution	Curve	β_m	c_m	Moment distribution	Curve	β_m	c_m
End moments only: contra-flexure & single end	a	0	0.6	End moments only: single curvature & single end	a	-1.0	1.0
	b	+0.5	0.4		b	-0.5	0.8
	c	+1.0	0.2		c	0	0.6
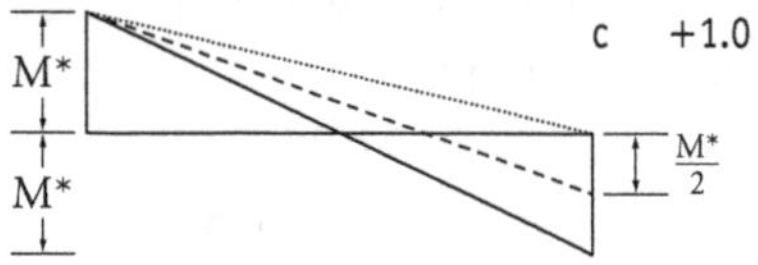				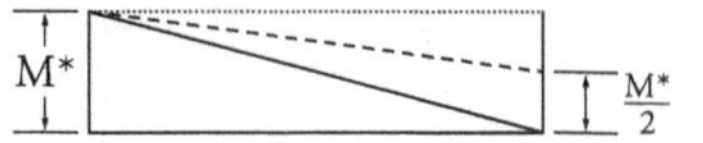			
UDL &/or end moments	a	-0.5	0.8	Mid-span concentrated force &/or equal end moments	a	-1.0	1.0
	b	-0.2	0.68		b	+0.5	0.4
	c	β	**		c	+1.0	0.2
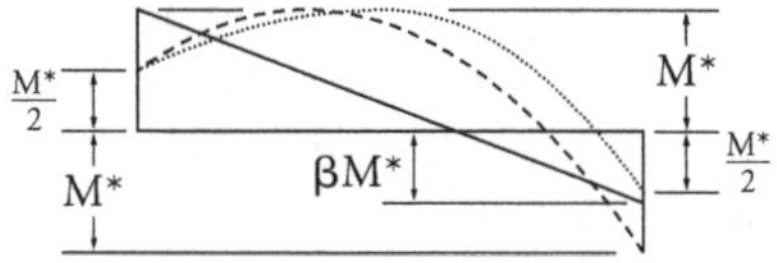				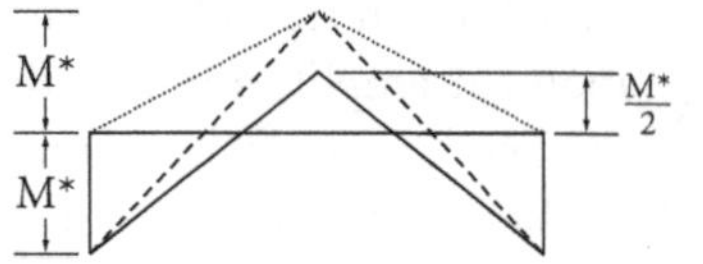			
UDL + equal end moments	a	-1.0	1.0	Mid-span concentrated force + single end moment	a	+0.4	0.44
	b	+0.2	0.52		b	0	0.6
	c	+0.6	0.36		c	+0.5	0.4
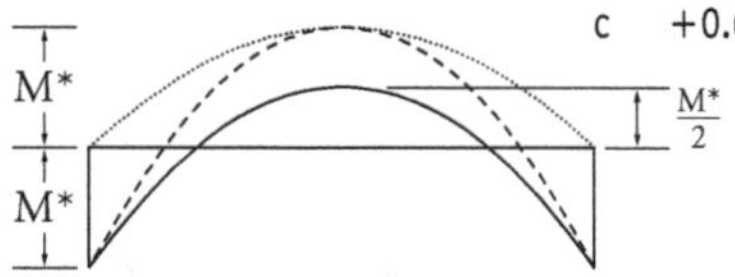				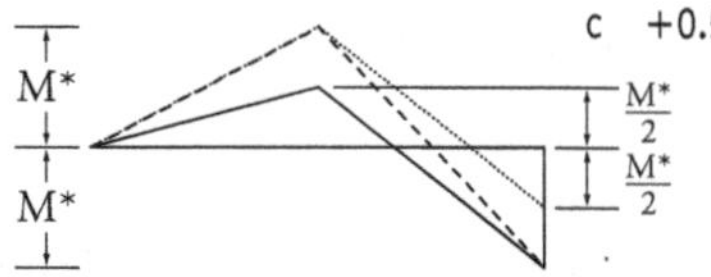			
UDL + unequal end moments	a	-0.4	0.76	Mid-span moment &/or end moments	a	+1.0	0.2
	b	+0.1	0.56		b	-0.4	0.76
	c	+0.7	0.32		c	-0.1	0.64
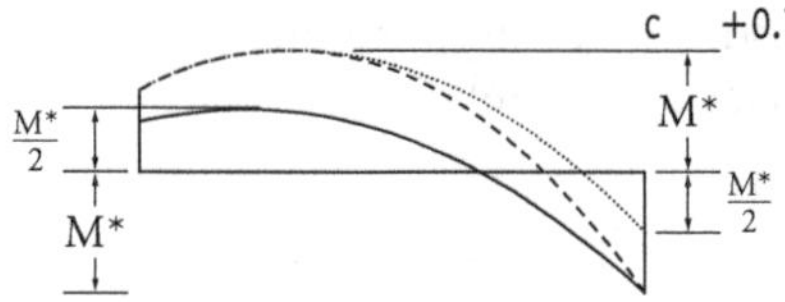				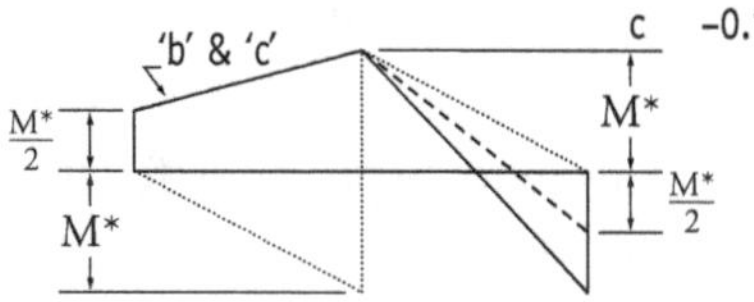			
UDL + single end moment	a	-0.5	0.8	Mid-span moment + end moments	a	-0.5	0.8
	b	+0.2	0.52		b	-0.1	0.64
	c	+0.2	0.52		c	+0.3	0.48
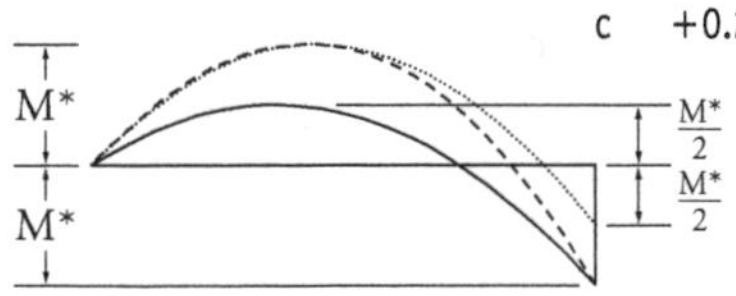				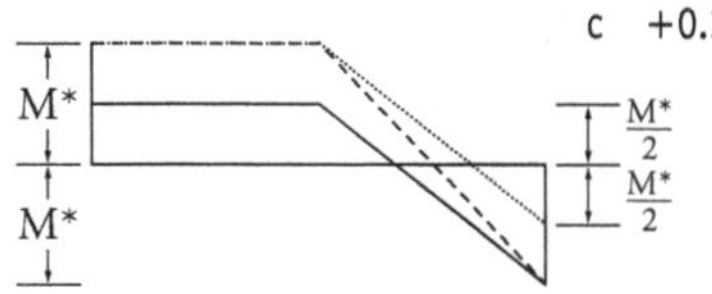			

Note: .... indicates moment distribution curve 'a'
- - - - - - - indicates moment distribution curve 'b'
———— indicates moment distribution curve 'c'
** indicates $c_m = 0.6 - 0.4\beta$ where β is positive when the member is in double curvature.

4.7 Examples

4.7.1 Example 4.1

Step	Description and calculations	Result	Units

Problem definition: Determine the design moments for the braced frame shown below by using the method of moment amplification.

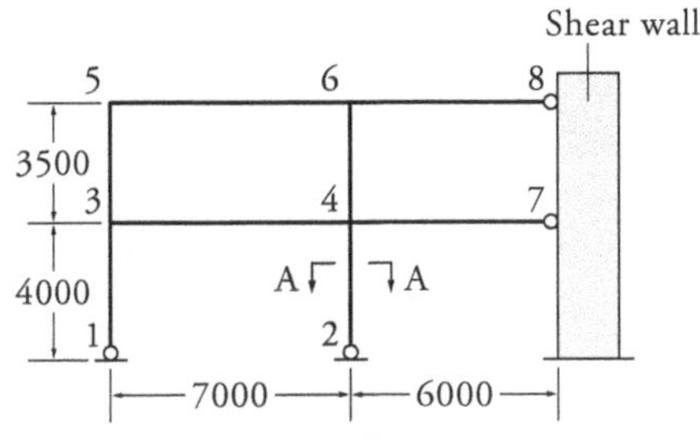

(a) Geometry

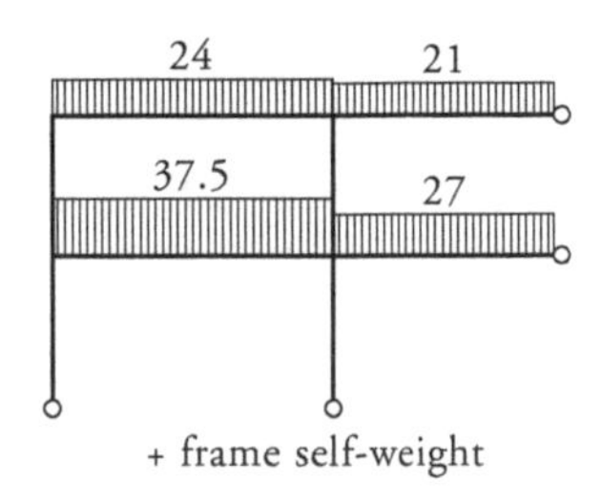

(b) Loads (kN/m)

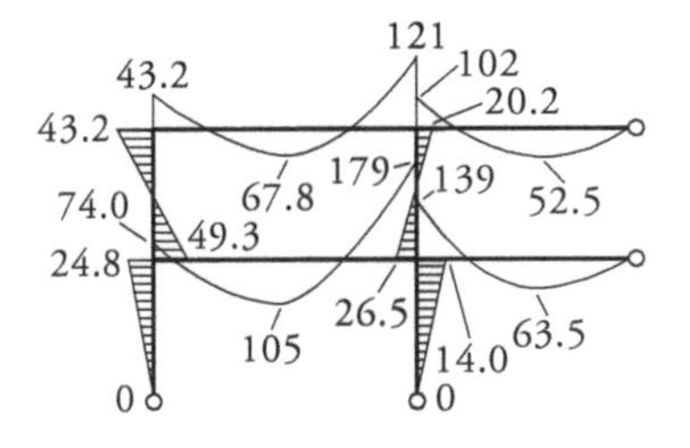

(c) Bending moments (kNm) from elastic first-order analysis (only column bending moment shown with cross-hatching for clarity)

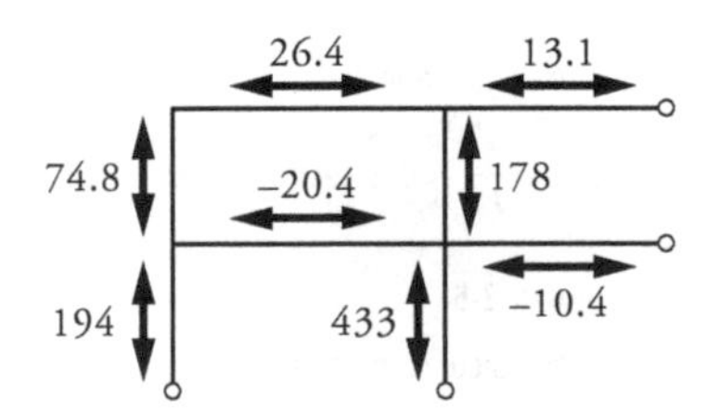

(d) Axial forces (kN) from elastic first-order analysis (negative values denote tension)

This example considers a braced frame and constituent members with out-of-plane behaviour prevented. The beams are also relatively lightly loaded in compression.

Note: Notional horizontal forces (Clause 3.2.4 of AS 4100 and Section 3.7 of this Handbook) are not considered in this instance as the frame is braced and the maximum lateral force is not significant (less than 1 kN for the first floor level).

Load factors and combinations are for the strength limit state.

1. Trial section properties used in the analysis:

Member	Section	I_x mm^4
1-3, 3-5	150 UC 30.0	17.6×10^6
2-4, 4-6	200 UC 46.2	45.9×10^6
Beams	310 UB 40.4	86.4×10^6

2. Determine the member elastic buckling loads:

(a) Column 1-3:

γ_1: pinned base (Clause 4.6.3.4(a) of AS 4100) $= 10.0$

$$\gamma_3 = \frac{\Sigma\left(\frac{I}{l}\right)_c}{\Sigma\left(\frac{\beta_e I}{l}\right)_b}$$

$$= \frac{\left(\frac{17.6}{4.0}\right) + \left(\frac{17.6}{3.5}\right)}{\left(1.0 \times \frac{86.4}{7}\right)} = 0.764$$

k_e from Fig. 4.5(a) $= 0.84$

$$l_e = k_e l$$

$$= 0.84 \times 4000 = 3360 \text{ mm}$$

$$N_{om} = \frac{\pi^2 EI}{l_e^2}$$

$$= \frac{\pi^2 \times 200 \times 10^3 \times 17.6 \times 10^6}{3360^2} \times 10^{-3} = 3080 \text{ kN}$$

$$\therefore \frac{N^*}{N_{om}} = \frac{194}{3080} = 0.0630$$

As $\frac{N^*}{N_{om}} < 0.1$ second-order effects can be neglected.

(b) Column 3-5:

γ_3: as calculated in (a) above $= 0.764$

$$\gamma_5 = \frac{\left(\frac{17.6}{3.5}\right)}{\left(1.0 \times \frac{86.4}{7}\right)} = 0.407$$

k_e from Fig. 4.5(a) $= 0.70$

$$l_e = k_e l$$

$$= 0.70 \times 3500 = 2450 \text{ mm}$$

$$N_{om} = \frac{\pi^2 \times 200 \times 10^3 \times 17.6 \times 10^6}{2450^2} \times 10^{-3} = 5790 \text{ kN}$$

$$\therefore \frac{N^*}{N_{om}} = \frac{74.8}{5790} = 0.0129$$

As $\frac{N^*}{N_{om}} < 0.1$ second-order effects can be neglected.

(c) Column 2-4:

γ_2: pinned base (Clause 4.6.3.4(a) of AS 4100) $= 10.0$

$$\gamma_4 = \frac{\left(\frac{45.9}{4}\right) + \left(\frac{45.9}{3.5}\right)}{\left(1.0 \times \frac{86.4}{7}\right) + \left(1.5 \times \frac{86.4}{6}\right)} = 0.724$$

k_e from Fig. 4.5(a) = 0.84

$$l_e = k_e l$$

$$= 0.84 \times 4000 = 3360 \text{ mm}$$

$$N_{om} = \frac{\pi^2 \times 200 \times 10^3 \times 45.9 \times 10^6}{3360^2} \times 10^{-3} = 8030 \text{ kN}$$

$$\therefore \frac{N^*}{N_{om}} = \frac{433}{8030} = 0.0539$$

As $\frac{N^*}{N_{om}} < 0.1$ second-order effects can be neglected.

(d) Column 4-6:

γ_4: as calculated in (c) above = 0.724

$$\gamma_6 = \frac{\left(\frac{45.9}{3.5}\right)}{\left(1.0 \times \frac{86.4}{7}\right) + \left(1.5 \times \frac{86.4}{6}\right)} = 0.386$$

k_e from Fig. 4.5(a) = 0.69

$$l_e = k_e l$$

$$= 0.69 \times 3500 = 2420 \text{ mm}$$

$$N_{om} = \frac{\pi^2 \times 200 \times 10^3 \times 45.9 \times 10^6}{2420^2} \times 10^{-3} = 15\,500 \text{ kN}$$

$$\therefore \frac{N^*}{N_{om}} = \frac{178}{15{,}500} = 0.0115$$

As $\frac{N^*}{N_{om}} < 0.1$ second-order effects can be neglected.

---

3. Moment amplification factors:

From the above preliminary evaluation of second-order effects, it appears that such effects can be neglected.

To illustrate the calculation of δ_b, a check will be made on the highest-loaded column in terms of $\frac{N^*}{N_{om}}$

– Column 1-3:

$$\beta_m = \frac{M_1}{M_2} = \frac{0}{24.8} = 0$$

$$c_m = 0.6 - 0.4\beta_m = 0.6$$

$$\delta_b = \frac{c_m}{\left(1 - \frac{N^*}{N_{om}}\right)}$$

$$= \frac{0.6}{\left(1 - \frac{194}{3080}\right)} = 0.640$$

Since δ_b <1.0 adopt δ_b = 1.0. Hence, no moment amplification needs to be applied to Column 1-3 (let alone any other column).

The beam members need not be assessed for second-order effects as their axial loads were relatively low (the highest N^*/N_{om} was 0.008) with the first-storey beams being in tension.

The design moments therefore equal the bending moments obtained from the elastic first-order analysis.

An elastic second-order analysis of the total structure indicated an 8% peak bending moment difference (for Column 1-3) to that from a first-order elastic analysis. This level of change in bending moments (<10%) indicates that second-order effects are negligible for the loaded structural frame.

Note: The example represents a typical frame in a low-rise building. Had the axial load in a column exceeded (say) $0.60N_{om}$, then the value of δ_b would have become larger than 1.0. This would be typical of columns in high-rise buildings.

4.7.2 Example 4.2

Step	Description and calculations	Result	Units

Problem definition: Determine the design moments for the sway frame shown below by using the method of moment amplification.

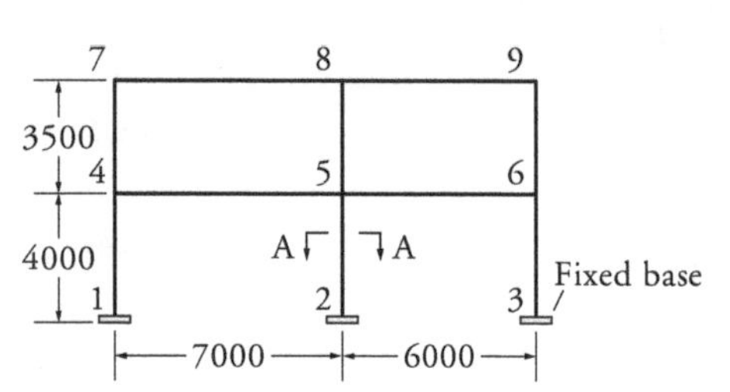

(a) Geometry

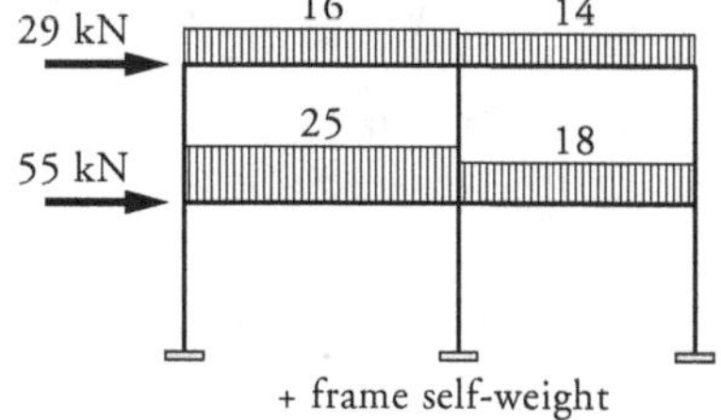

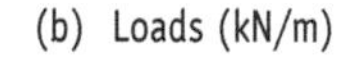

(b) Loads (kN/m)

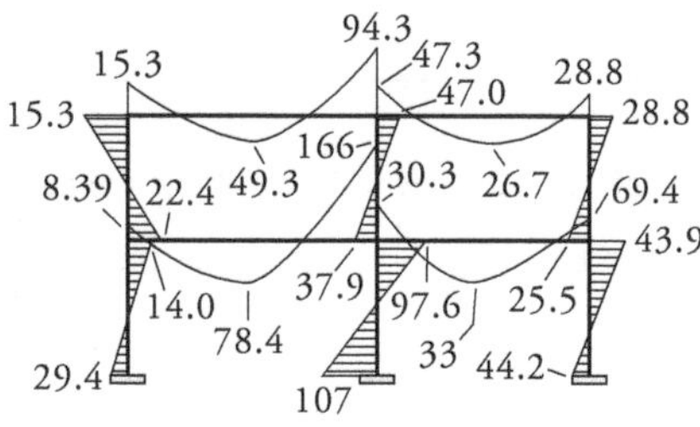

(c) Bending moments (kNm) from elastic first-order analysis (only column bending moment shown with cross-hatching for clarity)

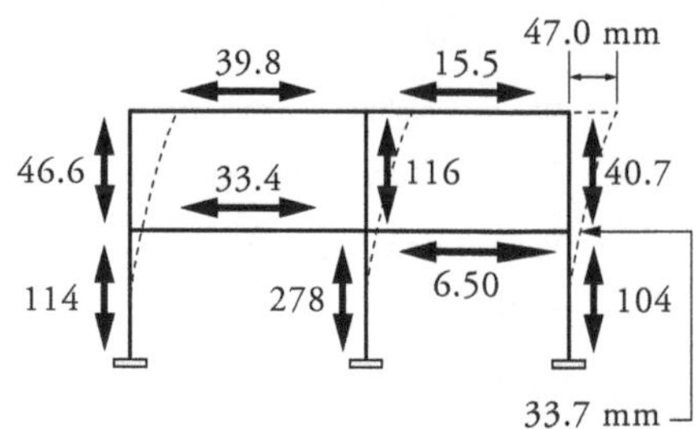

(d) Axial forces (kN) and sway deflections (mm) from elastic first-order analysis (all axial forces are compressive)

This example considers a sway frame and constituent members with out-of-plane behaviour prevented. The beams are also relatively lightly loaded in compression.

Note: Notional horizontal forces (Clause 3.2.4 of AS 4100 and Section 3.7 of this Handbook) are not considered, as there are other imposed lateral forces.

Load factors and combinations are for the strength limit state.

1. Trial section properties used in the analysis:

Member	Section	I_x mm^4
1-4, 4-7	150 UC 30.0	17.6×10^6
2-5, 5-8	200 UC 46.2	45.9×10^6
3-6, 6-9	150 UC 30.0	17.6×10^6
Beams	310 UB 40.4	86.4×10^6

2. Determine the elastic buckling loads (assuming it is a braced member):

(a) Column 1-4:

γ_1: rigid connection to base (Clause 4.6.3.4(b) of AS 4100) $= 0.6$

$$\gamma_4 = \frac{\Sigma\left(\frac{I}{l}\right)_c}{\Sigma\left(\frac{\beta_e I}{l}\right)_b}$$

$$= \frac{\left(\frac{17.6}{4.0}\right) + \left(\frac{17.6}{3.5}\right)}{\left(1.0 \times \frac{86.4}{7}\right)} = 0.764$$

k_e from Fig. 4.5(b) $= 1.22$

$$l_e = k_e l$$

$$= 1.22 \times 4000 = 4880 \text{ mm}$$

$$N_{om} = \frac{\pi^2 EI}{l_e^2}$$

$$= \frac{\pi^2 \times 200 \times 10^3 \times 17.6 \times 10^6}{4880^2} \times 10^{-3} = 1460 \text{ kN}$$

$$\therefore \quad \frac{N^*}{N_{om}} = \frac{114}{1460} = 0.078$$

As $N^*/N_{om} < 0.1$ second-order effects can be neglected.

(b) Column 2-5:

γ_2: rigid connection to base (Clause 4.6.3.4(b) of AS 4100) $= 0.6$

$$\gamma_5 = \frac{\left(\frac{45.9}{4}\right) + \left(\frac{45.9}{3.5}\right)}{\left(1.0 \times \frac{86.4}{7}\right) + \left(1.0 \times \frac{86.4}{6}\right)} = 0.919$$

k_e from Fig. 4.5(b) $= 1.24$

$$l_e = k_e l$$

$$= 1.24 \times 4000 = 4960 \text{ mm}$$

$$N_{om} = \frac{\pi^2 \times 200 \times 10^3 \times 45.9 \times 10^6}{4960^2} \times 10^{-3} = 3680 \text{ kN}$$

$$\therefore \quad \frac{N^*}{N_{om}} = \frac{278}{3680} = 0.076$$

As $\frac{N^*}{N_{om}} < 0.1$ second-order effects can be neglected.

(c) Column 3-6:

γ_3: rigid connection to base (Clause 4.6.3.4(b) of AS 4100) = 0.6

$$\gamma_6 = \frac{\left(\frac{17.6}{4}\right) + \left(\frac{17.6}{3.5}\right)}{\left(1.0 \times \frac{86.4}{6}\right)} = 0.655$$

k_e from Fig. 4.5(b) = 1.20

$$l_e = k_e l$$

$$= 1.20 \times 4000 = 4800 \text{ mm}$$

$$N_{om} = \frac{\pi^2 \times 200 \times 10^3 \times 17.6 \times 10^6}{4800^2} \times 10^{-3} = 1510 \text{ kN}$$

$$\therefore \frac{N^*}{N_{om}} = \frac{104}{1510} = 0.069$$

As $\frac{N^*}{N_{om}}$ <0.1 second-order effects can be neglected.

(d) Columns 4-7, 5-8 and 6-9:

These columns are less critically loaded and lower in effective lengths than their lower-storey counterparts and are not considered further.

---

3. Moment amplification factor for a 'braced' member, δ_b:

From the above preliminary evaluation of second-order effects, it appears that such effects can be neglected. However, to illustrate the calculation of δ_b, a check will be made on the lower-storey columns:

(a) Column 1-4:

$$\beta_m = \frac{M_1}{M_2} = \frac{14.0}{29.4} = 0.476$$

$$c_m = 0.6 - 0.4\beta_m = 0.6 - (0.4 \times 0.476) = 0.410$$

$$\delta_b = \frac{c_m}{\left(1 - \frac{N^*}{N_{om}}\right)}$$

$$= \frac{0.410}{\left(1 - \frac{114}{1460}\right)} = 0.445$$

Calculated $\delta_b < 1.0$, $\therefore \delta_b = 1.0$.

(b) Column 2-5:

$$\beta_m = \frac{M_1}{M_2} = \frac{97.6}{107} = 0.912$$

$$c_m = 0.6 - 0.4\beta_m = 0.6 - (0.4 \times 0.912) = 0.235$$

$$\delta_b = \frac{0.235}{\left(1 - \frac{278}{3680}\right)} = 0.254$$

Calculated δ_b <1.0, $\therefore \delta_b = 1.0$

(c) Column 3-6:

$\beta_m = \frac{M_1}{M_2} = \frac{43.9}{44.2}$ $= 0.993$

$c_m = 0.6 - 0.4\beta_m = 0.6 - (0.4 \times 0.993)$ $= 0.203$

$\delta_b = \frac{0.203}{\left(1 - \frac{104}{1510}\right)}$ $= 0.218$

Calculated $\delta_b < 1.0$, $\therefore \delta_b = 1.0$.

As $\delta_b < 1.0$ adopt $\delta_b = 1.0$ for all of the columns. Hence no 'braced' member moment amplification needs to be applied to the columns.

The beam members need not be assessed for second-order effects, as their axial loads were relatively low (the highest $\frac{N^*}{N_{om}}$ was 0.011).

---

4. Moment amplification factor for a sway member, δ_s:

The second-order effects from sway deformations is now checked by using the storey shear-displacement moment amplification method (Section 4.4.2.2):

(a) Lower storey:

$c_3 = \left(\frac{\Delta_s}{h_s}\frac{\Sigma N^*}{\Sigma V^*}\right)$

$= \frac{33.7}{4000} \times \frac{(114 + 278 + 104)}{(55 + 29)}$ $= 0.0497$

$\delta_s = \frac{1}{(1 - c_3)}$

$= \frac{1}{(1 - 0.0497)}$ $= 1.05$

$(\delta_m)_{ls}$ = moment amplification factor (overall) for lower storey

= max.[δ_b, δ_s] $= 1.05$

where the maximum value of δ_b is 1.0 for the columns (see Step 3 above).

(b) Upper storey:

$c_3 = \frac{(47.0 - 33.7)}{3500} \times \frac{(46.6 + 116 + 40.7)}{29}$ $= 0.0266$

$\delta_s = \frac{1}{(1 - 0.0266)}$ $= 1.03$

$(\delta_m)_{us}$ = moment amplification factor (overall) for upper storey

= max.[δ_b, δ_s] $= 1.03$

where the maximum value of δ_b is 1.0 for the columns (see Step 3 above).

---

5. Calculation of design bending moments:

For each storey, the bending moments from the elastic first-order analysis are multiplied by δ_m.

(a) Lower storey:

M^*_{ls} = amplified peak lower-storey moment

$= (M_m)_{ls} \times (\delta_m)_{ls}$

$= 107 \times 1.05$ $= 112$ kNm

This can similarly be done for the other columns in the storey.

(b) Upper storey:

M^*_{us} = amplified peak upper-storey moment

$= (M_m)_{us} \times (\delta_m)_{us}$

$= 47 \times 1.03$ $= 48.4$ kNm

This can similarly be done for the other columns in the storey.

Due to the structural and loading configuration used in this example it is seen that the overall second-order effects are less than 10% when compared with the results from the elastic first-order analysis. However, higher axial loads or larger deflections would have produced higher amplification factors. As noted in the Commentary to AS 4100, these braced and sway second-order effects could be neglected (δ_m <1.1). However, the above example assists to illustrate the braced and sway checks required for members in sway frames. The method used is valid when amplification factors do not exceed 1.4.

An elastic second-order analysis of the total structure indicated a 5% peak column bending moment difference (for Column 2-5) from that of a first-order elastic analysis, this indicating that the approximate amplification method gave a result close to the rigorous and more accurate non-linear method. However, in this instance, the level of change in bending moments (<10%) from first-order analysis further indicates that second-order effects are negligible for the loaded structural frame.

4.8 Summary

Interestingly, other literature on worked examples for in-plane second-order effects of rectangular steel-framed structures generally get moment amplifications of around 10%. This is particularly the case when suitable section stiffness is used for deflection constraints on column sway etc. One could then surmise that the above second-order effects are minimal for practical structures. This became evident to the authors when developing worked Examples 4.1 and 4.2, where changes of load magnitudes, section stiffness and base restraints for 'realistic' structures only produced second-order effects that were less than 10%. To try to trigger very significant second-order effects, there were some obvious changes tested when 'trialling' worked Example 4.2. These included:

- base restraint changed from rigid to pinned
- decreasing the sections by one size
- altering load magnitude

which then produced second-order effects in the range of 60%–70% (as noted from a second-order analysis program). However, the deformations—particularly the sway deflections—were inordinately excessive in this instance (e.g. storey sway to column height being 1/30 when refactored for serviceability loads). Even though the above suggests that second-order effects may not be significant in realistic rectangular-framed steel structures, the evaluation of these effects should not be dismissed, as they can become relevant for 'flexible' framing systems.

For the evaluation of second-order effects on pitched-roof portal frame buildings, Appendix CL in the Commentary to AS 4100 provides a simple approximation method for evaluating the elastic buckling load factor (λ_c), which then determines the sway amplification factor (δ_s). Two modes of in-plane portal frame buckling modes are considered—symmetrical and sway.

4.9 Further reading

- For additional worked examples see Chapter 4 of Bradford, et al. [1997].
- For a short summary on lower tier structural analysis in AS 4100 see Trahair [1992a]. This is also incorporated into Appendix CL of the AS 4100 Commentary.
- For hand calculation of manual moment amplification methods for specific framing configurations see Trahair [1992b,c,1993a]. These are also incorporated into Appendix CL of the AS 4100 Commentary.
- Additional references on the background to the structural analysis part of AS 4100 can be found in Bridge [1994], Hancock [1994a,b], Harrison [1990], Petrolito & Legge [1995], and Trahair & Bradford [1998].
- For a classical text on structural analysis of rigid frames see Kleinlogel [1973].
- For some authoritive texts on buckling see Bleich [1952], CRCJ [1971], Hancock [2007], Timoshenko [1941], Timoshenko & Gere [1961], Trahair [1993b] and Trahair & Bradford [1998] to name a few.

chapter 5

Beams & Girders

5.1 Types of members subject to bending

The term 'beam' is used interchangeably with 'flexural member'. In general, the subject matter covers most of the AS 4100 rules and some additional design situations. A simple classification of beams and girders is presented in Table 5.1. The purpose of the table is to give an overview of the various design considerations and many types of applications of beams in building and engineering construction. It also serves as a directory to subsections covering the particular design aspect.

Table 5.1 Design aspects covered for beam and girder design.

Aspect	Subgroup	Section
Section type:	Solid bars	5.3 & 5.4
	Hot-rolled sections: UB, UC, PFC	5.3–5.5, 5.7, 5.8
	Plate web girders	5.3–5.8
	Tubular sections: CHS, RHS and SHS	5.3-5.5, 5.7, 5.8
	Fabricated sections:	
	Doubly symmetrical	
	I-section / UB / UC	5.3–5.5, 5.7, 5.8
	Box section / tube	5.3-5.5, 5.7, 5.8
	Monosymmetrical	5.6
Design:	Flexure	5.2–5.6
	Shear	5.8
	Biaxial bending/combined actions	5.7
	Torsion	Appendix B
Loading:	Major plane bending	5.3–5.5
	Minor plane bending	5.3
	Combined actions	5.7
Lateral restraints:		5.4
Special design aspects:		
	Serviceability	5.10
	Economy	5.11

Typical section shapes used for flexural members are shown in Figure 5.1.

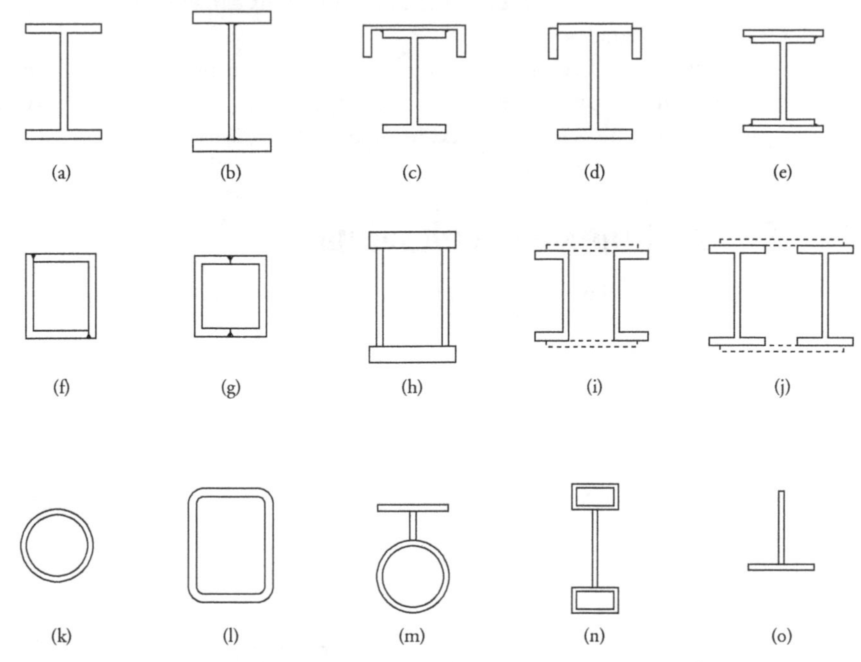

Figure 5.1 *Typical sections for beams: (a) hot-rolled (HR) universal section; (b) welded three (HR) plate I-section; (c) built-up HR universal section with HR channel (e.g. as in crane runway); (d) built-up HR universal section with flanges stiffened by HR flats or cut plate; (e) HR section with HR flange plates; (f) welded box section from HR angles; (g) welded box section from HR channels; (h) welded box section from HR plates; (i) compound HR channel section with intermittent ties (HR flats or plate); (j) compound HR universal section with intermittent ties (HR flats or plate); (k) cold-formed (CF) circular hollow section (CHS); (l) CF rectangular hollow section (RHS); (m) built-up CF hollow section with HR flats (e.g. in architectural applications); (n) built-up I-section using CF RHS flanges welded to HR plate (which can be flat or corrugated plate—e.g. Industrial Light Beam (ILB)); and (o) tee-section split from HR universal section or made from HR plate/flats.*

Depending on the type of end connections adopted in the design, structures can be classified into the following types:

(a) *Simple*—End connections are such that a relatively small degree of rotational restraint about the major axis is afforded to the beams. Consequently, it is assumed that no bending moments develop at the ends and the beams are designed as simply supported.

(b) *Rigid*—End connections are such that the rotational restraint of the beam ends tends to 100%. In practice it is acceptable if the restraint is at least 80%. Such connections are assumed in AS 4100 to possess enough rigidity to maintain the original included angle between the beam end and the connected members.

(c) *Semi-rigid*—The end connections are specifically designed to give a limited and controlled stiffness. Using the selected joint stiffness, the beam is designed as part of a frame with elastic nodes. These connection types are not commonly used, as a good understanding of the relationship between flexural restraint and load effects is required. However, these types of connections can readily be incorporated into elastic computer analysis methods.

It must be realised that secondary members and non-structural elements can exert a strong influence on the design. For example, a beam may form a part of a floor bracing system from which it receives additional axial loads.

5.2 Flexural member behaviour

It is helpful to consider the load deflection behaviour of a flexural member by means of Figure 5.2. A beam of compact section that is not subject to local buckling, and is restrained laterally and torsionally, would not fail until well after the onset of yielding, as shown by curve (a). Some beams fail before yielding, for one of four reasons:

(1) flexural (lateral) torsional buckling (b)

(2) local plate buckling of the compression flange or compression part of web (c)

(3) web shear yielding or shear buckling (d)

(4) web crushing (e).

Other types of failure can also prevent the full capacity of a beam from being reached, e.g. connection inadequacy, brittle tensile fracture, torsion and fatigue. The main objective of designing beams is to ensure that premature failures are ruled out as far as practicable by using appropriate constructive measures.

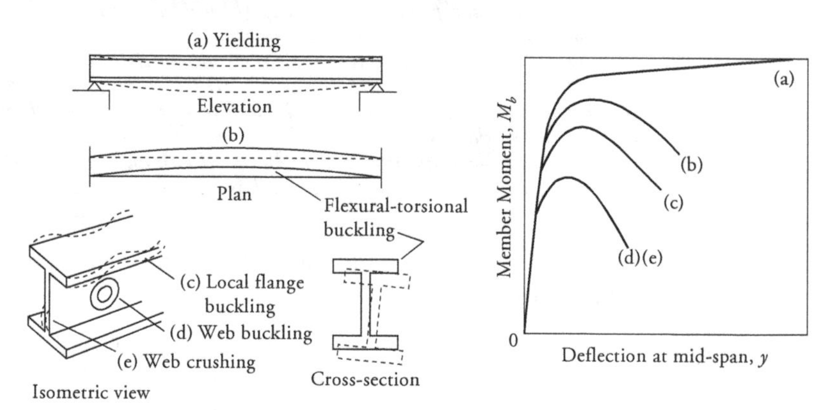

Figure 5.2 Modes of failure of an I-section beam (a) section yield; (b) flexural-torsional buckling; (c) local flange buckling; (d) web buckling; (e) web crushing

5.3 Bending moment capacity

The two bending moment capacities to be considered in design are:

- the nominal section moment capacity, and
- the nominal member moment capacity.

The nominal section moment capacity, M_s, refers to the flexural strength of a cross-section. The member moment capacity refers to the flexural-torsional capacity of the beam as a whole.

The nominal *section* moment capacity about the major and minor axis is given by:

$$M_{sx} = f_y Z_{ex}$$

$$M_{sy} = f_y Z_{ey}$$

The first design requirement is that at all sections of the beam must satisfy:

$$M_x^* \leqslant \phi M_{sx}$$

$$M_y^* \leqslant \phi M_{sy}$$

where f_y is the yield stress of steel, Z_{ex} and Z_{ey} are the effective section moduli, ϕ is the capacity reduction factor of 0.9, and M_x^*, M_y^* are the relevant design action effects.

The second, and often critical, requirement is that the member moment capacity be adequate. The value of the nominal *member* moment capacity, M_b, about the major axis is given by:

$$M_{bx} = \alpha_s \alpha_m M_{sx} \leqslant M_{sx}$$

making sure that in all segments and subsegments of the beam satisfy:

$$M_x^* \leqslant \phi M_{bx}$$

where α_s is the slenderness reduction factor and α_m is the moment modification factor (see Section 5.5). The slenderness reduction factor, α_s, varies with the ratio of M_{sx}/M_o, where M_o is termed the reference buckling moment. For equal-flanged I-beams and PFCs, M_o is given by

$$M_o = \sqrt{\left(\frac{\pi^2 EI_y}{l_e^2}\right)}\sqrt{GJ + \left(\frac{\pi^2 EI_w}{l_e^2}\right)}$$

The value of α_s varies between near 0.1, for very slender beams or beam segments, and 1.0, for stocky beams:

$$\alpha_s = 0.6\left\{\sqrt{\left[\left(\frac{M_{sx}}{M_o}\right)^2 + 3\right]} - \left(\frac{M_{sx}}{M_o}\right)\right\}$$

For unequal-flanged (monosymmetrical) beams, see Section 5.6.

For CHS, SHS and RHS sections and solid bars having large J values and $I_w = 0$:

$$M_o = \sqrt{\left(\frac{\pi^2 EI_y GJ}{l_e^2}\right)}$$

where E = 200 000 MPa, G = 80 000 MPa, J is the torsion constant, I_w is the warping constant (see Section 5.5), and l_e is the effective length for flexural-torsional buckling.

Flexural-torsional buckling does not occur in beams bent about their minor axis, and thus for such beams $M_{by} = M_{sy}$, except where the load is applied at a point higher than $1.0b_f$ above the centre of gravity (where b_f = flange width of the I- or channel section).

The factor α_m depends on the shape of the bending moment diagram, and its value ranges between 1.0 for constant moment (always safe) up to 3.5 for some variable moment shapes listed in Table 5.5 of this Handbook.

The product $\alpha_s\alpha_m$ must not exceed 1.0, otherwise the value of M_{bx} would become greater than M_s. For example, with $\alpha_m = 1.5$ and $\alpha_s = 0.8$, the product is 1.2 >1.0, thus α_m would effectively be reduced to:

$$\alpha_{mr} = \frac{1.0}{\alpha_s} = 1.25$$

As can be seen from the equation for M_o, an increase of the effective length has the effect of reducing the reference buckling moment M_o and thus the factor α_s. It is therefore advantageous to reduce the effective length by incorporating additional lateral and/or torsional restraints. If continuous lateral restraint is available the effective length can be taken as zero, making $\alpha_s = 1.0$, in which case $M_{bx} = f_y Z_{ex}$. In practice, beams are often divided into several segments so as to reduce the effective length. Beam segments are lengths of beams between full or partial restraints (described in Section 5.4). The effective length, l_e, is taken as:

$$l_e = k_t k_l k_r l$$

where l is the segment or subsegment length, k_t is the twist restraint factor, k_l is the load height factor, and k_r is the lateral rotation restraint factor. The values of k-factors specified in AS 4100 are summarised in Tables 5.2.1 and 5.2.2 of this Handbook.

5.4 Beam segments and restraints

5.4.1 Definitions

The term 'restraint' denotes an element or connection detail used to prevent a beam cross-section from lateral displacement and/or lateral rotation about the minor axis and/or twist about the beam centre line. Restraints at beam supports are often supplemented by additional restraints along the the span—see Figure 5.3.

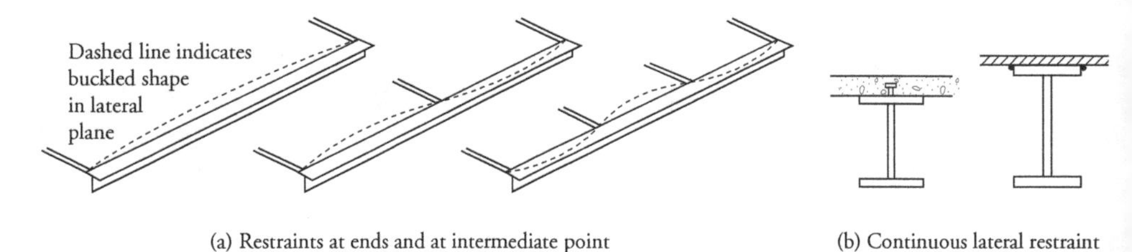

(a) Restraints at ends and at intermediate point

(b) Continuous lateral restraint

Figure 5.3 Arrangement of restraints

The following terms are used in AS 4100 and Trahair et al. [1993c,d] to describe cross-sectional restraints:

- *Full restraint* (F)—a restraint that prevents the lateral displacement of the critical flange of the cross-section and prevents twisting of the section.
- *Partial restraint* (P)—a restraint that prevents the critical flange of the cross-section from displacing laterally and partly prevents the section from twisting.
- *Lateral restraint* (L)—a restraint that prevents lateral displacement of the critical flange without preventing the twist of the section.

- *Nil restraint* (U)—a cross-section that does not comply with types F, P or L—i.e. unrestrained.
- *Continuous lateral restraint* (C)—a critical flange restraint provided continuously by a concrete slab, chequer plate or timber floor with the requirement that the segment ends are fully or partly restrained (practically resulting in $l_e = 0$). In this instance, any torsional buckling deformations do not occur even though lateral restraints are only applied. (See Figure 5.3(b)).
- *Lateral rotation restraint* (LR)—a restraint that prevents rotation of the critical flange about the section's minor axis.
- *Full lateral restraint*—a beam or beam segment with F, P or L restraints to the critical flange spaced (l_f) within the requirements of Clause 5.3.2.4 of AS 4100, such that the beam can be regarded as continuously restrained, resulting in $l_e = 0$. Typical spacing limits for equal-flanged I-sections are:

$$l_f \leqslant r_y \left[\left(80 + 50\beta_m \right) \sqrt{\left(\frac{250}{f_y} \right)} \right]$$

where β_m ranges from -1.0 to $+1.0$. Further expressions are given in Clause 5.3.2.4 of AS 4100 for other sections.

5.4.1.1 Additional terms

Critical flange—the flange that would displace laterally and rotate further if the restraints were removed. This is the compression flange of a simple beam and tension flange of a cantilever.

Critical section—the cross-section that governs the beam design. Clause 5.3.3 of AS 4100 notes this as the cross-section in a beam segment/subsegment with the largest ratio of M^* to M_s.

Segment—a portion of a beam between fully (F) or partially (P) restrained cross-sections. Restraint combinations (left and right) can be FF, PP or FP.

Segment length, l—length of the beam between restraints type F, P or L. For a beam having FF or PP end restraints and no mid-span restraints, the segment length is equal to the beam span. An additional lateral retraint at mid-span would result in a subsegment length of one-half span with end restraints of FL or PL.

Subsegment—a segment can be further subdivided into portions having at their ends at least the lateral (L) restraints to the critical flange. Restraint combinations can be FL, PL or LL.

Figure 5.4 in this Handbook, Figure 5.4.1 and 5.4.2 of AS 4100 and the connection diagrams in Trahair et al. [1993c,d] show examples of restraint designations.

The division of a beam into segments and subsegments (see Figure 5.5) does not affect the calculation of design bending moments and shears in the span—it affects only the calculation of the reference buckling moment M_o, the factor α_s and the breaking up of the bending moment distribution to respective beam segments and subsegments (for re-evaluation of α_m).

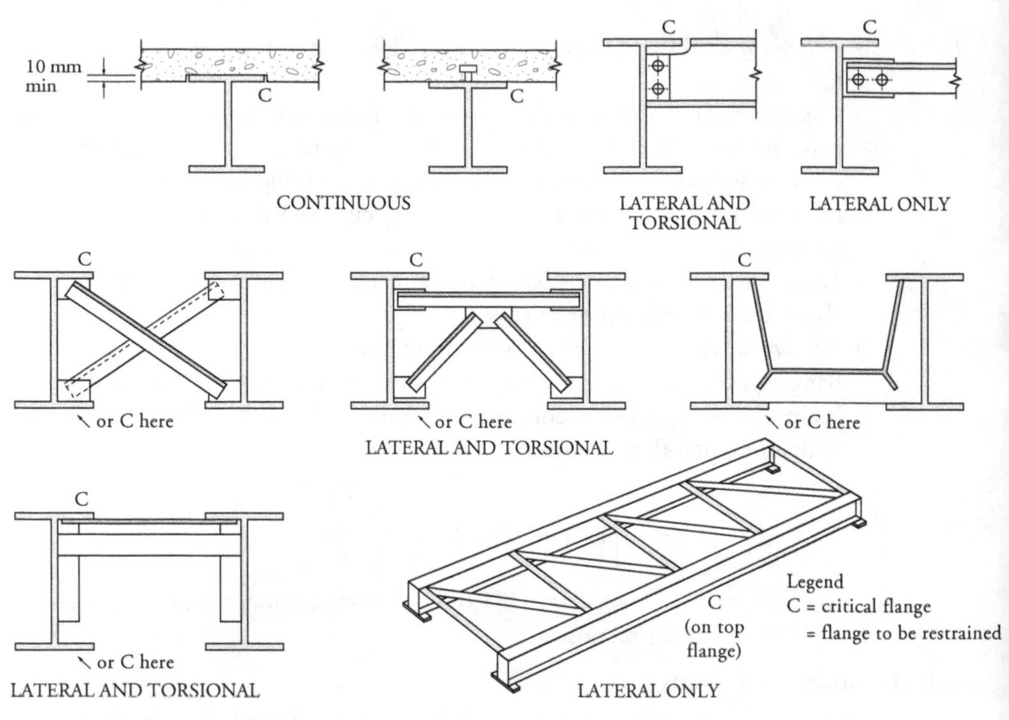

Figure 5.4 (a) *Restraining systems for prevention of flexural-torsional buckling failure of beams*

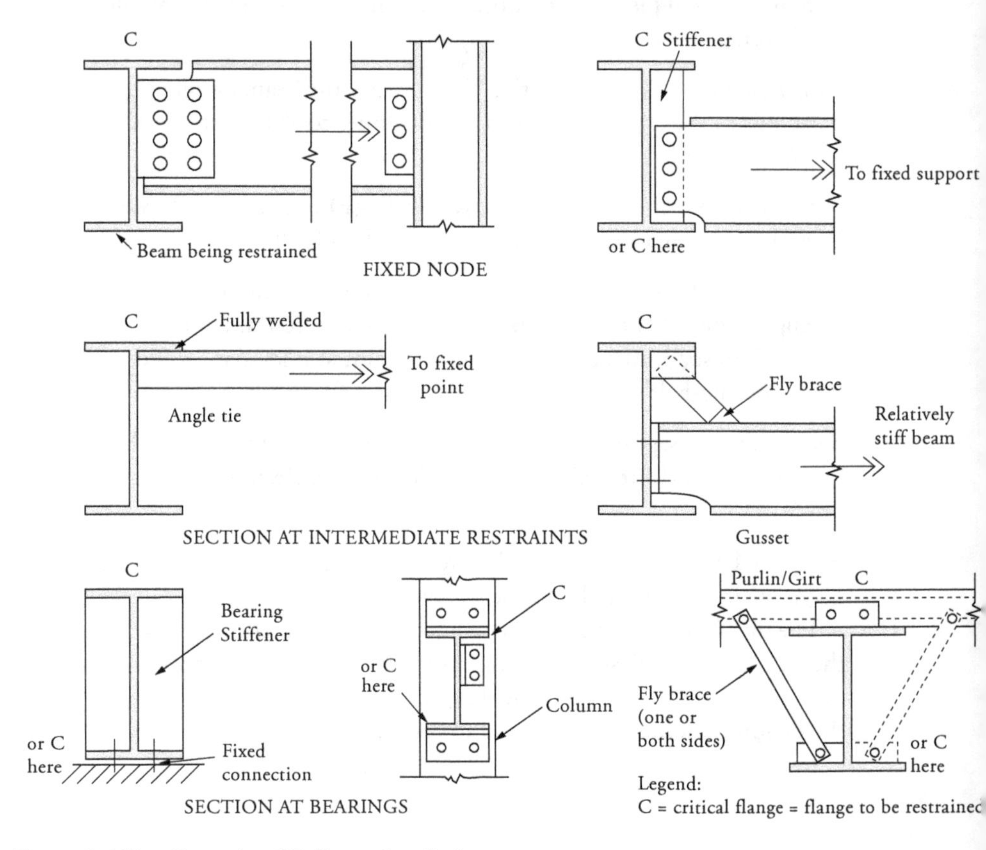

Figure 5.4(b) *Examples of Full restraints for beams*

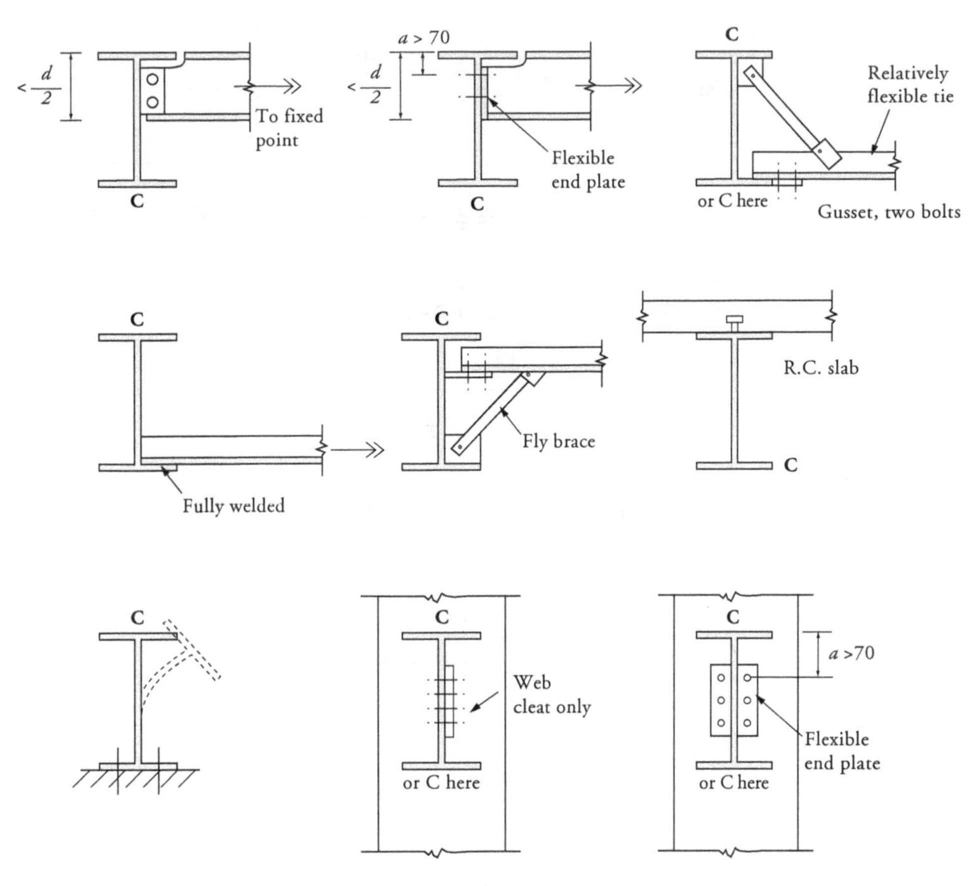

Figure 5.4(c) *Examples of Partial restraints for beams*

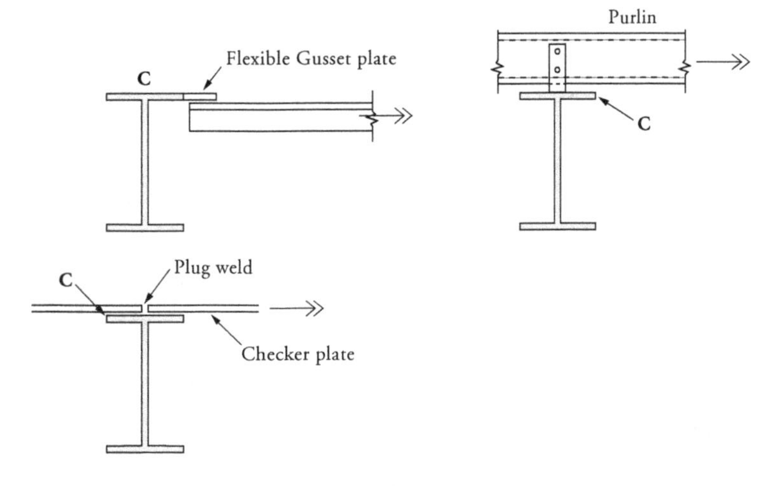

Figure 5.4(d) *Examples of Lateral restraints for beams*

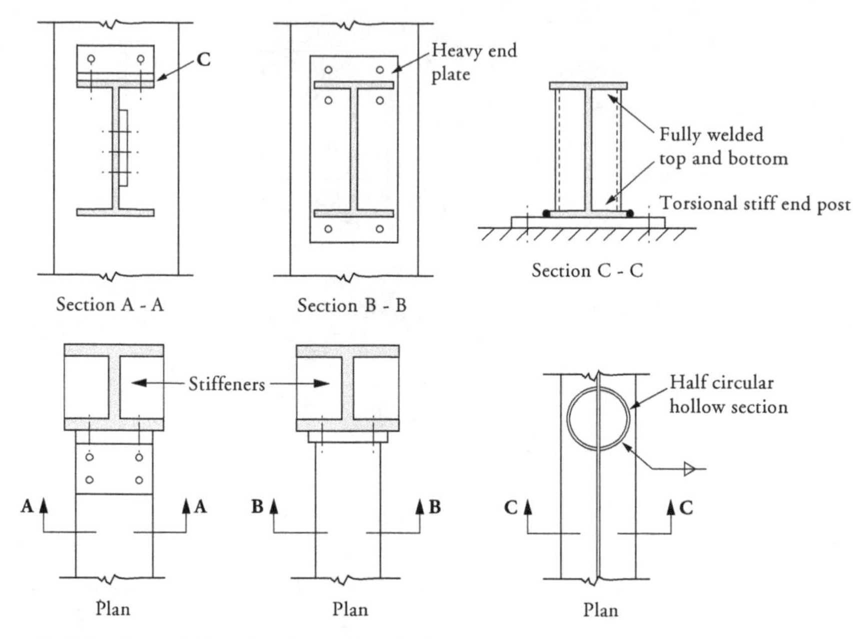

Figure 5.4(e) *Lateral Rotational restraints for beams*

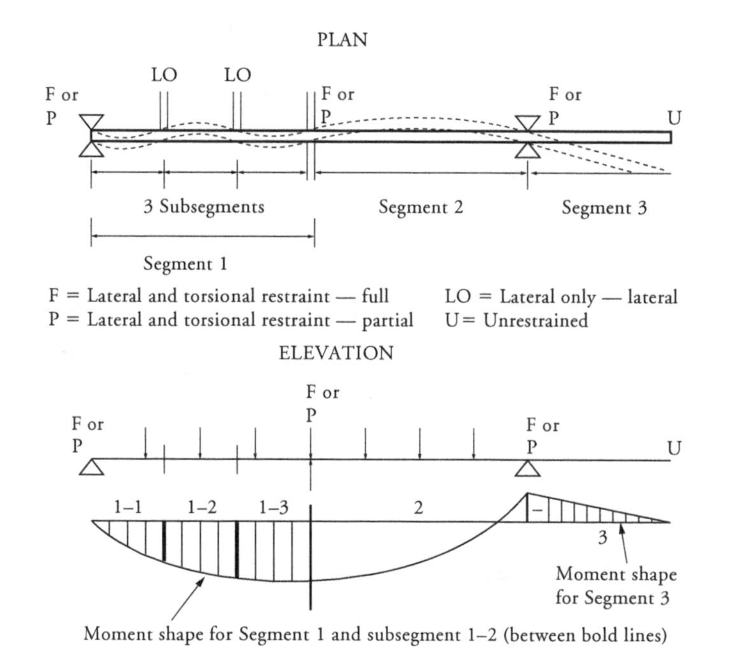

Figure 5.5 *The division of a beam into segments and subsegments*

5.4.2 Effective length factors

AS 4100 requires the effective length to be determined from the segment length and modified by three factors (Table 5.6.3 of AS 4100):

$$l_e = k_r k_t k_l l$$

k_t is the factor for twist distortion of the cross-section web with the values ranging between 1.0 for end cross-section restraint designations FF, FL, FU and above 1.0 for other restraint cases (see Table 5.6.3(1) of AS 4100).

k_r is the factor for the critical flange restraint against rotation about the minor principal axis, having values of:

- 0.70 for restraint designations FF, FP and PP, with both ends held against rotation about the y-axis
- 0.85 ditto, with one end only held
- 1.0 for negligible lateral rotation restraint.

k_l is a factor for the load application height above the shear centre and is equal to 1.0 where the (downward) load is applied at or below the shear centre, rises to 1.4 for the load applied to or above the beam top flange, and climbs to 2.0 for a cantilever.

Tables 5.2.1 and 5.2.2 give values of k_t, k_l, k_r for common beam applications.

5.4.3 Default values of k_t, k_l and k_r

Accepting at face value points of attachment of bracing, shear connectors and the like as lateral restraints (LR) for a beam, without going into their type, behaviour and effectiveness, simplifies calculations. The values of k_t, k_l and k_r are mostly on the conservative side. A more detailed calculation as described above could provide shorter effective lengths, and consequently more economical beams. However, at this simpler level of calculation, some important aspects are inherently more likely to be overlooked, such as that the lateral restraints may be badly located, resulting in an inferior or unsafe design.

Table 5.2.1 Default values of k_t, k_l and k_r to determine effective length, l_e of simple beams

Case	Direction of action/load, and flange on which it acts	k_t	k_l	k_r
1	Action/load acts downwards on top flange	1.1	1.4	1.0
2	Action/load acts downwards at shear centre or bottom flange	1.1	1.0	1.0

Table 5.2.2 Default values of k_t, k_l and k_r to determine effective length, l_e of cantilevers

Case	Direction of action/load, and flange on which it acts	k_t	k_l	k_r
3	Action/load acts downwards on top flange	1.1	2.0	1.0
4	Action/load acts downwards at shear centre or bottom flange	1.1	1.0	1.0

Notes: 1 k_t lies in the range 1.0 to 1.2+ for UB and UC, and 1.0 to 1.5+ for WB, WC and plate girders. The above-listed default values for k_t are mainly for UB and UC.

2 If all the actions/loads are located at the restraints (i.e. segment end), then k_l = 1.0 (except for cantilevers with unrestrained tip loads, k_l = 2.0).

continued

3 It may not be safer to use $k_r = 0.85$ or 0.7. If in doubt, use $k_r = 1.0$.

4 If loads are constrained to the plane of the minor (y) axis, use $k_l = 1.0$.

5 For downward (horizontal beam) or inward (vertical beam) acting loads in the above tables, the *top* or *outside* flange is the *critical* flange C.

6 For upward or outward loading the *bottom* or *inside* flange is the *critical* flange. The above tabulated values then apply when the "top flange" term is interchanged with "bottom flange" and vice versa, and the end restraint types are unchanged.

7 Change, for example in occupancy and loads, types of loads and reversals, and other aspects of different actions would require a more detailed determination of l_e.

5.4.4 Capacity of lateral restraint elements

The restraining elements at the ends of segments and subsegments, which may be end bearings, ties or floor joists, must be able to transfer a nominal action specified in Clause 5.4.3 of AS 4100. For a restraint required to prevent the lateral deflection of the critical flange, the requirement is:

$$N_r^* = 0.025 N_f^*$$

where N_r^* is the action to be resisted by the restraint, N_f^* is the segment flange force; for an equal-flanged section $N_f^* = M_m^*/h$, where h is the distance between flange centroids, and M_m^* is the maximum bending moment in either of the adjoining segments.

In a beam having many segments, each lateral restraint element must be designed for the specified restraining action except where the restraints are spaced more closely than is necessary (Clause 5.4.3.1 of AS 4100).

Where a restraining element continues over several parallel beams into a reaction point, it is necessary to use the 0.025 N_f^* force only for the most critical beam and halve the restraint force for the each of the remaining beams.

5.4.5 Capacity of twist restraint elements

The restraining element preventing the twisting of the section is subject to the application of the N_r^* force as for the lateral displacement restraint (Clause 5.4.3.2 of AS 4100).

5.5 Detailed design procedure

5.5.1 Effective section properties

5.5.1.1 General

As noted in Section 5.3, the basic capacity check for a flexural member is given by:

$$M^* \leq \phi\, \alpha_m \alpha_s Z_e f_y$$

provided that the value of M^* does not exceed:

$$M^* \leq \phi\, Z_e f_y$$

where α_m and α_s are factors respectively taking into account the distribution of the bending moment and the reduction of capacity due to flexural-torsional buckling effects. The value of factor α_m is in the range of 1.0 to 3.5, while α_s is in the range of 0.1 to 1.0.

The evaluation of these factors is given in Tables 5.5 to 5.8 inclusive.

The value of the effective modulus of section, Z_e, depends on the section geometry, or the section compactness: compact, non-compact or slender. For all standard hot-rolled sections, welded plate sections and structural steel hollow sections the designer can read off the Z_e values direct from ASI [2004, 2009a] and OneSteel [2011, 2012a,b].

For non-standard fabricated sections it is necessary to carry out section compactness checks. The general procedure for the determination of the effective section modulus, Z_e, is as follows:

(a) Calculate element slenderness values, λ_e, for each plate element carrying uniform or varying longitudinal compression stresses:

$$\lambda_e = \left(\frac{b}{t}\right)\sqrt{\left(\frac{f_y}{250}\right)}$$

(b) Find the ratio $\frac{\lambda_e}{\lambda_{ey}}$ for each element (ie flanges and webs)

(c) The whole section slenderness λ_s is taken to be equal to the λ_e value for the largest ratio of $\frac{\lambda_e}{\lambda_{ey}}$.

The values of the yield limit, λ_{ey}, can be read from Table 5.2 of AS 4100, reproduced here as Table 5.3. Several items of data must be assembled before entering the table:

- number of supported edges
- stress gradient between the plate edges: uniform or variable
- residual stress severity (see notes for Table 5.3).

Table 5.3 Slenderness limits for plate elements

Plate type and boundary condition	Stress distribution	Residual stress category	Slenderness limits λ_{ep}	λ_{ey}	λ_{ed}
Flat, One edge supported, other free	Uniform	SR	10	16	35
		HR	9	16	35
		LW, CF	8	15	35
		HW	8	14	35
	Gradient	SR	10	25	-
		HR	9	25	-
		LW, CF	8	22	-
		HW	8	22	-
Flat, Both edges supported	Uniform	SR	30	45	90
		HR	30	45	90
		LW, CF	30	40	90
		HW	30	35	90
	Gradient	SR, HR, LW, CF, HW	82	115	-
Circular hollow section		SR, HR, CF	50	120	-
		LW, HW	42	120	-

Notes: 1. SR = stress relieved
HR = hot-rolled (e.g. UB, UC, PFC, etc.)
CF = cold-formed (hollow sections)
LW = lightly welded
HW = heavily welded (WB, WC)

2. See Section 2.5 for further information.

continued

Example: Fabricated I-beam with 270 x 10 flanges and 530 x 6 web in Grade 250 plate steel (assume HW residual stress classification)

Element	λ_e	λ_{ep}	λ_{ey}	λ_e/λ_{ey}
Flange outstands				
1,2,3,4	13.2	8	14	0.943 ← Critical element
Web	88.3	82	115	0.768
Section slenderness:	$\lambda_s = 13.2$	$\lambda_{sp} = 8$	$\lambda_{sy} = 14$	(Section is non-compact)

For circular hollow sections, λ_s can be determined by:

$$\lambda_s = \left(\frac{d_o}{t}\right)\left(\frac{f_y}{250}\right).$$

For all section types, the effective section modulus, Z_e, is then determined from the following "compactness" classifications:

$\lambda_s \leq \lambda_{sp}$ compact section

$\lambda_{sp} < \lambda_s \leq \lambda_{sy}$ non-compact section

$\lambda_s > \lambda_{sy}$ slender section

5.5.1.2 Compact sections

In a compact section there is no possibility of local flange or web buckling (from longitudinal compression stresses) to prevent the attainment of full section plasticity, i.e.:

$$Z_e = S \leq 1.5\,Z$$

where S is the plastic section modulus determined for the fully plasticised section, i.e. using rectangular stress block; Z is the elastic section modulus calculated on the basis of linear variation of stress through the depth of section. Sectional property tables for standard rolled sections, welded sections and hollow sections give values of Z, S and Z_e (ASI [2004, 2009a] and OneSteel [2011, 2012a,b]. Typical values of these parameters for general sections are given in Table 5.4.

Table 5.4 Comparison of Z, S and Z_e values

SECTION	Z_x	S_x	$\frac{S_x}{Z_x}$	Z_{ex}
Square/ Rectangular (Flat) Bar	$\frac{td^2}{6}$	$\frac{td^2}{4}$	1.50	S_x
Round bar	$\frac{\pi d^3}{32}$	$\frac{d^3}{6}$	1.70(>1.50)	$1.5Z_x$
UB, UC	Prop.Tables	Prop.Tables	1.10 to 1.15	$(0.957 \text{ to } 1.00)S_x$
WB, WC	Prop.Tables	Prop.Tables	1.10 to 1.18	$(0.912 \text{ to } 1.00)S_x$
CHS	$\frac{\pi(d_o^4 - d_i^4)}{32d_o}$	$\frac{(d_o^3 - d_i^3)}{6}$	1.29 to 1.47	$(0.912 \text{ to } 1.00)S_x$
RHS	Prop.Tables	Prop.Tables	1.18 to 1.34	$(0.730 \text{ to } 1.00)S_x$
SHS	Prop.Tables	Prop.Tables	1.16 to 1.33	$(0.704 \text{ to } 1.00)S_x$

Notes: The above section listings are based on generally available sections (ASI [2004, 2009a] and OneSteel [2011, 2012a,b]).

Where sections feature relatively large holes, the section modulus must be reduced if the area reduction of either flange is:

$$\Delta A_f \geqslant \left(1 - \frac{f_y}{0.85 f_u}\right) A_f \text{ or } > 20\% \text{ flange area for Grade 300 steel.}$$

The reduced elastic and plastic section modulus may be calculated for the net area or, more conveniently, by:

$$Z_{red} = Z\left(\frac{A_n}{A_g}\right) \text{ and } S_{red} = S\left(\frac{A_n}{A_g}\right)$$

where A_n is the net area of the whole section and A_g is the gross area of the section.

When computing the reduced area, any deductions for fasteners should be treated as for connections of tensile members.

The above method of reduction of section moduli due to the presence of holes is also applicable to the calculation of Z and S for Non-compact and Slender sections.

5.5.1.3 Non-compact sections

Where one or more plate elements comprising the section are non-compact, that is λ_s exceeds λ_{sp} and is less than or equal to λ_{sy}, the section is deemed to be Non-compact. With Non-compact sections it is possible that some local buckling may take place before the attainment of section plasticity, i.e.:

$$Z_e = Z + c_z(Z_c - Z) \text{ where } c_z = \frac{(\lambda_{sy} - \lambda_s)}{(\lambda_{sy} - \lambda_{sp})}$$

where Z_c is the effective section modulus (Z_e) assuming the section is Compact (see Section 5.5.1.2).

Values of Z_e for standard rolled sections, welded I-sections and hollow sections can be read off directly from ASI [2004, 2009a] and OneSteel [2011, 2012a,b].

5.5.1.4 Slender sections

The term 'slender section' should not be confused with 'slender beam'. Where the slenderness of any plate element is more than the yield limit, λ_{ey}, the section is classified as slender. Normally it is best to avoid using slender sections, but it is sometimes necessary to check a section of this type. There are three situations to consider:

(a) Section elements having uniform compression (i.e. no stress gradient), e.g. flanges of UB or UC bent about the major principal axis:

- Method 1:

$$Z_e = Z\left(\frac{\lambda_{sy}}{\lambda_s}\right)$$

- Method 2:

  $Z_e = Z_r$, where Z_r is the elastic section modulus of a modified section obtained by removing the excess width of plates whose b/t exceeds the λ_{ey} limit (see Figure 5.6(d)).

(b) Sections with slenderness determined by a stress gradient in plate elements with one edge unsupported in compression, e.g. UB or UC bent about its minor principal axis:

$$Z_e = Z\left(\frac{\lambda_{sy}}{\lambda_s}\right)^2$$

(c) Circular hollow sections with $\lambda_s > \lambda_{sy}$

$$Z_e = \min(Z_{e1}, Z_{e2})$$

$$\text{where } Z_{e1} = Z\sqrt{\left(\frac{\lambda_{sy}}{\lambda_s}\right)} \quad \text{and} \quad Z_{e2} = Z\left(\frac{2\lambda_{sy}}{\lambda_s}\right)^2$$

Values of Z_e for standard rolled sections, welded I-sections and hollow sections can be read off direct from ASI [2004, 2009a] and OneSteel [2011, 2012a,b].

Figure 5.6 shows examples of sections having slender elements.

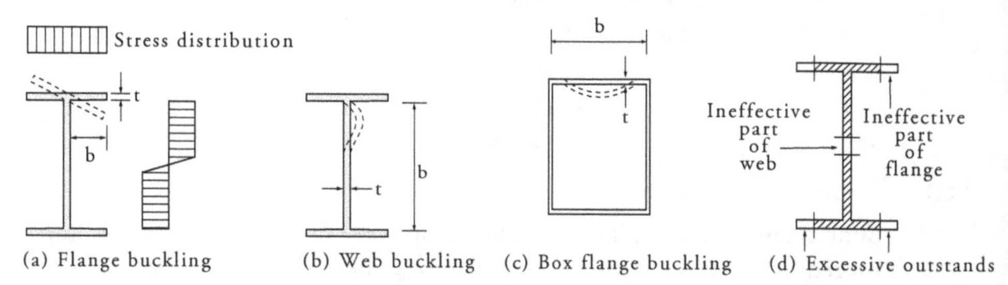

Figure 5.6 Cross-sections with slender elements

5.5.2 Continuously laterally restrained beams

This category embraces all beams bent about their minor axis and beams bent about the major axis and fully restrained against flexural-torsional buckling. To be fully restrained, the critical flange of the beam must be either continuously restrained in the lateral direction or restrained at close intervals, not exceeding $l_e/r_y = 20$. In addition, the end connections of the beam must be restrained against twisting and lateral displacement.

The spacing of effective lateral restraints must be such as to ensure that no capacity is lost on account of flexural-torsional buckling—see Section 5.4.1, ASI [2004, 2009a] or OneSteel [2012b].

Typical beams of this type are:

(a) beams carrying a concrete slab that engages the top (critical) flange or uses shear connectors at relatively close spacing (0.60 m for small sizes and 1.0–1.5 m for larger sizes)

(b) beams supporting chequer plate flooring and connected by intermittent welds

(c) purlins connected to roof sheeting by fasteners at every third sheeting ridge and loaded by dead load, live load plus wind pressure (with an overall inward effect).

The slenderness factor α_s is in this case equal to 1.0. The moment modification factor α_m also has a value of 1.0. The design moment capacity of a beam with full lateral restraint is simply:

$$M^* \leqslant \phi M_s$$

with $M_s = Z_e f_y$ the result is:

$$M^* \leqslant \phi Z_e f_y$$

Table 5.5 Values of α_m from Table 5.6.1 of AS 4100 for beam segments restrained at both ends.

Load Case	Beam segment between restraints	Moment Distribution	Particular β_m	α_m	Equation for α_m	Range of β_m and α_m
1.1	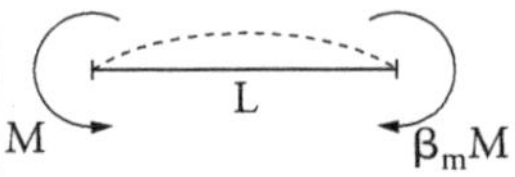 M, L, $\beta_m M$	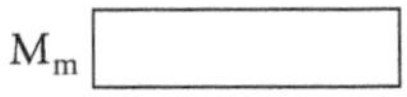 M_m	-1	1.00	$1.75 + 1.05\beta_m + 0.3\beta_m^2$	$-1 \leqslant \beta_m \leqslant 0.6$ $1.0 \leqslant \alpha_m \leqslant 2.5$
1.2	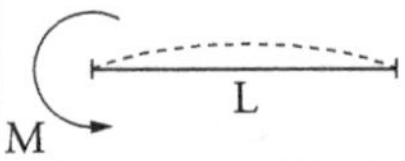 M, L	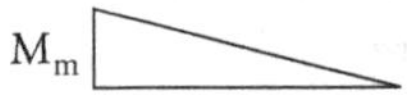 M_m	0	1.75		
1.3	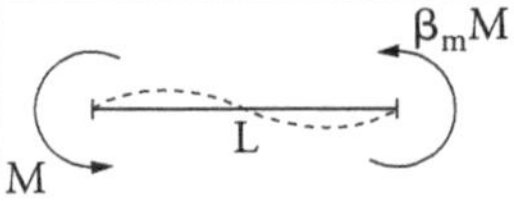 M, L, $\beta_m M$	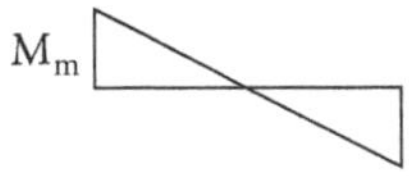 M_m	0.6 to 1.0	2.50	2.50	$0.6 < \beta_m \leqslant 1.0$ $\alpha_m = 2.50$
2.1	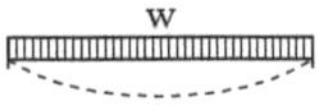 w	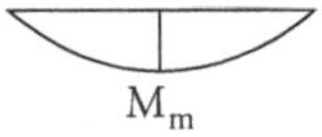 M_m	0	1.13	$1.13 + 0.12\beta_m$	$0 \leqslant \beta_m \leqslant 0.75$ $1.13 \leqslant \alpha_m \leqslant 1.22$
2.2	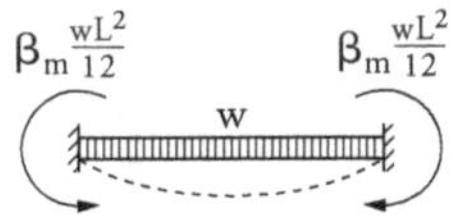 $\beta_m \frac{wL^2}{12}$, $\beta_m \frac{wL^2}{12}$	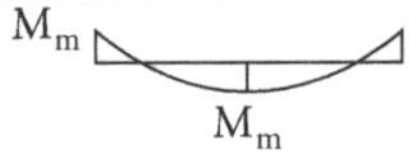 M_m, M_m	0.75	1.22		
2.3	w	 M_m	1	2.42	$-2.38 + 4.8\beta_m$	$0.75 \leqslant \beta_m \leqslant 1.0$ $1.22 \leqslant \alpha_m \leqslant 2.42$
3.1	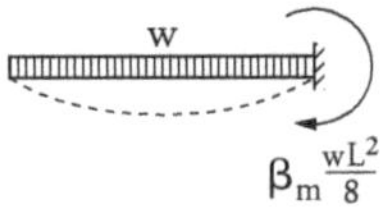 w	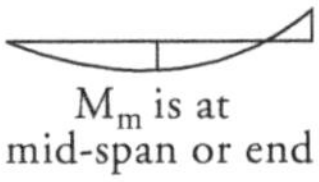 M_m is at mid-span or end	0.7	1.20	$1.13 + 0.10\beta_m$	$0 \leqslant \beta_m \leqslant 0.7$ $1.13 \leqslant \alpha_m \leqslant 1.20$
3.2	$\beta_m \frac{wL^2}{8}$		1	2.25	$-1.25 + 3.5\beta_m$	$0.7 \leqslant \beta_m \leqslant 1.0$ $1.20 \leqslant \alpha_m \leqslant 2.25$
4.1	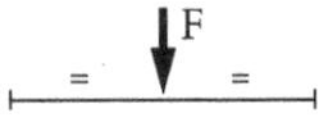 F, =, =	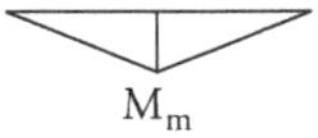 M_m	0	1.35	$1.35 + 0.36\beta_m$	$0 \leqslant \beta_m \leqslant 1.0$ $1.35 \leqslant \alpha_m \leqslant 1.71$
4.2	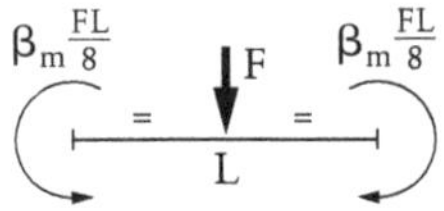 $\beta_m \frac{FL}{8}$, F, $\beta_m \frac{FL}{8}$, =, =, L	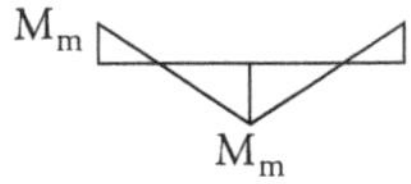 M_m, M_m	1	1.71		
5.1	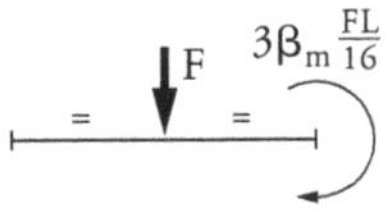 F, $3\beta_m \frac{FL}{16}$, =, =	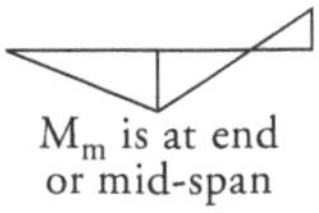 M_m is at end or mid-span	0	1.35	$1.35 + 0.15\beta_m$	$0 \leqslant \beta_m < 0.9$ $1.35 \leqslant \alpha_m < 1.5$
			1	1.80	$-1.20 + 3.0\beta_m$	$0.9 \leqslant \beta_m \leqslant 1.0$ $1.5 \leqslant \alpha_m \leqslant 1.8$
6.1	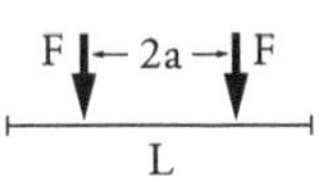 F ← 2a → F, L	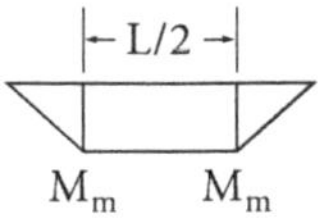 L/2, M_m M_m	—	1.09	← for $2a/L = 1/2$ $1.0 + 0.35(1 - 2a/L)^2$	$0 \leqslant 2a/L \leqslant 1.0$ $1.0 \leqslant \alpha_m \leqslant 1.35$
6.2	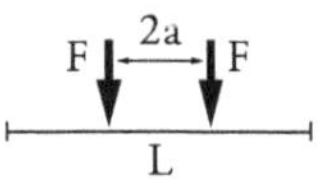 F 2a F, L	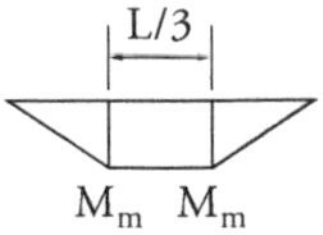 L/3, M_m M_m	—	1.16	← for $2a/L = 1/3$	

Notes:
1. Ends are Fully, Partially or Laterally restrained as in Clauses 5.4.2.1, 5.4.2.2 and 5.4.2.4 of AS 4100.
2. See Table 5.6.1 of AS 4100 for more cases.
3. See Table 5.6.2 of AS 4100 or Table 5.6(b) of this Handbook for segments unrestrained at one end.
4. The fourth column headed "Particular β_m - α_m" considers a specific β_m value and its related α_m.

For standard UB, UC and other sections the above procedure can be further reduced by referring to ASI [2004, 2009a] or OneSteel [2012b]. This is illustrated in Section 5.12.3.

5.5.3 Beams subject to flexural-torsional buckling

5.5.3.1 Description of the method

This method relies on the effective length concept for compression members with given end restraint conditions. This Section describes various tiered methods, which may be suitable for ordinary building structures. Where utmost economy in beam design is paramount it is best to use the buckling analysis method as noted in Clause 5.6.4 of AS 4100.

5.5.3.2 Evaluation of the moment modification factor α_m

A value of 1.0 can always be adopted for α_m, but this is very conservative in many situations. If a beam is continuously laterally restrained, $\alpha_m = 1.0$ ($= \alpha_s$) is automatically adopted as the section moment capacity is the maximum moment that can be obtained. Otherwise, there are two methods for determining the value of α_m:

(i) value obtained or interpolated from Table 5.6.1 or 5.6.2 of AS 4100.

(ii) value obtained from Clause 5.6.1.1(a)(iii) of AS 4100:

$$\alpha_m = \frac{1.7M_m^*}{\sqrt{[(M_2^*)^2 + (M_3^*)^2 + (M_4^*)^2]}} \text{ but not exceeding 2.5.}$$

The bending moment values of M_2 to M_4 correspond to the design bending moments at the quarter point, middle and third quarter point on the beam segment/subsegment being considered. M_m^* is the maximum design bending moment.

The first method is illustrated in Table 5.5 in this Handbook (based on Table 5.6.1 of AS 4100). Table 5.6 gives the values of α_m for simple loading patterns.

Table 5.6 Values of α_m for beams with simple loading patterns

(a) Simply supported beams—restrained at both ends

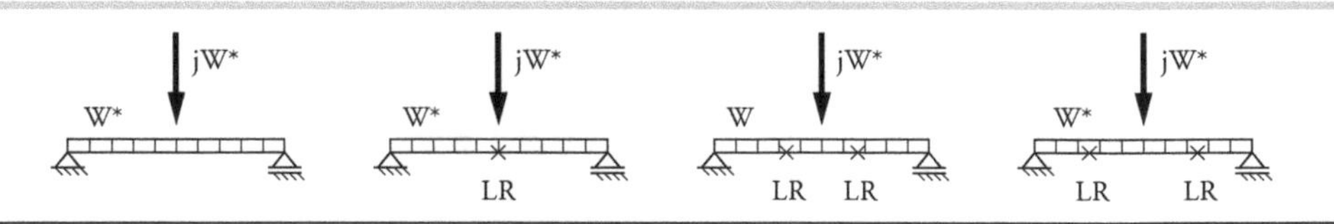

j	No LR	1 LR midspan	2 LRs at $\frac{1}{3}$ points	2 LRs at $\frac{1}{4}$ points
0	1.13	1.35	1.05	1.00
0.5	1.27	1.54	1.05	1.09
1	1.31	1.62	1.07	1.12
5	1.37	1.76	1.09	1.15
10	1.38	1.79	1.09	1.16

Notes:
1. W^* = Total uniformly distributed load.
2. jW^* = Concentrated load as a multiple of W^*.
3. LR = Full, Partial or Lateral restraint.
4. The α_m values for $j = 0$ are based on the more exact solutions from Table 5.6.1 of AS 4100 whereas the other values are calculated by superposition and Clause 5.6.1.1(a)(iii) of AS 4100.
5. The 2 LR cases only consider the critical middle segment as it has the highest moment.

continued

(b) Cantilevers—i.e. beam with one end unrestrained

j	α_m – **unrestrained tip, type U**
0	2.25
0.5	1.93
1	1.75
5	1.43
10	1.34
Tip load only	1.25
Tip moment only	0.25 uniform moment along span

Notes: 1. Notes 1 and 2 of Table 5.6(a) apply.
2. Lateral restraints are considered ineffective for cantilever spans (Trahair [1993d]).
3. The α_m values for $j = 0$, tip load only and tip moment only are from the more exact solutions listed in Table 5.6.2 of AS 4100 whereas the other values are calculated from the interpolation method noted in Section 3.3 of Trahair [1993d].

(c) Cantilevers—with tip restraint

j	α_m – **restrained tip, F or P but not L type.**
0	3.50
0.5	2.20
1	2.06
5	1.88
10	1.85

Notes: 1. Though this beam configuration may be a cantilever in a vertical support sense, it is designed in AS 4100 for flexural-torsional buckling to be a beam restrained at both ends.
2. Consequently, notes 1, 2 and 4 from Table 5.6(a) apply. See also note 2 of Table 5.6(b).

Where use is made of design capacity tables (ASI [2004, 2009a], OneSteel [2012b]), it is still necessary to evaluate α_m, as those tables assume a value of $\alpha_m = 1.0$:

$$M_R = \alpha_m (\phi M_b)_{DCT}$$

where the second term $(\phi M_b)_{DCT}$ is obtained from design capacity tables (ASI [2004, 2009a], OneSteel [2012b]), then check the following inequality with the appropriate α_m used in the above equation:

$$M^* \leq M_R$$

5.5.3.3 Evaluation of the slenderness reduction factor α_s

This section applies to beams or beam segments where the distance between restraints exceeds the limits given in Clause 5.3 of AS 4100 (see Section 5.4.1).

The purpose of the slenderness reduction factor α_s is to relate the actual capacity of a beam subject to flexural-torsional buckling to a fully restrained beam. The value of α_s is a function of the ratio M_s/M_o:

$$\alpha_s = 0.6\left\{\sqrt{\left[\left(\frac{M_s}{M_o}\right)^2 + 3\right]} - \frac{M_s}{M_o}\right\}$$

where M_s is the section moment capacity and M_o is the reference buckling moment:

$$M_o = \sqrt{\left(\frac{\pi^2 EI_y}{l_e^2}\right)\left(GJ + \frac{\pi^2 EI_w}{l_e^2}\right)}$$

where l_e is the effective length of the single segment beam or of a beam segment. It is determined from:

$l_e = k_t k_l k_r l$

G = shear modulus of steel (80 000 MPa)

J = torsion constant

I_w = warping constant

k_t, k_l, k_r are the components of the effective length factor k (see Section 5.4.2 and 5.4.3)

k_t = twist factor, which depends on the flexibility of the web

k_l = load position factor, normally 1.0, increasing to 1.2 or more when the load is applied at the level of the top flange

k_r = 'rotation' factor, actually a factor taking into account the resistance to rotation in plan of the flange at the end of the segment

As can be seen, M_o is a function of many variables: the 'column' term $\frac{\pi^2 EI_y}{l_e^2}$ is quite dominant, and torsional/warping terms play an important role. The equation gives a hint of how to increase the flexural-torsional buckling resistance of a beam:

(a) by decreasing l_e and by increasing I_y

(b) by increasing J and I_w.

Finally, the slenderness factor α_s used in calculating the nominal member moment capacity, is noted above.

Values of α_s are listed in Table 5.7 and plotted in Figure 5.7 for the purpose of illustration. As can be seen from the plot, the values of α_s are less than 1.0, except when the beam is 'stocky' and fails by yielding.

Table 5.7 Values of α_s

M_s/M_o	α_s
0.05	1.010
0.10	0.981
0.20	0.926
0.30	0.875
0.40	0.827
0.50	0.782
0.60	0.740
0.80	0.665
1.00	0.600
1.20	0.544
1.40	0.496
1.60	0.455
1.80	0.419
2.00	0.387

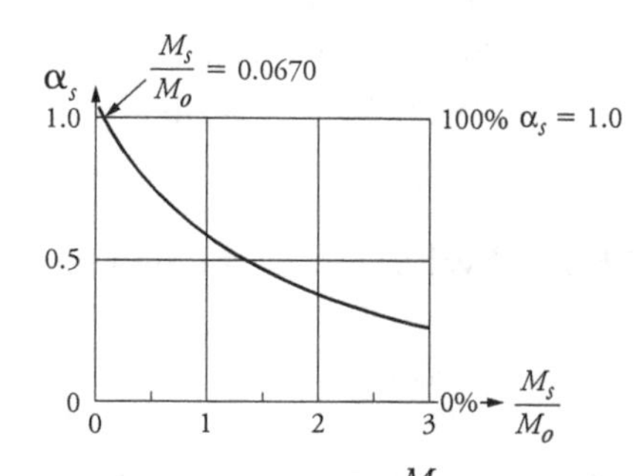

Figure 5.7 Plot of α_s vs $\frac{M_s}{M_o}$

Tables 5.8.1 and 5.8.2 give α_s values for UB and UC sections.

Table 5.8.1 Slenderness reduction factors (α_s) for UB Grade 300

Designation	Effective length l_e m												
	2	3	4	5	6	7	8	9	10	11	12	14	16
610UB 125	0.940	0.839	0.728	0.622	0.530	0.456	0.395	0.347	0.308	0.277	0.250	0.210	0.181
113	0.937	0.831	0.715	0.605	0.511	0.435	0.373	0.325	0.287	0.255	0.231	0.193	0.165
101	0.927	0.810	0.684	0.567	0.469	0.392	0.332	0.286	0.249	0.221	0.198	0.163	0.139
530UB 92.4	0.912	0.787	0.659	0.545	0.453	0.381	0.326	0.282	0.250	0.223	0.201	0.168	0.144
82.0	0.908	0.777	0.643	0.523	0.430	0.356	0.302	0.259	0.227	0.202	0.181	0.150	0.128
460UB 82.1	0.897	0.766	0.637	0.528	0.441	0.375	0.324	0.284	0.254	0.227	0.207	0.175	0.152
74.6	0.893	0.757	0.623	0.512	0.423	0.356	0.307	0.267	0.236	0.212	0.192	0.161	0.139
67.1	0.889	0.749	0.611	0.493	0.403	0.338	0.285	0.248	0.218	0.194	0.175	0.147	0.126
410UB 59.7	0.879	0.734	0.695	0.484	0.398	0.333	0.286	0.250	0.221	0.198	0.179	0.151	0.131
53.7	0.861	0.703	0.555	0.440	0.355	0.293	0.248	0.214	0.188	0.168	0.151	0.127	0.109
360UB 56.7	0.875	0.732	0.600	0.494	0.413	0.352	0.305	0.268	0.240	0.216	0.197	0.167	0.145
50.7	0.871	0.719	0.582	0.471	0.389	0.328	0.282	0.247	0.219	0.197	0.179	0.151	0.131
44.7	0.851	0.689	0.545	0.429	0.347	0.288	0.244	0.212	0.186	0.167	0.151	0.127	0.109
310UB 46.2	0.873	0.730	0.598	0.498	0.415	0.355	0.310	0.274	0.245	0.222	0.230	0.172	0.150
40.4	0.857	0.697	0.560	0.450	0.370	0.311	0.268	0.234	0.208	0.187	0.170	0.144	0.125
32.0	0.813	0.627	0.478	0.372	0.299	0.247	0.211	0.184	0.162	0.146	0.132	0.111	0.097
250UB 37.3	0.828	0.668	0.537	0.441	0.370	0.317	0.277	0.245	0.220	0.200	0.183	0.156	0.137
31.4	0.813	0.640	0.500	0.399	0.328	0.277	0.239	0.210	0.187	0.169	0.154	0.130	0.144
25.7	0.742	0.550	0.414	0.325	0.265	0.222	0.192	0.169	0.151	0.136	0.125	0.105	0.092
200UB 29.8	0.805	0.647	0.523	0.433	0.367	0.317	0.279	0.248	0.224	0.204	0.187	0.160	0.140
25.4	0.789	0.616	0.485	0.391	0.327	0.278	0.242	0.214	0.193	0.174	0.159	0.136	0.119
22.3	0.788	0.609	0.473	0.376	0.310	0.263	0.228	0.200	0.179	0.162	0.148	0.126	0.109
18.2	0.648	0.459	0.343	0.272	0.223	0.191	0.166	0.147	0.132	0.119	0.109	0.093	0.081
180UB 22.2	0.674	0.516	0.411	0.339	0.288	0.249	0.220	0.197	0.177	0.162	0.148	0.127	0.112
18.1	0.641	0.469	0.362	0.294	0.245	0.211	0.185	0.164	0.148	0.134	0.123	0.106	0.092
16.1	0.623	0.444	0.336	0.268	0.223	0.191	0.167	0.148	0.133	0.121	0.111	0.094	0.083
150UB 18.0	0.622	0.473	0.377	0.311	0.264	0.229	0.202	0.181	0.163	0.149	0.137	0.117	0.103
14.0	0.566	0.409	0.313	0.254	0.212	0.183	0.160	0.143	0.129	0.117	0.107	0.092	0.080

Table 5.8.2 Slenderness reduction factors (α_s) for UC Grade 300

Designation	Effective length l_e m														
	2	3	4	5	6	7	8	9	10	12	14	16	18	20	22
310UC 158	0.997	0.954	0.908	0.875	0.816	0.775	0.736	0.699	0.667	0.606	0.566	0.510	0.471	0.437	0.408
137	0.994	0.950	0.898	0.846	0.798	0.751	0.708	0.670	0.634	0.570	0.517	0.472	0.432	0.398	0.370
118	0.991	0.945	0.890	0.834	0.779	0.728	0.680	0.637	0.599	0.532	0.477	0.431	0.392	0.363	0.331
96.8	0.990	0.936	0.874	0.808	0.746	0.687	0.636	0.585	0.542	0.471	0.414	0.369	0.331	0.300	0.277
250UC 89.5	0.977	0.919	0.854	0.796	0.737	0.689	0.644	0.601	0.536	0.501	0.449	0.404	0.368	0.336	0.311
72.9	0.969	0.902	0.827	0.755	0.691	0.631	0.578	0.533	0.496	0.428	0.379	0.337	0.303	0.275	0.252
200UC 59.5	0.943	0.864	0.785	0.711	0.649	0.598	0.549	0.508	0.472	0.412	0.364	0.325	0.294	0.268	0.245
52.2	0.941	0.850	0.766	0.688	0.622	0.565	0.516	0.474	0.438	0.377	0.331	0.294	0.265	0.240	0.220
46.2	0.939	0.849	0.751	0.672	0.601	0.541	0.490	0.447	0.410	0.351	0.306	0.271	0.242	0.220	0.200
150UC 37.2	0.892	0.790	0.703	0.629	0.565	0.513	0.467	0.429	0.395	0.342	0.300	0.266	0.240	0.217	0.199
30.0	0.869	0.748	0.645	0.561	0.493	0.439	0.394	0.357	0.326	0.278	0.240	0.212	0.190	0.172	0.157
23.4	0.857	0.719	0.603	0.510	0.451	0.384	0.341	0.305	0.278	0.232	0.201	0.175	0.156	0.141	0.128
100UC 14.8	0.780	0.654	0.560	0.485	0.424	0.376	0.338	0.306	0.279	0.237	0.206	0.181	0.162	0.147	0.134

Additional information given in Appendix H of AS 4100 can be applied to the case where the load is applied below (or above) the centroid of the section. The same appendix gives a procedure for calculating the reference elastic buckling moment M_o for sections with unequal flanges (monosymmetrical sections). A typical section of this type is the top-hat section used for crane runway beams.

5.6 Monosymmetrical I-section beams

These beams are symmetrical about the minor axis and have unequal flanges. A typical example is the UB section combined with a downturned channel, used for crane runways. Figure 5.8 illustrates some typical monosymmetrical sections. The design of monosymmetrical sections is covered in Clause 5.6.1.2 and Appendix H of AS 4100.

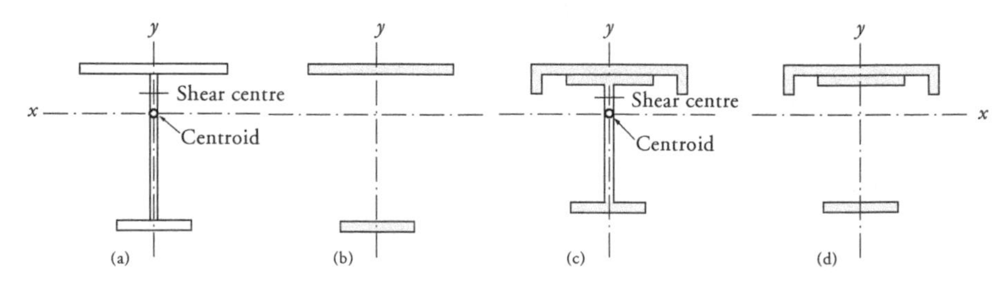

Figure 5.8 Monosymmetrical beams: (a) fabricated I-section; (b) section calculation for I_{cy} of (a)—only the compression flange is used (either above or below the x-axis, depending on the direction of bending); (c) UB section with downturned channel; and (d) section calculation for I_{cy} of (c)—one compression flange only.

The only difference in determining the nominal member moment capacity are the additional terms that appear in the equation for the reference buckling moment (see also Appendix H of AS 4100):

$$M_o = \sqrt{P_y}\left\{\sqrt{(GJ + P_w + 0.25\beta_x^2 P_y)} + 0.5\beta_x\sqrt{P_y}\right\}$$

where

$$P_y = \frac{\pi^2 EI_y}{l_e^2} \qquad P_w = \frac{\pi^2 EI_w}{l_e^2} \qquad \beta_x = 0.8d_f\left(\frac{2I_{cy}}{I_y} - 1\right)$$

$$I_w = \frac{I_y(d_f)^2}{4} \text{ for doubly-symmetric I-sections}$$

$$= I_{cy}d_f^2\left(1 - \frac{I_{cy}}{I_y}\right) \text{ for monosymmetrical I-sections.}$$

d_f = distance between flange centroids

I_{cy} = second moment of area of the compression flange about the section minor principal *y*-axis (see Figure 5.8).

5.7 Biaxial bending and bending with axial force

5.7.1 Biaxial bending

Biaxial bending occurs when bending moments are applied about both the major and minor principal axes. The method used in elastic design was to calculate the stresses about both axes and total them up. In limit states design, due to the non-linear methods adopted to optimise member economies, the method of superposition does not apply and design verification is done by the method of combined actions.

The capacity checks for biaxial bending without axial force are specified in Clauses 8.3 and 8.4 of AS 4100. The calculation procedure used is covered in Section 6.3 of this Handbook. Also the following, Section 5.7.2, discusses the capacity checks for generalised combined actions.

5.7.2 Bending combined with axial force

The flexural members subject to bending combined with a compressive axial force are termed 'beam-columns'. Typical beam-columns are columns and rafters of portal frames, beams doubling up as lateral bracing members and/or compression chords in wind trusses. Beam-columns in rigid and multistorey frames are dealt with in Chapter 6.

Unavoidably, the capacity calculations are somewhat complex to optimise member efficiencies. The following combined actions checks are required by AS 4100:

- reduced *section* moment capacity, M_{rx}, bending about *x*-axis
- reduced *section* moment capacity, M_{ry}, bending about *y*-axis
- biaxial *section* moment interaction
- in-plane *member* moment capacity, M_{ix} (about *x*-axis)
- in-plane *member* moment capacity, M_{iy} (about *y*-axis)
- out-of-plane *member* moment capacity, M_{ox} (about *x*-axis)
- biaxial *member* moment interaction.

The distinction between the section and member capacities is as follows:

- Section capacity is a function of section yield strength and local buckling behaviour
- Member capacity is a function of section capacity, flexural-torsional buckling resistance (beam action) and lateral buckling resistance (column action) of the member.

The verification method is covered in Section 6.3 which gives the general calculation procedure.

5.8 Web shear capacity and web stiffeners

5.8.1 General

Included in this Section are design methods for:

- web shear
- combined shear and bending
- web buckling
- web bearing.

With stiffened webs, the following aspects are relevant:

- transverse, intermediate stiffener proportioning
- end stiffeners
- end posts
- axial loads on stiffeners
- longitudinal stiffeners.

The main elements of plate web girders and hot-rolled I-sections to be verified for strength design are shown in Figure 5.12 and discussed as follows. Web thickness limits are given in Table 5.9.

Table 5.9 Minimum web thickness, $k_y = \sqrt{\left(\frac{f_y}{250}\right)}$

Arrangement	Minimum thickness t_w
Unstiffened web bounded by two flanges:	$k_y d_1/180$
Ditto for web attached to one flange (Tee):	$k_y d_1/90$
Transversely stiffened web:	
when $1.0 \leqslant s/d_1 \leqslant 3.0$ (See Note 4 also)	$k_y d_1/200$
$0.74 < s/d_1 \leqslant 1.0$	$k_y s/200$
$s/d_1 \leqslant 0.74$	$k_y d_1/270$
Web having transverse and one longitudinal stiffener:	
when $1.0 \leqslant s/d_1 \leqslant 2.4$	$k_y d_1/250$
$0.74 \leqslant s/d_1 \leqslant 1.0$	$k_y s/250$
$s/d_1 < 0.74$	$k_y d_1/340$
Webs having two longitudinal stiffeners and $s/d_1 \leqslant 1.5$	$k_y d_1/400$
Webs containing plastic hinges	$k_y d_1/82$

Notes:
1. The above limits are from Clauses 5.10.1, 5.10.4, 5.10.5 and 5.10.6 of AS 4100.
2. d_1 is the clear depth between the flanges
3. s is the spacing of transverse stiffeners
4. When $s/d_1 > 3.0$ the web panel is considered unstiffened.

5.8.2 Shear capacity of webs

The nominal shear capacity of a web subject to approximately uniform shear stress distribution is given in Clause 5.11.2 of AS 4100. For a single web bounded by two flanges, as in I-section beams and channels, the shear stress distribution is relatively uniform over the depth of the web. For such sections the nominal shear yield capacity of a web, V_w, is as follows:

$$V_w = 0.6 f_y A_w \quad \text{for} \quad \frac{k_y d_1}{t_w} \leq 82$$

where

$$k_y = \sqrt{\left(\frac{f_y}{250}\right)}$$

and A_w is the effective area of the web: $A_w = (d_1 - d_d)\,t_w$
with d_1 as the clear depth of the web, d_d as height of any holes up to a height of $0.1d_1$ for unstiffened webs ($0.3d_1$ if web is stiffened) and t_w is the web thickness. (Note: see Step 7 of Section 5.12.3 (Example 5.3) on the use of d_1 for hot-rolled sections such as UB, UC and PFC). If known, the design yield stress of the web is used for f_y in the above expression.

For $\frac{k_y d_1}{t_w} > 82$, V_w can be determined from V_b in Section 5.8.3. Non-uniform shear stress distribution occurs when the section being checked for shear has two webs (e.g. RHS/SHS and box sections), only one flange or has no flanges at all. This is covered by Clause 5.11.3 of AS 4100 as follows:

$$V_v = \frac{2V_w}{\left(0.9 + \frac{f^*_{vm}}{f^*_{va}}\right)} \quad \text{but not exceeding } V_w$$

where

V_w is noted above for approximately uniform shear stress distribution

f^*_{vm} = maximum shear stress

f^*_{va} = average shear stress

The ratio $\frac{f^*_{vm}}{f^*_{va}}$ is equal to 1.5 for a web without flanges, between 1.2 and 1.3 for webs attached to one flange, and 1.0 to 1.1 for two flanges. A solution of this ratio for structural Tees is given in ASI [2009a] and for RHS/SHS in ASI [2004] and OneSteel [2012b].

A special case is circular hollow sections, for which:

$$V_w = 0.36 f_y A_e \qquad \text{(see Clause 5.11.4 of AS 4100 for definition of } A_e\text{)}$$

For all relevant sections, the web adequacy check requires that:

$$V^* \leq \phi V_w \text{ or } V^* \leq \phi V_v \text{ where appropriate}$$

and ϕ = capacity reduction factor = 0.9.

5.8.3 Buckling capacity of unstiffened webs

Apart from web failure by shear yielding, the web can fail by shear buckling. AS 4100 requires that all webs having a web thickness, t_w, less than $\frac{k_y d_1}{82}$ should be checked for shear buckling capacity as follows:

$$V_b = \alpha_v (0.6 f_y A_w)$$

where

$$\alpha_v = \left(\frac{82 t_w}{d_1 k_y}\right)^2$$

$$k_y = \sqrt{\left(\frac{f_y}{250}\right)}$$

with d_1, f_y and A_w being described above in Section 5.8.2.

Finally, the overall check for shear buckling with ϕ = 0.9 is:

$$V^* \leq \phi V_b$$

The web may be stiffened if the web shear yielding or buckling capacity is inadequate. Figure 5.9 illustrates various methods of stiffening the webs. This is considered further in Sections 5.8.6 and 5.8.7.

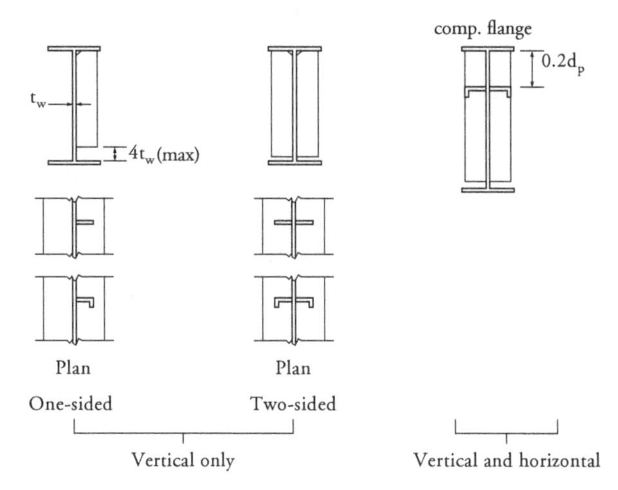

Figure 5.9 Methods of web stiffening

5.8.4 Shear and bending interaction for unstiffened webs

Shear capacity of webs in the locations of relatively high bending moment may have to be reduced, as specified in Clauses 5.12.2 and 5.12.3 of AS 4100—the latter clause being easier to manipulate and herein considered further. The reduction factor α_{vm} can be calculated from:

$$\alpha_{vm} = \left[2.2 - \frac{1.6M^*}{(\phi M_s)}\right] \qquad \text{for } 0.75\phi M_s < M^* \leq \phi M_s$$

For $M^* \leqslant 0.75\phi M_s$ then $\alpha_{vm} = 1.0$—i.e. no reduction in shear capacity. The nominal shear capacity due to interaction with bending moment is then:

$$V_{vm} = \alpha_{vm} V_v$$

where V_v and M_s are the nominal section capacities in web shear and moment, with V_v given as either V_w or V_v (whichever is relevant) as noted in Section 5.8.2 above.

No reduction is normally needed with floor beams loaded with UDL. Significant reduction of shear capacity can occur at the root of a cantilever, where maximum moment combines with maximum shear. Similar caution is needed with continuous beams and simply supported beams loaded with mid-span concentrated loads. Figure 5.10 shows beam configurations where interaction of shear and bending can occur and plots the values of reduction factor α_{vm}.

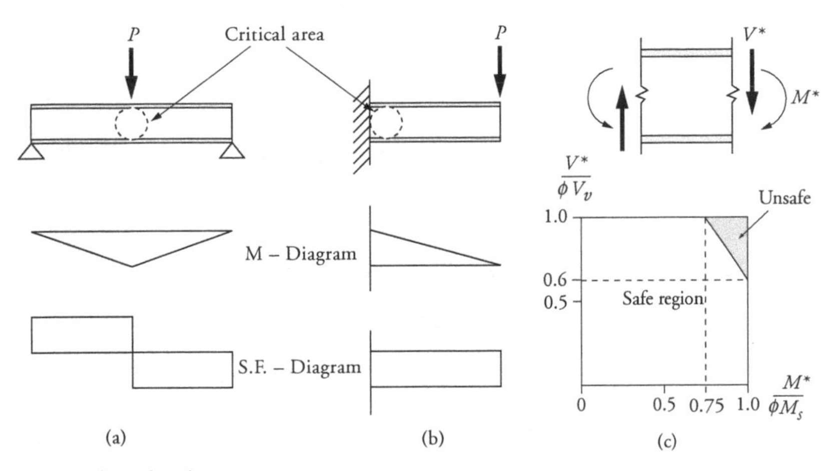

Figure 5.10 Shear-bending interaction

As noted in Figure 5.10(c), shear-bending interaction does not occur when $M^* \leqslant 0.75\phi M_s$ or $V^* \leqslant 0.6\phi V_v$ (where $V_v = V_w$ or V_v as noted in Section 5.8.2). Where these conditions are satisfied then there is no reduction in shear capacity.

5.8.5 Bearing capacity of unstiffened webs

5.8.5.1 General

Open section and RHS/SHS webs in the vicinity of a loading or reaction point must be checked for adequate capacity against yielding and buckling from localised bearing forces. Should the capacity not be adequate, the web should be provided with bearing stiffener(s). AS 4100 offers a simplified procedure based on the 'load dispersion' method. Figure 5.11 shows the dispersion lines, which slope at 1:1 in the web and at 1:2.5 through the flanges.

5.8.5.2 Yield capacity

Bearing yield capacity of the web (i.e. web crushing) at the junction of the web and the flange is computed on the assumption of a uniformly distributed bearing load over the distance b_{bf}:

$$R_{by} = 1.25 b_{bf} t_w f_y$$

and

$$R^* \leqslant \phi R_{by}$$

where t_w is the thickness of the relevant web of an open section (e.g.UB or WB), f_y is the design yield stress and R^* is the design bearing or reaction force.

The dispersion width is determined either as:

$$b_{bf} = b_s + 5t_f \text{, or}$$

$$b_{bf} = b_s + 2.5t_f + b_d \text{, whichever is the lesser applicable,}$$

where b_s is the length of the stiff bearing, t_f the flange thickness and b_d is the remaining distance to the end of the beam such that $b_d \leqslant 2.5t_f$ (see Figure 5.11(b)).

A different procedure is used for square and rectangular hollow sections:

$$R_{by} = 2.0\, b_b t f_y \alpha_p$$

and

$$R^* \leqslant \phi R_{by}$$

where α_p is a reduction factor as given in Clause 5.13.3 of AS 4100.

For an interior bearing with $b_d \geqslant 1.5d_5$:

$$\alpha_p = \frac{0.5}{k_s}\left[1 + (1 - \alpha_{pm}^2)\left(1 + \frac{k_s}{k_v} - (1 - \alpha_{pm}^2)\right)\left(\frac{0.25}{k_v^2}\right)\right]$$

For an end bearing with $b_d < 1.5d_5$:

$$\alpha_p = \sqrt{(2 + k_s^2)} - k_s$$

with typical values for α_p ranging between 0.25 and 0.65,

$$\alpha_{pm} = \frac{1}{k_s} + \frac{0.5}{k_v}$$

d_5 is the flat width of the RHS/SHS web depth and b_d is the distance from the stiff bearing to the end of the beam (see Figure 5.11(b) as an example). Also:

$$k_s = \left(\frac{2r_{ext}}{t}\right) - 1$$

$$k_v = \frac{d_5}{t}$$

r_{ext} = outside radius of section

t = RHS/SHS thickness.

This method for RHS/SHS is further explained with worked examples and tables in ASI [2004] and OneSteel [2012b].

5.8.5.3 Buckling capacity

This subsection applies to beam webs, which are not stiffened by transverse, longitudinal or load bearing web stiffeners. Webs subject to transverse loads can be verified for capacity by using a strut analogy. The nominal area of the "strut" is taken as $A_w = b_b t_w$ and other parameters are listed in Table 5.10.

Table 5.10 Web slenderness ratios for unstiffened webs

	I-section beam	**Hollow section**
Effective width	b_b	b_b
Interior bearing slenderness ratio	$\frac{l_e}{r} = \frac{2.5d_1}{t_w}$	$\frac{l_e}{r} = \frac{3.5d_5}{t_w}$ (for $b_d \geqslant 1.5d_5$)
End bearing slenderness ratio	$\frac{l_e}{r} = \frac{2.5d_1}{t_w}$	$\frac{l_e}{r} = \frac{3.8d_5}{t_w}$ (for $b_d < 1.5d_5$)

Notes: 1. d_1 = clear depth between flanges
2. t_w = thickness of web.
3. b_b = total bearing width dispersion at neutral axis (Figure 5.11(b))—see ASI [2004], OneSteel [2012b] or Figure 5.13.1.3 of AS 4100 to calculate this for RHS/SHS.
4. d_5 = flat width of RHS/SHS web.
5. As noted in Section 1.14(a), the amendment to AS 4100 (AS 4100 AMD 1 - see Appendix D) stipulates that for I- or C-sections that the web geometric slenderness ratio (l_e/r) must be $5.0d_1/t_w$ and not $2.5d_1/t_w$.

The effective width of the web section, b_b, is determined on the basis of the rule of dispersion (see Section 5.8.5.2 and Figure 5.11).

The next step is to obtain the value of the section slenderness factor α_c (see Clause 6.3.3 and Table 6.3.3(3) of AS 4100), assuming that the value of α_b is equal to 0.5 and $k_f = 1.0$. The web bearing buckling capacity is then given by:

$$\phi R_{bb} = \phi \alpha_c A_w f_y$$

where α_c is determined in Section 6 of AS 4100 and shown in Chapter 6 of this Handbook. Then $R^* \leqslant \phi R_{bb}$ must be observed.

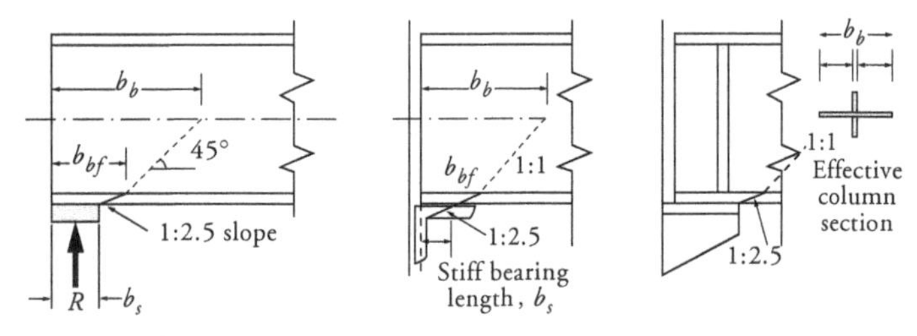

(a) Force dispersion at end bearing points.

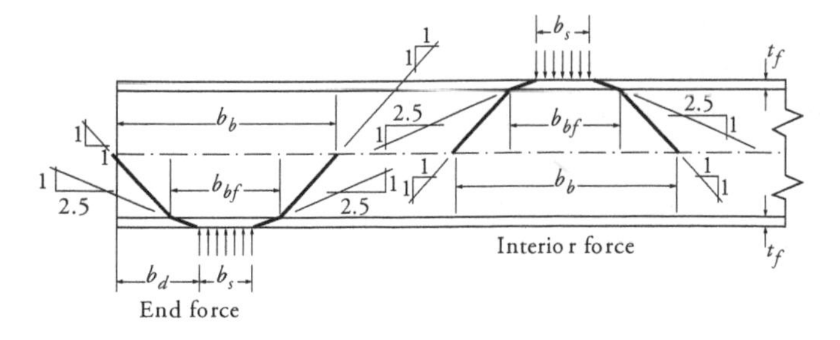

(b) General force dispersion in I-section flange and web

Figure 5.11 Web bearing and the load dispersion method

5.8.5.4 Combined bending and bearing

AS 4100 has additional rules for combined bending and bearing on square and rectangular hollow sections. The following interaction equation is used:

$$\frac{1.2\,R^*}{(\phi R_b)} + \frac{M^*}{(\phi M_s)} \leqslant 1.5$$

except when $\frac{b_s}{b} < 1.0$ and $\frac{d_5}{t} > 30$, the following inequality is used:

$$\frac{0.8\,R^*}{(\phi R_b)} + \frac{M^*}{(\phi M_s)} \leqslant 1.0$$

where b is the width of the hollow section, M^* is the design bending moment in the hollow section at R^*, the applied bearing force. In other words, the presence of bending moment reduces the bearing capacity ϕR_b and vice versa.

5.8.6 Webs with load bearing stiffeners

5.8.6.1 General

Where the web alone has insufficient capacity to carry the imposed concentrated loads and/or lateral forces it may be strengthened by bearing stiffeners directly adjacent to the load with full or partial contribution from the web—see Figure 5.12 and items 1 and 4 in Figure 5.13. Clauses 5.14.1 to 5.14.5 of AS 4100 provide design procedures for these types of web stiffeners (as in Example 5.5).

The ends of load bearing stiffeners are tightly fitted to bear snugly against the loaded flange. Sufficient welds or other fasteners are to be provided along the length of the stiffeners to distribute the reaction or load to the web.

5.8.6.2 Geometry and stiffness requirements

A geometric limitation for the stiffener is that its outstand from the face of the web, b_{es}, is such that:

$$b_{es} \leqslant \frac{15t_s}{\sqrt{\frac{f_{ys}}{250}}}$$

where t_s is the thickness and f_{ys} is the design yield stress of the flat stiffener without the outer edge continuously stiffened (e.g. the stiffener is from flat bar or plate and not an angle section).

5.8.6.3 Yield capacity

Clause 5.14.1 of AS 4100 notes the yield capacity of a load bearing stiffener, R_{sy}, to be a combination of the yield capacity of the unstiffened web, R_{by}, and the yield capacity of the stiffener, i.e.

$$R_{sy} = R_{by} + A_s f_{ys}$$

where R_{by} is determined from Section 5.8.5.2 and A_s is the cross-section area of the stiffener. The design bearing load or reaction force, R^*, must be less than or equal to ϕR_{sy} where ϕ = 0.9 (as in Example 5.6).

5.8.6.4 Buckling capacity

The buckling capacity of the web and load bearing stiffener combination, R_{sb}, are assessed to Clause 5.14.2 of AS 4100. This is similar to the method used for unstiffened webs where the web is analogised to a strut and designed to Section 6 of AS 4100. The strut effective cross-section, A_s, is taken as the stiffener area plus that of an effective length of the web taken as the lesser of:

$$\frac{17.5t_w}{\sqrt{\frac{f_y}{250}}} \text{ and } \frac{s}{2}$$

where t_w is the web thickness and s is the web panel width or spacing to the next web stiffener—if present. Calculations are then done to evaluate the web-stiffener second moment of area about the axis parallel to the web, I_s, and radius of gyration, r_s $\left(=\left[\frac{I_s}{A_s}\right]^{0.5}\right)$.

The effective length, l_e, of the stiffener-web strut is taken as either the clear depth between flanges, d_1, or $0.7d_1$ if the flanges are restrained by other structural elements in the plane of the stiffener against rotation.

Having evaluated l_e/r_s, the design capacity, ϕR_{sb}, is then calculated by the method noted in Section 5.8.5.3 to evaluate α_c (with $\alpha_b = 0.5$ and $k_f = 1.0$) and $\phi = 0.9$ such that:

$$\phi R_{sb} = \phi \alpha_c A_s f_y$$

ϕR_{sb} must be greater than or equal to the design bearing load or reaction force, R^*.

5.8.6.5 Torsional end restraints

Load-bearing stiffeners are also used to provide torsional restraint at the support(s). Clause 5.14.5 of AS 4100 requires the second moment of area of a pair of stiffeners, I_s, about the centreline of the web satisfies:

$$I_s \geqslant \frac{\alpha_t d^3 t_f R^*}{1000F^*}$$

where

$$\alpha_t = \frac{230}{\left(\frac{l_e}{r_y}\right)} - 0.60 \quad \text{and } 0 \leqslant \alpha_t \leqslant 4$$

d = depth of section

t_f = thickness of critical flange (see Section 5.4.1)

R^* = design reaction at the support/bearing

F^* = total design load on the member between supports

$\left(\frac{l_e}{r_y}\right)$ = load-bearing stiffener slenderness ratio noted in Section 5.8.6.4

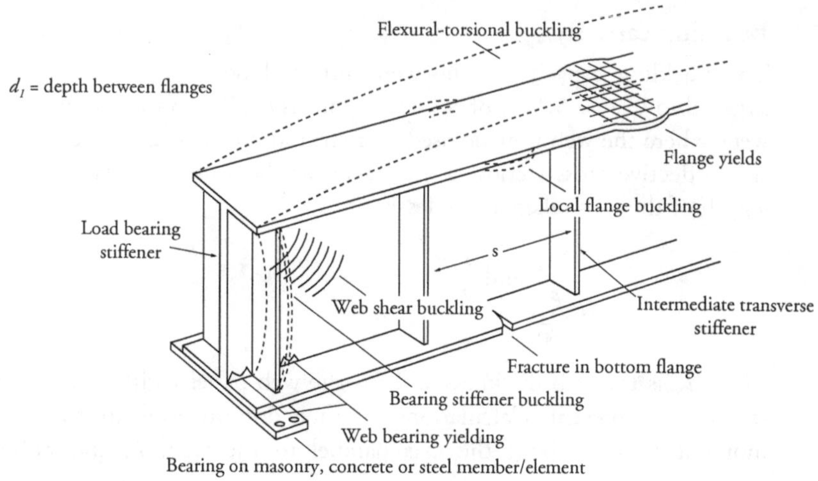

Figure 5.12 Critical areas for consideration of web stiffeners

5.8.7 Webs stiffened by intermediate transverse stiffeners

5.8.7.1 General

There are situations where webs should be made as light as possible or have shear and imposed forces in excess of the web capacity. Web stiffeners are employed to make the web perform better in these instances. The main benefit of intermediate transverse stiffeners (see Figure 5.12 and item 3 of Figure 5.13) is in the increase of the buckling resistance of the web. It should be realised that there is a cost involved in fitting the stiffeners and quite often a thicker, plain web will be a better choice. The design of the intermediate web stiffeners is covered in Clause 5.15 of AS 4100.

Intermediate web stiffeners are usually fillet-welded to the web. Intermittent fillet welds can be used for beams and girders not subjected to weather or corrosive environments, otherwise it is recommended that continuous fillet welds be used. The stiffeners should be in contact with the top flange but a maximum gap of $4t_w$ is recommended between the bottom (tension) flange and the end of the stiffener. Flat plate stiffeners are usually employed for beams and girders up to 1200 mm deep, and angle stiffeners with one leg outstanding are used for deeper girders so as to increase their stiffness. Intermediate stiffeners may be placed on one or both sides of a web. Requirements for such stiffeners are noted below.

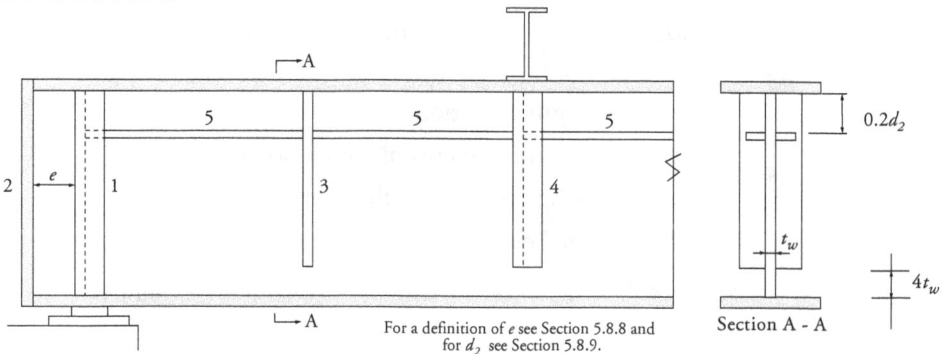

Figure 5.13 Web stiffeners: 1 - load bearing stiffener; 2 - end plate; 3 - intermediate transverse stiffener; 4 - load bearing stiffener; 5 - longitudinal stiffener.

5.8.7.2 Geometry and stiffness requirements

(a) The minimum area of an intermediate stiffener, A_s, should comply with Clause 5.15.3 of AS 4100:

$$A_s \geqslant 0.5\gamma A_w (1 - \alpha_v)\, c_k \left(\frac{V^*}{(\phi V_u)}\right)$$

where

$$c_k = k_s - \frac{k_s^2}{\sqrt{(1 + k_s^2)}}$$

$$k_s = \frac{s}{d_p} \quad (s = \text{spacing between stiffeners and } d_p = \text{depth of web})$$

γ = 2.4 for a single plate stiffener (one side of the web)
= 1.8 for a single angle stiffener
= 1.0 for a pair of stiffeners (one each side of the web)

$$\alpha_v = \left(\frac{82\, t_w}{d_p k_y}\right)^2 \left(\frac{a}{k_s^2} + b\right) \leqslant 1.0$$

$$k_y = \sqrt{\left(\frac{f_y}{250}\right)}$$

a = 0.75 and b = 1.0 for $1.0 \leqslant k_s \leqslant 3.0$

a = 1.0 and b = 0.75 for $k_s \leqslant 1.0$

(b) The outstand of intermediate transverse stiffeners must satisfy the provisions of Section 5.8.6.2.

(c) From Clause 5.15.5 of AS 4100, the minimum flexural stiffness depends on whether or not the stiffener receives applied loads and moments. A stiffener not so loaded must have a minimum stiffness of:

$$I_s = 0.75\, d_1 t_w^3 \quad \text{for } \frac{s}{d_1} \leqslant 1.41$$

$$= \frac{1.5\, d_1^3 t_w^3}{s^2} \quad \text{for } \frac{s}{d_1} > 1.41$$

An increase in the second moment of area of intermediate web stiffeners, I_s, is required by Clause 5.15.7 of AS 4100 when the stiffeners carry external imposed forces, shears and moments. This may arise from cross-beams and the like with their eccentric vertical loads to give such design action effects as $(M^* + F_p^* e)$ acting perpendicular to the web. The increased second moment of area, ΔI_s, is given by:

$$\Delta I_s = d_1^3 \frac{\left(2F_n^* + \dfrac{(M^* + F_p^* e)}{d_1}\right)}{\phi E t_w}$$

where

F_n^* = design force normal to the web

$M^* + F_p^* e$ = design moments normal to the web

F_p^* = design eccentric force parallel to the web

e = eccentricity of F_p^* from the web

t_w = thickness of the web

d_1 = depth between flanges.

An increase in strength of a intermediate transverse stiffener is necessary to carry imposed transverse loads parallel to the web such as the reaction from a cross-beam. In this instance the stiffener must be designed as a load bearing stiffener (see Section 5.8.6).

5.8.7.3 Yield capacity

Due to the nature of loading on intermediate transverse stiffeners, yield capacity checks are not undertaken.

5.8.7.4 Buckling capacity

Buckling capacity checks are undertaken on intermediate transverse stiffeners which must satisfy Clause 5.15.4 of AS 4100 as such:

$$V^* \leqslant \phi(R_{sb} + V_b)$$

where

ϕ = capacity reduction factor = 0.9

R_{sb} = nominal buckling capacity of the stiffener (see Section 5.8.6.4)

and V_b is the nominal shear buckling capacity:

$$V_b = \alpha_v \alpha_d \alpha_f (0.6 f_y A_w)$$

where

A_w = gross sectional web area = $d_1 \times t_w$

$$\alpha_v = \left(\frac{82 t_w}{d_p k_y}\right)^2 \left(\frac{a}{k_s^2} + b\right) \leqslant 1.0$$

t_w = web thickness

d_p = depth of web or deepest web panel

$$k_y = \sqrt{\left(\frac{f_y}{250}\right)}$$

$$k_s = \frac{s}{d_p}$$

s = horizontal spacing between stiffeners

a = 0.75 and b = 1.0 for $1.0 \leqslant k_s \leqslant 3.0$

a = 1.0 and b = 0.75 for $k_s \leqslant 1.0$

$$\alpha_d = 1 + \frac{1 - \alpha_v}{1.15 \alpha_v \sqrt{(1 + k_s^2)}} \text{ ; or}$$

= 1.0 for end web panels with end posts with specific shear buckling conditions (see Clause 5.15.2.2 of AS 4100)

α_f = 1.0 ; or

$$= 1.6 - \frac{0.6}{\sqrt{\left[1 + \left(\frac{40 b_{fo} t_f^2}{d_1^2 t_w}\right)\right]}}$$ with specific b_{fo} conditions (see below)

b_{fo} = flange restraint factor, taken as the least of the following:

$= \frac{12 t_f}{k_y}$; or

= distance from the mid-plane of the web to the nearest edge at a flange (or zero if no flange present); or

= half the clear distance between webs for two or more webs

with t_f being the nearest flange thickness and d_1 is the clear web depth between flanges.

The values of $\alpha_v \alpha_d$ range from 1.0 for stocky webs to 0.2 for slender webs. These values are plotted for ease of use in Figure 5.14. For the evaluation of buckling capacity of a stiffened web which contains other design actions (axial load, significant bending moment, patch loading on flange not necessarily on a stiffener, etc), reference should be made to Appendix I of AS 4100.

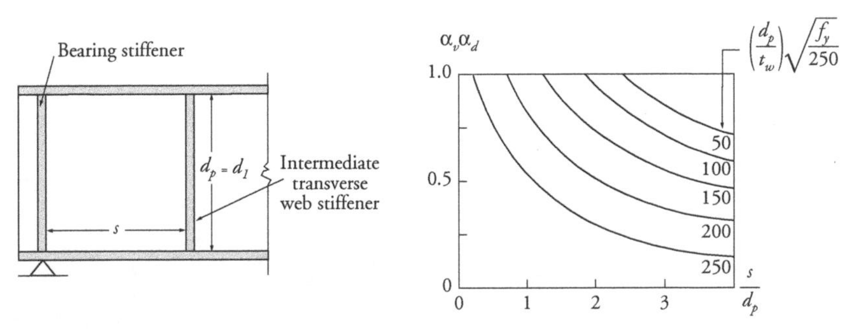

Figure 5.14 *Web buckling factor ($\alpha_v \alpha_d$)*

5.8.7.5 Connecting stiffeners to webs

Welds or other fasteners connecting each intermediate transverse stiffener not subject to external loads are required by Clause 5.15.8 of AS 4100 to transmit a minimum shear force in kN/mm of:

$$v_w^* = \frac{0.0008 t_w^2 f_y}{b_{es}}$$

where b_{es} is the stiffener outstand from the web face and t_w is the web thickness.

5.8.8 End posts

As noted in Clause 5.15.9 of AS 4100, end posts are required for end panels of beams and girders (also see Clause 5.15.2.2 of AS 4100) and are composed of a load bearing stiffener and a parallel end plate separated by a distance e (see items 1 and 2 of Figure 5.13). This stiffener-end plate combination also provides torsional restraint to the beam/girder end(s).

The design of the load bearing stiffener is done in accordance with Clause 5.14 of AS 4100 (see Section 5.8.6 of this Handbook) and should not be smaller than the end plate. The area of the end plate, A_{ep}, must also satisfy:

$$A_{ep} \geqslant \frac{d_1\left[\left(\frac{V^*}{\phi}\right) - \alpha_v V_w\right]}{8ef_y}$$

where α_v is given in Section 5.8.7.4 or Clause 5.11.5.2 of AS 4100 and V_w is the nominal shear yield capacity as noted in Section 5.8.2 or Clause 5.11.4 of AS 4100.

5.8.9 Longitudinal web stiffeners

Longitudinal (or horizontal) stiffeners can be used with advantage in very deep girders. For best results, they should be placed at a distance of $0.2d_2$ (see below for a definition of d_2) from the compression flange. Clause 5.16 of AS 4100 gives the requirements for this type of stiffener which is noted as item 5 in Figure 5.13.

These stiffeners are to be made continuous across the web or extend between transverse stiffeners and are connected by welds or other fasteners.

A longitudinal stiffener at a distance of $0.2d_2$ from the compression flange should have a second moment of area, I_s, about the face of the web not less than:

$$I_s = 4a_w t_w^2 \left\{\left(1 + \frac{4A_s}{a_w}\right)\left(1 + \frac{A_s}{A_w}\right)\right\}$$

where

$a_w = d_2 t_w$

A_s = area of the stiffener

d_2 = twice the clear distance from the neutral axis to the compression flange

A second horizontal stiffener, if required, should be placed at the neutral axis and should have an I_s not less than:

$$I_s = d_2 t_w^3$$

5.9 Composite steel and concrete systems

Composite beams require shear connectors to combine the compression flange of the steel beam to the reinforced concrete floor slab. The advantage of this type of construction is that the weight of the steel beam can be reduced because the concrete slab contributes to the capacity of the beam. The top flange of the beam can be designed as fully restrained against lateral deflection; thus the beam can be regarded as a stocky beam ($l_e = 0$). Deflections are also reduced because of an increased effective second moment of area.

The rules for the design of composite steel and concrete beams are given in AS 2327. The space in this Handbook does not permit treatment of composite action and design. Figure 5.15 illustrates the composite action. At the time of publication of this Handbook, Standards Australia had approved a project to further develop AS 2327 and consider additional items as composite slabs and columns.

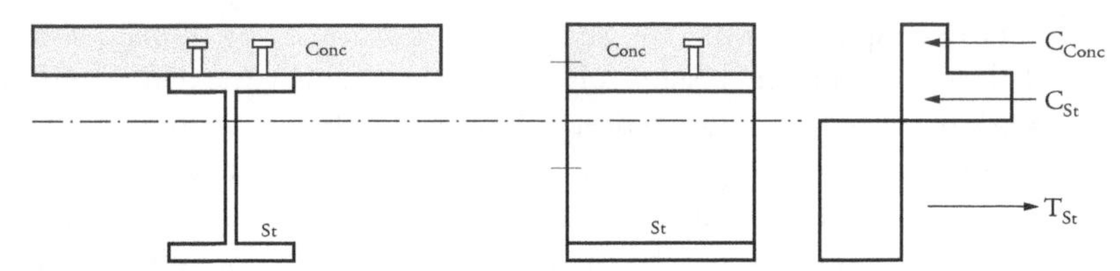

Figures 5.15 Composite steel and concrete beam

5.10 Design for serviceability

Flexural elements have to be checked for the serviceability limit state, which includes the check of deflections, bolt slip, vibration and corrosion. As the serviceability limit state occurs at 'working' loads rather than ultimate design loads, the structure can be regarded as behaving elastically. For a deflection check, it is sufficient to carry out a first-order elastic analysis (without amplification factors) based on 'working' (unfactored loads), use elastic section properties and compute the deformations. See Sections 1.8, 5.12, 5.13, 10.9, 10.11 and Appendices B and C of this Handbook for additional information on serviceability analysis/design and using elastic methods.

Deflections are computed as a part of structural analysis. The load cases for deflection calculations consist of nominal (unfactored) loads. Deflections at the ultimate limit state of strength are usually of no consequence and are therefore not subject to limitation. Some guidance on deflection limits are given in AS/NZS 1170.0 or Appendix B of AS 4100. Additional information can be found in Chapter 10 of this Handbook. Nevertheless, deflection limits should be based on dictates of true serviceability rather than adhering to some ratio of deflection to span.

5.11 Design for economy

Two main causes of uneconomical design in flexural members are the use of non-compact/slender sections and beams having excessive member slenderness in the lateral direction. Non-compact/slender sections utilise only a part of the cross-section, and therefore material is wasted. Also, increasing the size of the beam to meet the deflection limit can waste material. Often it is possible to introduce continuity or rigid end connections to overcome the deflection limit without wasting material.

Excessive slenderness can be measured by the magnitude of the slenderness reduction factor α_s (Section 5.3). A beam having a value of α_s less than 0.70 is regarded as uneconomical. Wherever possible, such beams should be redesigned in one of the following ways:

- by introducing more lateral restraints (i.e. shorter segment lengths) or
- by changing the section to one that exhibits better flexural-torsional resistance such as wide flange, or top-hat section or a hollow section.

Consideration should be given to using concrete floor slabs with some positive means of connection (e.g. shear studs) to the beam, or to providing positive connection between the steel floor plate and the beam by tack welding, etc.

Longer-span beams are sometimes governed by considerations of acceptable deflections. In many instances it is possible to satisfy the deflection limit without loss of section efficiency, choosing deeper beams of lighter section.

Finally, the detailing of end connections and web stiffeners needs appropriate attention so as to avoid costly cutting, fitting and welding.

Reference should also be made to Section 1.11 for other aspects of designing for economy.

5.12 Examples

5.12.1 Example 5.1

Step	Description and computation	Result	Unit
	(a) Select a trial section using Grade 400 steel for the simply supported beam shown. The actions/loads are applied to the top flange. Find a section with the ASI Design Capacity Tables (ASI [2009a]) using the properties of section and moment capacities listed. Assume the ends of the beam are securely held by bolts through the bottom flange anchored to the supports.		
	There are no intermediate lateral restraints between the supports.		
	(b) Check the trial section chosen in (a) for bending moment using AS 4100.		
	(c) Place one intermediate lateral restraint (LR) in the centre of the span. Comment on the difference between having none and one intermediate lateral restraint.		

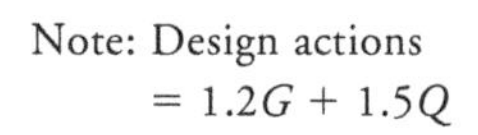

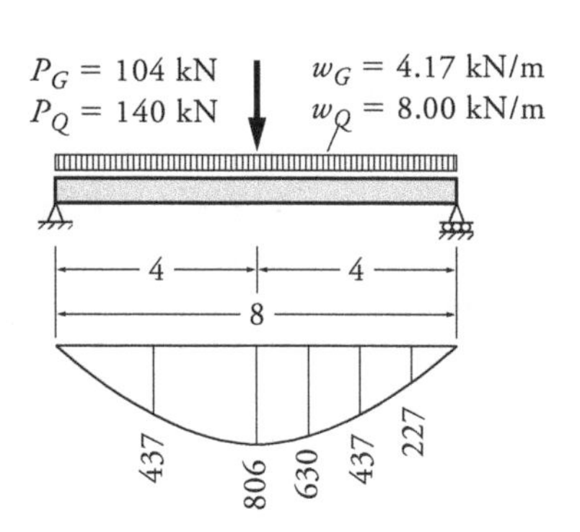

Step	Description and computation	Result	Unit
(a)	Select a trial section using ASI [2009a]		
1	Data		
1.1	Steel grade	400	MPa
1.2	Actions/loads (unfactored)		
	Uniformly distributed permanent action/dead load	4.17	kN/m
	Uniformly distributed imposed action/live load	8.00	kN/m
	Concentrated permanent action/dead load	104	kN
	Concentrated imposed action/live load	140	kN

Step	Description	Value	Unit
1.3	Design action effects (from equilibrium and load factors)		
	Design moment M^*	806	kNm
	Design shear V^*	235	kN
1.3.1	M^* for Step 1.4 and part (c) with its 1 intermediate LR to calculate α_m		
	M_m^* for Step 1.4 and part (c) = M_3^* for Step 1.4 =	806	kNm
	M_2^* for part (c) =	227	kNm
	M_3^* for part (c) = M_2^* and M_4^* for Step 1.4 =	437	kNm
	M_4^* for part (c) =	630	kNm
1.4	Moment modification factor α_m in AS 4100 Clause 5.6.1.1(a)(iii)		
	This calculates to α_m =	1.35	
2	Effective length of beam l_e in AS 4100 Clause 5.6.3		
2.1	Factors k_t, k_l and k_r assumed for first trial (see Table 5.2.1 for default values):		
	k_t = see Note 1 in Table 5.2.1 (for WB conservatively say) =	1.4	
	k_l = loads acting on top flange, acting downwards =	1.4	
	k_r = see Section 5.4.2 also =	1.0	
	l_s = distance between lateral restraints (LRs) = span when there are no intermediate LRs between the supports =	8.0	m
	$l_e = k_t\, k_l\, k_r\, l_s$ = 1.4 × 1.4 × 1.0 × 8.0 = 15.7 (round up for trial design, say)	16.0	m
3	Trial a section using ASI [2009a] which is conservatively based on $\alpha_m = 1.0$:		
	M^* is reduced by $\dfrac{1.0}{\alpha_m} = \dfrac{1.0}{1.35}$ to benefit from the higher α_m		
	$\alpha_m = 1.0$ listed in the ASI tables as follows:		
	$M_r^* = \dfrac{M^*}{\alpha_m} = \dfrac{806}{1.35} =$	597	kNm
	Enter ASI [2009a] Table 5.3-2 page 5-44 or Chart page 5-45 with l_e and M_r^*, choose **900WB218 Grade 400** with moment capacity 699 kNm calling this ϕM_{br} to satisfy the inequality:	**Answer** to **(a)** No int. LR	
	$M_r^* \leqslant \phi M_{br} \rightarrow 597 \leqslant 699 \rightarrow$ true	OK	
(b)	Check selection 900WB218 Grade 400 in (a) for bending using AS 4100		
4	Properties of 900WB218 Grade 400 is taken from ASI [2009a] Tables 3.1-1(A) and (B) pages 3-6 and 3-7		
	d_1 =	860	mm
	t_f =	25.0	mm
	t_w =	12.0	mm
	A_g =	27 800	mm^2
	I_x =	4060×10^6	mm^4
	I_y =	179×10^6	mm^4
	Z_{ex} =	9.84×10^6	mm^3
	r_y =	80.2	mm
	E = Young's modulus ... AS 4100 Notation ... =	200×10^3	MPa
	G = shear modulus = ... AS 4100 Notation ... =	80×10^3	MPa
	J = torsion constant =	4020×10^3	mm^4
	I_w = warping constant =	35.0×10^{12}	mm^6
	Section slenderness =	Non-compact about both *x*- and *y*-axis	
	f_{yf} =	360	MPa
	f_{yw} =	400	MPa

5	A better value of k_t to get l_e is now possible with the benefit of having established the trial section in (a): k_t = twist restraint factor. Bolts through the bottom flange securely fastening the end of the beam to the support provide the end with *partially restrained* cross-section of type P as in AS 4100 Figure 5.4.2.2. The top flange is the critical flange if loads act on it, and towards centre of section. Then AS 4100 Table 5.6.3(1) for restraint arrangement **PP** gives: $k_t = 1 + \left(\frac{2d_1}{l}\right)\left(\frac{0.5\,t_f}{t_w}\right)^3$ $= 1 + \left(2 \times \frac{860}{8000}\right)\left(0.5 \times \frac{25.0}{12.0}\right)^3 =$	1.24	
6	l_e = Effective length adjusted for actual k_t $= k_t\, k_l\, k_r\, l_s$ $= 1.24 \times 1.4 \times 1.0 \times 8000 =$	13 900	mm
7	M_o = reference buckling given by AS 4100 Equation 5.6.1.1(3): Equation is split into 3 parts *A*, *B* and *C* in these calculations:		
7.1	Let $A = \frac{\pi^2 E I_y}{l_e^2}$ $= \pi^2 \times 200 \times 10^3 \times 179 \times \frac{10^6}{13900^2} =$	1.83×10^6	N
7.2	Let $B = \frac{\pi^2 E I_w}{l_e^2}$ $= \pi^2 \times 200 \times 10^3 \times 35.0 \times \frac{10^{12}}{13900^2} =$	3.58×10^{11}	Nmm^2
7.3	Let $C = GJ$ $= 80 \times 10^3 \times 4020 \times 10^3 =$	322×10^9	Nmm^2
7.4	Equations are recombined into: $M_o = \sqrt{[A(C + B)]}$ $= \sqrt{[1.83 \times 10^6 \times (322 \times 10^9 + 3.58 \times 10^{11})]} =$ =	1.12×10^9 1120	Nmm kNm
8	$M_{sx} = M_s$ = nominal section moment capacity: $= f_y\, Z_{ex}$ AS 4100 Clause 5.2.1 $= 360 \times 9.84 \times 10^6 =$ = $\phi M_{sx} = 0.9 \times 3540$ ... AS 4100 Table 3.4 for ϕ ... =	3540×10^6 3540 3190	Nmm kNm kNm
9	α_s = slenderness reduction factor AS 4100 Equation 5.6.1.1(2) $= 0.6\left(\sqrt{\left[\frac{M_s^2}{M_o^2} + 3\right]} - \frac{M_s}{M_o}\right)$ $= 0.6\left(\sqrt{\left[\frac{3540^2}{1120^2} + 3\right]} - \frac{3540}{1120}\right) =$	0.266	

---

10 ϕM_b = member moment capacity AS 4100 Equation 5.6.1.1(1)

$= \phi\, \alpha_m\, \alpha_s\, M_s$

$= 0.9 \times 1.35 \times 0.266 \times 3540$ — 1140 kNm

$\leqslant \phi M_s = 3190$ — OK

10.1 $M^* \leqslant \phi M_b$ in AS 4100 Clause 5.1 is satisfied with:

$806 \leqslant 1140$ being true

900WB218 Grade 400 is OK with efficiency $\frac{806}{1140} = 0.71 = 71\%$ — **Answer** to **(b)**

Note: This may be considered excessive in terms of reserve of capacity however a check of all of the WBs smaller and lighter than 900WB218 Grade 400 will note they are not adequate for bending from the loading/restraint conditions. This is due to the significant influence of the PP end restraint conditions on k_t and subsequently l_e for deeper "girder" type (non-universal) members (see Table 5.2.1).

(c) Effect of additional LR ...using AS 4100 and ASI [2009a] ...

11 Improved value of α_m = moment modification factor with 1 LR from AS 4100 Clause 5.6.1.1(a)(iii) gives

$$\alpha_m = \frac{1.7 M_m^*}{\sqrt{[M_2^{*2} + M_3^{*2} + M_4^{*2}]}}$$

$$= \frac{1.7 \times 806}{\sqrt{[227^2 + 437^2 + 630^2]}} \leqslant 2.5 =$$ — 1.71

Note the earlier value of $\alpha_m = 1.35$ in Step 1.4

---

12 Much better/smaller effective length l_e from AS 4100 Clause 5.6.3 and assume same $k_t = 1.24$ gives

$l_e = k_t\, k_l\, k_r\, l_s$

$= 1.24 \times 1.4 \times 1.0 \times 4000 =$ — 6940 mm

Note previously from (b) in Step 6, $l_e = 13\,900$ mm.

---

13 $M_r^* = \frac{M^*}{\alpha_m}$ = reduced M^* to use with tables to gain benefit of $\alpha_m = 1.71$ being greater than the $\alpha_m = 1.0$ built into the tables:

$M_r^* = \frac{806}{1.71} =$ — 471 kNm

13.1 Select beam from ASI [2009a] Table 5.3-2 or Chart pages 5-44 and 45 for $l_e = 6.94$ m and $M_r^* = 471$ kNm gives:

700WB115 Grade 400 ...

which has ϕM_{br} for $l_e = 7.0$ m = — 472 kNm

to satisfy $M_r^* \leqslant \phi M_{br} \rightarrow 471 \leqslant 472 \rightarrow$ true — OK

and also AS 4100 Clause 5.2.1 requires

$$\phi M_{br} \leqslant \frac{\phi M_{sx}}{\alpha_m} \rightarrow 472 \leqslant \frac{1300}{1.71} \rightarrow 472 \leqslant 760 \rightarrow \text{true}$$ — OK

where $\phi M_{sx} = 1300$ kNm is also from ASI [2009a] Table 5.3.2 page 5-44

700WB115 Grade 400 satisfies bending moment capacity ϕM_b

13.2 Now recheck k_t and l_e (with improved PL restraint condition instead of the previous PP) due to the revised beam section.

Repeat Steps 5 & 12 to 13.1 with results only shown:

k_t = from AS 4100 Table 5.6.3(1) with a different formula — 1.08

l_e = $1.08 \times 1.4 \times 1.0 \times 4000 = 6050$ (round to) — 6000 mm

ϕM_{br} (from ASI [2009a] with inbuilt $\alpha_m = 1.0$) = — 577 kNm

	to satisfy $M_r^* \leq \phi M_{br}$ → 471 ≤ 577 → true	OK	
	$\phi M_{br} \leq \frac{\phi M_{sx}}{\alpha_m}$ → 577 ≤ 760 → true	OK	
	700WB115 Grade 400 satisfies ϕM_b	**Answer to (c)**	
14	**Comment** on benefit of introducing **1 LR at mid-span:** The additional lateral restraint provided the following beneficial effects: reduction in effective length and increase in moment modification factor, α_m, which resulted in a reduced beam size. The saving in the Grade 400 beam mass from parts (b) and (c) is:		
	(b) 900WB218 with no intermediate lateral restraint, mass =	218	kg/m
	(c) 700WB115 with one (1) intermediate lateral restraint, mass =	115	kg/m
	Beam mass saving = (218 − 115)/218 × 100 =	47%	
	Note: The above example utilised Partial (P) restraints for deeper "girder" type members which have a significant, non-readily apparent effect on bending design capacities. Had they not been present, then k_t = 1.0 (Table 5.6.3(1) of AS 4100) and no iteration would be required to calculate l_e based on the to-be-determined beam depth (d_1). For simple examples of beams with Full (F) restraint conditions see Section 5.3.6 of ASI [2004, 2009a] and OneSteel [2012b].		

5.12.2 Example 5.2

Step	Description and calculations	Result	Unit
	Find a UB in Grade 300 steel as a trial section for the beam shown. Each end is placed on a bearing plate 50 mm wide resting on top of a reinforced concrete wall and fastened to the wall by two bolts through the bottom flange. A structural tee 75CT11.7 Grade 300 steel brace acting as an intermediate lateral restraint (LR) is attached by two bolts to the top flange at mid-span. The other end of the brace is anchored by two bolts to the inside face of the reinforced concrete wall by an end plate welded to the brace. The beam supports reinforced concrete floor panels 83 mm thick, and which are removable and unattached to the beam. The beam must also carry an occasional short-term imposed action/live load of 10 kN at mid-span arising from industrial plant and equipment. The floor has an imposed action/live load of 3 kPa.		

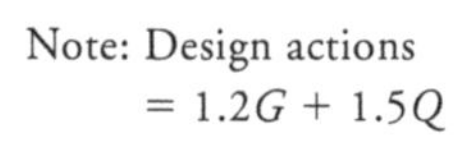

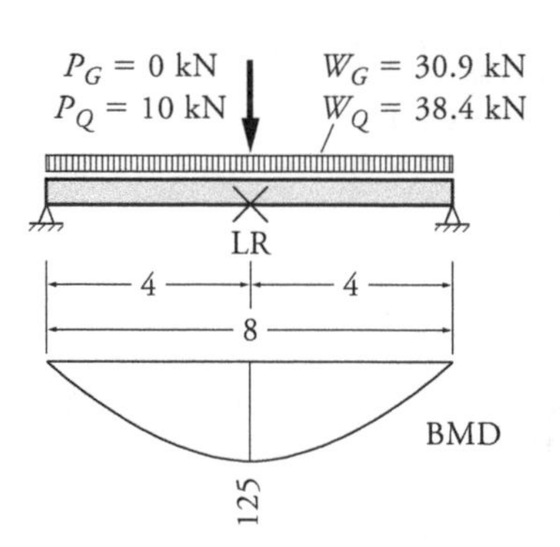

Step	Description and calculations	Result	Unit
1	Span and width: span = 8 m width of load tributary area = 1.6 m		

2	Nominal actions/loads.		
2.1	Uniformly distributed permanent actions/dead loads:		
	self-weight of steel beam initially estimated/guessed =	0.54	kN/m
	weight of 83 mm thick reinforced concrete floor panels at 25 kN/m^3		
	$= 0.083 \times 25 =$	2.08	kN/m^2
	W_G = *total* uniformly distributed dead load from self-weight of beam and removable concrete floor panels		
	$= 0.54 \times 8 + 2.08 \times 1.6 \times 8 =$	30.9	kN
2.2	Uniformly distributed imposed action/live loads =	3.0	kPa
	W_Q = *total* uniformly distributed imposed action/live load		
	$= 3.0 \times 1.6 \times 8 =$	38.4	kN
2.3	Point actions/loads:		
	P_G = point permanent action/dead load = none =	0	kN
	P_Q = point imposed action/live load =	10.0	kN
3	Strength design actions/loads and effects:		
3.1	Use combined factored actions/loads in AS/NZS 1170.0 Clause 4.2.2(b) which is $E_d = [1.2\,G, 1.5\,Q] = 1.2G + 1.5Q$ in the following:		
3.1.1	W^* = total uniformly distributed design action/load		
	$= 1.2W_G + 1.5W_Q$		
	$= 1.2 \times 30.9 + 1.5 \times 38.4 =$	94.7	kN
3.1.2	P^* = point design action/load		
	$= 1.2P_G + 1.5P_Q$		
	$= 1.2 \times 0 + 1.5 \times 10 =$	15.0	kN
3.2	Design action/load effects:		
	simple beam is loaded symmetrically, therefore:		
	R^* = design reaction at each support		
	$= 0.5W^* + 0.5P^*$		
	$= 0.5 \times 94.7 + 0.5 \times 15 =$	54.9	kN
	= design shear force		
	$= V^*$ = ... when there is no overhang ... =	54.9	kN
	M^* = design moment		
	$= \dfrac{W^*L}{8} + \dfrac{P^*L}{4}$		
	$= 94.7 \times \dfrac{8.0}{8} + 15 \times \dfrac{8.0}{4} =$	125	kNm
4	Effective length of beam l_e		
	$l_e = k_t\, k_l\, k_r\, l_s$ from AS 4100 Clause 5.6.3		
	Default $k_t\, k_l\, k_r$ values are given in Table 5.2.1 of this Handbook		
	l_s = distance between restraints		
	= distance between support and brace at mid-span =	4.0	m
	$l_e = 1.1 \times 1.4 \times 1.0 \times 4 =$	6.16	m
5	Size of beam-*conservative* Answer 1		
	ASI [2009a] Table 5.3-5 page 5-50 or Chart page 5-51 gives by linear interpolation:		
	410UB59.7 Grade 300 steel OK for bending with its design capacity of		
	126 kNm which is greater than or equal to the required 125 kNm	**Answer 1**	
	This is based on the conservative value of $\alpha_m = 1.0$ assumed from ASI [2009a] above. However, the 410UB59.7 will have a higher design capacity as the acutal α_m will be greater than 1.0. A more economical size is given in Answer 2 below.		

6	Moment modification factor α_m		
	Load ratio $\frac{P^*}{W^*} = \frac{15.0}{94.7} =$	0.158	
	Table 5.6(a), 1 LR mid-span with $j = 0.158$ interpolates to give α_m =	1.41	
7	Size of beam—more *economical* answer: making good use of new/better $\alpha_m = 1.41$ as follows:		
	Find reduced design moment $M_r^* = \frac{M^*}{\alpha_m} = \frac{125}{1.41} =$	88.7	kNm
	Enter ASI [2009a] Table 5.3-5 page 5-50 with its inbuilt $\alpha_m = 1.0$, and using l_e =	6.16	m
	Get a reduced ϕM_b for 360UB50.7 and call it ϕM_{br} =	91.9	kNm
	360UB50.7 Grade 300 is OK	**Answer 2**	
	because $M_r^* \leq \phi M_{br} \rightarrow 88.7 \leq 91.9 \rightarrow$ true, satisfied		

5.12.3 Example 5.3

Step	Description and calculations	Result	Unit
	Check the suitability of a 360UB50.7 Grade 300 for the beam in Example 5.2. This solution relies mainly on AS 4100 and minimally on ASI [2009a], for example, to get basic section properties. Omit deflection checks.		
	The next Example 5.3.1 is simpler because much more extensive use is made of ASI [2009a] as an aid, giving a shorter solution.		

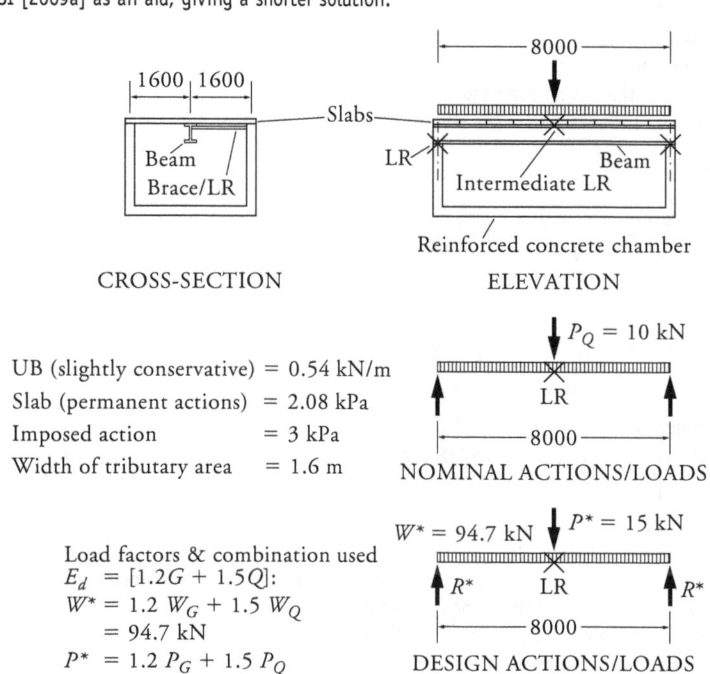

1	Design actions/loads and effects:		
	$W^* =$	94.7	kN
	$P^* =$	15.0	kN
	$R^* =$	54.9	kN
	$V^* =$	54.9	kN
	$M^* =$	125	kNm
2	Section properties of 360UB50.7 Grade 300: ASI [2009a] Tables 3.1-3(A) and (B) pages 3-10 and 11		
	$Z_{ex} =$	897×10^3	mm^3
	$I_x =$	142×10^6	mm^4
	$t_f =$	11.5	mm
	$t_w =$	7.3	mm
	$d_1 =$	333	mm
	$d =$	356	mm
	$f_{yf} =$	300	MPa
	$f_{yw} =$	320	MPa
3	Effective length l_e: $l_e = k_t\, k_l\, k_r\, l_s$ from AS 4100 Clause 5.6.3		
	Lateral restraint arrangement is overall symmetrical in which support end of each segment is **P**artially, and the other end at mid-span is **L**aterally, restrained from AS 4100 Figures 5.4.2.2 and 5.4.2.4 respectively.		
	For the restraint type of the segment with length between support and an intermediate lateral restraint, LR, at mid-span in this example, AS 4100 Clauses 5.3 to 5.6 notes this to be:	PL	
	k_t = twist restraint factor is given in AS 4100 Table 5.6.3(1): $= 1 + \frac{d_1}{l_s}\left(\frac{t_f}{2t_w}\right)^3$		
	$= 1 + \frac{333}{4000}\left(\frac{11.5}{2 \times 7.3}\right)^3 =$	1.04	
	k_l = load height factor relative to centre of beam cross-section =	1.4	
	... because top flange is critical flange as it is simply supported and in compression, and load acts directly on it to twist the beam more as it buckles (from AS 4100 Table 5.6.3(2)).		
	k_r = lateral rotation restraint factor =	1.0	
	Safer to use 1.0 most of the time or if uncertain. AS 4100 Table 5.6.3(3).		
	l_s = length of segment between lateral restraints LR = length between support and LR afforded by brace at mid-span =	4	m
	$l_e = 1.04 \times 1.4 \times 1.0 \times 4 =$	5.82	m
4	Moment modification factor α_m		
	Load ratio $= j = \frac{P^*}{W^*} = \frac{15}{94.7} =$	0.158	
	By interpolation in Table 5.6(a) for case with 1 LR at mid-span gives α_m = or use AS 4100 Clause 5.6.1.1(a)(iii)	1.41	
5	Slenderness reduction factor α_s (AS 4100 Clause 5.6.1.1(a)) From Table 5.8.1 for $l_e = 5.82$ get by interpolation:		
	$\alpha_s =$	0.404	
	or use AS 4100 Equations 5.6.1.1(2) and (3)		

6	Bending		
6.1	Section moment capacity ϕM_{sx} from AS 4100 Clause 5.2.1 is $\phi M_{sx} = \phi\ f_y\ Z_{ex}$ $= 0.9 \times 300 \times 897 \times \frac{10^3}{10^6} =$	242	kNm
	ϕ is given in AS 4100 Table 3.4		
6.2	Member moment capacity ϕM_{bx} $\phi M_{bx} = \alpha_m\ \alpha_s\ \phi M_{sx} \leqslant \phi M_{sx}$ AS 4100 Equation 5.6.1.1(1) $= 1.41 \times 0.404 \times 242 = 138 \leqslant \phi M_{sx} = 242 =$	138	kNm
6.3	$M^* \leqslant \phi M_{bx}$ in AS 4100 Clause 5.1 is OK because: 125 ≤ 138 requirement is true/satisfied	Member capacity ϕM_b **OK**	
7	Web shear capacity ϕV_v: $\phi V_v = \phi\ 0.6\ f_{yw}\ A_w = \phi\ 0.6\ f_{yw}\ d\ t_w$ $= 0.9 \times 0.6 \times 320 \times 356 \times \frac{7.3}{10^3} =$	449	kN
	$V^* \leqslant \phi V_v$ is satisfied because 54.9 ≤ 449 is true.	Web shear ϕV_v **OK**	
	In the absence of horizontal web stiffeners: Can use d_1 for d_p in web slenderness d_p/t_w which in hot-rolled I-sections such as UB and UC *all* satisfy AS 4100 Clause 5.11.2(a) meaning their webs are stocky to permit full shear yield $V_w = 0.6\ f_{yw}\ A_w$ given in Clause 5.11.4 to be used. Similarly for WC and PFC. A cursory examination of d_1/t_w in WB sections in tables shows a few possible exceptions for considering web shear buckling.		
	Note also that d was used instead of d_1 in the above ϕV_v calculation for hot-rolled (HR) sections as opposed to welded sections when d_1 would be used. This is due to HR sections such as UB and UC having full steel "flow" at the web-flange junction from the manufacturing process. However, the same can't be said for other types of fabricated sections. Part 5 of ASI [2009a] also notes this differentiation.		
	Shear and bending interaction need not be considered as the peak shear force ($V^* = 54.9$ kN) is less than 60% of the design shear capacity ($\phi V_v = 449$ kN) – see Section 5.8.4.		
8	Web bearing at supports		
8.1	Web bearing yield ... AS 4100 Clause 5.13.3 and Figure 5.13.1.1(b) inverted		
	b_s = length of stiff bearing = width of bearing plate =	50	mm
	b_{bf} = length of bearing between web and flange $= b_s + 2.5\ t_f$ AS 4100 Clause 5.13.1 $= 50 + 2.5 \times 11.5 =$	78.8	mm
	ϕR_{by} = web bearing yield capacity ... at support $= \phi 1.25\ b_{bf}\ t_w\ f_y$ AS 4100 Clause 5.13.3 ... $f_y = f_{yw}$ $= \frac{0.9 \times 1.25 \times 78.8 \times 7.3 \times 320}{10^3} =$	207	kN
	$R^* \leqslant \phi R_{by}$ is satisfied because 54.9 ≤ 207 is true ... yield part of AS 4100 Clause 5.13.2	Web bearing yield ϕR_{by} **OK**	
8.2	Web bearing buckling ... at support b_b = width of web notionally as a column loaded axially with R^* $= b_{bf} + \frac{d_2}{2}$		

	$= b_{bf} + \frac{d_1}{2}$		
	$= 78.8 + \frac{333}{2} =$	245	mm
	End portion of beam as a column with cross-section $t_w b_b$ with area A_n ... also $\alpha_b = 0.5$ and $k_f = 1.0$ ... AS 4100 Clause 5.13.4 ...		
	$A_n = t_w\, b_b$		
	$= 7.3 \times 245 =$	1790	mm^2
	End as column with geometric slenderness $\frac{l_e}{r}$		
	$\frac{l_e}{r} = \frac{2.5\, d_1}{t_w}$		
	$= 2.5 \times \frac{333}{7.3} =$	114	
	λ_n = modified slenderness to suit material properties ... AS 4100 Clause 6.3.3		
	$\lambda_n = \frac{l_e}{r} \sqrt{k_f} \sqrt{\left(\frac{f_y}{250}\right)}$		
	$= 114\sqrt{1.0} \sqrt{\left(\frac{320}{250}\right)} =$	129	
	α_b = member constant for residual stress distribution =	0.5	
	α_c = slenderness reduction factor to be applied to column section capacity ... AS 4100 Table 6.3.3(3)		
	$\alpha_c =$	0.345	
	ϕR_{bb} = web bearing buckling capacity ... at support ...		
	$= \phi\, \alpha_c\, f_{yw}\, A_n$		
	$= 0.9 \times 0.345 \times 320 \times \frac{1790}{10^3} =$	178	kN
	$R^* \leqslant \phi R_{bb}$ is satisfied because $54.9 \leqslant 178$ is true		
	... buckling part of AS 4100 Clause 5.13.2	Web bearing buckling $\phi \boldsymbol{R_{bb}}$ **OK**	
8.3	Web bearing in conclusion is ...	OK	
9	Deflection. See Example 5.3.1.	OK	

5.12.3.1 Example 5.3.1

A *simplified* check of the beam shown in Example 5.3 is done for moment, shear and web bearing capacity, and deflection.

Use aids from this Handbook and ASI [2009a]. A detailed check is done in Example 5.3.2.

Given:
Span l is 8 m.
The beam top flange is connected at mid-span by a brace from the RC wall giving lateral restraint (LR).
The beam ends receive lateral restraint from bolts anchoring the bottom flange to the supporting walls via 50 mm wide bearing plates.
Downward actions/loads act on the top flange.
Beam is 360UB50.7 Grade 300.

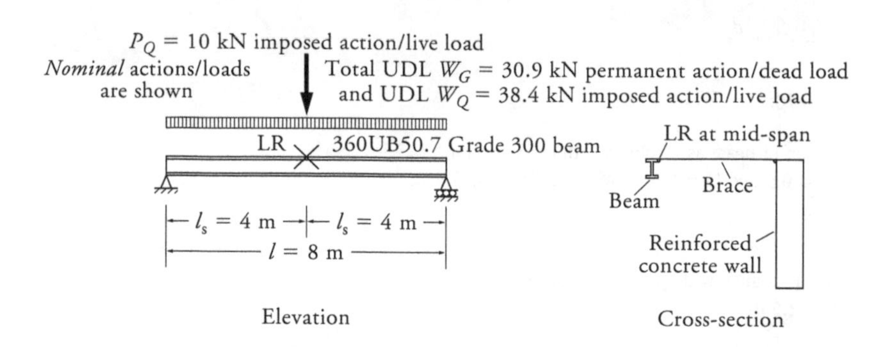

Step	Quantity	Result	Unit
	Data		
	360UB50.7 Grade 300		
	$I_x =$	142×10^6	mm^4
	$t_f =$	11.5	mm
	$t_w =$	7.3	mm
	$d_1 =$	333	mm
1	*Nominal* actions/loads are the same as in Examples 5.2 and 5.3. Loads when consolidated, reduce down to:		
1.1	*Total* uniformly distributed permanent action/dead load W_G including self-weight of beam and concrete panels.		
	$W_G =$	30.9	kN
1.2	*Total* uniformly distributed imposed action/live load W_Q		
	$W_Q =$	38.4	kN
1.3	Point permanent action/dead load P_G		
	$P_G =$	0	kN
1.4	Point imposed action/live load P_Q		
	$P_Q =$	10	kN
2	*Strength* design actions/loads and effects are calculated from *nominal* loads in Step 1. Briefly repeating from Example 5.2 gives:		
	$W^* = 1.2 \times 30.9 + 1.5 \times 38.4 =$	94.7	kN
	$P^* = 1.2 \times 0 + 1.5 \times 10 =$	15.0	kN
	$R^* = 0.5 \times (94.7 + 15) =$ ...symmetrical...=	54.9	kN
	$V^* = R^* =$ ...no overhang ...=	54.9	kN
	$M^* = 94.7 \times \frac{8.0}{8} + 15 \times \frac{8.0}{4} =$	125	kNm
3	Effective length of segment l_e between support and LR at mid-span-		
	$l_s =$	4.0	m
	k values are obtained from Table 5.2.1 of this Handbook		
	$l_e = k_t\, k_l\, k_r\, l_s$		
	$= 1.1 \times 1.4 \times 1.0 \times 4 =$	6.16	m
4	Moment modification factor α_m for the shape of the bending moment diagram		
	From Table 5.6(a) for 1 LR at mid-span, ...		
	or AS 4100 Clause 5.6.1.1(a)(iii) ... with		
	$j = \frac{P^*}{W^*} = \frac{15.0}{94.7} =$	0.158	
	Get $\alpha_m =$	1.41	

5	Blank (previous Example 5.3 required α_s. Its use is not overt in ASI [2009a] and does not need to be specifically calculated here).		
6	Moment capacity ϕM		
6.1	Section moment capacity ϕM_s		
	ASI [2009a] Table 5.3-5 page 5-50 or Table 5.2-5 page 5-38 ... get ϕM_s =	242	kNm
6.2	Member moment capacity ϕM_b		
	With l_e = 6.16 m in ASI [2009a] Table 5.3-5 page 5-50 or Chart page 5-51 get ϕM_{b1} for $\alpha_m = 1.0$:		
	$\phi M_{b1} =$	91.9	kNm
	Improve on ϕM_{b1} for $\alpha_m = 1.41$, getting		
	$\phi M_b = \alpha_m \phi M_{b1}$ $= 1.41 \times 91.9 =$	130	kNm
	Check section moment capacity ϕM_s is not exceeded by ϕM_b because of an overly large α_m ...		
	$\phi M_b \leqslant \phi M_s \rightarrow 130 \leqslant 242 \rightarrow$ true, satisfied ...	OK	
	and $\phi M_b =$	130	kNm
	(Step 6.2 of the last example notes ϕM_b (= ϕM_{bx}) = 138 kN, the slight difference being due to linear interpolation approximations and numerical rounding).		
	AS 4100 Clause 5.1 requires ...		
	$M^* \leqslant \phi M_b \rightarrow 125 \leqslant 130 \rightarrow$ true, satisfied ...	OK	
	360UB50.7 Grade 300 is satisfactory for bending ... and **Moment capacity**	ϕM_b OK	
7	Web shear capacity ϕV_v		
	ASI [2009a] Table 5.3-5 page 5-50		
	$\phi V_v =$	449	kN
	$R^* \leqslant \phi V_v \rightarrow 54.9 \leqslant 449$ = true ... and **Web shear**	ϕV_v OK	
8	Web bearing capacity ϕR_b (at end supports)		
	b_s = ... 50 wide bearing plate ... =	50	mm
	$b_{bf} = b_s + 2.5\ t_f$ — AS 4100 Clause 5.13.1 $= 50 + 2.5 \times 11.5 =$	78.8	mm
	$b_b = b_{bf} + \frac{d_1}{2}$ — AS 4100 Figure 5.13.1.1(b) inverted		
	$= 78.8 + \frac{333}{2} =$	245	mm
	$\frac{\phi R_{bb}}{b_h} = 0.725$ — ASI [2009a] Table 5.2-5 page 5-38		
	$\phi R_{bb} = 0.725 \times b_b$ $= 0.725 \times 245 =$	178	kN
	$\frac{\phi\ R_{by}}{b_{bf}} = 2.63$ — ASI [2009a] Table 5.2.5 page 5–38		
	$\phi R_{by} = 2.63 \times b_{bf}$ $= 2.63 \times 78.8 =$	207	kN
	$\phi R_b = \text{Min}(\phi R_{by}, \phi R_{bb})$ = Min (207,178) =	178	kN
	$R^* \leqslant \phi R_b \rightarrow 54.9 \leqslant 178$ is true, satisfied ... and **Web bearing**	ϕR_b OK	

9	Deflection. Refer to AS/NZS 1170.0 Clause 4.3 combining (a) and (b) with Table 4.1, and AS 4100 Appendix B, ASI [2009a] Table T5.3 page 5-19 or Appendix C of this Handbook.		
9.1	*Serviceability* design actions/loads for deflection:		
	$W^* = W_G + \psi_l W_Q = 30.9 + 0.6 \times 38.4 =$	53.9	kN
	$P^* = P_G + \psi_s P_Q = 0 + 1.0 \times 10 =$	10.0	kN
9.2	Actual deflection $\delta = \dfrac{5W^*L^3}{384EI} + \dfrac{P^*L^3}{48EI}$		
	$= \dfrac{5 \times 53.9 \times 10^3 \times 8000^3}{(384 \times 2 \times 10^5 \times 142 \times 10^6)} + \dfrac{10.0 \times 10^3 \times 8000^3}{(48 \times 2 \times 10^5 \times 142 \times 10^6)} =$	16.4	mm
9.3	Permissible deflection $\Delta = \dfrac{l}{250} = \dfrac{8000}{250} =$	32	mm
	$\delta \leq \Delta \rightarrow 16.4 \leq 32 \rightarrow$ true, satisfactory ... and **Deflection**	δ OK	
	Further deflection requirements are noted in Section 1.8.		
10	Summary		
	360UB50.7 Grade 300 is $\dfrac{125}{130} = 0.96$ or 96% effective in bending capacity. Web shear, bearing yield and buckling, and deflection are all satisfactory.	All **OK**	
Addendum	Shear–bending interaction		
	As an example, check a section of the beam 2 m from the support in which:		
	$V^* =$	31.2	kN
	$M^* =$	86.0	kNm
	ASI [2009a] Table 5.3-5 page 5-50 or Table 5.2-5 page 5-38 gives:		
	$\phi M_s =$	242	kNm
	$0.75\phi M_s =$	182	kNm
	From AS 4100 Clause 5.12.3		
	because $M^* \leq 0.75\phi M_s \rightarrow 86.0 \leq 182$ is true, satisfied		
	then $\phi V_{Vm} = \phi V_V =$	449	kN
	Thus $V^* \leq \phi V_V \rightarrow 31.2 \leq 449$ is true, satisfied.		
	Shear–bending interaction is	**OK**	

Note: Example 5.3.1 above involves the same beam and loads used in Examples 5.2 and 5.3. Example 5.2 established a 360UB50.7 Grade 300 as a possible trial section for the beam. Example 5.3 without aids and relying mainly on AS 4100 shows the selection is satisfactory for moment and web capacities.

Example 5.3.1 uses aids from this Handbook and ASI [2009a] to simplify the calculations to check moment and web capacities and deflection. It includes some comments and references.

5.12.3.2 Example 5.3.2

Example 5.3.1 is revisited showing only calculations.
Do a complete check of the beam shown.

Span l is 8 m.
Top flange has 1 lateral restraint (LR) at mid-span. Ends are partially restrained (P) by anchor bolts in the bottom flange. Bearing plates are 50 mm wide.

Actions/loads sit directly on the top flange, and act downwards.
Beam is 360UB50.7 Grade 300.

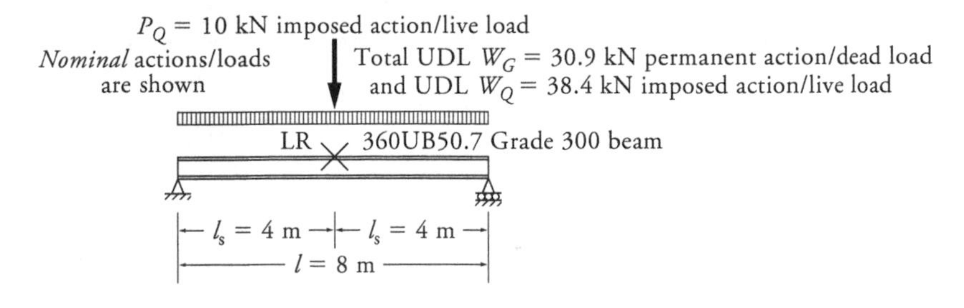

Elevation

Step	Quantity	Result	Unit
	Data		
	360UB50.7 Grade 300		
	I_x =	142×10^6	mm^4
	t_f =	11.5	mm
	t_w =	7.3	mm
	d_1 =	333	mm
	Z_{ex} =	897×10^3	mm^3
	f_{yf} =	300	MPa
	I_y =	9.60×10^6	mm^4
	J =	241×10^3	mm^4
	I_w =	284×10^9	mm^6
1	*Nominal* actions/loads are		
1.1	Total uniformly distributed permanent action/dead load including self-weight of beam and concrete panels W_G =	30.9	kN
1.2	Total uniformly distributed imposed action/live load W_Q =	38.4	kN
1.3	Point permanent action/dead load		
	P_G =	0	kN
1.4	Point imposed action/live load		
	P_Q =	10	kN
2	*Strength* design actions/loads and effects		
	$W^* = 1.2 \times 30.9 + 1.5 \times 38.4 =$	94.7	kN
	$P^* = 1.2 \times 0 + 1.5 \times 10 =$	15.0	kN
	$R^* = 0.5 \times (94.7 + 15) =$	54.9	kN
	$V^* = R^* =$	54.9	kN
	$M^* = 94.7 \times \frac{8.0}{8} + 15 \times \frac{8.0}{4} =$	125	kNm

3	Beam segment effective length based on default k values in Table 5.2.1 (could also use the more precise evaluation of k_t as noted in Step 3 of Example 5.3—but not in this instance) $l_e = 1.1 \times 1.4 \times 1.0 \times 4 =$	6.16	m
4	Section moment capacity, ϕM_s $M_s = 300 \times 897 \times \frac{10^3}{10^6} =$ $\phi M_s = 0.9 \times 269 =$	269 242	kNm kNm
5	Moment modification factor α_m from Table 5.6(a), 1 LR mid-span $j = \frac{P^*}{W^*} = \frac{15.0}{94.7} =$ $\alpha_m =$	0.158 1.41	
6	Slenderness reduction factor, α_s $A = \frac{\pi^2 \times 200 \times 10^3 \times 9.6 \times 10^6}{6160^2} =$ $B = \frac{\pi^2 \times 200 \times 10^3 \times 284 \times 10^9}{6160^2} =$ $C = 80 \times 10^3 \times 241 \times 10^3 =$ $M_o = \frac{\sqrt{499 \times 10^3 \times (19.3 \times 10^9 + 14.8 \times 10^9)}}{10^6} =$ $a_s = 0.6\left[\sqrt{\left[\left(\frac{269}{130}\right)^2 + 3\right]} - \left(\frac{269}{130}\right)\right] =$	499×10^3 14.8×10^9 19.3×10^9 130 0.378	N Nmm^2 Nmm^2 kNm
7	Member moment capacity $\phi M_b =$ $\phi M_b = 0.9 \times 1.41 \times 0.378 \times 269 =$ $\phi M_b \leq \phi M_s \rightarrow 129 \leq 242$ true, satisfied $M^* \leq \phi M_b \rightarrow 125 \leq 129$ true, satisfied. Moment capacity	129 ϕM_b **OK**	kNm
8	Web shear capacity (see step 7 of Example 5.3 for calculation of ϕV_v) $\phi V_v =$ $R^* \leq \phi V_v \rightarrow 54.9 \leq 449$ is satisfied.	449 ϕV_v **OK**	kN
9	Web bearing capacity ϕR_b (note: $\phi R_{by}/b_{bf}$ and $\phi R_{bb}/b_b$ can be evaluated from Step 8 of Example 5.3) $b_s =$ $b_{bf} = 50 + 2.5 \times 11.5 =$ $b_b = 78.8 + 333/2 =$ $\phi R_{bb} = 0.725 \times 245 =$ $\phi R_{by} = 2.63 \times 78.8 =$ $\phi R_b = \text{Min}(207,178) =$ $R^* \leq \phi R_b \rightarrow 54.9 \leq 178$ is satisfied. **Web bearing**	 50 78.8 245 178 207 178 ϕR_b **OK**	 mm mm mm kN kN kN
10	*Serviceability* design actions/loads for deflection: See Step 9 of Example 5.3.1 for a full calculation.		
11	Summary 360UB50.7 Grade 300 is 97% efficient in bending/moment capacity. Web shear, bearing yield and buckling, and deflection are also satisfactory.	 **OK**	

5.12.4 Example 5.4

Step	Description and calculations	Result	Unit

The welded girder shown has loads applied to the top flange. The girder is partially restrained at the supports and two intermediate points indicated by LR at the third points along the span. The flanges are thicker over the central portion of the span. The bottom flanges are 10 mm thinner than the top flanges. Check the moment capacity.

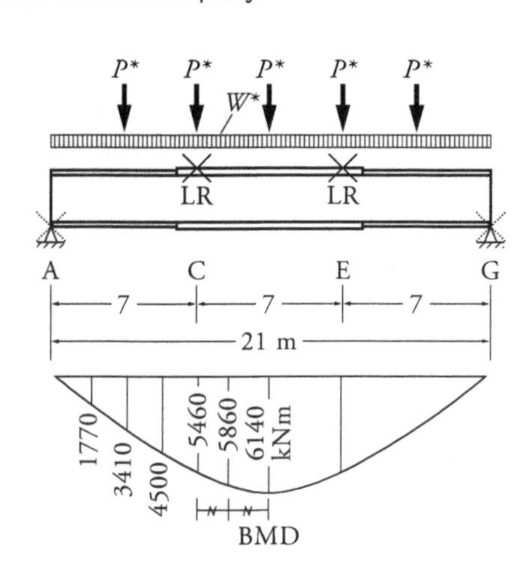

Note: End segment AC is the same as for EG.

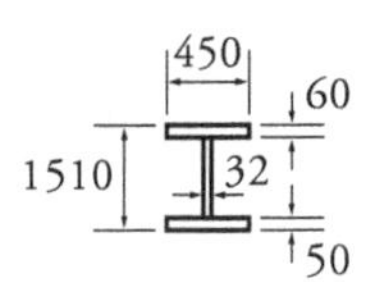

Full Section ... CE

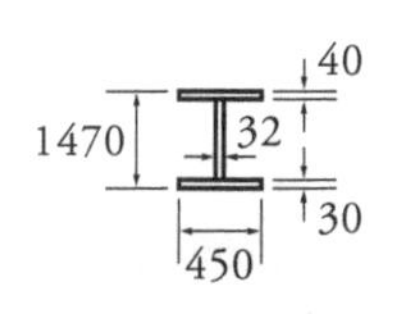

Reduced Section ... AC & EG

Step	Description and calculations	Result	Unit
	Data:		
	AS/NZS 3678 Steel grade (special grade used instead of standard Grade 250) =	300	
	Span of girder =	21 000	mm
	Design actions/loads and effects:		
	P^* =	250	kN
	W^* (including self-weight) =	840	kN
	R^* =	1045	kN
	M^* in segment CE at mid-span =	6140	kNm
	M^* in segments AC and EG at C and E	5460	kNm
1	Section **properties** of **full section** in middle segment CE (see Section 9.3.1 and Appendix B.3 also):		
	$d = 1510$ $d_1 = 1400$ $t_w = 32$		mm
	Top flange: $b_f = 450$ $t_f = 60$		mm
	Bottom flange: $b_f = 450$ $t_f = 50$		mm
	A_g =	94 300	mm^2
	I_x =	33.4×10^9	mm^4
	I_y =	839×10^6	mm^4
	$Z_{x\ top}$ =	46.1×10^6	mm^3
	$Z_{x\ bot}$ =	42.5×10^6	mm^3
	r_y =	94.3	mm
	Torsion constant: $J = \Sigma\, 0.333bt^3$ =	66.4×10^6	mm^4
	I_y of critical flange alone = I_{cy}:		
	$I_{cy} = \frac{60 \times 450^3}{12} =$	456×10^6	mm^4
	Distance between centroids of flanges, d_f:		
	$d_f = 1510 - 30 - 25 =$	1455	mm
	Warping constant: $I_w = I_{cy} d_f^2 \left(1 - \frac{I_{cy}}{I_y}\right)$ (AS 4100, Appendix H, Clause H4)		

Step	Calculation	Value	Unit
	$= 456 \times 10^6 \times 1455^2 \times \left(1 - \frac{456}{839}\right) =$	441×10^{12}	mm^6
	Equal area axis from bottom =	795	mm
	Plastic modulus of **full section** S_x =	51.5×10^6	mm^3
2	Section **properties** of **reduced section** in end segment AC and EG (see Section 9.3.1 and Appendix B.3 also):		
	$d = 1470$ $d_1 = 1400$ $t_w = 32$		mm
	Top flange: $b_f = 450$ $t_f = 40$		mm
	Bottom flange: $b_f = 450$ $t_f = 30$		mm
	A_g =	76 300	mm^2
	I_x =	23.4×10^9	mm^4
	I_y =	535×10^6	mm^4
	$Z_{x\ top}$ =	33.6×10^6	mm^3
	$Z_{x\ bot}$ =	30.3×10^6	mm^3
	r_y =	83.8	mm
	Torsion constant: $J = \Sigma\, 0.333bt^3$ =	28.9×10^6	mm^4
	I_y of critical flange alone = I_{cy}:		
	$I_{cy} = \frac{40 \times 450^3}{12} =$	304×10^6	mm^4
	Distance between centroids of flanges, d_f:		
	$d_f = 1470 - 20 - 15 =$	1435	mm
	Warping constant: $I_w = I_{cy} d_f^2 \left(1 - \frac{I_{cy}}{I_y}\right)$ (AS 4100, Appendix H, Clause H4)		
	$= 304 \times 10^6 \times 1435^2 \times \left(1 - \frac{304}{535}\right) =$	270×10^{12}	mm^6
	Equal area axis from bottom =	785	mm
	Plastic modulus of **reduced section** S_x =	38.1×10^6	mm^3
3	Section slenderness λ_s of **full section** in middle segment CE:		
3.1	Element slenderness λ_e from AS 4100 Clause 5.2.2:		
	$\lambda_e = \frac{b}{t}\sqrt{\left(\frac{f_y}{250}\right)}$		
3.2	Top flange element λ_e (assume lightly welded longitudinally (LW)):		
	$\lambda_e = \frac{209}{60}\sqrt{\left(\frac{280}{250}\right)} =$	3.69	
	$\frac{\lambda_e}{\lambda_{ey}} = \frac{3.69}{15} =$	0.246	
3.3	Bottom flange element λ_e (assume lightly welded longitudinally (LW)):		
	$\lambda_e = \frac{209}{50}\sqrt{\left(\frac{280}{250}\right)} =$	4.42	
	$\frac{\lambda_e}{\lambda_{ey}} = \frac{4.42}{15} =$	0.295	
3.4	Web element λ_e (assume lightly welded longitudinally (LW)):		
	$\lambda_e = \frac{1400}{32}\sqrt{\left(\frac{280}{250}\right)} =$	46.3	
	$\frac{\lambda_e}{\lambda_{ey}} = \frac{46.3}{115} =$	0.403	

3.5 Which element is the most slender is given by:

max (0.246, 0.295, 0.403) = 0.403

0.403 is identified with the web which is the critical element — *web* critical

3.6 Section slenderness λ_s:

Section as a whole has its slenderness λ_s controlled by λ_e of web
λ_e is analogous to strength of a chain = ... weakest link ...
with elements as links ...and $\lambda_s = \lambda_e$ of web = 46.3

With web critical. Web is supported along *two* edges by flanges, and bending stress *varies* from compression to tension

From AS 4100 Table 5.2, $\lambda_{sp} = \lambda_{ep}$ of web = 82

$\lambda_s \leqslant \lambda_{sp}$ = true because 46.3 ⩽ 82 = true ...

... to comply with AS 4100 Clause 5.2.3....
... and λ_s of **full section** in middle segment CE is ... Compact

---

4 Section slenderness λ_s of **reduced section** in end segment AC:
can show that the bottom flange governs with

$\lambda_{sp} = \lambda_{ep}$ = 8

so that $\lambda_s \leqslant \lambda_{sp}$ is true, satisfied.

λ_s of **reduced section** in end segment AC is ... Compact

---

5 Effective section modulus Z_{ex} of **full section** in middle segment CE:

Being compact: $Z_{ex} = S_x$ Then Z_{ex} of full section = 51.5×10^6 mm^3

Check: $S_x \leqslant 1.5\ Z_x$ OK

---

6 Effective section modulus Z_{ex} of **reduced section** in end segment AC:

Being compact: $Z_{ex} = S_x$ Then Z_{ex} of reduced section = 38.1×10^6 mm^3

Check: $S_x \leqslant 1.5\ Z_x$ OK

---

7 Effective length l_e of **middle segment** CE:

Downwards loads act on top flange, makes it *critical flange*
Restraints are all partial, giving arrangement PP

AS 4100 Table 5.6.3(1)

$$k_t = 1 + \frac{2d_w}{l_s}\left(\frac{0.5\ t_f}{t_w}\right)^3$$

$$= 1 + \left[2 \times \frac{1400}{7000}\right] \times \left(0.5 \times \frac{60}{32}\right)^3 =$$ 1.33

k_r = for unrestrained lateral rotation of flange about y-axis 1.0

k_l = for PP and load within segment = 1.4

$l_e = k_t\ k_l\ k_r\ l_s$
$= 1.33 \times 1.4 \times 1.0 \times 7000 =$ 13 000 mm

---

8 Effective length l_e of **end segment** AC:

Using assumptions from Step 7 above.

$$k_t = 1 + \left[2 \times \frac{1400}{7000}\right] \times \left(0.5 \times \frac{40}{32}\right)^3 =$$ 1.10

k_r = 1.0

k_l = 1.4

$l_e = 1.10 \times 1.4 \times 1.0 \times 7000 =$ 10 800 mm

The reason for the lower effective length is due to the reduced difference in stiffness between the web and the flange.

Step	Calculation	Result	Unit
9	Slenderness reduction factor α_s for **middle segment** CE requires: AS 4100 Clauses 5.6.1.1(a) and 5.6.1.2		
	Coefficient $\beta_x = 0.8\, d_f \left(\frac{2\, I_{cy}}{I_y} - 1\right)$		
	$= 0.8 \times 1455 \times \left(\frac{2 \times 456 \times 10^6}{839 \times 10^6} - 1\right) =$	101	mm
9.1	Let **A** $= \frac{\pi^2\, EI_y}{l_e^2}$		
	$= \frac{\pi^2 \times 200 \times 10^3 \times 839 \times 10^6}{13000^2} =$	9.80×10^6	N
9.2	Let **B** $= \frac{\pi^2\, EI_w}{l_e^2}$		
	$= \frac{\pi^2 \times 200 \times 10^3 \times 441 \times 10^{12}}{13000^2} =$	5.15×10^{12}	Nmm^2
9.3	Let **C** $= GJ$		
	$= 80 \times 10^3 \times 66.4 \times 10^6 =$	5.31×10^{12}	Nmm^2
9.4	M_o = reference buckling moment of **full section** in middle segment CE requires AS 4100 Clause 5.6.1.2:		
	$M_o = \sqrt{A}\left[\sqrt{[C + B + 0.25\, \beta_x^2\, A]} + 0.5\, \beta_x \sqrt{A}\right] =$	10.6×10^9	Nmm
	$=$	10 600	kNm
9.5	M_s = nominal section moment capacity of **full section** in middle segment CE:		
	$M_s = f_y\, Z_{ex}$		
	$= 280 \times 51.5 \times 10^6 =$	14.4×10^9	Nmm
	$=$	14 400	kNm
	$\phi M_s = 0.9 \times 14400 =$	13 000	kNm
9.6	α_s = slenderness reduction factor for **middle segment** CE (AS 4100 Equation 5.6.1.1(2)):		
	$= 0.6 \left\{\sqrt{\left[\frac{M_s^2}{M_o^2} + 3\right]} - \frac{M_s}{M_o}\right\}$		
	$= 0.6 \left\{\sqrt{\left[\frac{14400^2}{10600^2} + 3\right]} - \frac{14400}{10600}\right\} =$	0.506	
10	Moment modification factor α_m for **middle segment** CE (AS 4100 Clause 5.6.1.1(a)(iii)):		
	$\alpha_m = \frac{1.7\, M_m^*}{\sqrt{(M_2^{*2} + M_3^{*2} + M_4^{*2})}}$		
	$= \frac{1.7 \times 6140}{\sqrt{(5860^2 + 6140^2 + 5860^2)}} \leqslant 2.5 =$	1.01	
11	ϕM_b = member moment capacity of **full section** in middle segment CE:		
	$= \phi\, \alpha_m\, \alpha_s\, M_s$		
	$= 0.9 \times 1.01 \times 0.506 \times 14400 =$	6620	kNm
12	$M^* \leqslant \phi M_b$ requirement in AS 4100 Clause 5.1: $6140 \leqslant 6620$ true	OK	
	$M^* \leqslant \phi M_s$ is also good: $6140 \leqslant 13\,000$ true	OK	
	ϕM_b of **full section** in middle segment CE is adequate for moment capacity.	**Full section CE OK in bending**	

Step	Calculation	Result	Unit
13	Slenderness reduction factor α_s for **end segment** AC requires:		
	AS 4100 Clauses 5.6.1.1(a) and 5.6.1.2		
	Coefficient $\beta_x = 0.8\, d_f \left(\dfrac{2\, I_{cy}}{I_y} - 1\right)$		
	$= 0.8 \times 1435 \times \left(\dfrac{2 \times 304 \times 10^6}{535 \times 10^6} - 1\right) =$	157	mm
13.1	Let **A** $= \dfrac{\pi^2 EI_y}{l_e^2}$		
	$= \dfrac{\pi^2 \times 200 \times 10^3 \times 535 \times 10^6}{10800^2} =$	9.05×10^6	N
13.2	Let **B** $= \dfrac{\pi^2 EI_w}{l_e^2}$		
	$= \dfrac{\pi^2 \times 200 \times 10^3 \times 270 \times 10^{12}}{10800^2} =$	4.57×10^{12}	Nmm^2
13.3	Let **C** $= GJ$		
	$= 80 \times 10^3 \times 28.9 \times 10^6 =$	2.31×10^{12}	Nmm^2
13.4	$\boldsymbol{M_o}$ = reference buckling moment of **reduced section** in end segment AC requires AS 4100 Clause 5.6.1.2		
	$M_o = \sqrt{A}\left[\sqrt{[C + B + 0.25\, \beta_x^2\, A]} + 0.5\, \beta_x \sqrt{A}\right] =$	8.63×10^9	Nmm
	=	8630	kNm
13.5	$\boldsymbol{M_s}$ = nominal moment section capacity of **reduced section** in end segment AC:		
	$M_s = f_y\, Z_{ex}$		
	$= 280 \times 38.1 \times 10^6 =$	10.7×10^9	Nmm
	=	10 700	kNm
	$\phi M_s = 0.9 \times 10700 =$	9630	kNm
13.6	α_s = slenderness reduction factor for **end segment** AC:		
	$= 0.6\left[\sqrt{\left[\dfrac{M_s^2}{M_o^2} + 3\right]} - \dfrac{M_s}{M_o}\right]$		
	$= 0.6\left[\sqrt{\left[\dfrac{10700^2}{8630^2} + 3\right]} - \dfrac{10700}{8630}\right] =$	0.534	
14	α_m = moment modification factor for **end segment** AC		
	$= \dfrac{1.7\, M_m^*}{\sqrt{(M_2^{*2} + M_3^{*2} + M_4^{*2})}}$		
	$= \dfrac{1.7 \times 5460}{\sqrt{(1770^2 + 3410^2 + 4500^2)}} \leqslant 2.5 =$	1.57	
15	$\boldsymbol{\phi M_b}$ = member moment capacity of **reduced section** in end segment AC		
	$= \phi\, \alpha_m\, \alpha_s\, M_s$		
	$= 0.9 \times 1.57 \times 0.534 \times 10700 =$	8070	kNm
16	$M^* \leqslant \phi M_b$ requirement in AS 4100 Clause 5.1: $5460 \leqslant 8070$ true	OK	
	$M^* \leqslant \phi M_s$ is also good: $5460 \leqslant 9630$ true	OK	
	$\boldsymbol{\phi M_b}$ of **reduced section** in end segment AC is adequate for moment capacity.	**Reduced section AC OK in bending**	

5.12.5 Example 5.5

Step	Description and calculations	Result	Unit
	The welded girder shown has loads applied to the top flange. The girder is continuously laterally restrained along the span and at the supports. Web stiffeners are spaced at 1500 mm. Check the girder for moment capacity, and its web capacity to take shear and bearing. Stiff bearing length is 200 mm at supports. Dimensions and other details of the girder are tabulated below.		

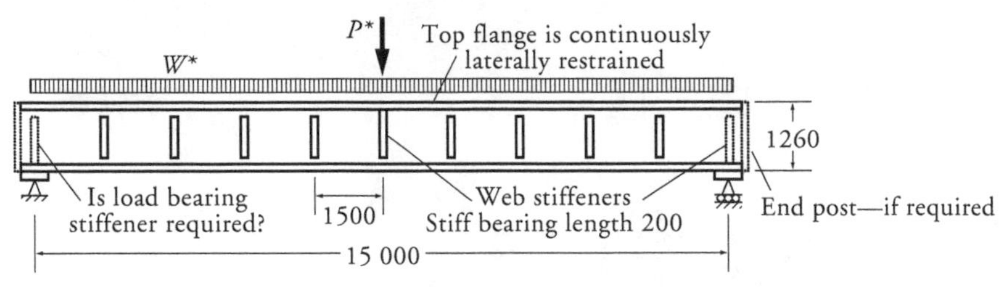

Elevation of plate girder

Step	Description and calculations	Result	Unit
1	Data		
	AS/NZS 3678 Steel grade (special grade used instead of standard Grade 250) =	300	
	Span of girder	15 000	mm
1.1	Cross-section dimensions:		
	$d =$	1260	mm
	$d_1 =$	1200	
	$b_f =$	500	mm
	$t_f =$	30	mm
	$t_w =$	8	mm
1.2	Properties of section:		
	$A_g =$	39 600	mm^2
	$I_x =$	12.5×10^9	mm^4
	$I_y =$	625×10^6	mm^4
	$Z_x =$	19.8×10^6	mm^3
	$r_y =$	126	mm
	Torsion constant: $J = \Sigma\, 0.333bt^3$	9.20×10^6	mm^4
	Warping constant: $I_w = \dfrac{I_y\, d_f^2}{4}$		
	$= \dfrac{625 \times 10^6 \times (1260 - 30)^2}{4} =$	236×10^{12}	mm^6
	Plastic modulus $S_x =$	21.3×10^6	mm^3
2	Actions/loads		
2.1	Nominal actions/loads:		
	Self-weight, uniformly distributed load	3.12	kN/m
	Permanent action/dead load, UDL	9.38	kN/m
	Imposed action/live load, UDL	18	kN/m
	Point imposed action/live load at mid-span	250	kN
2.2	Design actions/loads and effects		
	AS/NZS 1170.0 Clause 4.2.2(b)		
	$w^* = 1.2\, w_G + 1.5\, w_Q$		
	$= 1.2 \times (3.12 + 9.38) + 1.5 \times 18 =$	42.0	kN/m
	$W^* = w^* l$		
	$= 42.0 \times 15 =$	630	kN

	$P^* = 1.2\ P_G + 1.5\ P_Q$		
	$= 1.2 \times 0 + 1.5 \times 250 =$	375	kN
	Beam and actions/loads are symmetrical:		
	$R^* = 0.5 \times (630 + 375) =$	503	kN
	$V^* =$	503	kN
	$M^* = 630 \times \frac{15}{8} + 375 \times \frac{15}{4} =$	2590	kNm
3	Section slenderness at mid-span:		
3.1	Flange element slenderness λ_e AS 4100 Clause 5.2 ...		
	$\lambda_e = \frac{b}{t}\sqrt{\frac{f_y}{250}}$		
	$= \frac{246}{30}\sqrt{\frac{280}{250}} =$	8.68	
	$\frac{\lambda_e}{\lambda_{ey}} = \frac{8.68}{14} =$ (assume heavily welded longitudinally (HW))	0.620	
3.2	Web element slenderness λ_e		
	$\lambda_e = \frac{b}{t}\sqrt{\frac{f_y}{250}}$		
	$= \frac{1200}{8}\sqrt{\frac{320}{250}} =$	170	
	$\frac{\lambda_e}{\lambda_{ey}} = \frac{170}{115}$	1.48	
3.3	Worst element for slenderness:		
	max (0.620, 1.48) =	1.48	
	Web λ_e is closer to its λ_{ey} than flange λ_e (indeed it more than exceeds it)		
	Web is *critical* element to control section slenderness as a whole:		
	λ_s = section slenderness		
	= *web* element slenderness =	170	
	Note in the following reference to AS 4100 Table 5.2, *two* flanges support the *web* and the stress *varies* linearly from *compression to tension* (i.e. Flat plate element type, both longitudinal edges supported with compression at one edge, tension at the other).		
	Then $\lambda_{sy} = \lambda_{ey}$ of *web* =	115	
	Compare $\lambda_s = 170$ with limits in AS 4100 Table 5.2 and Clauses 5.2.3, 5.2.4 and 5.2.5 to find the category of section slenderness:		
	$\lambda_s > \lambda_{sy}$ is true and the section is ...	Slender	
	Effective section modulus Z_e:		
	$Z_e = Z\left(\frac{\lambda_{sy}}{\lambda_s}\right)$		
	$= 19.8 \times 10^6 \times \left(\frac{115}{170}\right) =$	13.4×10^6	mm^3
	The above is considered to be a conservative method of evaluating Z_e for slender sections. Clause 5.2.5 of AS 4100 also permits the use of calculating Z_e by establishing an effective section after omitting the portions of the section in excess of the width $\lambda_{sy}t$ (see Section 5.5.1.4). This may be more optimal in this instance as the flanges are fully effective and the web area would only be reduced. However, the following calculations will use the above conservative Z_e evaluation method for simplicity.		
4	Section moment capacity ϕM_s:		
	$\phi M_s = f_y\ Z_{ex}$		
	$= 280 \times 13.4 \times \frac{10^6}{10^6}$	3750	kNm

Step	Calculation	Result	Unit
5	Member moment capacity ϕM_b:		
	$\phi M_b = \alpha_m \alpha_s \phi M_s \leqslant \phi M_s$		
	α_m = 1.0 for a continuously restrained beam		
	α_s = 1.0 for a continuously restrained beam		
	$\phi M_b = 1.0 \times 1.0 \times 3750 = \phi M_s$ in this instance =	3750	kNm
	$M^* < \phi M_b$ is satisfied because 2590 < 3750 is true	member moment capacity ϕM_b OK	
6	Minimum web thickness:		
	$\frac{s}{d_1} = \frac{1500}{1200} = \frac{s}{d_p} =$	1.25	
	and $1 \leqslant \frac{s}{d_1} \leqslant 3.0$ in AS 4100 Clause 5.10.4(a) is satisfied	True	
	to give minimum web thickness t_w required:		
	minimum $t_w = \frac{d_1}{200}\sqrt{\frac{f_y}{250}}$		
	$= \frac{1200}{200}\sqrt{\frac{320}{250}} =$	6.79	mm
	$t_w = 8$ actual > 6.79	web $t_w = 8$ is OK	
7	Web shear capacity ϕV_v:		
7.1	Web slenderness =		
	$= \frac{d_p}{t_w} = \frac{1200}{8} =$	150	
	$\frac{d_p}{t_w} \geqslant \frac{82}{\sqrt{\frac{320}{250}}} = 72.5$ … and because 150 > 72.5	True …	
	… web can *buckle* instead of yielding in shear … from AS 4100 Clause 5.11.2(b)	$V_u = V_b$	
7.2	From AS 4100 Clauses 5.11.2(b) and 5.11.5.1, the nominal shear buckling capacity, V_b, of an unstiffened web is:		
	$V_b = \alpha_v V_w \leqslant V_w$		
	where		
	$\alpha_v = \left[\frac{82}{\left(\frac{d_p}{t_w}\right)\sqrt{\left(\frac{f_y}{250}\right)}}\right]^2$		
	$= \left[\frac{82}{\left(\frac{1200}{8}\right)\sqrt{\left(\frac{320}{250}\right)}}\right]^2 =$	0.233	
	V_w = nominal shear yield capacity		
	$= 0.6 f_y A_w = 0.6 f_y d_p t_w$		
	$= \frac{0.6 \times 320 \times 1200 \times 8}{10^3} =$	1840	kN
	Therefore		
	$\phi V_b = 0.9 \times 0.233 \times 1840 =$	386	kN
	$\leqslant \phi V_w$ ………. satisfies ϕV_b criterion		
	however,		
	$V^* > \phi V_b$ which is unsatisfactory as 503 > 386 and the web needs to be stiffened for shear buckling by intermediate transverse web stiffeners. This is understandable as the web is very slender.		

7.3 Nominal web shear buckling capacity in a web with (vertical) intermediate transverse stiffeners (as per the initial diagram to this worked example).
Assume intermediate transverse stiffeners are
2 No. 150 x 10 Grade 300 flat bars each side of the web.
As noted in AS 4100 Clause 5.15.4, the buckling capacity of the web/stiffener combination must satisfy:

$V^* \leqslant \phi(R_{sb} + V_b)$

where

R_{sb} = nominal buckling capacity of web/intermediate stiffener as noted in AS 4100 Clause 5.14.2 ...
with $l_e = d_1$ (as the flanges are not connected to the stiffener), $\alpha_b = 0.5$ and $k_f = 1.0$ ......

V_b = nominal shear buckling capacity for a stiffened web as noted in AS 4100 Clause 5.11.5.2...
with $\alpha_d = \alpha_f = 1.0$

l_{ewc} = effective length of web cross-section area on each side of stiffener for column action

$$= \text{min. of } \frac{17.5t_w}{\sqrt{\left(\frac{f_y}{250}\right)}} \text{ or } \frac{s}{2}$$

$$= \text{min. of } \frac{17.5 \times 8}{\sqrt{\left(\frac{320}{250}\right)}} \text{ or } \frac{1500}{2} = \text{min.}[124;\ 750\ ] = \qquad 124 \text{ mm}$$

A_{ws} = effective cross-section area of web/stiffener

$$= 2 \times (150 \times 10) + 2 \times (124 \times 8) = \qquad 4980 \text{ mm}^2$$

I_{ws} = second moment of area of web/stiffener taken about axis parallel to the web.

$$= \frac{10 \times (300 + 8)^3}{12} + \frac{2 \times 124 \times 8^3}{12} = \qquad 24.4 \times 10^6 \text{ mm}^4$$

$$r_{ws} = \sqrt{\frac{24.4 \times 10^6}{4980}} = \qquad 70.0 \text{ mm}$$

$\frac{l_{ws}}{r_{ws}}$ = web/stiffener compression member slenderness ratio

$$= \frac{d_1}{r_{ws}} = \frac{1200}{70.0} = \qquad 17.1$$

λ_n = web compression member modified slenderness ratio

$$= \left(\frac{l_{ws}}{r_{ws}}\right)\sqrt{(k_f)}\sqrt{\left(\frac{f_y}{250}\right)}$$

(note: where indicated, allow for the lower f_y of the stiffener i.e. from 320 → 310 MPa)

$$= 17.1 \times 1.0 \times \sqrt{\frac{310}{250}} = \qquad 19.0$$

α_c = web compression member slenderness reduction factor from AS 4100 Table 6.3.3(3), α_c = 0.972

R_{sb} = design bearing buckling capacity of a unstiffened web

$$= \alpha_c A_{ws} f_y$$

$= \dfrac{0.972 \times 4980 \times 310}{10^3} =$ 1500 kN

(note: where indicated, revert back to the higher f_y of the web as it controls shear buckling i.e. from 310 → 320 MPa).

From AS 4100 Clause 5.11.5.2:

$\dfrac{s}{d_p} = \dfrac{1500}{1200} =$ 1.25

For $1.0 \leqslant \dfrac{s}{d_p} \leqslant 3.0$:

$$\alpha_v = \left[\frac{82}{\left(\frac{d_p}{t_w}\right)\sqrt{\left(\frac{f_y}{250}\right)}}\right]^2 \left[\frac{0.75}{\left(\frac{s}{d_p}\right)^2} + 1.0\right] \leqslant 1.0$$

$$= \left[\frac{82}{150\sqrt{\left(\frac{320}{250}\right)}}\right]^2 \left[\frac{0.75}{(1.25)^2} + 1.0\right] =$$ 0.346

α_d = as noted above in V_b definition (in step 7.3) = 1.00

α_f = as noted above in V_b definition (in step 7.3) = 1.00

$V_b = \alpha_v \alpha_d \alpha_f V_w$

$= \dfrac{0.346 \times 1.00 \times 1.00 \times (0.6 \times 320 \times 1200 \times 8)}{10^3} =$ 638 kN

Then $\phi(R_{sb} + V_b) = 0.9 \times (1500 + 638) =$ 1920 kN

$V^* \leqslant \phi(R_{sb} + V_b)$ OK as 503 < 1920 kN — Intermediate stiffeners OK

Checks for stiffener minimum area (AS 4100 Clause 5.15.3), minimum stiffness (AS 4100 Clause 5.15.5) and outstand of stiffeners (AS 4100 Clause 5.15.6) show that this intermediate transverse stiffener configuration and loading type is also adequate. Additionally, as V^* is about 26% of the stiffened web design shear capacity, this difference can be optimised by re-sizing the stiffeners using an iterative procedure as noted in this step (7.3).

Shear and bending interaction need not be considered as the peak shear force (V^* = 503 kN) is less than 60% of the design shear capacity with the stiffened web (ϕV_v = 1920 kN) – see Section 5.8.4.

---

8 Web bearing at a support:

8.1 Bearing lengths

AS 4100 Figures 5.13.1.1 and 2 use notation for end force bearing (see Figure in Example 5.6 also):

b_s = stiff bearing length

= ... as specified in description of girder ... = 200 mm

b_{bf} = bearing length at junction of flange and web for yield

$= b_s + 2.5\, t_f$

$= 200 + 2.5 \times 30 =$ 275 mm

b_b = ... used later in step 8.3...

= length of web at mid – height ... width of "column"

$= b_{bf} + \dfrac{d_1}{2}$

$= 275 + \dfrac{1200}{2} =$ 875 mm

8.2	Bearing yield capacity ϕR_{by} (unstiffened):		
	$\phi R_{by} = \phi 1.25\, b_{bf}\, t_w\, f_y$		
	$= \dfrac{0.9 \times 1.25 \times 275 \times 8 \times 320}{10^3} =$	792	kN
	$R^* \leqslant \phi R_{by}$ OK because $503 \leqslant 792$ true	Web bearing yield OK	
8.3	Web buckling at support (unstiffened)		
8.3.1	Slenderness of web analogised as a column From AS 4100 Clause 5.13.4 -		
	α_b = web compression member section constant =	0.5	
	k_f = web compression member form factor =	1.0	
	A_{wc} = web compression member cross-section area		
	$= t_w b_b = 8 \times 875 =$	7000	mm^2
	(note: b_b is calculated above in Step 8.1 and further explained in Section 5.8.5.3).		
	$\dfrac{l_e}{r}$ = web compression member slenderness ratio		
	$= \dfrac{2.5 d_1}{t_w} = \dfrac{2.5 \times 1200}{8} =$	375	
	λ_n = web compression member modified slenderness ratio		
	$= \left(\dfrac{l_e}{r}\right)\sqrt{(k_f)}\sqrt{\left(\dfrac{f_y}{250}\right)}$		
	$= 375 \times 1.0 \times \sqrt{\dfrac{320}{250}} =$	424	
	α_c = web compression member slenderness reduction factor from AS 4100 Table 6.3.3(3), α_c =	0.0422	
	ϕR_{bb} = design bearing buckling capacity of an unstiffened web		
	$= \phi \alpha_c A_{wc} f_y$		
	$= \dfrac{0.9 \times 0.0422 \times 7000 \times 320}{10^3} =$	85.1	kN
	AS 4100 Clause 5.13.2 requires $R^* < \phi R_{bb}$ ... which is not satisfied because 503 < 85.1 is false	Web bearing buckling *fails*	
	Load bearing stiffeners are required for the web. Example 5.6 considers the design of such load-bearing stiffeners.		

5.12.6 Example 5.6

Step	Description	Result	Unit
	The plate girder in Example 5.5 shows load-bearing stiffeners are required at the ends terminating at the supports. Intermediate web stiffeners are placed at 1500 centres. Steel is Grade 300. The following is an example of load-bearing stiffener design.		
	Check the adequacy of a pair of 200 × 25 stiffeners placed within the stiff bearing length b_s, which is given as 200 mm.		

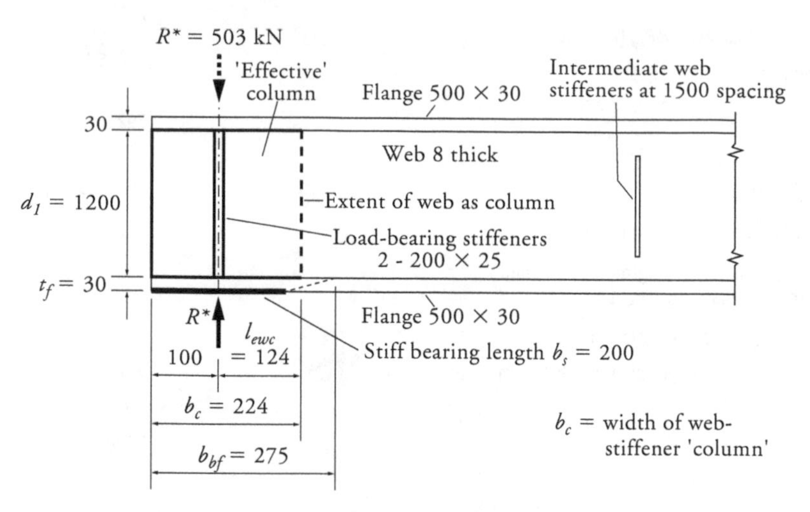

Elevation of left end of girder

1	Design reaction R^* is calculated in Example 5.5:		
	$R^* =$	503	kN
2	Girder cross-section dimensions:		
	$d = 1260$ $b_f = 500$ $t_f = 30$ $t_w = 8$ $d_1 = 1200$		mm
3	Stiffeners, Area A_{st} one pair, 2 No. 200 × 25mm:		
	$A_{st} = 2 \times 200 \times 25 =$	10 000	mm^2
4	Material properties:		
	Grade 300 steel		
	Design yield stress values from Table 2.3 or AS 4100 Table 2.1 for AS/NZS 3678 Grade 300 plate:		
	Web with $t_w = 8$ gives $f_{yw} =$	320	MPa
	Stiffeners with $t_s = 25$ gives $f_{ys} =$	280	MPa
5	Stiffener outstand, check:		
	Stiffener outstand b_{es} in AS 4100 5.14.3 is limited to:		
	$b_{es} \leqslant \dfrac{15t_s}{\sqrt{\left(\dfrac{f_{ys}}{250}\right)}}$		
	$= \dfrac{15 \times 25}{\sqrt{\left(\dfrac{280}{250}\right)}} =$	354	mm
	Actual outstand = 200 ⩽ 354	OK	
6	Stiffener spacing s:		
	s = spacing between stiffeners =	1500	mm
	$\dfrac{s}{2} = \dfrac{1500}{2}$	750	mm

Step	Calculation	Result	Unit
7	Yield capacity of the web/load-bearing stiffener The web and load-bearing stiffener combine to act as an axially loaded column (see note at the end of the example). This must satisfy AS 4100 Clause 5.14.1: $R^* \leqslant \phi R_{sy}$ where R^* = design bearing or reaction force R_{sy} = nominal yield capacity of the stiffened web $= R_{by} + A_s f_{ys}$		
	$R_{by} = 1.25 b_{bf} t_w f_{yw} = \frac{1.25 \times 275 \times 8 \times 320}{10^3} =$	880	kN
	(note: b_{bf} is calculated in Step 8.1 of Example 5.5 and further explained in Section 5.8.5.2). A_s = area of stiffener in contact with the flange		
	$= 2 \times (200 \times 25) =$	10 000	mm^2
	(note: allow for the lower f_y of the stiffener not the web i.e. from 320 → 280 MPa) then		
	$\phi R_{sy} = 0.9 \times \left(880 + \frac{10000 \times 280}{10^3}\right) =$	3310	kN
	$R^* < \phi R_{sy}$ ... which is satisfied because $503 < 3310$	Web/Stiffener yielding is OK	
8	Buckling capacity of the web/load-bearing stiffener The web and load-bearing stiffener combine to act as an axially loaded column (see note at the end of the example) This must satisfy AS 4100 Clause 5.14.2: $R^* \leqslant \phi R_{sb}$ where R_{sb} = nominal buckling capacity of web/load-bearing stiffener with $l_e = d_1$ (or $0.7d_1$ if the flanges are rotationally restrained by other structural elements in the plane of the stiffener), $\alpha_b = 0.5$, and $k_f = 1.0$ l_{ewc} = effective length of web cross-section area on each side of stiffener for column action $= \text{min. of } \frac{17.5 t_w}{\sqrt{\left(\frac{f_y}{250}\right)}} \text{ or } \frac{s}{2}$		
	$= \text{min. of } \frac{17.5 \times 8}{\sqrt{\left(\frac{320}{250}\right)}} \text{ or } \frac{1500}{2} = \text{min.}[124; 750\ ] =$	124	mm
	A_{ws} = effective cross-section area of web/stiffener (see Figure at the beginning of the example)		
	$= 2 \times (200 \times 25) + (100 + 124) \times 8 =$	11 800	mm^2
	I_{ws} = second moment of area of web/stiffener taken about axis parallel to the web		

$$= \frac{25 \times (400 + 8)^3}{12} + \frac{(100 + 124) \times 8^3}{12} =$$ 142×10^6 mm^4

$$r_{ws} = \sqrt{\frac{142 \times 10^6}{11800}} =$$ 110 mm

$\frac{l_{ws}}{r_{ws}}$ = web/stiffener compression member slenderness ratio

$$= \frac{d_1}{r_{ws}} = \frac{1200}{110} =$$ 11.0

(Note: had the flanges been restrained against torsion (in the plane of the stiffener) then l_{ws} could have been $0.7d_1$ instead of d_1).

λ_n = web/stiffener compression member modified slenderness ratio

$$= \left(\frac{l_{ws}}{r_{ws}}\right)\sqrt{(k_f)}\sqrt{\left(\frac{f_y}{250}\right)}$$

(note: where indicated, allow for the lower f_y of the stiffener i.e. from 320 → 280 MPa)

$$= 11.0 \times 1.0 \times \sqrt{\frac{280}{250}} =$$ 11.6

α_c = web compression member slenderness reduction factor from AS 4100 Table 6.3.3(3), α_c = 0.997

ϕR_{sb} = design bearing buckling capacity of the web/stiffener

$$= 0.9\alpha_c A_{ws} f_y$$

$$= \frac{0.9 \times 0.997 \times 11800 \times 280}{10^3} =$$ 2960 kN

$R^* \leqslant \phi R_{sb}$ OK as $503 < 2960$ kN — Load-bearing stiffeners OK

---

9 Requirement for an End Post

AS 4100 Clauses 5.15.9 and 5.15.2.2 note that an end post is required if the end web panel width, s_{end}, does not satisfy the following criteria:

(a) $V^* \leqslant \phi V_b$ from AS 4100 Clause 5.11.5.2 with $\alpha_d = 1.0$, and

(b) does not undergo any shear and bending interaction effect (from AS 4100 Clause 5.12)

Additionally, AS 4100 Clause 5.14.5 notes the following:

(c) a minimum stiffness of load-bearing stiffeners if they are the sole means of providing torsional end restraints to the member supports.

Checking the above:

(a) $\frac{s_{end}}{d_p} = \frac{1500}{1200} =$ 1.25

From Step 7.3 in Example 5.5 with $\alpha_d = 1.0$:

α_v = 0.346

α_f = 1.0

$\phi V_b = 0.9 \times 638 =$ 574 kN

$V^* \leqslant \phi V_b$ OK as $503 < 574$ kN at the supports

(b) There is no shear and bending interaction as the end supports see the highest shear force and the lowest moment.

(c) If the load-bearing stiffener is the sole means of beam support torsional restraint, the second moment of area of a pair of stiffeners, I_s, about the web centreline must satisfy:

$$I_s \geq \left(\frac{\alpha_t}{1000}\right)\left(\frac{d^3 t_f R^*}{F^*}\right)$$

where

$\frac{l_e}{r_y}$ = web/stiffener compression member slenderness ratio as calculated in Step 8 =	11.0	
$\alpha_t = \frac{230}{\left(\frac{l_e}{r_y}\right)} - 0.60$ (for $0 \leqslant \alpha_t \leqslant 4.0$) =	4.0	
R^* =	503	kN
$F^* = 2 \times 503$ =	1006	kN

then minimum I_s $(=I_{smin})$

$I_{smin} = \left(\frac{4.0}{1000}\right)\left(\frac{1260^3 \times 30 \times 503}{1006}\right)$ =	120x10^6	mm^4

From Step 8 above, actual I_s is

I_s =	142x10^6	mm^4

The minimum stiffness for the load-bearing stiffener is satisfied as $I_s > I_{smin}$ → $142 \times 10^6 > 120 \times 10^6$ — Load-bearing stiffener OK for torsional stiffness

Hence no End plate is required. — No End plate required.

If the design end shear force or the stiffener spacing increases an end plate designed to AS 4100 Clause 5.15.9 may have to be provided.

Note 1 Buckling web and stiffeners are analogised to a column buckling about the horizontal axis along the web. Buckling about the vertical axis along the stiffeners is prevented by the continuity of the web beyond the extent of the section used in the calculations.

Comment The calculations detailed above for web and load-bearing stiffener buckling, plus the check in Example 5.5 to see whether load-bearing stiffeners are required or not, are lengthy to do manually in repetitive calculations. Conclusions can be drawn from the results, and short-cuts judiciously applied. This is more so if a decision is made at the outset to provide stiffeners for whatever reason.

5.13 Further reading

- For additional worked examples see Chapter 5 of Bradford, et al [1997].
- For bending moment/shear force distribution and deflection of beams see Syam [1992], Young et al. [2012] or Appendix C of this Handbook.
- Rapid beam sizing with industry standard tables and steel sections can be found in ASI [2004, 2009] and OneSteel [2012b].
- For some authoritative texts on buckling see Bleich [1952], CRCJ [1971], Hancock [2007], Timoshenko [1941], Timoshenko & Gere [1961], Trahair [1993b] and Trahair & Bradford [1998] to name a few.
- For typical beam connections also see ASI [2003, 2009b], Hogan & Munter [2007a–h], Hogan & Van der Kreek [2009a–e], Syam & Chapman [1996] and Trahair, et al [1993c,d].

- Clause 5.3.1 of AS 4100 notes that beams/beam segments must have at least one end with a F or P restraint. The only departure from this is for beam sub-segments which have LL—i.e L at both ends. This does increase the member moment capacity, however, in this instance the beam sub-segment must be part of an overall beam/beam segment that has an F or P restraint to react against any twisting of the critical flange. To establish the link between F, P, L and U beam restraint categories to practical connections see Trahair, et al [1993c].
- Watch out for designing cantilevers (in a lateral and torsional restraint sense) where, even though the actual length is used in effective length calculations, the α_m is different to that used for beam segments with both ends restrained—see Trahair [1993d] for further details.
- An excellent reference on composite steel-concrete construction behaviour, design and systems is Oehlers & Bradford [1995, 1999]. From an Australian design and construction perspective, Ng & Yum [2008] and Durack & Kilmister [2009] are also very good publications in the area.

chapter 6

Compression & Beam-Column Members

6.1 Types of compression members

Compression members used in building structures can be divided broadly into columns, beam-columns, struts and compression members in trusses. A more detailed classification system is given in Table 6.1.

Table 6.1 Classification of compression & beam-column members

Aspect	Subdivision	Section
Member type	Solid shafts:	
	• Single (element) shaft	6.2-6.4
	• Uniform section	6.2-6.5
	• Variable section	6.3.3
	Compound shafts:	
	• Latticed members	6.5
	• Battened members	6.5
Construction	Steel alone:	
	• Rolled sections	6.2-6.4
	• Welded sections	6.2-6.4
	• Thin-walled sections	NC
	Composite steel and concrete:	
	• Externally encased	6.6
	• Concrete filled	6.6
Loading	Axial without/with bending:	
	• Axial	6.2
	• Axial load with uniaxial bending	6.3
	• Axial load with biaxial bending	6.3
Restraints	Restraint position	
	• End restraints only	6.7
	• Intermediate restraints	6.7

NC = Not covered by this text.

6.2 Members loaded only axially

6.2.1 Failure modes

A member subject to an axial compressive load can fail in one of four modes:

- compressive yielding (squashing)
- local buckling
- column buckling in the elastic range
- column buckling in the inelastic range.

The prominence of each of these failure modes is dependent on several factors including: section slenderness, member slenderness, strength, influence of restraints and connections, level of material imperfections, level of geometric imperfections and residual stresses.

6.2.2 Compressive yielding

Only very 'stocky' compression members fail by yielding. The ultimate load at which such a member uniformly yields is often called the 'squash' capacity, and is given by:

$$N_s \quad = A_n f_y$$

where f_y is the design yield stress and A_n is the net area of the section. The gross cross-section area, A_g, may be used if the unfilled cross-section holes are relatively small. To qualify as 'stocky' the column would have a slenderness ratio, l/r, of less than 25, approximately. Some bearing blocks and stocky struts fall into this category.

6.2.3 Local buckling

Steel sections in compression are considered to be generally composed of flat plate elements. The only departure from this model are Circular Hollow Sections (CHS) which are composed of curved (i.e. circular) elements. Regardless, these elements may be 'stocky', slender or somewhere in between. For steel sections with slender (or near-slender) elements subject to compression stresses, the possibility of a short wavelength buckle (i.e localised "rippling") may develop before the section yields. When this occurs, the section is considered to have undergone local buckling. This is different to overall member buckling where the buckle half-wavelength is nearly the length of the member. AS 4100 considers this behaviour by modifying the 'squash' capacity (see Section 6.2.2) with a local buckling form factor, k_f.

The phenomena of local buckling is also encountered for sections subject to bending (Section 5). In this instance, the section elements subject to compression stresses (either uniform or varying from tension to compression) categorise the section as either compact, non-compact or slender. However, unlike bending where not every element may have compression stresses, compression members need to have every element assessed for the possibility of local buckling so as to determine the overall section behaviour.

6.2.4 Buckling in the elastic range

Failure by pure elastic (column) buckling can occur only in slender compression members. The theoretical elastic buckling load, N_{om}, is given by the classic Euler equation:

$$N_{om} = \frac{\pi^2 EI_{min}}{l^2}$$

For columns with the same end restraint conditions about both principal axes and noting that

$$r^2_{min} = \frac{I_{min}}{A_n} \quad \text{one obtains:}$$

$$N_{om} = \frac{\pi^2 EA_n}{\left(\frac{l}{r_{min}}\right)^2}$$

where E is Young's modulus of elasticity (E = 200 000 MPa), I_{min} is the second moment of area about the minor principal axis, l the column length and A_n the net cross-section area. This equation has validity only with perfectly straight columns, free of residual stresses, loaded at the centre of gravity and knife-edge supports. Such ideal conditions can be achieved only in laboratory conditions.

The Euler equation overestimates the column capacity, increasingly as l/r drops below 200, and thus it cannot be used for column capacity evaluation. The expression for N_{om} does, however, provide a useful notion of the bifurcation buckling load. It also shows that the capacity of a column is inversely proportional to the square of the slenderness ratio l/r, and proportional to the section area.

6.2.5 Failure in the inelastic range

In the real world, compression members have imperfections, and consequently the design buckling capacity of practical columns is less than predicted by the theoretical elastic buckling load, N_{om}. The main reasons for the discrepancy are:

- initial lack of straightness (camber)
- initial eccentricity of axial loads
- residual stresses induced during manufacture and fabrication.

It is difficult to fabricate columns having a camber less than $l/1000$, approximately. Thus there is always a small eccentricity of load at column mid-length. The effect of a camber of $l/1000$ (indeed $l/500$ for manufactured sections) has been included in the design standard, with an understanding that it should be checked for excessive camber before erection.

Residual stresses are a result of manufacture and fabrication. Rolled sections develop residual stress fields as a result of some non-uniform cooling during manufacture. Welded sections develop residual stresses as a result of weld shrinkage forces. Residual stresses for hollow sections arise from cold-forming. Further sources of residual stresses are cold straightening and hot-dip galvanizing.

Verification of strength needs to be carried out for:

- critical cross-sections (combined axial and bending capacity)
- member as a whole (member buckling capacity).

Exhaustive testing programs have been carried out overseas and in Australia to ascertain the precise influence of the initial camber and the residual stresses. AS 4100 is based on this research, as discussed in Sections 6.2.9 and 6.2.10.

Figure 6.1 shows a comparison between 'squash' capacity, elastic buckling capacity, and the design axial capacity for a typical UB section.

6.2.6 Glossary of terms

Area, effective, A_e The effective area, calculated from the sum of effective section elements.

Area, gross, A_g The calculated nominal area of the total cross-section.

Area, net, A_n The area of the section less the areas lost by penetrations and unfilled holes, as determined in accordance with Section 6.2.1 of AS 4100.

Biaxial bending capacity Capacity of a member subject to bending moments about both the major and the minor principal axes, combined with axial load (if present).

Capacity reduction factor, ϕ A knockdown factor for nominal strength (=0.9 for compression members).

Form factor, k_f Ratio of the effective to the gross area of a cross-section.

In-plane capacity, M_{ix}, M_{iy} Member bending moment capacity, where bending and buckling takes place in the same plane. Examples are CHS, SHS sections bent about any axis, I-sections and PFC sections bent about their minor axis.

Member elastic buckling load, N_{om} Critical buckling load of an idealised elastic column.

Member slenderness, λ The ratio of the effective length of the member to the respective radius of gyration.

Nominal section capacity, N_s Compression capacity based on net section times the form factor times the section yield stress.

Out-of-plane capacity, M_{ox} Member bending moment capacity, where bending occurs in the major plane and buckling takes place in the lateral direction (flexural-torsional and column buckling). Examples are I-sections and PFC sections bent about their major axis.

Plate element slenderness, λ_e The ratio of b/t times a yield stress adjustment factor, used in the calculation of the net area of the section.

Uniaxial bending capacity Capacity of a member subject to a single principal axis bending moment combined with axial load.

The elastic buckling load of a compression member is given by:

$$N_{om} = \frac{\pi^2 EI}{(k_e l)^2} = \frac{\pi^2 EA_n}{\left(\frac{k_e l}{r}\right)^2}$$

where E is the Young's modulus of elasticity equal to 200 000 MPa, I is the second moment of area about the axis of buckling, A_n is the net area of the section, k_e is the member effective length factor and l is the member length.

The member effective length factor k_e is equal to 1.0 for columns with pins at both ends, and varies for other end restraint conditions as shown in Figures 4.3 and 4.4. The ratio $\frac{k_e l}{r}$ is termed the slenderness ratio.

A comparison between the column elastic buckling load and the design member capacity is shown in Figure 6.1.

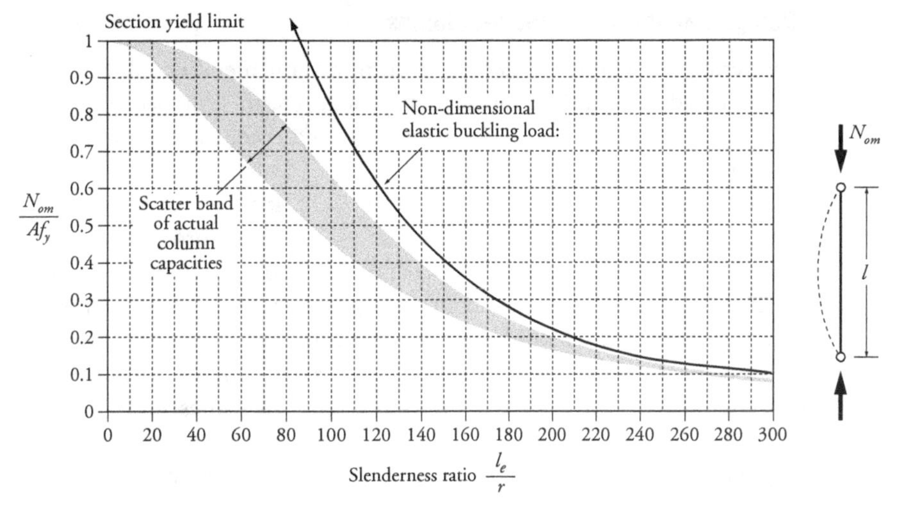

Figure 6.1 Plot of non-dimensionalised compression capacity

6.2.7 Concentrically loaded compression members

AS 4100 makes a distinction between section and member capacities. The nominal *section* capacity is the yield (squash) capacity of the net/effective section. The *member* capacity is concerned with resistance to column buckling and flexural-torsional buckling (the latter buckling mode is applicable when combined actions are present).

The relevant design capacities are:

(a) Nominal *section* capacity:

$$N_s = k_f A_n f_y$$

where k_f is the section form factor:

$$k_f = \frac{A_e}{A_g}$$

checking that the following inequality is satisfied:

$$N^* \leqslant \phi N_s$$

(b) Nominal *member* capacity is given by:

$$N_c = \alpha_c N_s = \alpha_c k_f A_n f_y \leqslant N_s$$

checking that the following inequality is satisfied:

$$N^* \leqslant \phi N_c$$

where A_n is the net section area, α_c is the slenderness reduction factor computed in accordance with Clause 6.3.3 of AS 4100. The value of α_c cannot exceed 1.0 and can be as low as 0.020 for very slender members.

The evaluation of the net/effective section properties and slenderness reduction factor is dealt with in Sections 6.2.9 and 6.2.10.

6.2.8 Design capacities of beam-columns

The term 'beam-column' denotes a compression member subjected to bending action effects in addition to compressive axial load. Beam-columns are prevalent in practice because the effects of frame action and induced eccentricity, result in bending moments transmitted to the columns.

The design bending moments in beam-columns need to be amplified, as detailed in Chapter 4, unless they are obtained from a second-order or buckling analysis. The section properties and slenderness reduction factors are determined in the same way for columns subject only to axial compression, and for beams subject only to bending/moments. The uniaxial moment capacity of a beam-column is reduced in the presence of the axial force. Additionally, biaxial bending further reduces the moment capacity. The design verification of beam-columns is presented in Section 6.3.

6.2.9 Section capacity and properties of columns & beam-columns

In general, many manufactured standard steel sections are considered as 'stocky'. Nevertheless, some standard sections are classified as slender. ASI [2004, 2009a] and OneSteel [2011, 2012a,b] clearly indicate those sections. The main concern is that local buckling of slender plate elements can occur before the attainment of the section capacity. From Clause 6.2 of AS 4100 the nominal section capacity in compression is:

$$N_s = k_f A_n f_y$$

The net section area is computed as follows:

$$A_n = A_g - A_d$$

where A_d is the sum of area deductions for holes and penetrations. No deductions need to be made for filled holes for the usual 16 to 24 mm bolt diameters and penetrations that are smaller than specified in Clause 6.2.1 of AS 4100. The form factor, k_f, is determined from:

$$k_f = \frac{A_e}{A_g}$$ where $A_g = \Sigma(b_i t_i)$ in this instance. (See Figure 6.2 also).

The effective cross-sectional area is computed from:

$$A_e = \Sigma(b_{ei} t_i) \leqslant A_g$$

$$b_{ei} = \lambda_{eyi} t_i \sqrt{\left(\frac{250}{f_{yi}}\right)} \leqslant b_i$$

where b_{ei} is the effective width of the i-th plate element of the section.

λ_{eyi} is the plate yield slenderness limit, obtained from Table 6.2, b_i is the clear width of a plate element having two plates supporting it longitudinally from crumpling (the web between the flanges), or the outstand of the element supported along one longitudinal edge only (the flange of an open section), and t_i and f_{yi} are respectively the thickness and design yield stress of the plate element being considered (see Figure 6.2).

Table 6.2 Plate element yield slenderness limits (from Table 6.2.4 of AS 4100).

Plate element slenderness	Long. edges support	Residual stresses	Yield limit λ_{ey}
Flange or web:	One only	SR, HR	16
		LW, CF	15
		HW	14
	Both	SR, HR	45
		LW, CF	40
		HW	35
Circular hollow section:		All	82

Legend: SR = stress-relieved, CF = cold-formed, HR = hot-rolled or hot-finished,
LW = lightly welded (longitudinally), HW= heavily welded (longitudinally).

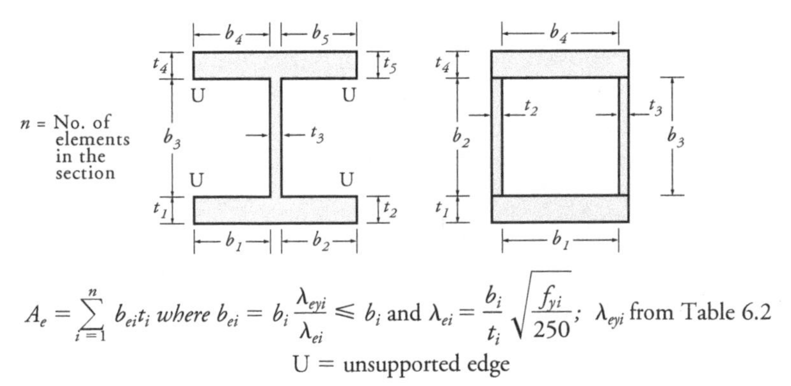

$$A_e = \sum_{i=1}^{n} b_{ei} t_i \; where \; b_{ei} = b_i \frac{\lambda_{eyi}}{\lambda_{ei}} \leqslant b_i \text{ and } \lambda_{ei} = \frac{b_i}{t_i}\sqrt{\frac{f_{yi}}{250}}; \; \lambda_{eyi} \text{ from Table 6.2}$$

U = unsupported edge

Figure 6.2 Effective area calculation for compression members. Note that 'clear' widths are used for flat plate elements—see also Example 6.2 (Section 6.9.2). For circular elements see Example 6.1 (Section 6.9.1).

Another section property required for column calculations is the member slendernes ratio, λ_n (see Section 6.2.10). This is dependent on the radii of gyration, r, defined as:

$$r_x = \sqrt{\left(\frac{I_x}{A_n}\right)}$$

$$r_y = \sqrt{\left(\frac{I_y}{A_n}\right)}$$

Section property tables (ASI [2004, 2009a] and OneSteel [2011, 2012a,b]) list the values of r_x and r_y for all standard sections. For a preliminary estimate, the radius of gyration can be approximately guessed from the depth of section, as shown in Table 6.3.

Table 6.3 Approximate values of radius of gyration

Section	r_x	r	r_y
UB/WB	0.41d		0.22b
UC/WC	0.43d		0.25b
PFC	0.40d		0.32b
CHS		0.34d_o	
Box, RHS, SHS	0.37d		0.40b
Solid square bar		0.29a	
Solid round bar		0.25d	

6.2.10 Member capacity and slenderness reduction factor for columns & beam-columns

From Clause 6.3.3 of AS 4100, the expression for nominal member capacity, N_c, is given by:

$$N_c = \alpha_c N_s = \alpha_c k_f A_n f_y \leqslant N_s$$

where α_c is the slenderness reduction factor, which depends on the slenderness ratio $\frac{l_e}{r}$, form factor k_f, yield stress of steel f_y and section constant α_b.

The evaluation of the effective length is described in Chapter 4 (specifically Section 4.5). Some special cases are described in this Section. This method relies on the effective length concept, and attempts to reduce a column with various end restraint conditions to an equivalent pin-ended column of modified length l_e, such that its capacity closely matches the capacity of the real column. The modified compression member slenderness, λ_n, is a means used to reduce the number of tables:

$$\lambda_n = \left(\frac{l_e}{r}\right)\sqrt{\left(\frac{k_f f_y}{250}\right)}$$

Using the modified slenderness ratio λ_n only a single graph or table is required for all values of yield stress, as shown in Table 6.6, to determine α_c.

Effective lengths of columns with idealised (though typically considered) end restraints are shown in Figure 6.3. Members in latticed frames are given in Table 6.4 and members in other frames are shown in Section 4.5.3.

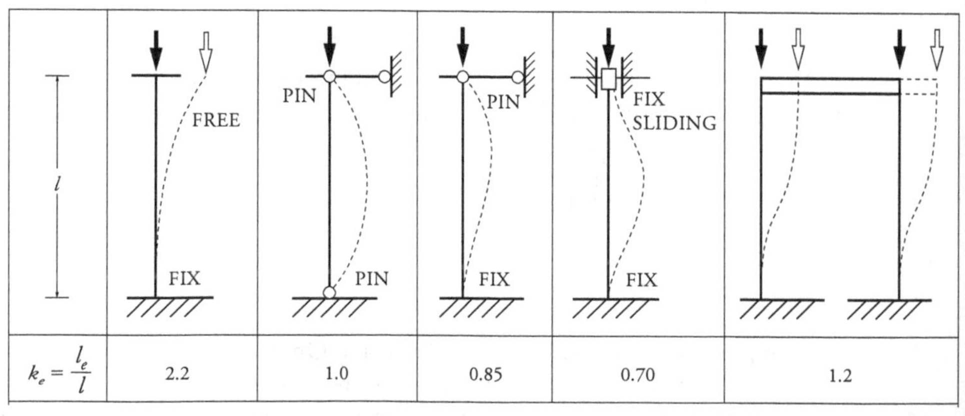

Figure 6.3 Effective length factors for idealised columns (see also Figure 4.3).

Table 6.4 Effective lengths for compression members in trusses

Member	Buckling plane	Effective length l_e
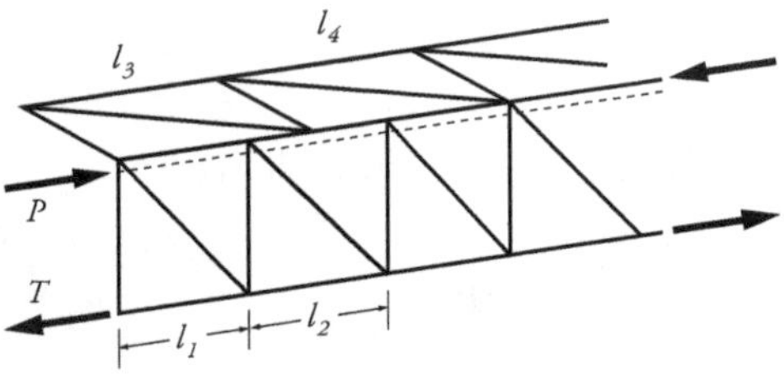	In plane of truss	$1.0\ l_1$ or l_2
	Out of truss plane	$1.0\ l_3$ or l_4
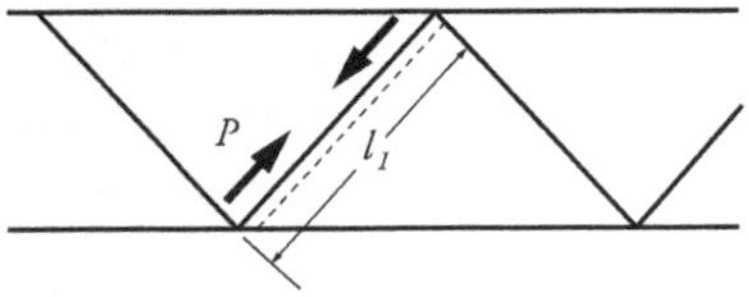	In plane of truss	$1.0\ l_1$
	Out of plane	$1.0\ l_1$
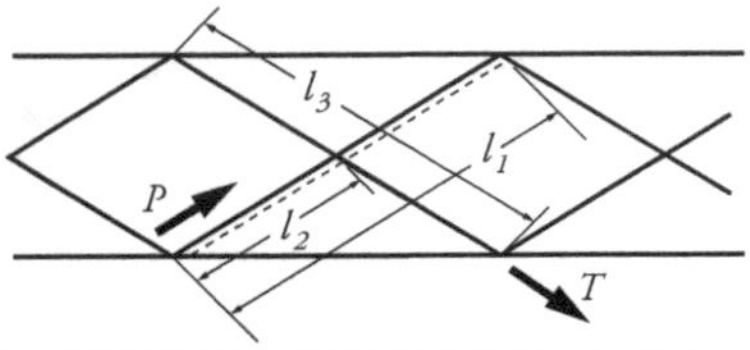	In plane of truss	$1.0\ l_2$
	Out of plane	$l_1\sqrt{1-\frac{3Tl_1}{4Pl_3}}$ but not less than $0.7l_3$
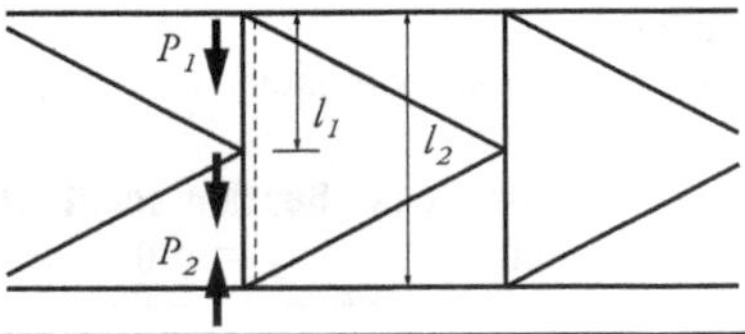	In plane of truss	$1.0\ l_1$ if $P_1 = P_2$ $0.75\ l_1$ if $P_2 = 0$ or tensile
	Out of plane	$1.0\ l_2$
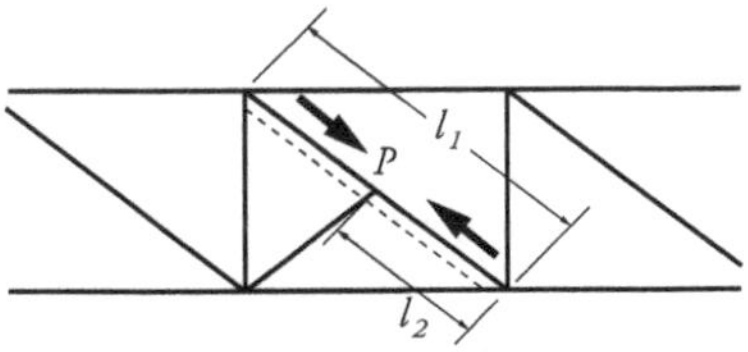	In plane of truss	$1.0\ l_2$
	Out of plane	$1.0\ l_1$

Two values of λ_n (see above) should normally be calculated, one for buckling about the major principal axis and the other about the minor principal axis.

The manual procedure for calculating the slenderness reduction factor α_c is given in AS 4100 as follows:

Factor α_a:

$$\alpha_a = \frac{2100\,(\lambda_n - 13.5)}{(\lambda_n^2 - 15.3\lambda_n + 2050)}$$

Member section constant α_b = (see Table 6.5).
Combined slenderness:

$$\lambda = \lambda_n + \alpha_a\alpha_b$$

Imperfection factor:

$$\eta = 0.00326\,(\lambda - 13.5),\ \text{but}\ \eta \geqslant 0$$

Factor ξ:

$$\xi = \frac{\left[\left(\frac{\lambda}{90}\right)^2 + 1 + \eta\right]}{2\left(\frac{\lambda}{90}\right)^2}$$

The slenderness reduction factor, α_c, is then calculated as follows:

$$\alpha_c = \xi\left\{1 - \sqrt{\left[1 - \left(\frac{90}{\xi\lambda}\right)^2\right]}\right\} \leqslant 1.0$$

Values of α_c can be calculated from the above equations or be readily evaluated by linear interpolation in Table 6.6 of this Handbook (or Table 6.3.3(3) of AS 4100).

The influence of the residual stress pattern on the capacity of a column is represented by the section constant α_b. For the selection of the constant α_b the following attributes are needed:

- section type
- method of manufacture and fabrication giving rise to residual stresses
- thickness of the main elements
- form factor k_f.

Table 6.5 lists the values of the section constant α_b.

Table 6.5 Values of compression member section constant α_b
k_f = section form factor (k_f = 1.0 for stocky (i.e. "compact") sections)

Section type	Manufacturing method	Thickness mm	Section constant, α_b, for $k_f = 1.0$	$k_f < 1.0$
RHS, SHS, CHS	Hot-formed, or cold-formed and stress-relieved	Any	-1.0	−0.5
	Cold-formed, not stress-relieved	Any	-0.5	−0.5
UB, UC	Hot-rolled	<40*	0	0
		⩾40*	+1.0	+1.0
Channels	Hot-rolled	Any	+0.5	+1.0
Angles, T-sections	Hot-rolled or flame-cut ex UB/UC	Any	+0.5	+1.0
Plate web I/H girders	Flame-cut edges	Any	0	+0.5
	As-rolled plate edges	⩽40*	+0.5	+0.5
	" "	>40*	+1.0	+1.0
Box section	Welded sections	Any	0	0
Other sections		Any	+0.5	+1.0

Note: 1) 'Any' means any practical thickness.
2) * indicates flange thickness.

The plot of the slenderness reduction factor, α_c, against the modified slenderness ratio, λ_n, in Figure 6.4 shows five column buckling curves, one for each α_b value of −1.0, −0.5, 0, +0.5 and +1.0, representing the different levels of residual stress and imperfections. The −1.0 curve is associated with sections having the lowest imperfections (e.g. hollow sections) and residual stress. The value of α_c can also be readily read off Table 6.6.

Table 6.6 Values of compression member slenderness reduction factor α_c

λ_n = modified slenderness, $\left(\frac{l_e}{r}\right)\sqrt{(k_f)}\left(\sqrt{\left(\frac{f_y}{250}\right)}\right)$

α_b = section constant as per Table 6.5

λ_n	Value of section constant α_b				
	−1.0	−0.5	0	+0.5	+1.0
10	1.000	1.000	1.000	1.000	1.000
15	1.000	0.998	0.995	0.992	0.990
20	1.000	0.989	0.978	0.967	0.956
25	0.997	0.979	0.961	0.942	0.923
30	0.991	0.968	0.943	0.917	0.888
40	0.973	0.940	0.905	0.865	0.818
50	0.944	0.905	0.861	0.808	0.747
60	0.907	0.862	0.809	0.746	0.676
70	0.861	0.809	0.748	0.680	0.609
80	0.805	0.746	0.681	0.612	0.545
90	0.737	0.675	0.610	0.547	0.487
100	0.661	0.600	0.541	0.485	0.435
110	0.584	0.528	0.477	0.431	0.389
120	0.510	0.463	0.421	0.383	0.348
130	0.445	0.406	0.372	0.341	0.313
140	0.389	0.357	0.330	0.304	0.282
150	0.341	0.316	0.293	0.273	0.255
160	0.301	0.281	0.263	0.246	0.231
170	0.267	0.251	0.236	0.222	0.210
180	0.239	0.225	0.213	0.202	0.192
190	0.214	0.203	0.193	0.184	0.175
200	0.194	0.185	0.176	0.168	0.161
210	0.176	0.168	0.161	0.154	0.148
220	0.160	0.154	0.148	0.142	0.137
230	0.146	0.141	0.136	0.131	0.127
240	0.134	0.130	0.126	0.122	0.118
250	0.124	0.120	0.116	0.113	0.110
260	0.115	0.111	0.108	0.105	0.102
270	0.106	0.103	0.101	0.098	0.096
280	0.099	0.096	0.094	0.092	0.089
290	0.092	0.090	0.088	0.086	0.084
300	0.086	0.084	0.082	0.081	0.079

Notes:

1) Linear interpolation permitted.
2) More intermediate values of λ_n can be found in Table 6.3.3(3) of AS 4100.

Steel grade	$\sqrt{\left(\frac{f_y}{250}\right)}$
200	0.894
250	1.00
300	1.10
350	1.18
450	1.34
500	1.41
600	1.55
700 (f_y = 650 MPa to AS 3597)	1.61
700 (f_y = 690 MPa to AS 3597)	1.66
700 (f_y = 620 MPa to AS 3597)	1.57

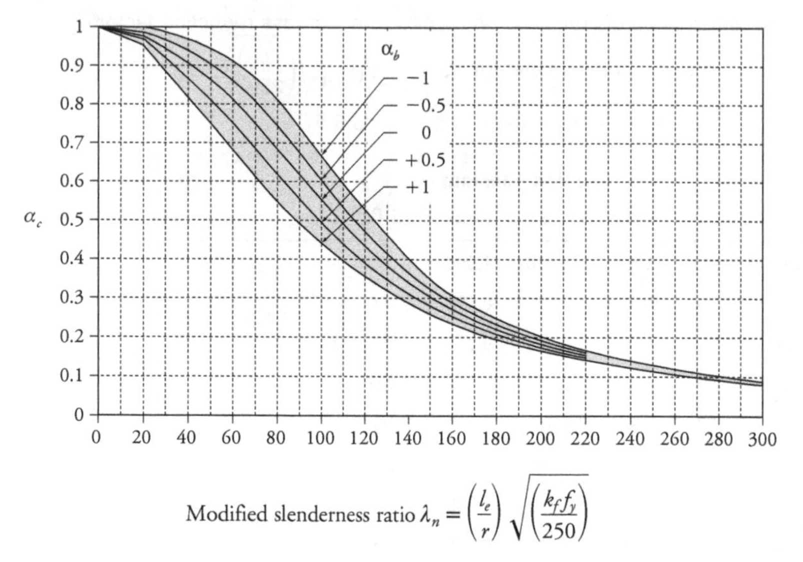

Figure 6.4 *Values of compression member slenderness reduction factor α_c under the influence of λ_n and section constant α_b (see Table 6.5 to see the relationship between α_b and the relevant steel section).*

Finally, the nominal member capacity, N_c, is calculated as follows:

$$N_c = \alpha_c k_f A_n f_y \leq N_s$$

where A_n is the net area of the section, that is $A_n = A_g - A_d$, and if there are no deductions for unfilled holes, $A_n = A_g$.

Though not explicitly evident in the above series of equations to develop the member axial compression capacity, Section 6 of AS 4100 only considers flexural buckling (i.e. not twisting, etc) to be the predominant buckling mode for the member stability design of columns. Hence, open sections such as some angles, tees and short cruciforms may not be adequately handled by AS 4100 as they have a propensity to twist rather than essentially flexurally buckle. Refer to Clause C6.3.3 of the AS 4100 Commentary for further guidance on the buckling design of these sections under axial compression loadings.

As noted in Section 1.14(a), the amendment to AS 4100 (AS 4100 AMD 1—see Appendix D) now has design provisions for compression member design based on Euler-type (flexural) buckling modes and the additional differentiation of members which are subject to torsional or flexural-torsional buckling and those members not subject to those buckling modes.

Additionally, most structural steel members that are loaded in compression generally buckle about either principal axis. Therefore, like flexural members, it is necessary to calculate section/member properties (e.g. slenderness ratios) and capacities about the x-and y-axis for column design.

The tedium of manual calculation of the member design capacities for relatively straightforward columns can be avoided by the use of ASI [2009a] for hot-rolled and welded open sections and OneSteel [2012b] or ASI [2004] for hollow sections.

6.3 Design of beam-columns

6.3.1 Concepts

The term 'beam-column' arises from the members subject to combined actions of axial force and bending. Design of beam-columns is covered by Section 8 of AS 4100. The method of calculation of the design compression capacity N_c is described in Section 6.2 of the Handbook applies equally to beam-columns.

Compression members are almost always subject to combined compression and bending. The bending moments may arise from the eccentric application of the load, from lateral loads applied to columns, and from the overall frame action. Light trusses having only insignificant node eccentricities can be designed neglecting the induced secondary bending moments.

Interaction between the axial compression force and bending moments produces three effects:

- Column buckling and bending interaction amplifies bending moments, as discussed in Chapter 4 of the Handbook.
- Axial compression force reduces the bending capacity of the beam-column.
- Bending about both the major and minor axes (biaxial bending) reduces axial member and bending capacity (see Figure 6.5).

The design bending capacity of a beam-column is determined in the same way as for beams covered in Chapter 5. The procedure adopted in AS 4100 is to divide the verification of combined action capacity into:

- *section* capacity verification concerned with checking that no section is loaded beyond its bending capacity
- *member* capacity verification concerned with the interaction between buckling and bending both *in plane* and *out of plane*
- *biaxial bending interaction* (see Figure 6.5).

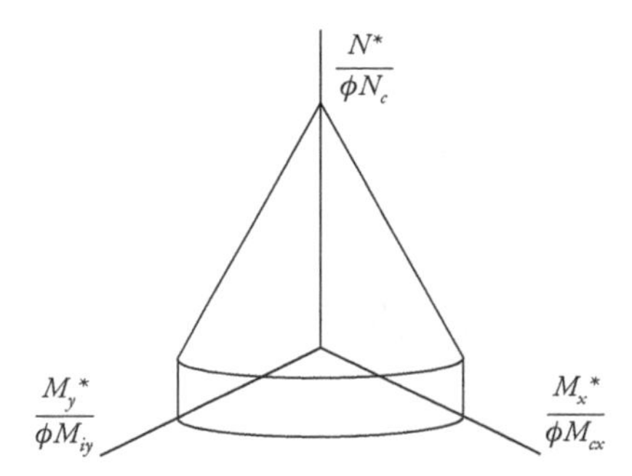

Figure 6.5 Biaxial bending and axial load interaction space

It is important to distinguish between three types of beam-column *member* behaviour modes:

(a) *in-plane*, where member deformations from imposed loads ***and*** subsequent buckling deformations occur in the same plane. This includes:

- members bending about their minor principal axis (y-axis)—that is, load *and* buckling deformations are only in the x-z plane (e.g. Figure 6.6(a)). The buckling is primarily due to column action and, in this instance, members bending about their minor axis cannot undergo flexural-torsional buckling.
- members bending about their major principal axis (x-axis) and only constrained to buckle about this axis also—that is, load *and* buckling deformations are only in the y-z plane (e.g. Figure 6.6(b)). The buckling is due to column action as flexural-torsional buckling, which can only occur about the minor principal axis, is suppressed.

(b) *out-of-plane*, where member deformations from imposed loads do ***not*** occur in the same plane as the buckling deformations—e.g. Figure 6.6(c). Practically, this occurs for beam-columns subject to bending moments about the major principal axis (x-axis) with buckling deformations about the minor principal axis (y-axis). The interaction between column buckling and flexural-torsional buckling needs to be considered as the latter buckling mode is not suppressed.

(c) *biaxial bending*, where bending occurs about both principal axis (x- and y-axis) with or without axial loads. The loading and buckling deformations present in this situation are a combination of (a) and (b) with subsequent interaction effects.

The procedures from Chapter 5 on bending apply to all the above three beam-column design situations—of particular note for (b) in determining flexural-torsional buckling behaviour. It should also be noted that, in general, sections with a large ratio of M_{sx}/M_{sy} and bending moments acting about their x-axis are prone to *out-of-plane* behaviour. Sections bent about their y-axis are subject to *in-plane* buckling only, except where the imposed loads are applied high above the centre of gravity.

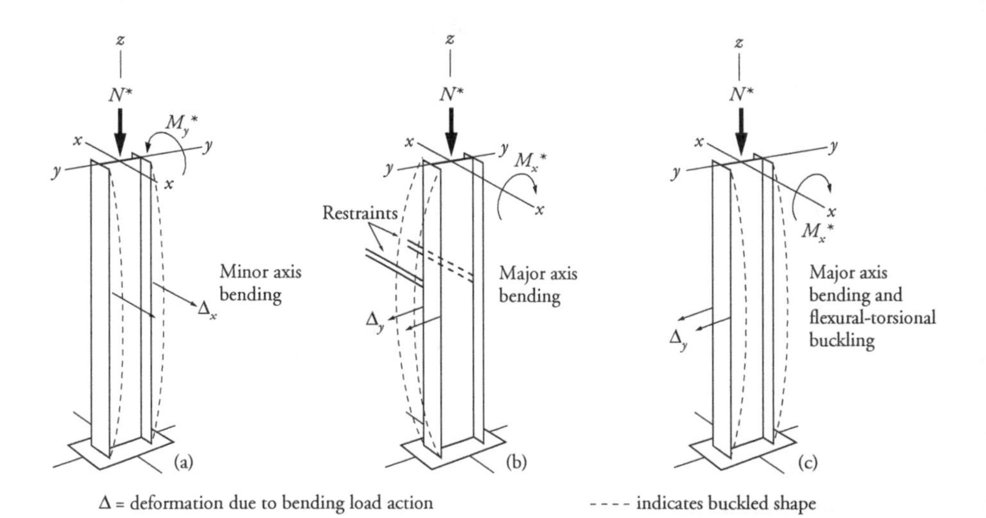

Figure 6.6 *Combined axial compression and bending: (a) in-plane bending about weak axis with column buckling; (b) in-plane bending about strong axis with column buckling; (c) out-of-plane bending to column & flexural-torsional buckling*

6.3.2 Beam-column verification procedure

The following procedures begin with a lower tier approach, which may be suitable for ordinary building structures but tends to be conservative. Thereafter, more economical higher tier design methods are highlighted. The procedure for designing beam-columns is given as follows.

Step 1: Frame analysis
Analyse the frame for design actions and their combinations giving the most adverse action effects. Carry out a second-order analysis or moment amplification procedure with first-order elastic analysis as described in Chapter 4. Determine the effective lengths of the column using the methods described in Chapter 4 and, from these, determine the nominal buckling loads and moment amplification factors (if required).

Step 2: Data assembly
The complete set of data required for manual or computer calculations is as follows:

Design actions (loads) and design action effects:

N^* = design axial load

M^*_{xi} = design bending moment about the x-axis: at end A, end B, and intermediate points of high bending moments, i = 1, 2, 3 ...

M^*_{yi} = ditto, about the y-axis.

Design bending moments are bending moments obtained from a second-order analysis or a first-order analysis factored by an amplification factor (see Chapter 4).

Member and section data required:

l = actual length between full, partial and lateral restraints for bending about the x-axis

l_x = actual length between restraints for column buckling about the x-axis

l_y = ditto, about the y-axis

f_y = design yield stress

A_n, k_f, Z_{ex}, Z_{ey}, I_x, I_y, J, I_w, r_x, r_y

and other sectional data such as compactness, flange thicknesses, etc. (see ASI [2004, 2009a] and OneSteel [2011, 2012a,b]).

Step 3: Effective section properties
Calculate the net section by deducting larger holes and cut-outs but not filled bolt holes, as described in Section 6.2.9.

Step 4: Effective lengths—column action
For a simple method of calculating column effective lengths:

$$l_{ex} = k_{ex} l_x$$

$$l_{ey} = k_{ey} l_y$$

where k_{ex} and k_{ey} are the effective length factors obtained from Figure 6.3 for simple framed buildings or from Table 6.4 for triangulated frames. For further reading on column effective lengths, see Section 4.5.

Step 5: Column slenderness reduction factors α_{cx} and α_{cy}
The term 'slenderness reduction factor' used in Clause 6.3.3 of AS 4100 refers to the reduction of the section capacity on account of column slenderness.

Step 6: Axial compression capacity
As noted in Section 6.2.7, the axial compression capacity is:

$$N_s = k_f A_n f_y$$

$$N_c = \alpha_c N_s = \alpha_c k_f A_n f_y \leq N_s$$

checking that $N^* \leq \phi N_s$ and $N^* \leq \phi N_c$ with $\phi = 0.9$.

Step 7: Moment capacity
Chapter 5 gives full details and aids for calculating the beam slenderness reduction factor, α_s, and the reference buckling moment, M_o. The beam slenderness reduction factor is given by:

$$\alpha_s = 0.6\left[\sqrt{\left[\left(\frac{M_s}{M_o}\right)^2 + 3\right]} - \frac{M_s}{M_o}\right]$$

where the section moment capacity M_s is obtained from $M_s = Z_e f_y$.

The reference buckling moment for a doubly symmetric section is given by:

$$M_o = \sqrt{\left[\left(\frac{\pi^2 EI_y}{l_e^2}\right)\left(GJ + \frac{\pi^2 EI_w}{l_e^2}\right)\right]}$$

where the effective length for bending, l_e, about the x-axis is noted in Section 5.4.2 and given by

$$l_e = k_r k_t k_l l$$

The nominal member moment capacities are given by:

$$M_{bx} = \alpha_m \alpha_s M_{sx} \leq M_{sx}$$

where α_m is the moment modification factor (see Section 5.5.3.2), and

$$M_s = f_y Z_e$$

The design moment capacity checks with $\phi = 0.9$ are:

$$M_x^* \leq \phi M_{bx}$$

$$M_y^* \leq \phi M_{sy}$$

The beam-column design process is slightly more complex and may involve trial-and-error steps. The designer can modify the outcome by exercising control over the position, type and stiffness of the lateral restraints to achieve overall economy.

Step 8: Reduced *section* moment capacity due to combined actions
As noted in Clauses 8.3.2 and 8.3.3 of AS 4100, this is basically a reduction in bending capacity of a section subject to combined axial compression and uniaxial bending. Only elastically designed members are considered here.

Using the lowest tier, for any section bent about the *major* principal axis:

M_{rx} = nominal section moment capacity, bending about the x-axis, reduced by axial force (tension or compression)

$$= M_{sx}\left(1 - \frac{N^*}{\phi N_s}\right)$$

Less conservatively, for compact doubly symmetrical I-sections, and rectangular and square hollow sections (for tension members and compression members with $k_f = 1$):

$$M_{rx} = 1.18\, M_{sx}\left(1 - \frac{N^*}{\phi N_s}\right) \leqslant M_{sx}$$

A relatively accurate check can be applied for compression members with $k_f < 1.0$:

$$M_{rx} = M_{sx}\left\{1 - \frac{N^*}{\phi N_s}\right\}\left\{1 + 0.18\left(\frac{\lambda_1}{\lambda_2}\right)\right\} \leqslant M_{sx}$$

where $\lambda_1 = (82 - \lambda_e)$ and $\lambda_2 = (82 - \lambda_{ey})$, refer to the plate slenderness of the web (see Table 5.3).

For any section bent about the *minor* principal axis (lowest tier):

M_{ry} = nominal section moment capacity, bending about the y-axis, reduced by axial force (tension or compression)

$$M_{ry} = M_{sy}\left(1 - \frac{N^*}{\phi N_s}\right)$$

Alternatively, for compact doubly symmetrical I-sections:

$$M_{ry} = 1.19\, M_{sy}\left(1 - \left[\frac{N^*}{\phi N_s}\right]^2\right) \leqslant M_{sy}$$

and for compact RHS and SHS hollow sections:

$$M_{ry} = 1.18\, M_{sy}\left(1 - \left[\frac{N^*}{\phi N_s}\right]\right) \leqslant M_{sy}$$

Final checks required (with ϕ = capacity factor = 0.9):

$$M_x^* \leqslant \phi M_{rx}$$

$$M_y^* \leqslant \phi M_{ry}$$

Step 9: Section capacity under *biaxial* bending

As noted in Clause 8.3.4 of AS 4100, the following interaction inequality can conservatively be used for any type of section:

$$\frac{N^*}{\phi N_s} + \frac{M_x^*}{\phi M_{sx}} + \frac{M_y^*}{\phi M_{sy}} \leqslant 1.0$$

Alternatively, for compact doubly symmetrical I-sections, rectangular and square hollow sections:

$$\left[\frac{M_x^*}{\phi M_{rx}}\right]^\gamma + \left[\frac{M_y^*}{\phi M_{ry}}\right]^\gamma \leqslant 1.0$$

where $\gamma = 1.4 + \dfrac{N^*}{\phi N_s} \leq 2.0$, M_{rx} and M_{ry} are values calculated for uniaxial bending.

Step 10: In-plane member moment capacity
Having determined the *section* moment capacities at critical section(s) of the member, it is necessary to determine the capacities of the *member* as a whole. This time the compression member capacity, N_c, is used instead of the section capacity, N_s. *In-plane* buckling, entails bending and column buckling, both occurring in the same plane (see Section 6.3.1(a)).

The design procedure for checking this mode of failure is given in Clause 8.4.2.2 of AS 4100 for elastically analysed/designed compression members. Plastically analysed/designed members are covered by Clause 8.4.3 of AS 4100. This is the only clause to specifically consider plastic methods under combined actions and gives limits on section type (i.e. only doubly symmetric I-sections), member slenderness, web slenderness and plastic moment capacities.

For a general section analysed elastically with compression force, the in-plane member moment capacity, M_i, is given by:

$$M_{ix} = M_{sx}\left(1 - \frac{N^*}{\phi N_{cx}}\right)$$

$$M_{iy} = M_{sy}\left(1 - \frac{N^*}{\phi N_{cy}}\right)$$

where N_c is the member axial capacity for an effective length factor $k_e = 1.0$ for both braced and sway members, unless a lower value of k_e can be established for braced members. All this is premised on using the appropriate l_e for $N^* \leq \phi N_c$ when the member is designed for compression alone.

The above two expressions for M_i can be slightly confusing as, depending on the axis of bending, only M_{ix} **<u>or</u>** M_{iy} needs to be evaluated for uniaxial bending with axial compression force. As an example, for M_y^* acting with N^* only, M_{iy} is only evaluated. M_{ix} is not evaluated as there is no bending about that axis. The reverse applies for x-axis bending. M_{ix} and M_{iy} will need to be calculated if both M_x^* and M_y^* are present with axial force, N^* — see Step 12 for member capacity under biaxial bending.

Alternatively, for doubly symmetrical, compact I-sections, rectangular and square hollow sections bending about the x-axis, with $k_f = 1.0$:

$$M_{ix} = M_{sx}[(1 - c_2)\, c_3 + (1.18\, c_2 c_3^{0.5})] \leq M_{rx}$$

where

$$c_2 = (0.5 + 0.5\beta_m)^3$$

$$c_3 = 1 - \frac{N^*}{\phi N_{cx}}$$

$$\beta_m = \frac{M_{1x}}{M_{2x}}$$ = ratio of smaller to larger end bending moments (positive in reverse curvature)

Similarly, the in-plane member moment capacity for bending and buckling about the y-axis, M_{iy}, can be evaluated by changing the x subscript into y for the above higher tier equation for M_{ix}.

For the final check (with ϕ = capacity factor = 0.9):

$$M_x^* \leqslant \phi M_{ix} \text{ or } M_y^* \leqslant \phi M_{iy} \text{ as appropriate.}$$

Step 11: *Out-of-plane member* moment capacity
This check is for flexural-torsional buckling in conjunction with column buckling. In such situations the axial compression force aggravates the flexural-torsional buckling resistance of columns bent about the major principal axis. As an example, the typical portal frame column is oriented such that its major principal plane of bending occurs in the plane of the frame, with buckling ocurring out of plane. From Clause 8.4.4 of AS 4100, the value of the out-of-plane member moment capacity, M_{ox}, is obtained for any section, conservatively by:

$$M_{ox} = M_{bx}\left(1 - \frac{N^*}{\phi N_{cy}}\right)$$

where N_{cy} is the nominal member capacity calculated for out-of-plane buckling (about the y-axis), M_{bx} is the nominal member moment capacity of a member without full lateral restraint and having a moment modification factor (α_m) reflective of the moment distribution along the member (see Section 5.5.3.2) or conservatively taken as $\alpha_m = 1.0$.

There is an alternative expression for M_{ox} given in Clause 8.4.4.1 of AS 4100, which should give a less conservative solution at the expense of some extra computational effort. This alternative method has the following limitations: sections must be compact, doubly symmetrical I-sections, having $k_f = 1.0$. Additionally, both ends of the segment must be at least partially restrained:

$$M_{ox} = \alpha_{bc} M_{bxo} \sqrt{\left[1 - \frac{N^*}{\phi N_{cy}}\right]\left[1 - \frac{N^*}{\phi N_{oz}}\right]} \leqslant M_{rx}$$

where $$\alpha_{bc} = \frac{1}{\left\{(0.5 - 0.5\beta_m) + (0.5 + 0.5\beta_m)^3\left[0.4 - \frac{0.23\, N^*}{\phi N_{cy}}\right]\right\}}$$

M_{bxo} = nominal member moment capacity without full lateral restraint with $\alpha_m = 1.0$

and, $$N_{oz} = \frac{A\left(GJ + \frac{\pi^2 E I_w}{l_z^2}\right)}{(I_x + I_y)}$$

where l_z is the distance between partial or full torsional restraints and β_m defined in Step 10.

Step 12: Member capacity under *biaxial* bending
Where a compression member is subject to the simultaneous actions of N^* (if present), M_x^* and M_y^*, the following check is required:

$$\left[\frac{M_x^*}{\phi M_{cx}}\right]^{1.4} + \left[\frac{M_y^*}{\phi M_{iy}}\right]^{1.4} \leqslant 1.0$$

M_{cx} in the first term of the inequality should be the lesser of M_{ix} and M_{ox} (see Step 10 and 11 above). M_{iy} is also noted in Step 10.

Step 13: The influence of self-weight-induced moment
Self-weight of members made of rolled or tube sections can reduce the compression capacity quite significantly. Self-weight should be included with the design actions applied to the member.

As an example, Woolcock et al. [2011] provide a rigorous method (and design capacity tables) for compression bracing and self-weight.

Step 14: Design review
The adequacy, efficiency and economy of the solutions should be subjected to thorough appraisal. A 'what if' analysis should be carried out, questioning in particular the section make-up, steel grade and feasibility of added restraints.

6.3.3 Variable section columns

Variable section or non-prismatic beam-columns are sometimes used in portal frames as tapered columns and rafters. There are a number of published methods (Bleich [1952], CRCJ [1971], Lee et al. [1972]) for the evaluation of elastic critical buckling load, N_{om}, of such columns. A conservative method of verification of the compression capacity of variable section columns is presented in Clause 6.3.4 of AS 4100. The first step after computing the value of N_{om} is to compute the section capacity N_s of the smallest section in the column (conservative). Based on this, compute the modified slenderness using:

$$\lambda_n = 90\sqrt{\left(\frac{N_s}{N_{om}}\right)}$$

and proceed as discussed in Sections 6.2.10 and 6.3.2.

6.4 Struts in triangulated structures

The web members in simple trusses are often connected eccentrically, as shown in Figure 6.7. This may be due to geometric considerations, however, in most instances, fabrication economy will dictate connection eccentricity for easier cutting, fitting and welding.

Eccentricity may be evident in two planes at a connection. The eccentricity may be in the plane of the truss (Figure 6.7(a)) and/or in the orthogonal plane to the truss (Figure 6.7(b)). Lighter trusses using angle sections generally have this double eccentricity and the design provisions for such connections can be found in Clause 8.4.6 of AS 4100. Although AS 4100 and its Commentary do not specifically mention the evaluation of, and limits for, an in-plane eccentricity, it may be used as a good starting point for determining connection capacity for such trusses.

In large trusses carrying considerable loads, the members should be connected concentrically (i.e. all member centrelines coincide at a point in the connection thereby negating secondary effects as bending moments at the connection). If this significantly infringes on fabrication economics (which it will in many instances) then allowance must be made for the second-order moments from bending eccentricities. Due to the nature of the sections used (e.g. I-section chords), the inherent eccentricity to be considered is in-plane. Member design would then follow the normal method of design for combined actions. Connection loading and design would be done separately (see Chapter 8).

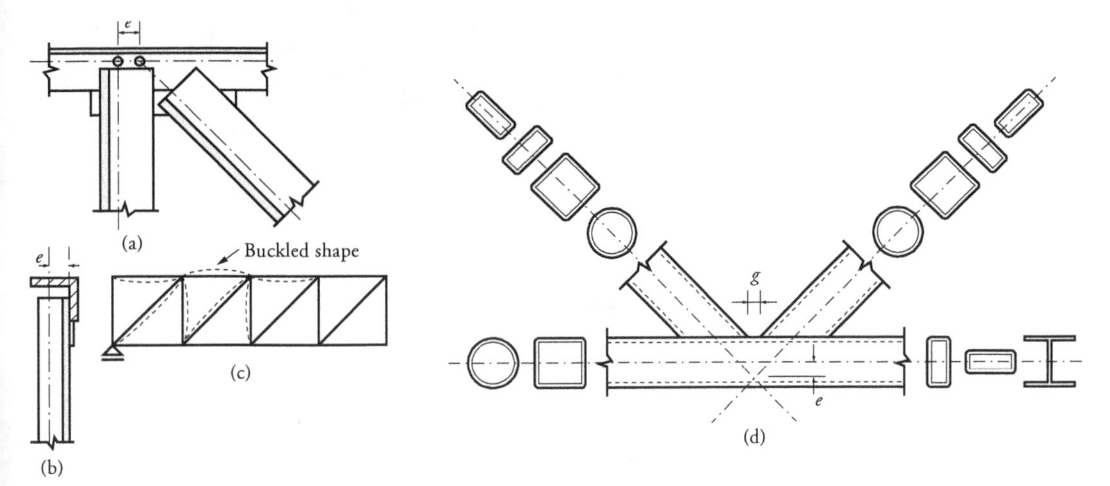

Figure 6.7 *The configuration and effect of eccentric node connections in trusses: (a) elevation of in-truss plane eccentricity in angle trusses; (b) section of out-of-plane eccentricity in angle trusses; (c) in-plane buckled shape of members in trusses; and (d) elevation of in-plane eccentricity in tubular trusses.*

Tubular trusses are gaining much popularity, and these types of trusses can be readily designed for connection eccentricity to satisfy aesthetics and fabrication economies. Again, due to the nature of the members, the only eccentricity to consider is in-plane (Figure 6.7(d)). There has been much research work done on these connections which can provide ready design solutions for adequate connection strength and behaviour (CIDECT [1991, 1992, 2008, 2009], Syam & Chapman [1996], Packer & Henderson [1997]). Such connection design models permit significant eccentricities (within limits) such that secondary bending moments can be neglected. Connection loading and design is part of the connection model, and member design would follow the normal method of design for combined actions.

Of note, tubular connections as shown in Figure 6.7(d) should have a gap between the two diagonal members with clear separation between adjacent welds. Unless there are other mitigating factors, overlap diagonal member connections should be avoided for arguments of fabrication economy.

6.5 Battened and laced struts

Compression members can be composed of two or more main components tied together by "lacing" or "battens" to act as one compression member. "Back-to-back" members also fall into this category (see items *l, m, n, o* in Figure 6.13). Special provisions for battening, lacing and shear connections are given in Clauses 6.4 and 6.5 of AS 4100. When these rules are complied with, the compound struts can be verified for capacity as if they were one column shaft (see Figures 6.8 and 6.9 and 6.10).

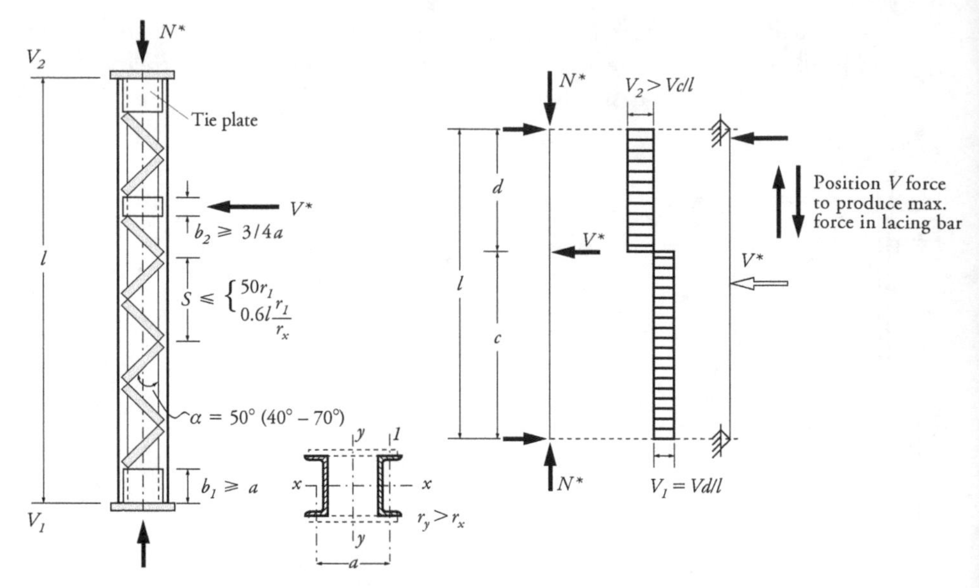

Figure 6.8 Laced compression members

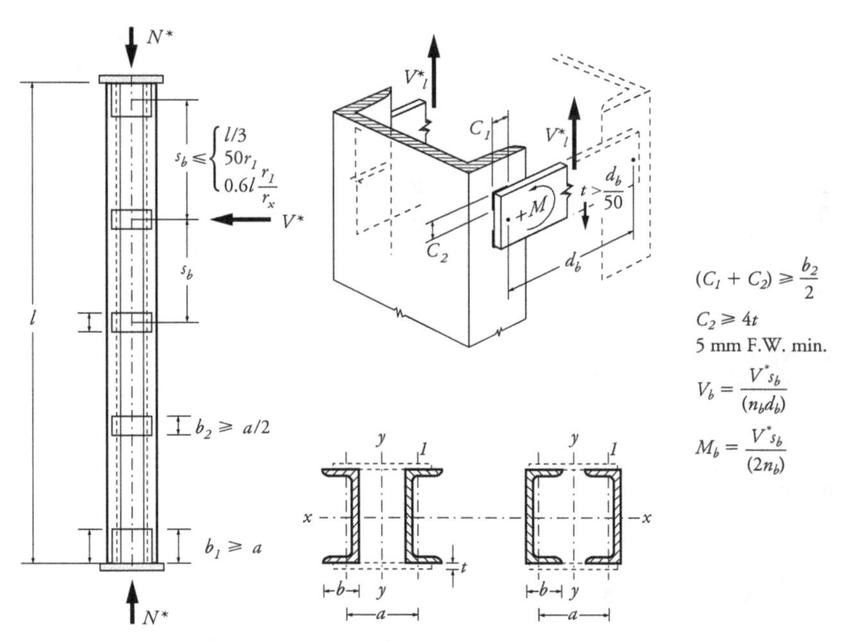

Figure 6.9 Battened compression members

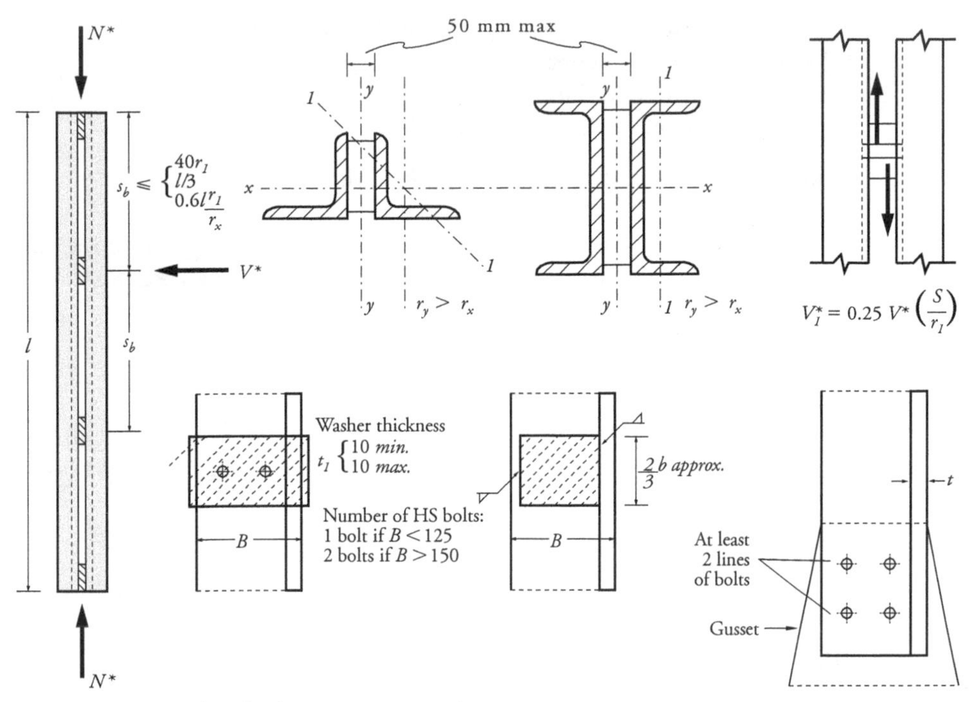

Figure 6.10 Back-to-back compression members

The function of the battens or lacing is to prevent main components from acting as separate columns and thus reduce the capacity of the battened member. The capacity of the compound column without effective battens or lacing would decrease to just twice the capacity of each main component. The battens or lacing members must be designed to resist the action of a lateral force V^* applied at any position of the member:

$$V^* = \frac{\pi N^*\left(\frac{N_s}{N_c} - 1\right)}{\lambda_n} \geqslant 0.01N^* \text{ (Clause 6.4.1 of AS 4100)}$$

where

N_s = nominal section capacity of the compression member (Section 6.2.9)

N_c = nominal member capacity (Section 6.2.10)

N^* = design axial force applied to the compression member

λ_n = the modified member slenderness (Clauses 6.4.3.2 and 6.3.3 of AS 4100)

AS 4100 also requires the following items to be satisfied for laced (Clause 6.4.2), battened (Clause 6.4.3) and back-to-back (Clause 6.5) compression members:

- configuration requirements for back-to-back members
- maximum slenderness ratio of main component
- slenderness ratio calculation of the overall composite member
- lacing angle (where relevant)
- effective length of a lacing/batten element
- slenderness ratio limit of a lacing/batten element

- mutual opposite side lacing requirement (where relevant)
- tie plates (for lacing systems)
- minimum width and thickness of a batten (for batten systems)
- minimum design loads (for batten and back-to-back systems)

Figures 6.8 to 6.10 illustrate the configuration and principal design rules for these members.

Battens and their connections (Figure 6.9) have to be designed for the effects of force V^* applied laterally at any point. The actions on the battens are as follows:

(a) Shear parallel to the axis of the main components:

$$V_l^* = \frac{V^* s_b}{n_b d_b}$$

where n_b is the number of battens resisting the shear, d_b is the lateral distance between the centres of weld groups or bolts, and s_b is the batten spacing (see Figures 6.9 and 6.10).

(b) Design bending moment:

$$M^* = \frac{V^* s_b}{2n_b}$$

6.6 Composite steel and concrete columns

Combining steel and concrete has advantages over bare steel columns where large axial loads are encountered, as for example in high-rise construction. The basic principle is that the load is shared between the steel shaft and the concrete. Composite columns may be constructed in a variety of ways:

- concrete-filled tubular columns (sometimes called externally reinforced concrete columns)
- concrete-encased I-section columns
- concrete-encased latticed columns (which are not common).

Typical composite column sections are shown in Figure 6.11. The benefits of these forms of column construction include higher load carrying capacities, increased fire resistance and faster construction times. An Australian Standard for designing these types of columns is soon to be under preparation (at the time of publication of this Handbook). In lieu of a local Standard, many structural designers have been using AS 5100.6, Eurocode 4 (EC 4) and CIDECT [1994,1998] for suitable guidance in this area.

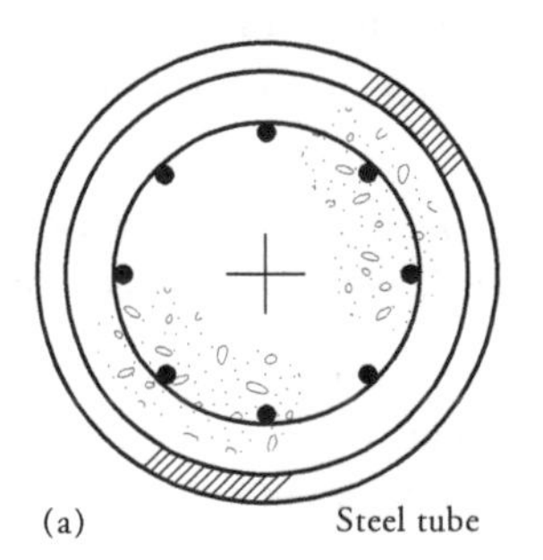

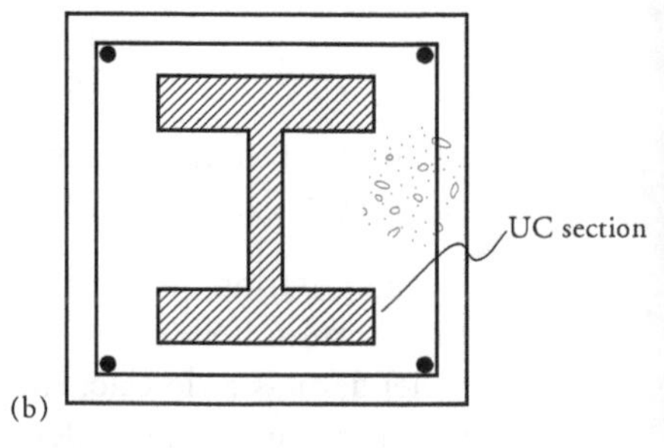

Figure 6.11 Composite steel and concrete column sections: (a) concrete filled tube (may/may not have internally placed reinforcing; (b) concrete encased I-section.

6.7 Restraining systems for columns and beam-columns

End restraints are usually not a problem with beam-columns, because there is usually adequate torsional and lateral stiffness at the bases and at the beam-to-column connections. The connections should, however, be checked for their ability to prevent twisting of the beam-column section, particularly if any flexible or web cleat connections are employed. A concrete floor would provide full torsional restraint to the beam-column shaft at the floor levels.

Intermediate lateral-torsional restraints are often used with open sections to reduce the effective length, thus increasing the beam-column member capacity. The stiffness and capacity of the restraints are important. Restraints should be designed so that the buckling by twisting is prevented. UB and UC sections need both the flanges of beam-columns to be restrained. This can be achieved with restraint braces of sufficient flexural and/or axial stiffness and detailing that ensures that both column flanges are restrained against flexural or torsional buckling. Hollow sections are easy to restrain because they have relatively high flexural or torsional capacity and are not susceptible to torsional buckling. Some details of flexural-torsional restraints are shown in Figure 6.12.

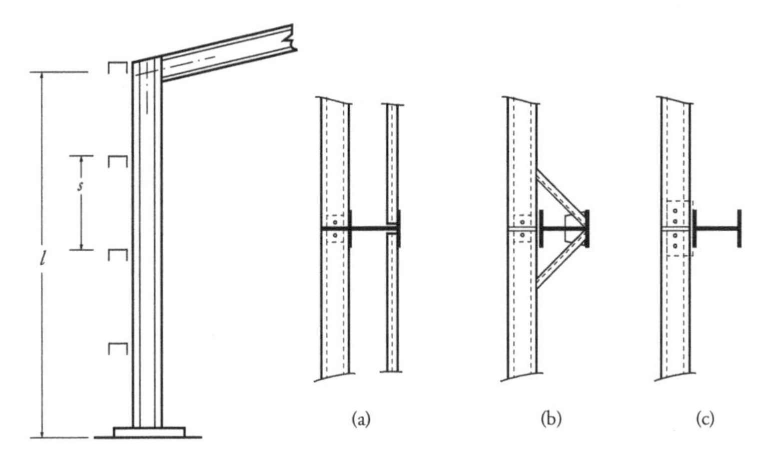

Figure 6.12 *Beam-columns braced by girts on one flange only: (a) with ties attached to the opposite flange; (b) with fly braces stabilising the other flange; (c) with rigid connection using extended girt cleats and more (H.S.) bolts.*

For columns, Clause 6.6 of AS 4100 requires some minimum loads to be resisted by compression member restraints and associated connections. These restraining members should generally be designed to transmit the greater of the following:

- any notional horizontal forces (see Section 3.7) required to be transmitted by the column system and any other restraint forces prevailing from the design loads to reaction points, or
- 2.5% of the maximum member axial compression force at the restraint position (this may be reduced on a group restraint basis if the restraint spacing is more closely spaced than is required for the member to attain its full section compression capacity).

For a series of parallel compression members being restrained by a line of restraints, AS 4100 permits a reduction in the accumulation of restraint forces for parallel members beyond the connected member. This reduction is from 2.5% to 1.25% and applies for the situation of no more than seven parallel members being restrained by the line of restraints.

Interestingly, Clause 9.1.4 of AS 4100 which considers the minimum design actions on connections, requires a minimum load of 0.03 (3%) times the capacity of the tension/compression member. Hence for restraint connections, the 3% rule should be used instead of 2.5% for restraint elements.

6.8 Economy in the design

The cost of construction dictates that sections be efficiently designed. The simple principle is to design columns in such a way that the ratio of radius of gyration to the section depth or width is as high as possible, and that the α_c (for columns) and α_s (for beam-columns) values are as high as possible. Some column sections that are usually employed in practice are shown in Figure 6.13. If possible, compound sections should be avoided in the interest of economy. However, where heavy loads are to be resisted by the column, the use of compound sections may be the only feasible solution.

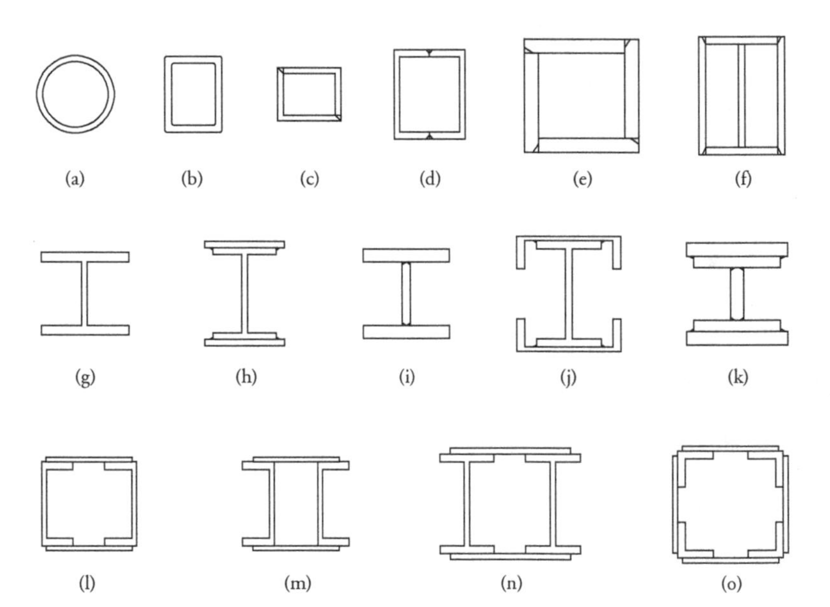

Figure 6.13 *Typical compression member sections: (a) Circular Hollow Sections (CHS); (b) Rectangular/Square Hollow Sections (RHS/SHS); (c) welded box from angles; (d) welded box from channels; (e) welded box from plates; (f) boxed-off I-section; (g) Universal Column (UC) or Universal Beam (UB) section; (h) flange plated UC/UB; (i) Welded Column (WC) or welded 3-plate column; (j) hybrid universal section with channel stiffening of flanges; (k) flange reinforced welded 3-plate column section; (l) laced/battened channel section column—toes inwards; (m) laced/battened channel section column—toes outwards; (n) laced/battened I-section column; (o) laced/battened angle section column.*

As can be seen from Figure 6.4, the slenderness ratio, l_e/r, gives an indication of how efficient the column will be in resisting axial compression. It follows that the spacing of lateral restraints should be chosen so as to minimise the slenderness ratio or, in practical terms, to aim for an l/r_y of less than 100 for columns.

Open sections such as UC or UB sections have a large ratio of r/d about the strong axis (where d = section depth), and the designer can utilise this beneficial feature by orienting the column section so that its web is in the plane of the frame action. This, however, necessitates closer spacing of restraints in the weak plane.

Tubular sections (CHS, SHS and RHS) have relatively large ratios of r/d and are less sensitive to combined actions, and thus also make economical compression members, as can be seen from Table 6.7.

Table 6.7 compares column capacities between various steel sections to obtain an overall ranking of column efficiencies based on these sections. The sections vary from cross-section areas totally away from the section centre (e.g. hollows sections), to steel elements crossing/near the section centre (UC and EA), to those with the cross-section essentially located on the section centre (Round Bar). Though the parameters of l_e/r, r/d and α_c are a useful indication of relative efficiency within a section type, they do not fully take into account the strength and mass of the section. The last parameter is a direct indication of the plain steel cost. Consequently, Table 6.7 has been formatted to include the ratio of design compression member capacity to its mass and is expressed in kN/(kg/m)—or much how capacity can we "squeeze" out of every kilogram of the relevant section. Obviously, the higher the ratio the better.

Interestingly, hollow sections stand out over the other sections and occupy the first two positions in the ranking. The SHS is placed in the top ranking as it harnesses the benefits of higher strength in this application. CHS and SHS may switch about in ranking depending on the overall slenderness of the column. Regardless, from Table 6.7, it is evident that tubular sections are more efficient than "open" type sections in column applications. Also quite noteworthy are the significant differences with much less efficient column sections like angle and solid sections—particularly the latter with a very low kN/(kg/m) value. The only missing information is the $/tonne or $/m cost for each section type to get a good estimate of relative value and efficiency between section types. Due to the variability of such information, it may be obtained from steel suppliers, Watson et al [1996] or ASI [2009b] which provides some basic information on this area. However, based on current general costings and price differentials (at time of publication), Table 6.7 provides a good reflection of efficient column section ranking in this method of assessment.

Table 6.7 Comparison and ranking of compression capacity efficiency of various sections with N^* = 950 kN, l_e= 3500 mm for axis of buckling as noted. (Rank 1 = most efficient, 7 = least efficient).

Rank	Section	Grade	Mass kg/m	r mm	l_e/r	r/d	α_c	$\frac{\phi N_c}{(kg/m)}$ kN/(kg/m)
1	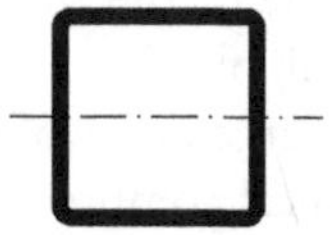 150 × 150 × 6.0 SHS	C450L0	26.2	58.2	60.1	0.388	0.741	**38.2**
2	 219.1 × 6.0 CHS	C350L0	31.5	75.4	46.4	0.344	0.885	**35.6**
3	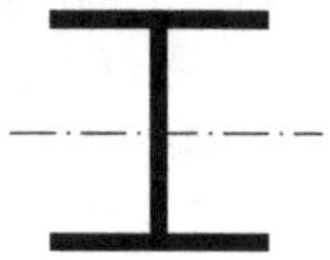 150UC37.2	300	37.2	68.4	51.2	0.422	0.830	**28.5**
4	 200UC46.2	300	46.2	51.0	68.6	0.251	0.714	**24.6**
5	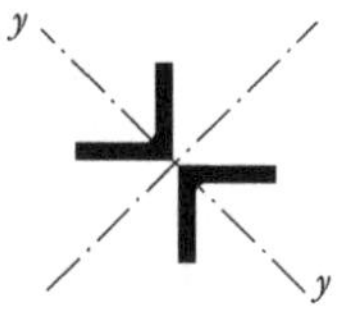 125 × 125 × 12.0EA	300	45.0	48.3	72.5	0.273	0.617	**21.2**
6	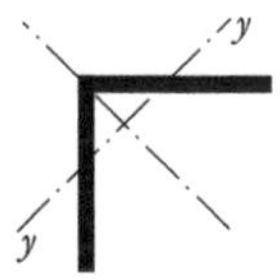 200 × 200 × 20.0EA	300*	60.1	39.3	89.1	0.256	0.520	**16.7**
7	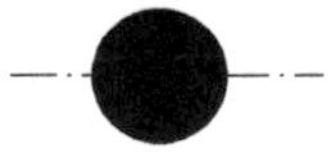 120 mm dia. Round Bar	300*	88.8	30.0	117	0.250	0.367	**11.8**

Note: *f_y = 280 MPa. Also, it is assumed that the sections with angles will be controlled by minor axis buckling.

6.9 Examples

6.9.1 Example 6.1

Step	Description and calculations	Result	Unit
	Verify the capacity of the tubular compression member subjected to axial load only. The member end connections are pinned to prevent any moment developing at these ends. The two ends of the member are braced against lateral sway in both directions.		
	Data		
	Design axial load $N^* =$	1030	kN
	Member length $l =$	3800	mm
1	Guess the trial section		
	219.1 x 6.4 CHS Grade C350L0 (OneSteel [2012b] Table 3.1-2(1) page 3-8) with properties:		
	$A_g =$	4280	mm^2
	$r =$	75.2	mm
	$f_y =$	350	MPa
2	Section slenderness λ_e, form factor k_f, net area A_n		
	$\lambda_e = \left(\frac{d_o}{t}\right)\left(\frac{f_y}{250}\right)$ ... AS 4100 Clause 6.2.3...		
	$= \left(\frac{219.1}{6.4}\right)\left(\frac{350}{250}\right) =$	47.9	
	$\lambda_{ey} =$ ... AS 4100 Table 6.2.4 ...=	82	
	$d_{e1} = d_o\sqrt{\left(\frac{\lambda_{ey}}{\lambda_e}\right)} \leqslant d_o$ ... AS 4100 Clause 6.2.4		
	$= 219.1\sqrt{\left(\frac{82}{47.9}\right)} \leqslant 219.1$		
	$= 287 \leqslant 219.1 =$	219.1	mm
	$d_{e2} = d_o\left(\frac{3\,\lambda_{ey}}{\lambda_e}\right)^2$ ... AS 4100 Clause 6.2.4		
	$= 219.1\left(\frac{3 \times 82}{47.9}\right)^2 =$	5780	mm
	$d_e = \min(d_{e1}, d_{e2})$ = effective outside diameter		
	$= \min(219.1, 5780) =$	219.1	mm
	d_i = internal diameter		
	$= d_e - 2t$		
	$= 219.1 - 2 \times 6.4 =$	206.3	mm
	Effective area $A_e = \frac{\pi d_e^2}{4} - \frac{\pi d_i^2}{4}$		
	$= \frac{\pi \times 219.1^2}{4} - \frac{\pi \times 206.3^2}{4} =$	4280	mm^2

which obviously equals the gross cross-section area, A_g

	$k_f = \frac{A_e}{A_g}$		
	$= \frac{4280}{4280} =$	1.0	
	Area of holes A_h = ... no holes in section ... =	0	mm^2
	Net area $A_n = A_e - A_h$		
	$= 4280 - 0 =$	4280	mm^2
3	Nominal section capacity N_s ... AS 4100 Clause 6.2 ...		
	$N_s = k_f A_n f_y$		
	$= \frac{1.0 \times 4280 \times 350}{10^3} =$	1500	kN
4	Effective length of the member l_e		
	Member is effectively pinned at both ends:		
	$l_e = k_e l$ ... Figure 6.3 or AS 4100 Figure 4.6.3.2 ...		
	$= 1.0 \times 3800$	3800	mm
	$\frac{l_e}{r} = \frac{3800}{75.2} =$	50.5	
	Modified slenderness λ_n		
	$\lambda_n = \frac{l_e}{r}\sqrt{\left(\frac{k_f f_y}{250}\right)}$ ... AS 4100 Clause 6.3.3 ...		
	$= 50.5 \times \sqrt{\left(\frac{1.0 \times 350}{250}\right)} =$	59.8	
5	Member section constant α_b for shape of section ...residual stress		
	Cold-formed (non-stress relieved) CHS category, and $k_f = 1.0$: AS 4100 Table 6.3.3(1):		
	$\alpha_b =$	-0.5	
6	Axial member capacity ϕN_c		
	From Table 6.6 or AS 4100 Table 6.3.3(3) interpolate to get slenderness reduction factor α_c		
	$\alpha_c =$	0.863	
	$N_c = \alpha_c N_s$		
	$= 0.863 \times 1500 =$	1290	kN
	$N_c \leq N_s$ = true as $1290 \leq 1500$ = true	OK	
	$\phi N_c = 0.9 \times 1290 =$	1160	kN
	$N^* \leq \phi N_c$ = true as $1030 \leq 1160$ = true	OK	
	219.1 × 6.4 CHS Grade C350L0 is satisfactory for *member* axial capacity	**Answer (@ 33.6 kg/m)**	
	The same result would be obtained if the above compression member is a diagonal in a truss with its ends welded to a gusset and connected concentrically. The effective length would then be equal to the geometric length of the diagonal member.		

6.9.2 Example 6.2

Step	Description and calculations	Result	Unit
	Using Example 6.1, check the adequacy of a Square Hollow Section (SHS) being used instead of the previously selected CHS. (This example shows the calculation method to be used for sections in which the effective area does not equal the gross cross-section area for compression member calculations).		
	Data: As noted in Example 6.1 –		
	Design axial compression load, $N^* =$	1030	kN
	Member length, $l =$	3800	mm
1	Select a trial section		
	200 x 200 x 5.0 SHS Grade C450L0 (OneSteel [2012b] Table 3.1-6(1)) page 3-15 with properties –		
	$b = d =$	200	mm
	$t =$	5.0	mm
	$A_g =$	3810	mm^2
	$r = r_x =$	79.1	mm
	$f_y =$	450	MPa
2	Section slenderness, λ_e, effective area, A_e, form factor, k_f and net area, A_n		
	For uniform compression ***all*** four cross-section elements are the same and their element slenderness, λ_e, can be expressed as		
	$\lambda_e = \frac{b_{cw}}{t}\sqrt{\left(\frac{f_y}{250}\right)}$ ... AS 4100 Clause 6.2.3 ...		
	where, b_{cw} = clear width $= b - 2t =$	190	mm
	$= \frac{190}{5} \times \sqrt{\left(\frac{450}{250}\right)} =$	51.0	
	Effective width, b_e, for each compression element		
	$b_e = b_{cw}\left(\frac{\lambda_{ey}}{\lambda_e}\right) \leqslant b_{cw}$ ... AS 4100 Clause 6.2.4 with flat ...		
	... element both long edges, CF residual stress classification...		
	$= 190 \times \left(\frac{40}{51.0}\right) =$	149	mm
	clearly the elements are not fully effective as $b_e < b_{cw}$		
	Effective area, A_e, is then		
	$A_e = 4b_et = 4 \times (149 \times 5.0) =$	2980	mm^2
	Form factor, k_f		
	$k_f = \frac{A_e}{A_g} \approx \frac{A_e}{4b_{cw}t} = \frac{2980}{4 \times 190 \times 5.0} =$	0.784	
	Area of holes, A_h		
	$A_h =$ ... no holes in section ...	0	mm^2
	Net area of cross-section, A_n		
	$A_n = A_g - A_h = 3810 - 0 =$	3810	mm^2
3	Nominal section capacity, N_s .... AS 4100 Clause 6.2		
	$N_s = k_fA_nf_y = 0.784 \times 3810 \times 450/10^3 =$	1340	kN

4	Effective length of the compression member, l_e		
	l_e = ... as noted in Example 6.1 ...	3800	mm
	$\frac{l_e}{r} = \frac{3800}{79.1} =$	48.0	
	Modified slenderness, λ_n ... AS 4100 Clause 6.3.3 ...		
	$\lambda_n = \frac{l_e}{r}\sqrt{\left(\frac{k_f f_y}{250}\right)} = 48.0 \times \sqrt{\left(\frac{0.784 \times 450}{250}\right)} =$	57.0	
5	Member section constant, α_b , ... From AS 4100 Table 6.3.3(2) ... with Cold-formed (non-stress relieved) RHS category and $k_f < 1.0$		
	α_b =	-0.5	
6	Axial member capacity, ϕN_c		
	From Table 6.6 or AS 4100 Table 6.3.3(3) interpolate to get the slenderness reduction factor, α_c		
	α_c =	0.876	
	$N_c = \alpha_c N_s = 0.876 \times 1340 =$	1170	kN
	$N_c \leqslant N_s$ = true as 1170 ⩽ 1340 = true	OK	
	$\phi N_c = 0.9 \times 1170 =$	1050	kN
	$N^* \leqslant \phi N_c$ = true as 1030 ⩽ 1050 = true	OK	
	200 × 200 × 5.0 SHS Grade C450L0 is also satisfactory for *member* axial design capacity.	**Answer (@ 29.9 kg/m)**	

6.9.3 Example 6.3

Step	Description and calculations	Result	Unit
	Verify the capacity of the beam-column shown below. Use a UC section in Grade 300 steel. The column is fixed at the base and braced as shown. The top of the column is pinned and laterally restrained by braces. Beams B1 and B2 are connected to the column using simple construction to AS 4100 Clause 4.3.4.		
	Beam reactions *R* act at eccentricities *e* to the column. The values of *R* and *e* are shown in the data below.		

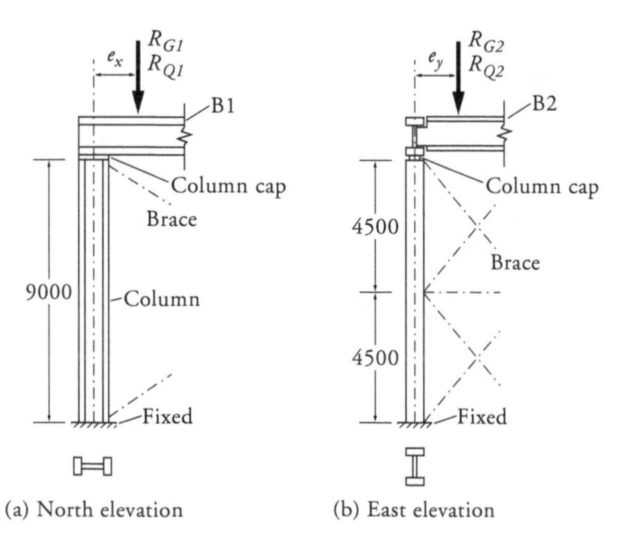

(a) North elevation (b) East elevation

Step	Description	Value	Units
	Data		
	Loads due to beam reactions R:		
	Beam B1: Permanent action/dead load R_{G1} =	260	kN
	Imposed action/live load R_{Q1} =	190	kN
	Beam B2: Permanent action/dead load R_{G2} =	78.1	kN
	Imposed action/live load R_{Q2} =	60.0	kN
	Column self-weight (say):	8.50	kN
	Minimum eccentricities from AS 4100 Clause 4.3.4...		
	... more details are given in Step 2:		
	e_x =	200	mm
	e_y =	80	mm
1	Trial section 250UC89.5 Grade 300 properties from ASI [2009a] Table 3.1-4(A) and (B) pages 3-12 and 13:		
	$A_n = A_g$ (No holes)	11 400	mm^2
	Z_{ex} =	1230×10^3	mm^3
	Z_{ey} =	567×10^3	mm^3
	I_x =	143×10^6	mm^4
	I_y =	48.4×10^6	mm^4
	r_x =	112	mm
	r_y =	65.2	mm
	f_{yf} =	280	MPa
	f_{yw} =	320	MPa
	k_f =	1.0	
	Section slenderness type for bending (about both principal axis)	Compact	
	Section is doubly symmetric I-section		
2	Eccentricity e of beam reactions:		
	Note minimum eccentricities AS 4100 Clause 4.3.4.		
	Column cap extends 200 mm from column centre line in north elevation, thus:		
	e_x =	200	mm
	Column cap extends 80 mm from column centre line in east elevation, thus:		
	e_y =	80	mm
3	Design axial load N^* and moments M^* (AS/NZS 1170.0 Clause 4.2.2(b)):		
	$N^* = 1.2 \times (R_{G1} + R_{G2} + 8.50) + 1.5 \times (R_{Q1} + R_{Q2})$		
	$= 1.2 \times (260 + 78.1 + 8.50) + 1.5 \times (190 + 60) =$	791	kN
	Unamplified design bending moments M_m^* due to eccentricity:		
	(a) Top of column:		
	$M_{mxt}^* = (1.2 \times 260 + 1.5 \times 190) \times 0.200 =$	119	kNm
	$M_{myt}^* = (1.2 \times 78.1 + 1.5 \times 60) \times 0.080 =$	14.7	kNm
	(b) Other column moments:		
	The following is assumed for initial design purposes –		
	i) beam-column x-axis bending/buckling		
	– undergoes double curvature bending		
	– at mid-height *lateral* restraint, assume a contraflexure point, then M_{xbr}^* =	0	kNm
	– at fixed base, $M_{xb}^* = 0.5M_{mxt}^* = 0.5 \times 119 =$	59.5	kNm
	ii) beam-column y-axis bending/buckling		
	– undergoes single curvature bending from *either end* to the *mid-height brace*		
	– at mid-height bracing, $M_{ybr}^* = 0.5M_{myt}^* = 0.5 \times 14.7 =$	7.35	kNm
	– at fixed base, $M_{yb}^* = 0.25M_{myt}^* = 0.25 \times 14.7 =$	3.68	kNm

If required, this can be refined in subsequent analysis after the preliminary section sizes are determined.

4 Member elastic buckling loads N_{omb} from AS 4100 Clause 4.6.2 and Figure 4.6.3.2:

k_{ex} (fixed base, pin top) = 0.85

k_{ey} (free to rotate at the two ends of upper half of beam-column) = 1.00

k_{ey} for the beam-column bottom half is not considered as it has reduced design moments and a smaller effective length (due to the fixed base).

For a braced frame:

$$N_{ombx} = \frac{\pi^2 EI_x}{(k_{ex} l_x)^2}$$

$$= \frac{\pi^2 \times 2 \times 10^5 \times 143 \times 10^6 \times 0.001}{(0.85 \times 9000)^2} = \quad 4820 \text{ kN}$$

$$N_{omby} = \frac{\pi^2 EI_y}{(k_{ey} l_y)^2}$$

$$= \frac{\pi^2 \times 2 \times 10^5 \times 48.4 \times 10^6 \times 0.001}{(1.0 \times 4500)^2} = \quad 4720 \text{ kN}$$

5 Moment amplification factors δ for a braced member/column to AS 4100 Clause 4.4.2.2:

$\beta_{mx} = \frac{59.5}{119}$ = ... one beam-column segment from base to tip ... = 0.500 (double curvature)

$\beta_{my} = \frac{-7.35}{14.7}$ = ... for beam-column segment above mid-height brace ... = −0.500 (single curvature)

$c_{mx} = 0.6 - 0.4\,\beta_{mx} = 0.6 - (0.4 \times 0.5) =$ 0.400

$c_{my} = 0.6 - 0.4\,\beta_{my} = 0.6 - [0.4 \times (-0.5)] =$ 0.800

$$\delta_{bx} = \frac{c_{mx}}{\left(1 - \frac{N^*}{N_{ombx}}\right)}$$

$$= \frac{0.400}{\left(1 - \frac{791}{4820}\right)} = \quad 0.479$$

$$\delta_{by} = \frac{c_{my}}{\left(1 - \frac{N^*}{N_{omby}}\right)}$$

$$= \frac{0.800}{\left(1 - \frac{791}{4720}\right)} = \quad 0.961$$

As the amplification factors δ_b are less than 1.0, use $\delta_b = 1.0$, giving $M^* = \delta_b M^*_m = 1.0 \times M^*_m = M^*_m$ in Step 3

(a) Top of column:

$M^*_{xt} = M^*_{mxt} =$ 119 kNm

$M^*_{yt} = M^*_{myt} =$ 14.7 kNm

(b) Other column moments:

$M^*_{xbr} =$ 0.0 kNm

$M^*_{xb} =$ 59.5 kNm

$M^*_{ybr} =$ 7.35 kNm

$M^*_{yb} =$ 3.68 kNm

(Note: *br* = at brace; *b* = at base)

(c) Maximum design moments:

$M_x^* = M_{mx}^* = \max.(119, 59.5)$ — 119 kNm

$M_y^* = M_{my}^* = \max.(14.7, 7.35, 3.68)$ — 14.7 kNm

For bending/buckling about the *y*-axis this is applicable to the segment above the beam-column mid-height brace.

---

6 *Section* capacities of 250UC89.5 Grade 300:

6.1 Axial *section* capacity N_s to AS 4100 Clause 6.2.1:

$N_s = k_f A_n f_y$

$= 1.0 \times 11400 \times 280 \times 0.001 =$ — 3190 kN

$\phi N_s = 0.9 \times 3190 =$ — 2870 kN

6.2 *Section* moment capacities M_s to AS 4100 Clause 5.2.1:

$M_{sx} = f_y Z_{ex}$

$= \dfrac{280 \times 1230 \times 10^3}{10^6} =$ — 344 kNm

$\phi M_{sx} = 0.9 \times 344 =$ — 310 kNm

$M_{sy} = f_y Z_{ey}$

$= \dfrac{280 \times 567 \times 10^3}{10^6} =$ — 159 kNm

$\phi M_{sy} = 0.9 \times 159 =$ — 143 kNm

6.2.1 This step is required to calculate the intermediate values of nominal section moment capacities reduced by axial force, M_i (see AS 4100 Clauses 8.3.2 and 8.3.3) prior to Step 6.3.2 for the biaxial bending check. It also illustrates the use of the combined actions *section* capacity check for the case of axial load with only uniaxial bending. The following is restricted to doubly symmetrical sections which are compact. If the beam-column section does not satisfy this criteria then 1.0 is used instead of 1.18 and 1.19 for M_{rx} and M_{ry} respectively below and the $N^*/\phi N_s$ term in the M_{ry} equation is not squared.

Reduced *section* moment capacities for compact, doubly symmetric I-sections:

In the presence of axial compression, and $k_f = 1.0$, can use AS 4100 Clause 8.3.2(a), in which:

$$M_{rx} = 1.18\, M_{sx} \left[1 - \frac{N^*}{(\phi N_s)}\right] \leq M_{sx}$$

$= 1.18 \times 344 \times \left[1 - \dfrac{791}{2870}\right] =$ — 294 kNm

$M_{rx} \leq M_{sx}$ is satisfied as $294 \leq 344$ is true — OK

If there was only uniaxial bending about the major principal *x*-axis with axial compression, then $M^* \leq \phi M_{rx}$ need only be satisfied for the combined actions *section* capacity check.

Section is doubly symmetrical I, and compact. Then can use AS 4100 Clause 8.3.3(a), in which:

$$M_{ry} = 1.19\, M_{sy} \left[1 - \frac{N^{*2}}{(\phi N_s)^2}\right] \leq M_{sy}$$

$= 1.19 \times 159 \times \left[1 - \dfrac{791^2}{2870^2}\right] =$ — 175 kNm

$M_{ry} \leq M_{sy}$ is not satisfied as $175 \leq 159$ is false — No good

M_{ry} is reduced down to ... = — 159 kNm

If there was only uniaxial bending about the minor principal y-axis with axial compression, then $M^* \leqslant \phi M_{ry}$ need only be satisfied for the combined actions *section* capacity check.

6.3 Biaxial *section* capacity in AS 4100 Clause 8.3.4 allows for:

6.3.1 *Any* type of section:

$$\frac{N^*}{\phi N_s} + \frac{M_x^*}{\phi M_{sx}} + \frac{M_y^*}{\phi M_{sy}} \leqslant 1$$

$$\frac{791}{2870} + \frac{119}{310} + \frac{14.7}{143} \leqslant 1$$

$0.762 \leqslant 1$ = true, satisfied — OK

Biaxial section capacity with conservative check gives 76% "capacity" usage. The following: Step 6.3.2 is more economical if applicable/used.

6.3.2 *Compact doubly symmetrical* I-section only:

$$\gamma = 1.4 + \frac{N^*}{\phi N_s} \leqslant 2.0$$

$$= 1.4 + \frac{791}{2870} \leqslant 2.0 =$$ 1.68

$$\left[\frac{M_x^*}{(\phi M_{rx})}\right]^{\gamma} + \left[\frac{M_y^*}{(\phi M_{ry})}\right]^{\gamma} \leqslant 1$$ from AS 4100 Clause 8.3.4

$$\left[\frac{119}{(0.9 \times 294)}\right]^{1.68} + \left[\frac{14.7}{(0.9 \times 159)}\right]^{1.68} =$$ 0.283

As $0.283 \leqslant 1$ = true, section OK for *biaxial section* capacity. **Answer1**

Biaxial section capacity with higher tier check gives 28% use of "capacity" compared with 76% in Step 6.3.1.

Above is *not* applicable if non-compact/asymmetrical I-section, but can also be used for compact RHS/SHS to AS/NZS 1163:

---

7 *Member* capacities of 250UC89.5 Grade 300 to AS 4100 Clauses 6.3.2 and 6.3.3:

7.1 Axial *member* capacity N_c :

$l_{ex} = 0.85 \times 9000 = \ldots$ 0.85 is taken from Step 4 ... = 7650 mm

$l_{ey} = 1.0 \times 4500 = \ldots$ 1.0 is taken from Step 4 ... = 4500 mm

$$\frac{l_{ex}}{r_x} = \frac{7650}{112} =$$ 68.3

$$\frac{l_{ey}}{r_y} = \frac{4500}{65.2} =$$ 69.0

Modified slenderness:

$$\lambda_{nx} = \frac{l_{ex}}{r_x}\sqrt{\frac{k_f f_y}{250}}$$

$$= 68.3 \times \sqrt{\frac{1.0 \times 280}{250}} =$$ 72.3

$$\lambda_{ny} = \frac{l_{ey}}{r_y}\sqrt{\frac{k_f f_y}{250}}$$

$$= 69.0 \times \sqrt{\frac{1.0 \times 280}{250}} =$$ 73.0

UB categorised HR, and $k_f = 1.0$ to AS 4100 Table 6.3.3(1):

α_b = member section constant = 0

Table 6.6 or AS 4100 Table 6.3.3(3) for slenderness reduction factor α_c:

α_{cx} = 0.733

	$\alpha_{cy} =$	0.728	
	N_{cx} and N_{cy} are $< N_s$ as α_{cx} and α_{cy} are < 1.0		
	$N_{cx} = \alpha_{cx} N_s$		
	$= 0.733 \times 3190 = \ldots N_s$ from Step 6.1 ... =	2340	kN
	$N_{cy} = \alpha_{cy} N_s$		
	$= 0.728 \times 3190 =$	2320	kN
7.2	In-plane *member* moment capacity, ϕM_i , to AS 4100 Clause 8.4.2.2:		

7.2.1 This is sometimes a confusing part of AS 4100 and may require some further explanation. The fundamental premise of combined action interaction checks is to reduce the relevant moment capacity due to the presence of an axial load working against the column capacity. Hence, for a particular beam-column segment, the three (3) nominal moment capacities that require reduction when an axial compressive load (N^*) is present, are:

(a) *section* moment capacity about the major principal axis, M_{sx}

(b) *section* moment capacity about the minor principal axis, M_{sy}

(c) *member* moment capacity for bending about the *x*-axis, M_{bx} (there cannot be the equivalent of this for bending about the *y*-axis).

The behaviour of **(a)** and **(b)** is mainly constrained to their respective ***in-plane*** interaction effects as loading and deformations occur in the same plane. Therefore, for combined actions, M_{sx} is reduced by N^* and the in-plane column member capacity N_{cx}. This also applies to M_{sy} which is reduced by N^* and N_{cy}.

However, **(c)** has loading effects and (buckling) deformations in mutually orthogonal planes – i.e. **out-of-plane.** From a member interaction perspective, M_{bx} must then be reduced by N^* and N_{cy} as the buckling deformations from M_{bx} and N_{cy} are in the same plane. The out-of-plane interaction check is considered in Step 7.3.

7.2.2 M_{ix} : AS 4100 Clause 8.4.2.2 with *x*-axis as the principal axis gives –

$$M_{ix} = M_{sx}\left[1 - \frac{N^*}{\phi N_{cx}}\right]$$

$= 344 \times \left[1 - \dfrac{791}{0.9 \times 2340}\right] =$	215	kNm
$\phi M_{ix} = 0.9 \times 215 =$	194	kNm
$M_x^* \leqslant \phi M_{ix}$ is true as $119 \leqslant 194$ is true. $M_x^* = 119$ is in the summary at the end of Step 5.	OK	

7.2.3 M_{iy} : AS 4100 Clause 8.4.2.2 with *y*-axis as the principal axis gives –

As for Step 7.2.2 but replacing *x* with *y*.

$$M_{iy} = M_{sy}\left[1 - \frac{N^*}{\phi N_{cy}}\right]$$

$= 159 \times \left[1 - \dfrac{791}{0.9 \times 2320}\right] =$	98.8	kNm
$\phi M_{iy} = 0.9 \times 98.8 =$	88.9	kNm
$M_y^* \leqslant \phi M_{iy}$ is true as $14.7 \leq 88.9$ is true. $M_y^* = 14.7$ is from Step 5.	OK	

7.3 Out-of-plane *member* moment capacity, ϕM_{ox}, to AS 4100 Clause 8.4.4.1:

7.3.1 As noted in Step 7.2.1, M_{bx} must be reduced by N^* and N_{cy} .

7.3.2 Effective length, l_e:

Both column flanges are connected to wind bracing at mid-height of the column.

l_e is the effective length of the beam part in a beam-column member. See AS 4100 Clause 5.6.3. For the top (more critical) part of the beam-column with FF restraint condition (see (b) East elevation):

$l_e = k_t\, k_l\, k_r\, l_s = \ldots$ with no transverse loads ... =	4.5	m

7.3.3 Moment modification factor, α_m:

For the beam-column bending about the *x*-axis, there are two segments subject to out-of plane flexural-torsional buckling and *y*-axis column buckling. These segments are above and below the mid-height brace point with the upper segment having the larger moment and subject to further investigation.

From Step 5, the ratio of the smaller to the larger bending moment, β_m (positive if in double curvature) for the upper segment beam-column is:

$\beta_m = \beta_{mx} = \dfrac{M^*_{xbr}}{M^*_{xt}} = \dfrac{0.0}{119} =$ 0.0

From Load Case 1.2 of Table 5.5 (or AS 4100 Table 5.6.1 Case1):

$\alpha_m = 1.75 + 1.05\beta_m + 0.3\beta_m^2$

$= 1.75 + 0.0 + 0.0 =$ 1.75

7.3.4 *Member* moment capacity about the (strong) *x*-axis, ϕM_{bx}, decreases with increasing flexural-torsional buckling. Use $l_e = 4.5$ m from Step 7.3.2 to read the value of ϕM_{bx} directly as follows:

ϕM_{bx1} = *member* moment capacity with $\alpha_m = 1.0$ from ASI [2009a] Table 5.3-6 page 5-52 for a 250UC89.5 Grade 300:

ϕM_{bx1} = ... top segment of beam-column ... = 255 kNm

ϕM_{bx} = moment capacity with $\alpha_m = 1.75$ is:

$\phi M_{bx} = \alpha_m \phi M_{bx1} \leqslant \phi M_{sx}$ ($\phi M_{sx} = 310$ kNm from Step 6.2)

$= 1.75 \times 255 =$ 446 kNm

$\phi M_{bx} \leqslant \phi M_{sx}$ = false as $446 \leqslant 310$ is false $\rightarrow \phi M_{bx} =$ 310 kNm

$M^*_x \leqslant \phi M_{bx}$ is satisfied because $119 \leqslant 310$ is true OK

$M_{bx} = \dfrac{\phi M_{bx}}{\phi}$

$= \dfrac{310}{0.9} =$ 344 kNm

M_{ox} = out-of-plane moment *member* capacity from AS 4100 Clause 8.4.4.1:

$M_{ox} = M_{bx}\left[1 - \dfrac{N^*}{\phi N_{cy}}\right]$

$= 344 \times \left[1 - \dfrac{791}{0.9 \times 2320}\right] =$ 214 kNm

7.4 Design moment *member* capacity about *x*-axis, ϕM_{cx} is given by:

M_{cx} = lesser of moments M_{ix} and M_{ox}

$= \min(M_{ix}, M_{ox})$

$= \min(215, 214) =$ ... with 215 from step 7.2.2 ... = 214 kNm

$\phi M_{cx} = 0.9 \times 214 =$ 193 kNm

7.5 Biaxial bending *member* moment capacity:

Step 5 gives $M^*_x =$ 119 kNm

$M^*_y =$ 14.7 kNm

Step 7.2.3 gives $\phi M_{iy} =$ 88.9 kNm

AS 4100 Clause 8.4.5.1 requires compliance with interaction inequality:

$$\left(\frac{M^*_x}{\phi M_{cx}}\right)^{1.4} + \left(\frac{M^*_y}{\phi M_{iy}}\right)^{1.4} \leqslant 1$$

$$\left[\frac{119}{193}\right]^{1.4} + \left[\frac{14.7}{88.9}\right]^{1.4} = 0.589 \leqslant 1$$

$0.589 \leqslant 1$ = true, satisfied OK

	250UC89.5 Grade 300 is adequate for *member* moment capacities in biaxial bending. Only 59% of member "capacity" is used.	**Answer2**	

6.9.4 Example 6.4

Step	Description and calculations	Result	Unit
	Verify the capacity of the beam-column shown below. The section is a 150UC30.0. Use Grade 300 steel. The rigid jointed frame is unbraced in the plane of the main frame action (north elevation). In-plane bending moments M_x^* have been determined by second-order analysis. Out-of-plane action is of the 'simple construction' type, and bracing is provided at right angles to the plane of the main frame. The main frame beams are 310UB40.4 at spacing of 7000 centre-to-centre of columns.		

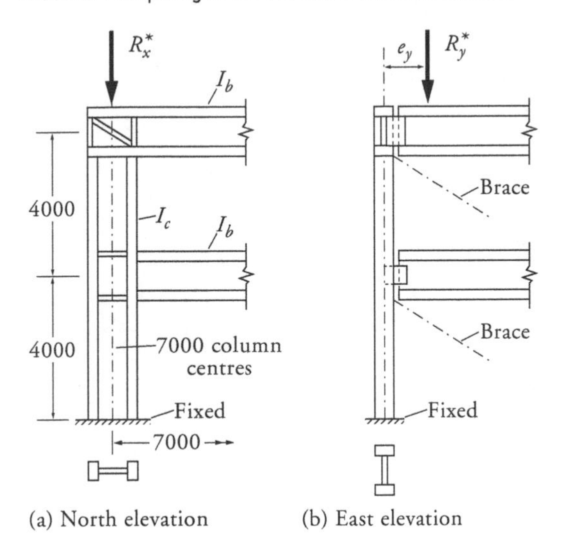

(a) North elevation (b) East elevation

Note: R_x^* is the vertical load exerted by the top beam onto the beam-column in the North elevation.

Step	Description and calculations	Result	Unit
	Data		
	R_x^* =	69.0	kN
	R_y^* =	47.0	kN
	Simple construction ... AS 4100 Clause 4.3.4 ...		
	e_y =	80	mm
	Column self-weight & other contributing dead loads =	8.10	kN
	In-plane frame bending moments on the beam-column from a second-order analysis (which requires no further amplification)		
	$M_{x\_top}^*$ =	35.0	kNm
	$M_{x\_mid}^*$ = ... at mid-height beam ... =	3.50	kNm
	Design axial load, N^*		
	N^* = 69.0 + 47.0 + 8.10 =	124	kN
	NOTE: This worked example will focus on the upper beam-column segment as it is more critically loaded.		

Step	Description	Result	Unit
1	Properties of trial section using 150UC30.0 Grade 300:		
	ASI [2009a] Tables 3.1.4(A) and (B) pages 3–12 and 13:		
	Geometry of section = doubly symmetrical I =	DSI	
	Slenderness of section about x-axis =	Compact x	
	Slenderness of section about y-axis =	Compact y	
	k_f =	1.0	
	$A_n = A_g$ ... No holes ... =	3860	mm^2
	Z_{ex} =	250×10^3	mm^3
	Z_{ey} =	110×10^3	mm^3
	I_x =	17.6×10^6	mm^4
	I_y =	5.62×10^6	mm^4
	r_x =	67.5	mm
	r_y =	38.1	mm
	f_{yf} =	320	MPa
	f_{yw} =	320	MPa
	Beam 310UB40.4 (from previous page of ASI [2009a] to that above)		
	I_x =	86.4×10^6	mm^4

2 Effective length of column l_e:

2.1 Effective length of column l_{ex} in plane of unbraced frame (northern elevation) AS 4100 Clause 4.6.3.4:

$$\gamma_1 = \frac{\Sigma\left(\frac{I}{l}\right)_c}{\Sigma\left(\frac{\beta_e I}{l}\right)_b}$$

$$= \frac{\left(\frac{17.6}{4.0} + \frac{17.6}{4.0}\right)}{\left(\frac{1.0 \times 86.4}{7.0}\right)} = \text{ ... at the joint with the middle beam ... } \quad 0.713$$

$$\gamma_2 = \frac{\left(\frac{17.6}{4.0}\right)}{\left(\frac{1.0 \times 86.4}{7.0}\right)} = \text{ ... at the joint with the top beam ... } = \quad 0.356$$

where β_e and the other terms are explained in Sections 4.5.3 and 4.7

k_{ex} = chart: AS 4100 Figure 4.6.3.3(b) for sway members = 1.17

$l_{ex} = k_{ex} l$

$= 1.17 \times 4000 =$ 4680 mm

2.2 Effective length of braced column l_{ey} in plane perpendicular to frame. Beam to column connection uses simple construction to accord with AS 4100 Figure 4.6.3.2:

$l_{ey} = k_{ey}\, l$

$= 1.0 \times 4000 =$ 4000 mm

3 Member elastic buckling loads N_{om}:

$$N_{omx} = \frac{\pi^2 E I_x}{l_{ex}^2}$$

$$= \frac{\pi^2 \times 2 \times 10^5 \times 17.6 \times 10^6 \times 0.001}{4680^2} = \quad 1590 \text{ kN}$$

$$N_{omby} = \frac{\pi^2 EI_y}{{l_{ey}}^2}$$

$$= \frac{\pi^2 \times 2 \times 10^5 \times 5.62 \times 10^6 \times 0.001}{4000^2} = \quad 693 \text{ kN}$$

---

4 Second-order effects on bending moments on beam-column

4.1 In-plane of frame (north elevation)

No further assessment required as results are from a second-order analysis – hence use moments noted in Data.

4.2 Out-of-plane to frame (east elevation)

4.2.1 First-order elastic analysis

It is assumed the beam-column undergoes double-curvature bending with the following –

$M^*_{y\_top}$ = moment eccentricity from top of column beam reaction

$= 47.0 \times 0.080 =$ 3.76 kNm

$M^*_{y\_mid} =$ 1.0 kNm

4.2.2 Moment amplification for a braced member

The top segment is considered to be in single curvature with no transverse loads. From AS 4100 Clause 4.4.2.2:

β_{my} = ratio of the smaller to larger bending moment (negative for single curvature)

$= \frac{-1.0}{3.76} =$ −0.266

$c_{my} = 0.6 - 0.4\beta_{my} = 0.6 - 0.4 \times (-0.266) =$ 0.706

N_{omby} = Elastic flexural buckling load about y-axis

$= \pi^2 EI_y / l_{ey}^2$ … from Step 3 … = 693 kN

$$\delta_b = \frac{c_m}{\left(1 - \frac{N^*}{N_{omby}}\right)} = \frac{0.706}{\left(1 - \frac{124}{693}\right)} = \quad 0.860$$

$=$ … as $\delta_b \leqslant 1.0$ then $\delta_b =$ 1.0

Hence, no moment amplification required and the first-order analysis results are sufficient.

4.2.3 Maximum design moments:

$M^*_x = M^*_{mx} = \text{max.}(35.0, 3.50)$ 35.0 kNm

$M^*_y = M^*_{my} = \text{max.}(3.76, 1.0)$ 3.76 kNm

---

5 *Section* capacities:

5.1 Axial *section* capacity N_s:

$N_s = k_f A_n f_y$

$= 1.0 \times 3860 \times 320 \times 0.001 =$ 1240 kN

5.2 *Section* moment capacities M_s in *absence* of axial load:

$M_{sx} = f_y Z_{ex}$

$$= \frac{320 \times 250 \times 10^3}{10^6} \quad 80.0 \text{ kNm}$$

$M_{sy} = f_y Z_{ey}$

$$= \frac{320 \times 110 \times 10^3}{10^6} \quad 35.2 \text{ kNm}$$

6	Combined actions *section* capacity		
	Reduced moment *section* capacity M_r, due to *presence* of axial load N^*. Top segment/half of column is critical as it has largest moment M^*.		
6.1	Reduced *section* moment capacity about *x*-axis, M_{rx}:		
	Section is DSI, compact *x* and $k_f = 1.0$ Therefore can use: AS 4100 Clause 8.3.2(a)		
	$M_{rx} = 1.18\, M_{sx}\left(1 - \frac{N^*}{\phi N_s}\right)$		
	$= 1.18 \times 80.0 \times \left(1 - \frac{124}{0.9 \times 1240}\right) =$	83.9	kNm
	$M_{rx} \leqslant M_{sx}$ = false as $83.9 \leqslant 80.0$ = false $\rightarrow M_{rx} =$	80.0	kNm
6.2	Reduced *section* moment capacity about *y*-axis, M_{ry}:		
	Section is DSI, compact *y* and $k_f = 1.0$. Therefore, can use: AS 4100 Clause 8.3.3(a)		
	$M_{ry} = 1.19\, M_{sy}\left[1 - \left(\frac{N^*}{\phi N_s}\right)^2\right]$		
	$= 1.19 \times 35.2 \times \left[1 - \left(\frac{124}{0.9 \times 1240}\right)^2\right] =$	41.4	kNm
	$M_{ry} \leqslant M_{sy}$ = false as $41.4 \leqslant 35.2$ = false $\rightarrow M_{ry} =$	35.2	kNm
6.3	Biaxial bending *section* capacity for *compact doubly symmetrical* I-section:		
	$\gamma = 1.4 + \frac{N^*}{\phi N_s}$		
	$= 1.4 + \frac{124}{(0.9 \times 1240)}$		
	$= 1.51 \leqslant 2.0 \rightarrow$	1.51	
	$\left[\frac{M_x^*}{(\phi M_{rx})}\right]^{\gamma} + \left[\frac{M_y^*}{(\phi M_{ry})}\right]^{\gamma} \leqslant 1$ ...AS 4100 Clause 8.3.4		
	$\left[\frac{35.0}{(0.9 \times 80.0)}\right]^{1.51} + \left[\frac{3.76}{(0.9 \times 35.2)}\right]^{1.51} =$	0.377	
	As $0.377 \leqslant 1$ = true, section **OK for biaxial bending *section* capacity**	**Answer1**	
7	*Member* capacities:		
7.1	Axial *member* capacity:		
7.1.1	Slenderness ratio $\frac{l_e}{r}$		
	$\frac{l_{ex}}{r_x} = \frac{4680}{67.5} =$	69.3	
	$\frac{l_{ey}}{r_y} = \frac{4000}{38.1} =$	105	
	Modified slenderness:		
	$\lambda_{nx} = \frac{l_{ex}}{r_x}\sqrt{\frac{k_f f_y}{250}}$		
	$= 69.3 \times \sqrt{\frac{1.0 \times 320}{250}} =$	78.4	
	$\lambda_{ny} = \frac{l_{ey}}{r_y}\sqrt{\frac{k_f f_y}{250}}$		

	$= 105 \times \sqrt{\dfrac{1.0 \times 320}{250}} =$	119	
7.1.2	Slenderness reduction factor α_c: UB categorised HR, and $k_f = 1.0$: AS 4100 Table 6.3.3(1) gives: α_b = ... member section constant ... =	0	
	Table 6.6 or AS 4100 Table 6.3.3(3), interpolate to get member slenderness reduction factor α_c:		
	α_{cx} =	0.692	
	α_{cy} =	0.426	
7.1.3	Axial member capacity ϕN_c :		
	$N_{cx} = \alpha_{cx} N_s$		
	$= 0.692 \times 1240 =$	858	kN
	$\phi N_{cx} = 0.9 \times 858 =$	772	kN
	$N_{cy} = \alpha_{cy} N_s$		
	$= 0.426 \times 1240 =$	528	kN
	$\phi N_{cy} = 0.9 \times 528 =$	475	kN
	ϕN_{cx} and ϕN_{cy} are $< \phi N_s$ as α_{cx} and α_{cy} are < 1.0		
7.2	In-plane *member* moment capacity, ϕM_i:		
7.2.1	M_{ix} — see explanation in Step 7.2.1 and 7.2.2 of Example 6.3:		
	$M_{ix} = M_{sx}\left[1 - \dfrac{N^*}{(\phi N_{cx})}\right]$		
	$= 80.0 \times \left[1 - \dfrac{124}{772}\right] =$	67.2	kNm
	$\phi M_{ix} = 0.9 \times 67.2 =$	60.5	kNm
	$M_x^* \leqslant \phi M_{ix}$ in AS 4100 Clause 8.4.2.2:		
	As ... $35.0 \leqslant 60.5$ is true, then $M_x^* \leqslant \phi M_{ix}$ = true, satisfied	OK	
7.2.2	M_{iy} — see explanation in Step 7.2.1 and 7.2.3 of Example 6.3:		
	$M_{iy} = M_{sy}\left[1 - \dfrac{N^*}{(\phi N_{cy})}\right]$		
	$= 35.2 \times \left[1 - \dfrac{124}{475}\right] =$	26.0	kNm
	$\phi M_{iy} = 0.9 \times 26.0 =$	23.4	kNm
	As $3.76 \leqslant 23.4$ true, then $M_y^* \leqslant \phi M_{iy}$ = true, satisfied	OK	
7.3	Out-of-plane *member* moment capacity, ϕM_{ox} : See explanation in Step 7.2.1 and 7.3 of Example 6.3.		
7.3.1	Bending effective length of beam-column segment l_e: Both column flanges are connected to wind bracing at mid-height of the column.		
	l_e = effective length of beam-column segment: $= k_t\, k_l\, k_r\, l_s$		
	$= 1.0 \times 1.0 \times 1.0 \times 4.0 =$ ... with no transverse loads ...	4.0	m
7.3.2	From the Data section of this worked example, the ratio of the smaller to larger bending moment, β_m, (positive if in double curvature) for the upper segment beam-column is:		
	$\beta_m = \beta_{mx} = \dfrac{-3.5}{35.0} =$	−0.10	

From Load Case 1.1 & 1.2 of Table 5.5 (or AS 4100 Table 5.6.1 Case 1):

α_m = moment modification factor

$= 1.75 + 1.05\beta_m + 0.3\beta_m^2$ ... (for $-1 \leqslant \beta_m \leqslant 0.6$) ...

$= 1.75 + [1.05 \times (-0.10)] + [0.3 \times (-0.10)^2] =$ 1.65

7.3.3 Nominal member moment capacity about *x*-axis, M_{bx}, in *absence* of axial load.

For a 150UC30.0 Grade 300 with $\alpha_m = 1.0$ and $l_e = 4.0$m from:

ϕM_{bx1} = *member* moment capacity for $\alpha_m = 1.0$ using ASI [2009a] Table 5.3-6 page 5-52:

$\phi M_{bx1} =$ 46.4 kNm

$M_{bx1} = \frac{\phi M_{bx1}}{\phi} = \frac{46.4}{0.9} =$ 51.6 kNm

For a 150 UC30.0 Grade 300 with $\alpha_m = 1.65$ then:

$M_{bx} = \alpha_m M_{bx1}$

$= 1.65 \times 51.6 =$ 85.1 kNm

$M_{bx} \leqslant M_{sx}$ = false as $85.1 \leqslant 80.0$ = false → $M_{bx} =$ 80.0 kNm

7.3.4 Out-of-plane *member* moment capacity in *presence* of axial load, M_{ox}: AS 4100 Clause 8.4.4.1:

$$M_{ox} = M_{bx}\left[1 - \frac{N^*}{(\phi N_{cy})}\right]$$

$$= 80.0 \times \left[1 - \frac{124}{475}\right] =$$ 59.1 kNm

7.3.5 Critical member moment capacity about *x*-axis, M_{cx}

$M_{cx} = \min.[\,M_{ix}, M_{ox}\,]$ from AS 4100 Clause 8.4.5.1

$= \min.[67.2, 59.1] =$ ...... 67.2 from Step 7.2.1..... = 59.1 kNm

$\phi M_{cx} = 0.9 \times 59.1 =$ 53.2 kNm

7.4 Biaxial bending *member* moment capacity

From AS 4100 Clause 8.4.5.1:

$$\left[\frac{M_x^*}{\phi M_{cx}}\right]^{1.4} + \left[\frac{M_y^*}{\phi M_{iy}}\right]^{1.4} \leqslant 1$$

$$\left[\frac{35.0}{53.2}\right]^{1.4} + \left[\frac{3.76}{23.4}\right]^{1.4} = 0.634 \leqslant 1$$

$0.634 \leqslant 1$ = true ... satisfactory ... → OK

150UC30.0 Grade 300 is adequate for *member* moment capacity in biaxial bending. Only 63% of member "capacity" used. **Answer 2**

6.10 Further reading

- For additional worked examples see Chapter 6 and 8 of Bradford, et al. [1997].
- Rapid column/beam-column sizing with industry standard tables and steel sections can be found in ASI [2004, 2009a] and OneSteel [2012b].
- Compression bracing members should allow for the combined actions of compression load and self-weight. Woolcock, et al. [2011] provides some excellent guidance and design tables for this situation.
- For some authoritative texts on buckling see Bleich [1952], CRCJ [1971], Hancock [2007], Timoshenko [1941], Timoshenko & Gere [1961], Trahair [1993b] and Trahair & Bradford [1998] to name a few.
- For typical column/compression member connections also see ASI [2003, 2009b], Hogan [2011], Hogan & Munter [2007a–h], Hogan & Van der Kreek [2009a–e], Syam & Chapman [1996] and Trahair, et al. [1993c].

chapter 7

Tension Members

7.1 Types of tension members

Tension members are predominantly loaded in axial tension, although inevitably they are often loaded in combined tension and bending. The bending moments may arise from eccentricity of the connections, frame action and self-weight of the members. A simple classification of tension members is presented in Table 7.1. This table gives an overview of the many types of tension member applications in building construction; it also serves as a directory to subsections covering the particular design aspects.

Table 7.1 Classification of tension members

Aspect	Subgroup	Section
(a) Type of construction		
Section type:		
Rigid	I-sections, hollow sections	7.2–7.5
	Angles and channels	7.2–7.5
Flexible	Plates, bars	7.4
	Steel rods	7.6
	Steel wire ropes	7.7
Construction	Single section	7.2–7.5
	Compound sections	7.4.4
(b) By position of restraints		
	End restraints at connections	7.4–7.5
	End and intermediate restraints	7.4.4
(c) By type of loads		
	Axial tension only	7.4.1
	Combined tension and bending	7.4.2
(d) By load fluctuation		
	Predominantly static loads	7.3–7.5
	Dynamic loads	7.2, 7.5.4
	Impact	–

7.2 Types of construction

Tubular tension members are increasingly used as they offer high capacities coupled with relatively high bending stiffness. Additionally, tubular members have superior resistance to axial compression loads, which is important for axially loaded members subject to reversal of loads.

Flexible members such as rods and steel wire ropes are often used in structures exposed to view, where the architect may prefer the member to be of the smallest possible size. Where load reversal can occur, the flexible members should be used in a cross-over arrangement such that one member is in tension while the other is allowed to buckle under compression load, unless both members are pretensioned to a level where compression will not occur. Steel wire ropes are also used for guying purposes. The main advantage of high-strength bars and steel strand is that they exhibit superior tension capacity at a minimum weight.

Tubular and angle tension members are typically used for tensile web members, and tension chords in trusses, wall and roof bracing members, hangers, stays and eaves ties. Where the load on the member changes from tension to compression, the member should be checked for compression capacity as well as tension capacity. Where the tension members are slender ($l/r > 200$) and cross one another in the braced panel, it can be assumed that the compression member will buckle, with the result that the tension member must be designed to resist 100% of the applied panel shear force. Figure 7.1 illustrates several types of situations in which tension members are used.

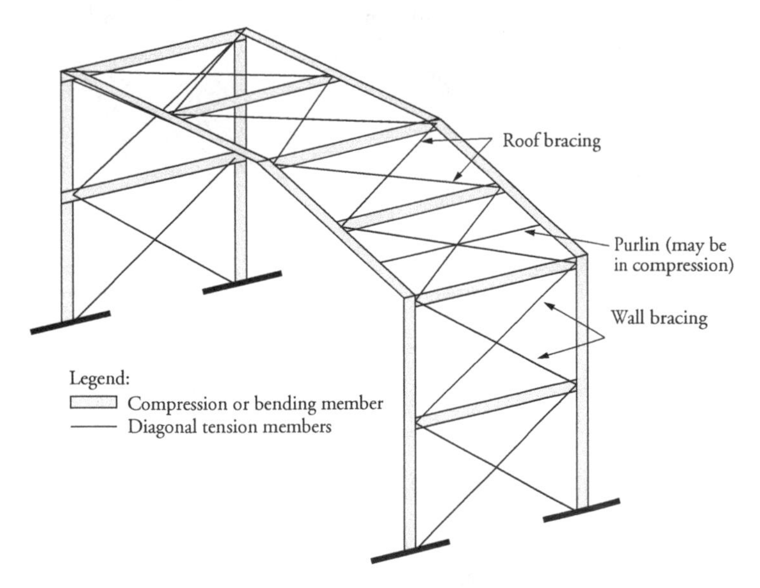

Figure 7.1 Typical tension member application

Tension members composed of two or more sections can be used where the tension member is relatively long. Compound members provide greater bending rigidity than single members. Therefore they show less sag (i.e. the 'take up' of tension load is not that rapid when there is excessive sag) and are less likely to vibrate under fluctuating load. They can resist quite large bending moments and can act in either tension or compression. Another benefit of compound members is that the eccentricity of connections can be eliminated altogether when they are tapered towards the ends. The disadvantage, however, is the increased cost of fabrication resulting from additional components comprising the compound ties: end battens, intermediate battens etc. Starred angles provide an economical solution with small eccentricities and need only a few short battens. Typical compound members are shown in Figure 7.2.

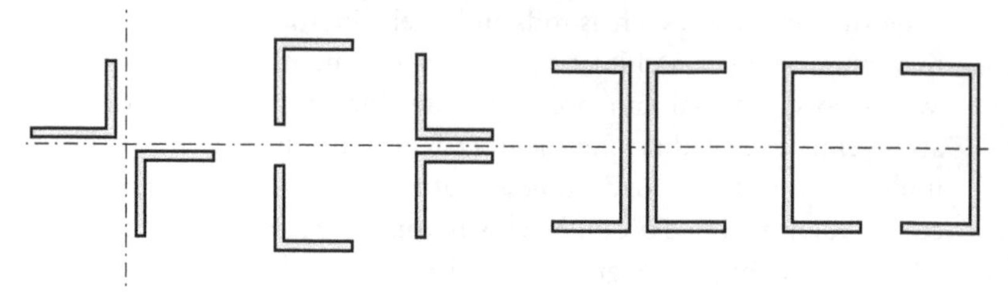

Figure 7.2 Typical compound section tension members

Bending restraint of the member end has an effect on the design of the member. The ends of the tension members can be rigidly, flexibly or pin constrained. Rigid restraint in the plane of the truss or frame is obtained by welding or by bolting, either directly to the framing members or to a substantial gusset (12 mm or thicker). Flexible end restraint is more common and occurs when the tension member is connected to the framing by means of a relatively thin gusset (less than 10 mm thick). The pin-type connection can provide bending restraint only in the plane of the pin. (Refer also to Section 8.10.2.) Figure 7.3 shows typical end connections for tensile members.

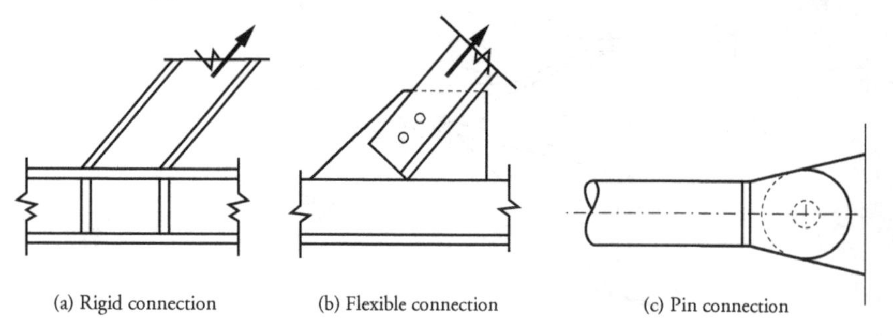

Figure 7.3 Typical end connections for tension members

7.3 Evaluation of load effects

Tensile forces and bending moments acting on a tensile member are determined by analysis, either by simple manual calculations or by computer frame analysis. Depending on the bending rigidity and the function of the tension member, the following situations may arise:

- Flexurally rigid tension members carrying large applied transverse loads or resisting significant bending moments because of frame action behave as beams subject to bending with axial tension force.
- Semi-flexible members with no applied transverse load behave as members subject to tension with some secondary bending.
- Flexible members (e.g. rods and cables) behave purely as ties.

Consequently, different design procedures need to be applied to various types of tension members. Typical examples of flexurally rigid members are rigid bracing systems in heavy frames. These members should be designed as flexural members carrying axial forces. Where load reversal can occur, the compression capacity usually governs.

Some examples of semi-flexible members are tubular and light, hot-rolled sections used in trusses and braces for wall and roof bracing systems. These members typically have l/r ratios of between 100 and 300. With these members, tension load predominates but bending due to self-weight may be significant (Woolcock & Kitipornchai [1985] and Woolcock et al. [2011]). There is a need to assess the effects of end connection eccentricity where the member is bolted or welded by a lap-type connection, or when the connection centroid is offset to the tie centroid, or when not all the tie cross-section elements are connected. For this case, AS 4100 provides a simplified method of assessment of the member capacity (see Section 7.4).

Flexible tension members, such as guy cables, rod braces, bow girder ties and hangers, act predominantly in pure tension. With longer members it is necessary to check the amount of sag of the cable and the longer-term resistance to fatigue.

7.4 Verification of member capacity

7.4.1 Tension capacity

7.4.1.1 General

The nominal axial capacity of a member loaded in tension, N_t, is calculated in accordance with Clause 7.2 of AS 4100 as follows:

$$N_t = 0.85 k_t A_n f_u, \text{ or}$$

$$N_t = A_g f_y, \text{ whichever is the lesser.}$$

The section is adequate in tension if:

$$N^* \leqslant \phi N_t$$

where

N^* = design axial tension force

A_g = gross area of cross-section.

A_n = the net cross-sectional area: $A_n = A_g - A_d + A_a \leqslant A_g$

A_d = the area lost by holes

A_a = the allowance for staggered holes, for each side step on ply thickness t (see Clause 9.1.10.3 of AS 4100):

$$= 0.25 \frac{s_p^{\,2} t}{s_g}$$

f_y = the yield stress used in design

s_p = staggered pitch, the distance measured parallel to the direction of the force (see Clause 9.1.10.3(b) of AS 4100)

s_g = gauge, perpendicular to the force, between centre-to-centre of holes in consecutive lines (see Clause 9.1.10.3(b) of AS 4100)

f_u = the tensile strength used in design

ϕ = the capacity reduction factor
= 0.9 for sections other than threaded rods
= 0.8 for threaded rods (assumed to behave like a bolt)
= 0.3–0.4 recommended for cables (not given in AS 4100)

k_t = a correction factor for end eccentricity and distribution of forces (see Table 7.3.2 of AS 4100 or Table 7.2 in this Handbook).

The factor k_t is taken as 1.0 if the member connection is designed for uniform force distribution across the end section, which is achieved when:

- each element of the member section is connected
- there is no eccentricity (i.e. the connection centroid coincides with the member centroid)
- each connected element is capable of transferring its part of the force.

As long as the ratio of $\frac{A_n}{A_g}$ is larger than $\frac{f_y}{(0.85k_t f_u)}$, the failure can be expected to be ductile gross yielding; otherwise, failure will occur by fracture across the weakest section. The factor of 0.85 provides additional safety against the latter event.

7.4.1.2 Members designed as 'pinned' at the ends

The rigorous computation of bending moment due to eccentricity caused by connecting only some elements of the section can be avoided by the use of the correction factor k_t (see Bennetts et al. [1986]). A value of k_t less than 1.0 applies where some elements of the section are not effectively connected, or where minor connection eccentricity exists, so that a non-uniform stress distribution is induced (see Figure 7.4 and Table 7.2). The method is convenient for designing building bracing systems and truss web members.

Table 7.2 The correction factor for distribution of forces in tension members, k_t

Configuration	k_t	Note
One-sided connection to:		
Single angle	0.75	Unequal angle—connected by short leg
	0.85	Otherwise
Twin angles on same side of gusset/plate	0.75	Unequal angle—connected by short leg
	0.85	Otherwise
Channel	0.85	Connected by web only
Tee (from UB/UC)	0.90	Flange connected only
	0.75	Web connected only (suggested value)
Back-to-back connection:		
Twin angles	1.0	On opposite sides of gusset/plate
Twin channels	1.0	Web connected on opposite sides of gusset/plate
Twin tees	1.0	Flange connected on opposite sides of gusset/plate
Connections to flanges only		
UB, UC, WB, WC and PFC	0.85	See note below

Note: The length of the flange-to-gusset connection is to be greater than the depth of the section. See Clause 7.3.2 of AS 4100 for further information on the above.

If a bending moment is applied to the member because of gross eccentricity or frame bending action, the member should be designed for combined tension and bending (see Section 7.4.2).

Particular care should be given to avoiding feature or details that may give rise to brittle fracture, especially when the member has to operate at relatively low ambient temperatures. Notches can be introduced by poor thermal cutting practices, micro-cracking around punched holes, welding defects or damage during handling and erection.

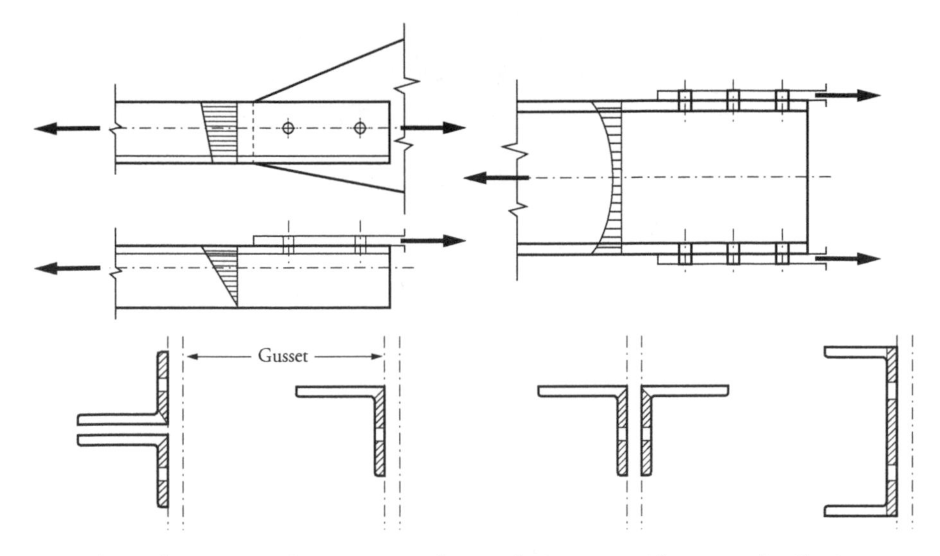

Figure 7.4 End connections for tension members producing non-uniform stress distribution

7.4.2 Combined actions – tension and bending

This section covers the situations where member axial tension forces, N^*, are present with bending moments from frame action (end moments), applied transverse forces or gross connection eccentricity. The design provisions for these combined actions follow the same method and terminology for beam-columns (Chapter 6) where, except for out-of-plane checks, the section/member moment capacity is reduced by the presence of axial load.

(a) *Section* capacity check
   Clause 8.3 of AS 4100 notes either of the following should be satisfied –
   (i) Uniaxial bending about the major principal (x-) axis with tension:

$$M_x^* \leqslant \phi M_{rx}$$

   where

$$M_{rx} = M_{sx}\left(1 - \frac{N^*}{\phi N_t}\right) \quad \text{or}$$

   for a higher tier assessment of compact, doubly symmetric I-sections, RHS and SHS

$$M_{rx} = 1.18 M_{sx}\left(1 - \left[\frac{N^*}{\phi N_t}\right]\right) \leqslant M_{sx}$$

   (ii) Uniaxial bending about the minor principal (y-)axis with tension:

$$M_y^* \leqslant \phi M_{ry}$$

where

$$M_{ry} = M_{sy}\left(1 - \frac{N^*}{\phi N_t}\right) \qquad \text{or}$$

for a higher tier assessment of compact, doubly symmetric I-sections, RHS and SHS

$$M_{ry} = H_T M_{sy}\left(1 - \left[\frac{N^*}{\phi N_t}\right]^n\right) \leq M_{sy}$$

where $H_T = 1.19$ & $n = 2$ for I-sections and $H_T = 1.18$ & $n = 1$ for RHS/SHS.

(iii) Biaxial bending with tension:

$$\frac{N^*}{\phi N_t} + \frac{M_x^*}{\phi M_{sx}} + \frac{M_y^*}{\phi M_{sy}} \leq 1.0 \qquad \text{or}$$

for a higher tier assessment of compact, doubly symmetric I-sections, RHS and SHS

$$\left(\frac{M_x^*}{\phi M_{rx}}\right)^\gamma + \left(\frac{M_y^*}{\phi M_{ry}}\right)^\gamma \leq 1.0$$

where

$$\gamma = 1.4 + \left(\frac{N^*}{\phi N_t}\right) \leq 2.0$$

and M_{rx} and M_{ry} are evaluated as noted as above.

(b) *Member* capacity check

For elastic design, Clause 8.4 of AS 4100 notes the following should be satisfied—

(i) In-plane capacity—for uniaxial bending with tension: use where relevant either (a)(i) or (a)(ii) above. For biaxial bending with tension: use (a)(iii) above.

(ii) Out-of-plane capacity – for bending about the major principal (x-)axis with tension which may buckle laterally:

$$M_x^* \leq \phi M_{ox}$$

where

$$M_{ox} = M_{bx}\left(1 + \frac{N^*}{\phi N_t}\right) \leq M_{rx}$$

where M_{rx} is noted in (a)(i) above and M_{bx} is the nominal member moment capacity for the member when subjected to bending (see Section 5.3). It is interesting to note that axial tension has a beneficial effect on the member when under x-axis bending and designing for flexural-torsional buckling. This is seen by the addition of the $N^*/\phi N_t$ term in the above equation for M_{ox}—as opposed to a subtraction which is noted for compression loadings with bending (see Step 11 of Section 6.3.2). Also, as noted in Chapter 6, there are no out-of-plane capacities to be checked for minor principal (y-)axis bending with axial tension.

(iii) Out-of-plane capacity – for biaxial bending with tension

$$\left(\frac{M_x^*}{\phi M_{tx}}\right)^{1.4} + \left(\frac{M_y^*}{\phi M_{ry}}\right)^{1.4} \leq 1.0 \qquad \text{where}$$

M_{tx} is the lesser of M_{rx} (see (a)(i) above) and M_{ox} (see (b)(ii) above), and M_{ry} is noted in (a)(ii) above.

(c) Quick combined action checks: For bracing systems and ties, tension members with combined actions generally have major principal (x-)axis bending moments. Based on this a conservative check for all section types is to use the lower tier provisions of (a)(i) and (b)(ii) above. Where noted, the higher tier provisions may be used for compact, doubly symmetric I-sections and RHS/SHS which can have significant design capacity increases. For other types of bending (e.g. y-axis or biaxial) the relevant parts are used from above. Hence, it is seen that not all the above provisions are used for each situation – perhaps only less than a third of them.

7.4.3 Tension and shear

Tensile capacity is reduced in the presence of shear across the section. Based on extensive tests on bolts, the interaction curve for tension and shear is an elliptical function, and the verification of capacity can be carried out by using the following interaction formula:

$$\left(\frac{N_t^*}{0.8N_t}\right)^{2.0} + \left(\frac{V^*}{0.8V_v}\right)^{2.0} \leqslant 1.0$$

It should be noted that the above interaction inequality was developed specifically for bolts which are unique in terms of their method of concentrated loading (with respect to the overall element) and are exposed to different boundary conditions than that encountered by typical structural elements (e.g. confinement effects from the bolt hole onto the bolt, etc.).

Unlike bolts, there appears to be no significant work or design provisions on the interaction effects of tension and shear on structural members. However, if required for design purposes, and in lieu of any other advice, the above inequality may be used as a starting point for considering this type of loading interaction. If used, then a rough rule of thumb is if the shear force is less than 60% of the member shear capacity then no interaction need be considered with tension capacity (and vice versa). It should be noted that situations where the shear force is greater than 60% of the shear capacity are rare.

7.4.4 Compound members

Tension members can be composed of two or more sections where it is necessary to increase their lateral stiffness as, for example, when members are alternately loaded in tension and compression. Clause 7.4 of AS 4100 specifies the minimum requirements for battens and lacing plates. The effective steel area of a compound tension member equals the sum of section areas provided that battening complies with the provisions of AS 4100, as illustrated in Figure 7.5.

7.5 End connection fasteners and detailing

7.5.1 General

Wherever possible, end connections should be so designed that the centroidal axes of the member and gusset coincide with one another and every part of the member section is connected. It is not always practical to eliminate eccentricity at the connections but it is

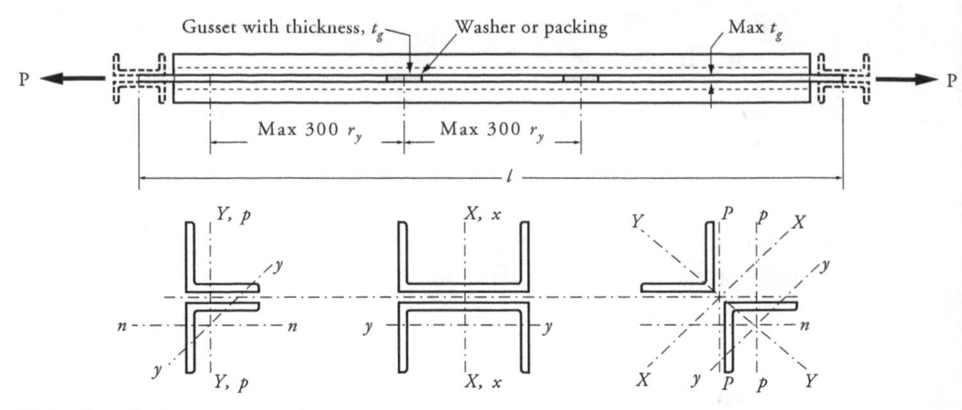

(a) Back-to-back tension members

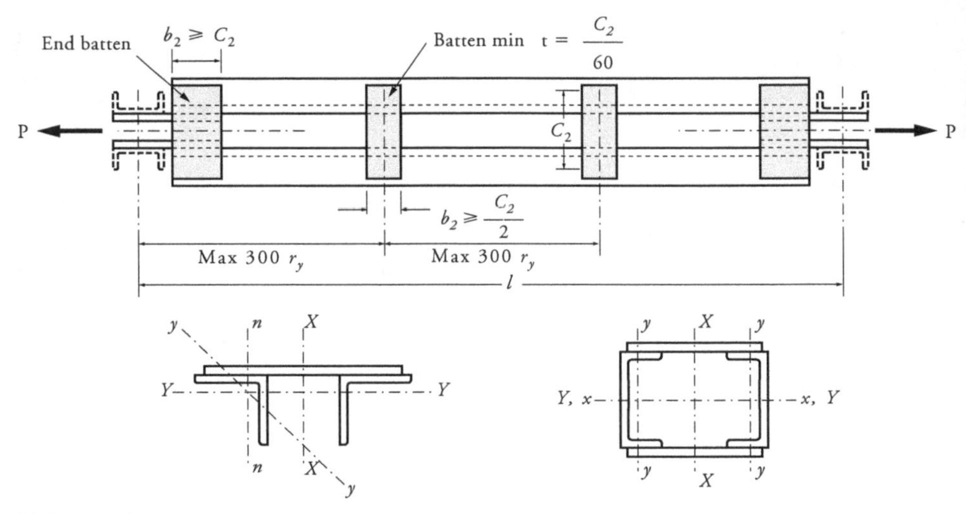

(b) Battened tension members

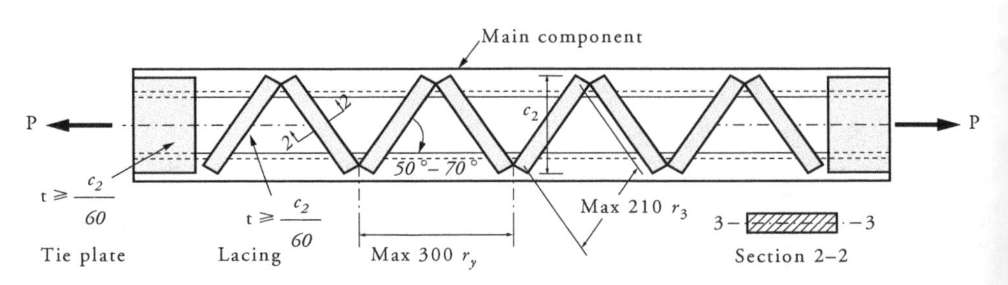

(c) Laced tension members (cross sections are the same as for battened tension members)

Figure 7.5 *Compound tension members (Note: x & y are local principal axes for the main component, n & p are local non-principal (i.e. "rectangular") axes for the main component, and; X & Y are global principal axes for the compound member).*

essential that the effects of eccentricity be taken into account in the design. A typical example is a truss node where heavy (or large) truss members are used with centroidal axes of web members not meeting at a common node. In such a case the bending moment resulting from the offset should be distributed to all members meeting at a node, in proportion to their stiffness. However, as noted in Section 6.4, there are many situations where this can be either neglected or readily considered in design e.g. for tubular and light to medium trusses.

7.5.2 Simple detailing method

Single- and double-angle members are often connected through one leg of the angle, web of a channel or flanges only of the UB section or a channel. A method of evaluation of members with such connections is outlined in Section 7.4.1. The actual connection design (bolting, welding, etc.) is covered in Chapter 8. Typical connection details of this type are illustrated in Figure 7.4.

7.5.3 Detailing of rigid and semi-rigid connections

Connections of this type can be made either direct to the main member or to a centrally situated gusset. In either case it is good practice to avoid significant eccentricities. The important point in detailing the connection is that the axial force and any bending moments are transferred in such a manner that the capacity of each element of the connection is maintained.

There are situations where it is not feasible (or economical) to totally eliminate the in-plane connection eccentricity, and in such cases the member should be checked for combined actions (see Chapter 5 and 6 of this Handbook and Section 8 of AS 4100).

Bolted connections are preferred by most steel erectors, and they should be used unless the number of bolts required becomes too large. Welded connections are used mostly in the fabrication of subassemblies such as trusses and bow girders. Their use in the erection is restricted to special cases, such as large connections where the number of bolts would be excessive and too costly.

7.5.4 Balanced detailing of fasteners

Wherever practicable, the centroid of the fastener group should be detailed to coincide with the centroid of the member. It is sometimes not practical to achieve this ideal without added complications, e.g. as in truss nodes. Some eccentricity of the fastener group with respect to the member centroid can be tolerated in statically loaded structures, and this is confirmed by tests carried out in the USA. A typical example is given in Figure 7.6. However, it must be stressed that unbalanced connections have been found to have an inferior performance in fatigue-loaded structures.

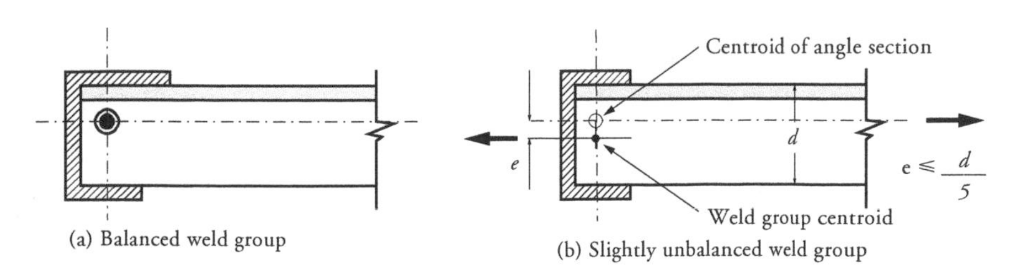

Figure 7.6 Welded connections

Splices in tension members should be checked for a minimum connection force, which should not be less than 30% of the member capacity. It is recommended that the splice detail and the fasteners be arranged to be balanced with respect to the tension member centroid so as to avoid introducing bending moments at the midpoint of the member. For design of connections and fasteners, see Chapter 8.

7.6 Steel rods

Steel rods are often used as tension members, particularly in lightweight steel structures. Typical uses are for wind bracing systems, underslung or bow girders, mast stays and hangers. Their main advantage is in compactness, especially when higher-strength rods are employed.

Threaded sockets usually terminate rod ends. A lower-cost end termination is by threading the ends of the rods through a bracket so a nut can be used to secure them in place. Threads made by a thread-cutting tool suffer from a drawback that the effective section area is reduced by 10%–14%. A better way is to form the threads by a roll-forming process, where the metal is merely deformed to produce the grooves so that the gross bar area can be used in capacity computations.

High-strength steel rods have the advantage of keeping the rod diameter as small as possible. The types often used are the VSL Stress Bar and MACALLOY 80 rods. They achieve steel strengths up to 610 MPa in smaller diameters. They are factory-threaded, using a rolled thread to avoid stress raisers. Couplers are used to form longer stays. Fitting the bar into the structure can be difficult without using turnbuckles or threaded forks.

The amount of sag in rods is a function of the span and the tension stress in the rod. Woolcock et al. [2011] suggest the following expression for the tensile stress in the rod under catenary action, f_{at}:

$$f_{at} = 9.62 \times 10^{-6} \left(\frac{l^2}{y_c} \right) \text{MPa}$$

where y_c is the mid-length deflection (sag) of the tie. Suggested upper limits for l/y_c are 100 for industrial buildings and 150 for institutional buildings.

High-strength rods are often used in lightweight structures where larger forces need to be transmitted with as small a size as possible.

A measure of initial tension is essential to keep the rods reasonably straight. Over-tightening in the field can be a problem for rods and connections, and a minimum sag should also be specified on the drawings. See Woolcock et al. [2011] for further information.

7.7 Steel wire ropes

Steel wire ropes or cables are used where relatively large tensions are being resisted. The individual wires in the cable are produced by repeated 'drawing' through special dies, such that after each draw the wire is reduced in diameter. This is repeated several times until the desired diameter is reached. The drawing process has the virtue of increasing the yield strength and the ultimate strength of the wire. The typical value of breaking tensile strength, f_u, is 1770 MPa for imported Bridon [undated] strands.

The following types of cables are used in steel construction:

- *Spiral strand:* Consists of wires laid out in the form of a spiral. Used for applications where relatively high values of Young's modulus of elasticity, *E*, is important.
- *Parallel strand:* Uses cables consisting of parallel wires (bridge design). Possesses relatively high *E* values.
- *Locked coil strand:* Similar to spiral strand, but with outer wires specially shaped to make a compact cable.
- *Structural steel rope:* Made of small wire strands wound spirally.

It should be borne in mind that the Young's modulus of elasticity of the cables can be considerably lower than for rolled-steel sections. Table 7.3 illustrates this.

Table 7.3 Young's modulus of elasticity, *E*, for various tendon types

Tendon type	Diameter mm	Young's modulus, *E* GPa	% of rolled of steel *E*
Steel rods:		200	100
Spiral strand cable:	≤ 30	175	87
	31 to 45	170	85
	45 to 65	165	83
	66 to 75	160	80
	> 76	155	78
Parallel wire strand:	all dia	195	98
Structural rope:	all dia	125	63

Note: Extract from the Bridon [undated] catalogue.

For breaking strength of cables, see Table 7.4.

Table 7.4 Properties of steel wire ropes for guying purposes

Strand dia, mm	Breaking strength, kN
13	171
16	254
19	356
22	455
25	610
30	864

Note: Cables up to 100 mm nominal diameter have been used.

While these capacities look quite impressive, it should be realised that larger 'safety factors' are necessary: that is, lower-capacity factors, ϕ, are used because of such unknowns as dynamic behaviour, fatigue, corrosion damage or rigging mistakes.

Thus capacity factors are in the range of 0.3 to 0.4, which is equivalent to saying that the utilisation factor is as low as 27%–36%, compared with rods having a utilisation factor of better than 90%, although associated with lower tensile strength.

The end terminations or sockets (forks) transfer the tensile action from the tendon to the framing gusset(s). The cables are secured in the sockets by molten zinc/lead alloy capable of withstanding 120% of the cable breaking load (see Figure 7.7).

Where adjustment of cable length is required, use is made of turnbuckles or rigging screws. These are proprietary items designed to overmatch the cable capacity. Typical end terminations are also shown in Figure 7.7.

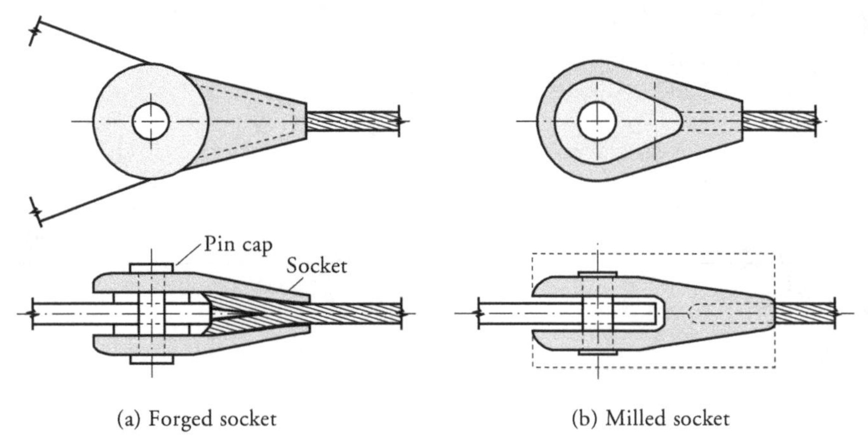

Figure 7.7 End terminations for rods and cables

Corrosion protection of the cables is an important consideration. Cable manufacturers use special procedures to provide uniform thickness of galvanizing. Often a special pliable coat of paint incorporating aluminium flakes or other proprietary system further protects the cables. It is important to inspect the coating at 3 to 5-year intervals and repair any damage as soon as possible (Lambert [1996]).

Angles and hollow section tension members develop some bending stresses as a consequence of the sag due to self-weight. Woolcock et al. [2011] have demonstrated that pretensioning would not be very effective for these sections. In the stricter sense the bending moment caused by self-weight does reduce the tensile (section) capacity of the tension member, even though the reduction may be only 10%. Hangers from the purlins or other stiff members at points along the span may be used to reduce the sag. Another method is to attach to the member a 'sag eliminator' catenary rod of relatively small diameter.

The magnitude of the mid-span bending moment for a member subject to tension and bending due to self-weight can be determined by the following approximate formula (Timoshenko [1941]):

$$M_m^* = \frac{0.125 w_s l^2}{(1 + 0.417 z^2)}$$

Where M_m^* is the mid-span bending moment of a pin-ended tensile member with length l, w_s is the UDL due to self-weight, $z^2 = \dfrac{N^* l^2}{4EI}$ and N^* the tension force. The value of the M_m^* moment will be slightly reduced by a member not connected by pins.

7.8 Examples

7.8.1 Example 7.1

Step	Description and calculations		Result	Unit
	Check the capacity of a 2 m long tension member used to suspend an overhead crane runway beam. It consists of a 250UC89.5 Grade 300 I-section with end connections by its flanges only as described in AS 4100 Clause 7.3.2(b). The vertical actions/loads N with their eccentricities e, and the horizontal action H are tabulated at the start of the calculations:			

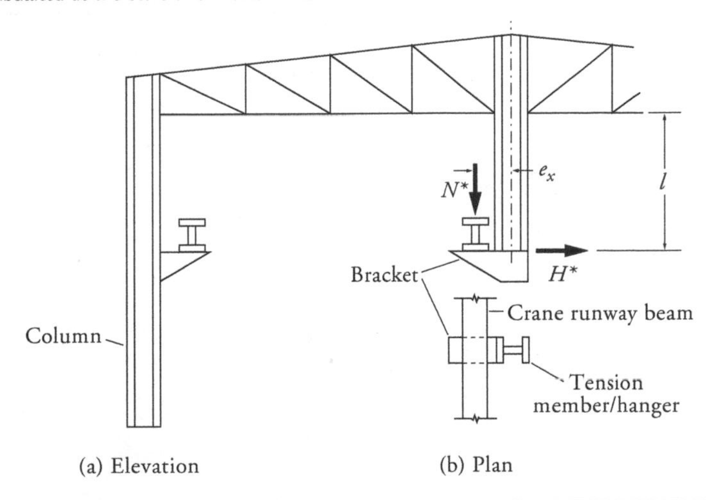

Step	Description and calculations		Result	Unit
1	Action/loads N, H			
	Permanent action/dead load	$N_G =$	99.0	kN
	Imposed action/live load	$N_Q =$	210	kN
	Imposed action/live load	$H_Q =$	18.0	kN
	Eccentricities			
	$e_x =$		280	mm
	$e_y =$		0	mm
	$l =$		2.0	m
2	Properties of 250UC89.5 Grade 300			
	ASI [2009a] Tables 3.1-4(A) and (B) pages 3-12 and 13			
	$A_n = A_g = \ldots$ no holes to deduct $\ldots =$		11 400	mm^2
	$Z_{ex} =$		1230×10^3	mm^3
	$Z_{ey} =$		567×10^3	mm^3
	$f_{yf} =$		280	MPa
	$f_{yw} =$		320	MPa
	Table 2.3 or ASI [2009a] Table T2.1 page 2-3 ...			
	$f_u =$		440	MPa

Step	Description	Result	Unit
3	Design action/load effects from runway reactions N^*, H^* and M_x^* ...		
	AS/NZS 1170.0, Clause 4.2.2(b) ...		
	Design axial tensile action/load N^*		
	$N^* = 1.2\,N_G + 1.5\,N_Q$		
	$= 1.2 \times 99.0 + 1.5 \times 210 =$	434	kN
	Design horizontal transverse action/load H^*		
	$H^* = 1.2\,H_G + 1.5\,H_Q$		
	$= 1.5 \times 18.0 =$	27.0	kN
	Design bending moment M_x^*		
	$M_x^* = H^*l + N^*e_x$		
	$M_x^* = 27.0 \times 2 + 434 \times 0.280 =$	176	kNm
4	Axial section capacity ϕN_t ...		
	AS 4100 Clause 7.2 ... tension failure: yield or fracture?		
4.1	Gross yielding N_{ty}		
	$N_{ty} = A_g f_y$		
	$= \dfrac{11400 \times 280}{1000} =$	3190	kN
4.2	Fracture N_{tf}		
	k_t = correction factor for distributon of forces = 0.85 (Table 7.2 or AS 4100 Clause 7.3.2(b))		
	$N_{tf} = 0.85\,k_t A_n f_u$		
	$= \dfrac{0.85 \times 0.85 \times 11400 \times 440}{1000} =$	3620	kN
4.3	$N_t = \min(N_{ty}, N_{tf})$		
	$= \min(3190, 3620) =$	3190	kN
4.4	$\phi N_t = 0.9 \times 3190 =$	2870	kN
	... check $N^* \leq \phi N_t \rightarrow 434 \leq 2870 \rightarrow$ true	OK	
5	***Section*** moment capacity ϕM_{sx} and ϕM_{rx}		
	AS 4100 Clause 5.2.1 gives ...		
5.1	$M_{sx} = f_{yf} Z_{ex}$		
	$= \dfrac{280 \times 1230 \times 10^3}{10^6} =$	344	kNm
	$\phi M_{sx} = 0.9 \times 344 =$	310	kNm
5.2	ϕM_{rx} = reduced ϕM_{sx} by presence of axial action/load N^*		
	... AS 4100 Clause 8.3.2 ... in symmetrical compact I- ...		
	$\phi M_{rx} = 1.18\,\phi M_{sx}\left(1 - \dfrac{N^*}{\phi N_t}\right)$		
	$= 1.18 \times 310 \times \left(1 - \dfrac{434}{2870}\right) =$	310 but ...	kNm
	... check $\phi M_{rx} \leq \phi M_{sx} \rightarrow 310 \leq 310 \rightarrow$ true, satisfied	OK	
	and $\phi M_{rx} =$	310	kNm
5.3	$M_x^* \leq \phi M_{rx}$ ...does this satisfy AS 4100 Clause 8.3.2? As $176 \leq 310 \rightarrow$ true, it does satisfy reduced ***section*** moment **capacity** ϕM_{rx} for combined axial tensile load and uniaxial bending about *x*-axis. Note $\phi M_{rx} = \phi M_{ix}$ *in-plane* moment capacity, which is different from out-of-plane moment capacity in Step 6.	**OK**	

6	***Member*** moment capacity ϕM_{bx} in which bending about x-axis undergoes flexural-torsional buckling about y-axis ... also called *out-of-plane* moment capacity ϕM_{ox} ... AS 4100 Clauses 5.6 and 8.4.4.2 ...		
6.1	On face value, one could assume that the member in question is a combination of a tension member and a cantilever subject to a tip moment and (horizontal) tip load. However, it would also be fair to assume that the crane runway beam via its support bracket actually supplies lateral (i.e. parallel to the runway beam) and torsional restraint to the cantilever tip. Based on the top of the tension member being connected to the roof truss it would be reasonable to further assume that even though the tension member is a cantilever in the plane of loading, it actually has flexural-torsional buckling restraints at both segment ends and can be designated as FF.		
	The effective length, l_e, calculation is straightforward (from AS 4100 Table 5.6.3) except that one may take $k_r = 0.85$ as the tip is rotationally restrained (about its y-axis) by the crane beam and its support bracket. However, the more conservative value of 1.0 will be used. Consequently,		
	l_e = "beam" effective length		
	$= k_t k_l k_r l = 1.0 \times 1.0 \times 1.0 \times 2.0 =$	2.0	m
	From Table 5.6.1 of AS 4100 or Table 5.5 here, the moment modification factor, α_m, is notionally between that of an end moment that produces a constant bending moment along the member length ($\alpha_m = 1.0$) and a transverse force at the end (with linearly varying moment along the length - i.e. $\alpha_m = 1.75$). A reasonably accurate value of α_m can be obtained from the parabolic approximation method of AS 4100 Clause 5.6.1.1(a)(iii) with the superposition of the two bending moment diagrams. However, initially try the more conservative approach of -		
	$\alpha_m =$	1.0	
6.2	Use ASI [2009a] Table 5.3-6 page 5-52 as an aid to get ϕM_{bx} of 250UC89.5 Grade 300 ... with $\alpha_m = 1.0$ and $l_e = 2.0$m ...		
	Then ϕM_{bx} = ... in absence of axial load N^* ... =	302	kNm
	and $M_{bx} = \dfrac{\phi M_{bx}}{\phi}$		
	$= \dfrac{302}{0.9} =$	336	kNm
6.3	AS 4100 Clause 8.4.4.2 gives M_{ox} in beneficial presence of axial tensile action/load N^* mitigating flexural-torsional buckling		
	$M_{ox} = M_{bx}\left[1 + \dfrac{N^*}{(\phi N_t)}\right] \leqslant M_{rx}$		
	$= 336 \times \left[1 + \dfrac{434}{2870}\right] =$	387	kNm
	$\phi M_{ox} = 0.9 \times 387$		
	= ... in presence of axial load N^* ... =	348	kNm
	... check $\phi M_{ox} \leqslant \phi M_{rx} \rightarrow 348 \leqslant 310 \rightarrow$ false, then $\phi M_{ox} =$	310	kNm
6.4	$M_x^* \leqslant \phi M_{ox}$ requirement in AS 4100 Clause 8.4.4.2 is		
	$176 \leqslant 310 \rightarrow$ true, satisfied	**OK**	
	Member out-of-plane moment **capacity** ϕM_{ox} in presence of axial load N^* is adequate. ***Member*** in-plane moment **capacity** ϕM_{ix} was satisfied in Step 5.3	**OK**	
6.5	Summary: Both section and member capacities are satisfied by 250UC89.5 Grade 300	**OK**	

7.8.2 Example 7.2

Step	Description and calculations	Result	Unit
	Select a section for a diagonal tension member of a truss. Verify its capacity. Use one equal angle in Grade 300 steel with one line of M20/S fasteners in one leg. Axial actions/loads are given at the start of the calculations.		

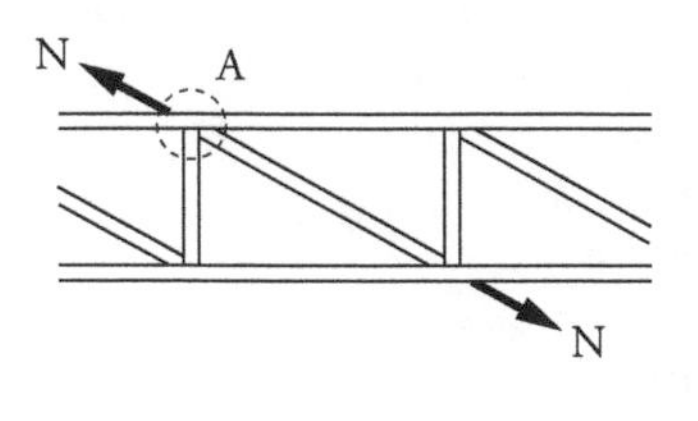

(a) Elevation

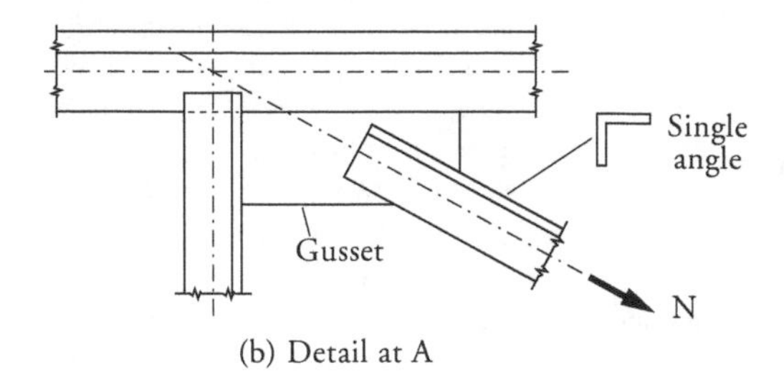

(b) Detail at A

Step	Description and calculations	Result	Unit
1	Axial action/loads, N		
	Permanent action/dead load $N_G =$	99.0	kN
	Imposed action/live load $N_Q =$	121	kN
2	Design action/load N^*		
	AS/NZS 1170.0, Clause 4.2.2(b) ...		
	$N^* = 1.2\,N_G + 1.5\,N_Q$		
	$= 1.2 \times 99.0 + 1.5 \times 121 =$	300	kN
3	Trial section 1 - 125×125×8 EA Grade 300 with one leg attached ... Properties		
	ASI [2009a] Tables 3.1-9(A)-1 and 3.1-9(B)-1 pages 3-20 and 21 ...		
	$A_g =$	1900	mm^2
	$A_n = 1900 - (20 + 2) \times 8$	1720	mm^2
	Table 2.3 or ASI [2009a] Table T2.1 page 2-3 ...		
	$f_y =$	320	MPa
	$f_u =$	440	MPa
4	*Section* capacity ϕN_t ... AS 4100 Clause 7.2 ... tension failure: gross yielding or fracture?		
4.1	Gross yielding N_{ty}		
	$N_{ty} = A_g\,f_y$		
	$= \dfrac{1900 \times 320}{1000} =$	608	kN
4.2	Fracture N_{tf}		
	$N_{tf} = 0.85\,k_t\,A_n\,f_u$		
	$= \dfrac{0.85 \times 0.85 \times 1720 \times 440}{1000} =$	547	kN
	where $k_t = 0.85$ is from Table 7.2 or AS 4100 Table 7.3.2 Case(i)		

4.3	$N_t = \min(N_{ty}, N_{tf})$		
	$= \min(608, 547) =$	547	kN
	$\phi N_t = 0.9 \times 547 =$	492	kN
4.4	$N^* \leqslant \phi N_t \rightarrow 300 \leqslant 492 \rightarrow$ true, satisfied	OK	
	1 - 125×125×8 EA Grade 300 with 1 line of M20/S is adequate	**Answer**	

Note: 1. A tension member in a truss under gravity actions/loads may sometimes sustain a reversal of action to compression when wind uplift occurs. If so, the member should also be checked as a strut (e.g. see Section 6.4).

7.9 Further reading

- For additional worked examples see Chapter 7 and 8 of Bradford, et al. [1997].
- Rapid tension member sizing with industry standard tables and steel sections can be found in ASI [2004, 2009a] and OneSteel [2012b].
- For typical tension member connections also see ASI [2003, 2009b], Hogan [2011], Hogan & Munter [2007a–h], Hogan & Van der Kreek [2009a–e] and Syam & Chapman [1996].
- For other references on tension bracing see Woolcock, et al. [2011], Woolcock & Kitipornchai [1985] and Kitipornchai & Woolcock [1985].

chapter 8

Connections

8.1 Connection and detail design

8.1.1 General

This chapter covers the design of elements whose function is to transfer forces from a member to the footings, from one member to another, or from non-structural building elements to the structure. Typical connections used in constructional steelwork are listed below:

- Shop connections and fixtures: direct connections between members; for example, in welded trusses, welded or bolted subframes, brackets and fixtures.
- Site connections and splices: column and beam splices, beam-to-beam and beam-to-column connections, column bases-to-footing connections.

Site connections are usually bolted for speed of erection, but in special circumstances welding is used. The reasons for the popularity of bolted site connections are:

- low sensitivity to unavoidable dimensional inaccuracies in fabrication, shop detailing or documentation

Table 8.1 Connections chapter contents

Subject	Subsection
Types of connections & design principles	8.1
Bolted connections	8.2 to 8.3
Bolt installation and tightening	8.2.5 to 8.2.7
Design of bolted connections	8.3
Connected plate elements	8.4
Welded connections	8.5 to 8.8
Types of welded joints	8.6
Design of a connection as a whole	8.9 to 8.10
Hollow section connections	8.10.1
Pin connections	8.10.2

- simplicity and speed of installation
- low demand on skills of workers
- relatively light and portable tools.

Design of bolted connections is covered in Sections 8.2 to 8.4 and design of welded connections in Sections 8.5 to 8.8 and connections as a whole in Sections 8.9 to 8.10. Pin connections are in Section 8.10.2. See Table 8.1 for the connection item index.

8.1.2 Design principles

While a relatively small percentage of the total steel mass is attributed to connections, in terms of cost they play a prominent part in the economy of steelwork (ASI [2009b], Watson et al [1996]). The making of connections and details is a labour-intensive process, because many large and small pieces have to be manufactured and fitted within specified tolerances. In designing details it is important to keep in mind the following principles:

- Design for strength:
  - direct force-transfer path;
  - avoidance of stress concentrations;
  - adequate capacity to transfer the forces involved.
- Design for fatigue resistance:
  - avoidance of notches;
  - careful design of welded joints.
- Design for serviceability:
  - avoidance of features that can cause collection of water;
  - ease of application of protective (and other) coatings;
  - absence of yielding under working load.
- Design for economy:
  - simplicity;
  - minimum number of elements in the connection;
  - reducing the number of members meeting at the connection.

8.1.3 Types of connections

The choice of the connection type must be related to the type of frame:

- Rigid framework: Figure 8.1(d) to (g)
  - rigidly constructed, welded connections;
  - non-slip bolted connections;
  - heavy bolted end plates;
  - assumed to have sufficient stiffness to maintain the original angles between members during loading.
- Simple framework: Figure 8.1(a) to (c)
  - flexible end plate connection;
  - slippage-permitted on web and flange cleats;
  - assumes the ends of the connected members do not develop bending moments.
- Semi-rigid framework:
  - connections designed for controlled rotational deflections and deformations;
  - behaves somewhere between rigid and simple connections.

The type of framing adopted is often governed by the optimum connection type for that particular project. For example, if the framework for an office building will not be exposed to view, then all-bolted field connections may be chosen purely due to the need for rapid erection; the frame will thus be of simple construction, possibly braced by the lift core. In contrast, if the steel framework of a building will be exposed to view, then the choice may be braced framework with pin-type connections or, alternatively, a rigid frame system with welded connections.

Connections can also be categorised by the fastener type used:

- Welded connections using:
  - butt welds;
  - fillet welds;
  - compound (butt-fillet) welds.
- Bolted connections using:
  - snug-tight bolting denoted by "/S";
  - controlled post-tensioned bolting:
    - working by bearing contact denoted by "/TB";
    - working by the friction grip principle denoted by "/TF".
- Pins and pinned connections.

Another useful way to categorise connections is by the construction stage:

- workshop connections: mostly welded;
- site connections: mostly bolted.

Various types of beam-column connections are shown in Figure 8.1.

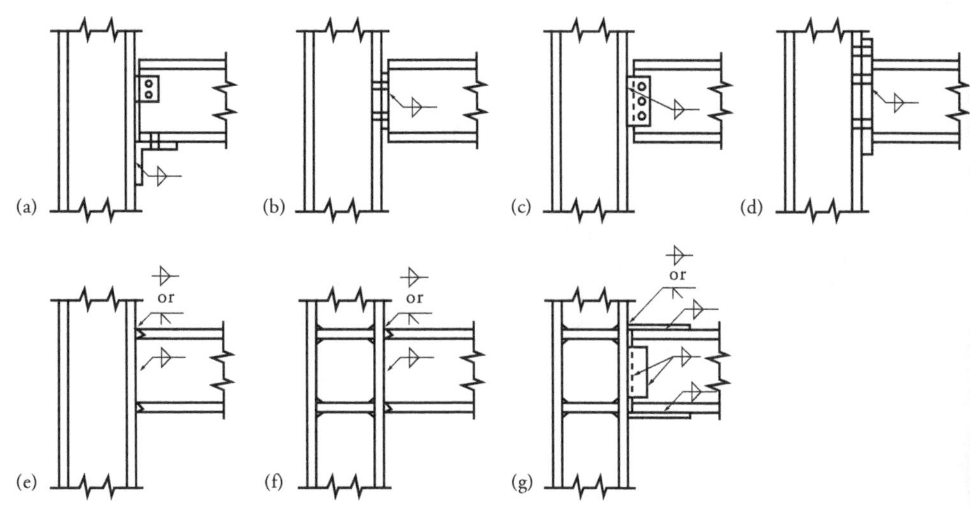

Figure 8.1 *Types of field connections (a) to (d) bolted; (e) to (g) welded on site (or in the shop if the beam is a stub piece).*

8.1.4 Minimum design actions

In recognition of the crucial importance of connection design, AS 4100 requires connections to be designed for forces at times larger than the member design actions. When the size of the member is larger than required for strength design, the connection design actions must not be less than the specified minimum. Table 8.2 summarises these code requirements.

Table 8.2 Minimum design actions on connections as specified in Clause 9.1.4 of AS 4100

Type and location	Minimum actions N^*_{min} or V^*_{min}	M^*_{min}	Notes
Tension members (excl. threaded rod bracing)			
Splices and end connections	$0.3\phi N_t$	0	N_t = nom. member tension capacity
Threaded rod bracing member with turnbuckles	ϕN_t		
Compression members			
End connections	$0.3\phi N_c$	0	N_c = nom. member compression capacity
Splices, at restrained points	$0.3\phi N_c$	0	
Splices, at unrestrained points	$0.3\phi N_c$	$0.001\ \delta\ N^* l_s$	δ = amplification factor δ_b or δ_s (Section 4.4 or Clause 4.4 of AS 4100) l_s = distance between points of lateral restraint
Fasteners in splices prepared for full bearing contact	$0.15\phi N_c$	0	
Flexural members			
Splices in flexural members	0	$0.3\ \phi M_b$	M_b = nom. member moment capacity
Ditto, but acting in shear only	V^*	M^*_v	V^* = design shear force M^*_v = moment due to V^* eccentricity on the connector group
Beam connections in			
Rigid construction	0	$0.50\ \phi M_b$	
Simple construction	min[$0.15\phi V_v$, 40 kN]	0	V_v = nominal shear capacity
Member subject to combined actions	simultaneously satisfy the above relevant requirements		

Note: ϕ is generally taken as 0.9 as per member design. There are separate design action effects and behaviour requirements for earthquake load combinations—see Section 13 of AS 4100 and AS 1170.4.

8.1.5 Block shear failure of components

As noted in Section 1.14(a), the amendment to AS 4100 (AS 4100 AMD 1—see Appendix D) stipulates that a phenomena called "Block Shear" failure be considered in design. Such a failure can occur when Connection "Components" (i.e. plate connection elements such as cleats, gusset plates, brackets, etc.) are subject to a design shear force and/or design tension force.

Unlike other national/international Standards, this failure mode had not been specifically noted in AS 4100 and in predecessor Standards, but was generally undertaken during connection design with separate connection design models (which are not listed in AS 4100). Prior to the release of AS 4100 AMD 1 (see Appendix D), it would be prudent to consider such a failure mode in connection design. Background information for this failure mode is noted in Hogan & Munter [2007a] and the application of such a failure mode is considered in Hogan & Munter [2007d-f]. See also Appendix E for further consideration and application of this failure mode in AS 4100.

8.2 Bolted connections

8.2.1 General

Bolted connections used in steel construction are of several types, described as follows:

- Simple framework (minimal rotational restraint to member ends):
  - flexible end plate connection;
  - web and flange cleat—slippage-permitted;
  - pin-type connection
- Semi-rigid framework (connections designed for controlled deformations)
- Rigid framework (maximum rotational restraint to the member end).

8.2.2 Definition of bolting terms

Bearing-type joint A bolted joint designed for maximum utilisation of HS structural bolts, which is achieved by allowing joint slip to take place and thus bring the bolt shank to bear on the walls of the holes; resistance to joint shear is derived by a combination of frictional resistance and bearing resistance (though design provisions only consider the latter type of resistance).

Commercial bolts Bolts of Property Class 4.6 steel, manufactured to medium tolerances.

Direct-tension indication (DTI) device Generally described as a flat washer with protrusions on one side of the washer. The protrusion side of the washer is placed under the bolt head and squashes under a pre-determined load. It is used as a primary means of installation for full tensioning of bolts. Over the last few years, "squirter" type DTI washers have been used in many projects. In this instance, when the correct bolt tension is reached, a calibrated amount of (typically) orange silicon is squirted from the DTI washer. These types of DTIs are considered to be compliant with the inspection of bolted connections provisions of AS 4100 (Clause 15.4.1(b))—e.g. see <www.hobson.com.au> for further information.

Edge distance The distance from the centre of a bolt hole to a free edge (or an adjacent hole).

Effective clamping force Net clamping force in a joint subject to tension parallel to the bolt axes; the applied tensile force has the effect of reducing the pre-compression of the joint plies.

Effective cross-section Area used in stress calculations being equal to the gross cross-sectional area less deductions for bolt or other holes and for excessive plate width (if applicable).

Faying surface Surfaces held in contact by bolts (the mating surfaces).

Friction grip bolts An obsolete term (see friction-type joint).

Friction-type joint A bolted joint using HS structural bolts designed so that the shear between the plies is lower than the safe frictional resistance of the pre-compressed mating surfaces, resulting in a non-slip joint. Attention to ply surface preparation around the bolt area is required for such joints.

Full tensioning A method of bolt tensioning capable of imparting to the bolt a minimum tension of at least 75% of the bolt proof stress. This permits a "stiffer" joint in terms of load-deflection behaviour. Two categories of bolting can be derived from such bolt installations—fully tensioned bearing (designated as "/TB") and fully tensioned friction (designated as "/TF").

Grip Total thickness of all connected plies of a joint.

Gross cross-sectional area Area computed from the cross-sectional dimensions without regard to any deductions.

High-strength interference body bolts These bolts, made of high-strength steel, are designed for a driving fit (interference fit).

High-strength structural bolts Bolts made of high-strength steel (typically Property Class 8.8) to commercial tolerances, and suitable for applications requiring high tightening torques.

HS High strength—as in high strength structural bolts.

Lapped connection Same as shear connection.

Lap-type moment connection Resistance against rotation due to moment acting in the plane of the mating surface and is achieved by bolts stressed in shear.

Load factor Safety factor against slip in friction-type joints.

Mild steel bolts An obsolete term (see Commercial bolts).

Minimum bolt tension Minimum tensile force induced in the bolt shank by initial tightening, approximately equal to the bolt proof load, see Full tensioning also.

Moment connection Flange-type or end plate moment connections are subjected to moments acting in a plane perpendicular to the mating surfaces.

Part-turn tightening Relies on the relationship between the bolt extension and the induced tension (see Section 9 of AS 4100). It is used as a primary means of installation for full tensioning of bolts.

Pin An unthreaded rod permitting the rotation of plies around the rod axis.

Pitch The distance between bolt centres along a line—also referred to as spacing.

Ply A single thickness of steel component (plate, flat bar, section flange/web, etc.) forming the joint.

Precision bolts Bolts, available in several grades of carbon steel, are manufactured to a high tolerance and are used mainly in mechanical engineering.

Proof load Bolt tension at proportional limit.

Prying force Additional forces on bolts subject to tension induced by the flexing of the plies.

Shear connection In this type of connection the forces are parallel to the mating surfaces, and the bolts are stressed in shear.

Slip factor Coefficient of friction between the mating surfaces/plies.

Snug-tight bolts Bolts tensioned sufficiently to bring into full contact the mating surfaces of the bolted parts. Designated in Australia and New Zealand as "/S".

Stress area Cross-sectional area of bolt used in verifying stresses in bolts subject to tension; it is approximately 10% larger than the core area (area at the root of thread).

Tension connection In this type of connection the mating surfaces are perpendicular to the direction of the applied tensile force, and the bolts are stressed in axial tension.

Tension and shear connection Applied tensile force is inclined to the mating surfaces, thus the bolts are subjected to combined shear and tension.

Torque-control tightening Method of tightening using either a hand-operated torque wrench or a power-operated tool; the method relies on the relationship involving friction between the bolt and nut threads and the applied torque. This is not a primary bolt installation method in AS 4100 (used for inspection only).

8.2.3 Types of bolts and installation category

Bolts used in steel construction are categorised as follows:

- Property Class 4.6 commercial bolts conforming to AS 1111
  These bolts are made of low carbon steel similar to Grade 250 steel and generally used in structural steel construction.
- Property Class 8.8, high-strength structural bolts conforming to AS/NZS 1252
  These bolts are made of medium carbon steel using quenching and tempering to achieve enhanced properties. Consequently they are sensitive to high heat input, for example welding.
- Property Class 8.8, 10.9 and 12.9 precision bolts
  These bolts are manufactured to very close tolerances, suitable for mechanical assembly. They can find their use as fitted bolts in structural engineering. Bolts marketed under the brand name 'Unbrako' and 'Huck' are in this category.

Higher-grade structural bolts, designation 10.9, are used in special circumstances where very large forces are transmitted and space is limited. One variety of these bolts incorporates a self-limiting initial tension feature (Huckbolt).

Property Class 10.9 and 12.9 bolts are susceptible to hydrogen pick-up, possibly leading to delayed brittle fracture. Hydrogen pick-up can occur from the pickling process used in galvanizing or from rust formation in service. Specialist advice should be sought before specifying these bolts.

Property Class 4.6 commercial bolts are suitable only for snug-tight installation designated as 4.6/S bolting category.

Property Class 8.8 structural bolts are capable of being highly tensioned. Their designation is 8.8/TB or 8.8/TF depending on whether they are used in bearing mode or friction mode connections respectively. They are also typically used for snug-tight installation designated as 8.8/S bolting category.

The details of commonly used bolts are given in Table 8.3 with other pertinent details in Tables 8.4(a) to (c).

Table 8.3 Characteristic properties of structural bolts

Property Class	Standard	Min. tensile strength, MPa	Min. yield stress, MPa
4.6	AS 1111	400	240
8.8	AS/NZS 1252	830	660

Property Class 4.6 and 8.8 bolts are available in nominal sizes of M12, M16, M20, M24, M30 and M36 as either untreated or galvanized. The larger-size bolts, M30 and M36, can be fully tensioned only by using special impact wrenches and should therefore not be used indiscriminately.

The use of washers is subject to the following rules:

- Use one hardened washer under the head or nut, whichever is used for tightening.
- Use a second washer where the bolt holes have clearance over 3 mm.
- Use thicker and larger washers with slotted holes.
- Use tapered washers where the flanges are tapered.

An excellent reference for bolting of steel structures is Hogan & Munter [2007a,c].

Table 8.4(a) Bolt dimensions for **Property Class 4.6** bolts to AS 1111
Tensile strength: 400 MPa. Yield stress: 240 MPa

Shank dia, mm	Tensile stress area, mm^2	Thread pitch mm	Bolt head/Nut width across flats, mm	Bolt head/Nut width across corners, mm
12	84.3	1.75	18	20
16	157	2.0	24	26
20	245	2.5	30	33
24	353	3.0	36	40
30	561	3.5	46	51
36	817	4.0	55	61

Table 8.4(b) Bolt dimensions for **Property Class 8.8** bolts to AS/NZS 1252
Tensile strength 830 MPa. Yield stress: 660 MPa

Shank dia, mm	Tensile stress area, mm^2	Thread pitch mm	Bolt head/Nut width across flats, mm	Bolt head/Nut width across corners, mm
16	157	2.0	27	31
20	245	2.5	34	39
24	353	3.0	41	47
30	561	3.5	50	58
36	817	4.0	60	69

Table 8.4(c) Main attributes of **Property Class 8.8** bolts to AS/NZS 1252

Bolting/Torque	**8.8/S**	**8.8/TB**	**8.8/TF**
Attribute	Snug-tight	Fully tensioned, bearing type	Fully tensioned, friction type
Shear joint			
Tightening	Snug-tight	Torque control	Torque control
Contact surfaces	Any	Bare metal	Bare metal (critical)
Joint slippage	Possible	Some slippage	Unlikely
Getting loose	Possible	Unlikely	Unlikely
Tightening tool	Hand wrench	Torque wrench	Torque wrench
Tension joint			
Tightening	Snug tight	Torque control	Same as 8.8/TB
Contact surfaces	Any	Painting OK	Same as 8.8/TB
Joint slippage	Unlikely	Zero	Same as 8.8/TB
Getting loose	Unlikely	Zero	Same as 8.8/TB
Compression joint			
Contact surfaces	Any condition	Any condition	—
Joint slippage	Some possible	Zero	—
Getting loose if vibrating	Unlikely	Not possible	—
Cost of installation	Low	Medium to high	High

Notes: The torque control is either by "load indicating washer" or "turn of the nut" method. Torque wrench/measurement is only used for inspection purposes in AS 4100.

8.2.4 Modes of force transfer

Two characteristic modes of force transfer are used:

- bearing mode for joints where connection is allowed to slip until the bolts come in bearing contact;
- friction or 'friction grip' mode for joints intended not to slip under limit loads.

Further discussion on the bolting types is given in Sections 8.2.5, 8.2.6, 8.3 and 8.4. Figure 8.2 illustrates the modes of force transfer.

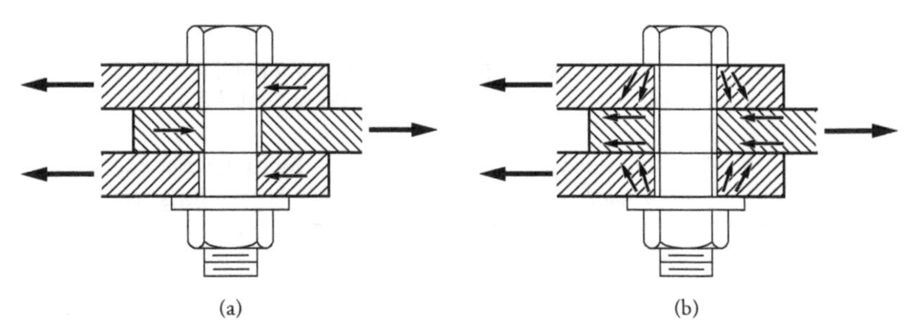

Figure 8.2 *Modes of force transfer at a bolted lap joint (a) bearing and shear on bolt; (b) friction grip.*

Some confusion can occasionally arise when special tight-fitting bolts (interference body bolts) are specified, because they are often called 'high-strength bearing bolts'. The shanks of these bolts are manufactured with small, sharp protrusions so that bolts can be driven into a slightly undersized hole for a very tight fit. These bolts permit the transmission of very high shear forces without slip but at a cost penalty. Their use is restricted to bridge and heavy construction.

The following points are pertinent for selection:

- magnitude of loads
- intended condition of mating surfaces (whether they are to be painted, galvanized or left "as rolled")
- maximum joint-slip permitted
- presence of load fluctuations or fatigue.

In order to comply with AS 4100 high-strength bolting requirements, only the following surface treatments can be used: flame-cleaned plain steel, shot-blasted plain steel, hot zinc sprayed and sandblasted, and inorganic zinc-rich paint. By modifying the slip factor, μ, which is assumed as 0.35 in AS 4100, it is also possible to use grit-blasted or wire-brushed hot-dip galvanized steel.

It is often desirable and economical to paint the steel in the fabricator's painting facility before shipping it to the site. When the steelwork is prepainted it is necessary to apply masking to the areas of the mating surfaces where friction bolting is used. Using inorganic zinc silicate paint is advantageous, as a relatively high friction coefficient can be achieved and masking may not be necessary.

8.2.5 Bolts in snug-tight connections (4.6/S & 8.8/S)

When bolts are installed as snug-tight (without controlled tension), they are assumed to act like pins (see Figure 8.2(a)). Shear forces transferred through bolts depend entirely on the shear capacity of the bolts and the bearing capacity of the ply. Tensile forces larger than the initial tension, transferred through the bolts, can open up the joint between the bolted plates because of the elongation of the bolt.

Under conditions of load reversal, snug-tight bolts should not be used for transfer of shear, as joint movement would result. This could cause damage to the protective finishes, and to the faying surfaces, and nuts might undo themselves. Shear and tension connections subject to high-cycle load fluctuations should be specified for fully tensioned, Property Class 8.8 bolts.

Both the Property Class 4.6 and 8.8 bolts can be installed as snug-tight: that is, tensioned just sufficiently to bring the bolted elements into full contact. The usual specification for snug-tight bolting is: the snug-tight condition is achieved when the full effort of an averagely fit worker, using a standard hand wrench (podger spanner) or a few impacts of an impact wrench, where the bolt or nut will not turn any further.

The behaviour of snug-tight bolts is shown in Figure 8.3. For the strength limit state, design capacities for snug-tight bolts (i.e. 4.6/S or 8.8/S) can be calculated as described in Section 8.3.1.1, or read from Table 8.5(a). Due to the inherent extra slip in its load-deformation characteristics, snug-tight bolts are generally used in simple or flexible connections (see Figure 8.1(a) to (c)).

The possibility of ply crushing and tearout may also need to be evaluated (see Sections 8.3.1.1(d) and (e), 8.4.1 and 8.4.2) for snug-tight bolted connections. Generally, snug-tight bolts are not used/permitted in dynamic loading (fatigue) situations.

8.2.6 Property Class 8.8 bolts, tension-controlled

Property Class 8.8 bolts, or high-strength structural bolts, installed under controlled tensioning procedures, can be designed to act as:

- bolts in bearing mode, 8.8/TB
- bolts in friction mode, 8.8/TF.

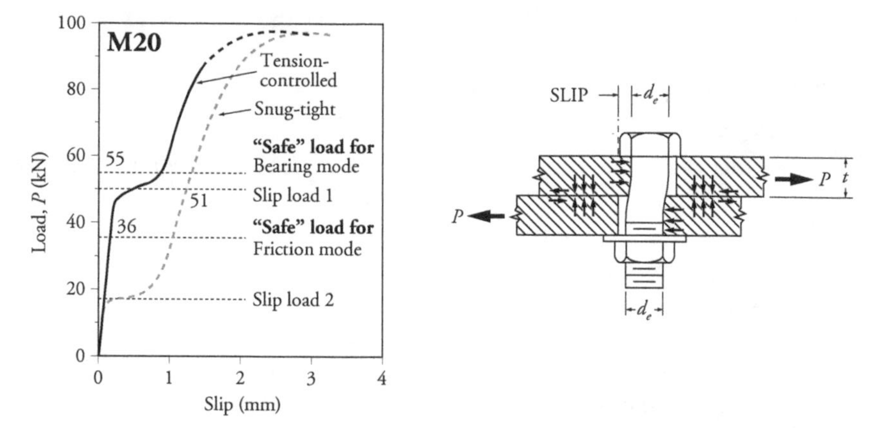

Figure 8.3 Behaviour of high-strength structural bolts. Slip load 1 applies to tension-controlled HS structural bolts using a slip factor of 0.35. Slip load 2 applies to snug-tight bolts.

There is no physical difference between the bolts themselves used in either of these modes. The only difference is that the treatment of the mating surface of friction-bolted joints (designation /TF) should be such as to ensure a high coefficient of friction. Property Class 8.8 bolts have heavier heads and nuts than Property Class 4.6 bolts because they have to resist larger initial tension and applied tensile loads. Bolting category 8.8/TB requires no special preparation of the mating surfaces, while category 8.8/TF depends on the value of the friction coefficient of the surfaces in contact.

When bolts carry tension only, it is acceptable to leave the paint coat covering the mating surfaces, and bolting category 8.8/TB is appropriate.

The behaviour of fully-tensioned bolts is shown in Figure 8.3.

8.2.6.1 Property Class 8.8 bolts, tension-controlled in bearing mode connections (8.8/TB)

Bolts designed for bearing mode of action rely mainly on a dowel-type action to transfer shear forces (see Figure 8.2(a)). As the load increases, the bolted elements will slip sufficiently to bring the bolt shank into contact with the wall of the bolt hole. This has the effect of stressing the bolt shank in shear and in bearing, and thus the term 'bearing-type joint' is used.

At the worst, the slip may become as large as the bolt hole clearance, say 2–3 mm for standard bolt holes. Bolt slips approaching these values have occurred in practice, but only when the bolts were incorrectly tensioned. A joint slip of at least 1 mm should be assumed in design, because the bolts are fully tensioned, and a part of the load is transferred by friction-grip action.

The fundamental difference between tension-controlled bearing mode (/TB) and snug-tight (/S) bolts is seen by the behaviour of these bolt types in Figure 8.3. Clearly, for the same Property Class, /S bolts slip much earlier than /TB bolts during the (shear) loading process. So the basic advantage of /TB bolts is the extra stiffness supplied over /S bolts by delaying slip from the pre-tensioning process. This advantage of extra stiffness to the overall joint is significantly used in rigid type connections—where snug–tight connections are generally not used (or permitted). Much research and testing has been done with /TB bolts in rigid and semi-rigid connections to correlate behaviour to design assumptions for this joint type.

For the strength limit state, design capacities for tension-controlled Property Class 8.8 bolts designed for bearing mode (i.e. 8.8/TB) can be calculated as described in Section 8.3.1.1, or read from Table 8.5(b). The value of the capacity reduction factor, ϕ, is taken as 0.8.

Interestingly, these capacities are the same as for the snug-tight Property Class 8.8 bolts (i.e. 8.8/S). This is due to both bolt categories have the same limiting condition for the strength limit state. The only difference is the extra rigidity offered by the /TB bolts that affects the load-slip performance which is very important for some connection types—e.g. rigid and semi-rigid connections (see Figure 8.1(d) and, if bolted, (g)).

For 8.8/TB bolts, like 8.8/S bolts, ply crushing and tearout capacities also have to be evaluated—more so in this instance as bolt strength is higher (see Sections 8.3.1.1(d) and (e), 8.4.1 and 8.4.2). Also, /TB bolts are used in dynamic loading (fatigue) situations when friction mode type bolts are seen to be uneconomical.

8.2.6.2 Property Class 8.8 bolts, tension-controlled in friction mode connections (8.8/TF)

The principal mode of action of friction-grip bolts is to use the friction resistance developed under initial tensioning of the bolts. As long as the working shear force is less than the frictional resistance on the preloaded joint, there is no slippage and the joint may be regarded as behaving elastically. The term applicable to this type of joint is 'friction-type joint', but a more descriptive term sometimes used is 'friction-grip joint'. The physical contact between the bolt shank and the bolt hole is not essential in this mode of load transfer, but should be anticipated in case of joint slippage. Each bolt can transfer a force equal to the frictional resistance of the mating surfaces surrounding the bolt.

Table 8.5(a) Design capacities for snug-tightened (category /S) bolts

Bolt size	Property Class 4.6 bolts (4.6/S)		Property Class 8.8 bolts (8.8/S)	
mm	Axial tension ϕN_{tf} (kN)	Single Shear * ϕV_f (kN)	Axial tension ϕN_{tf} (kN)	Single Shear* ϕV_f (kN)
M12	27.0	15.1 [22.4]	-	-
M16	50.2	28.6 [39.9]	104	59.3 [82.7]
M20	78.4	44.6 [62.3]	163	92.6 [129]
M24	113	64.3 [89.7]	234	133 [186]
M30	180	103 [140]	373	214 [291]
M36	261	151 [202]	541	313 [419]

Notes: 1. *Single plane shear values are for bolts with threads *included* in the shear plane, and shear values in [ ] are for bolts with threads *excluded* from the shear plane.

Table 8.5(b) Design bolt capacities for category 8.8/TB bolts

Bolt size	Axial tension	Single shear	
		Threads included	Threads excluded
mm	ϕN_{tf} (kN)	ϕV_f (kN)	ϕV_f (kN)
M16	104	59.3	82.7
M20	163	92.6	129
M24	234	133	186
M30	373	214	291
M36	541	313	419

Note: Threads included/excluded refer to threads included in or excluded from the shear plane.

Table 8.5(c) Design capacities for category 8.8/TF bolts installed in round holes (k_h = 1.0)

Bolt size	Bolt tension, kN at installation	Axial tension, kN	Single shear, kN		
			μ = 0.25	μ = 0.30	μ = 0.35
	N_{ti}	ϕN_{tf}	ϕV_{sf}	ϕV_{sf}	ϕV_{sf}
M16	95	66.5	16.6	20.0	23.3
M20	145	101	25.4	30.5	35.5
M24	210	147	36.8	44.1	51.5
M30	335	234	58.6	70.4	82.1
M36	490	343	85.7	103	120

Note: Axial tension design capacity (ϕN_{tf}) is only used for shear-tension interaction checks for /TF bolts.

As mentioned above, snug-tight bolts (/S) and tensioned-controlled in bearing mode bolts (/TB) are typically designed for the strength limit state. Unlike, /S and /TB bolts, tension-controlled friction mode bolts (/TF), are generally designed for the serviceability limit state as slip (or deflection) is being limited.

For the serviceability limit state, design capacities for tension-controlled Property Class 8.8 bolts designed for friction mode (i.e. 8.8/TF) can be calculated as described in Section 8.3.1.2, or read from Table 8.5(c). The design shear capacities in Table 8.5(c) are given for three values of coefficient of friction. The value of the "special" serviceability capacity reduction factor, ϕ, is taken as 0.7 as noted in Clause 3.5.5 of AS 4100. The /TF bolts can be used in dynamic loading (fatigue) situations though are somewhat penalised by their lower load carrying capacity. If used in this situation, fatigue checks do not have to be undertaken on /TF bolts (unless the bolt has prying forces under tension loads—see Table 11.5.1(3) of AS 4100).

Even though /TF bolts are designed for the serviceability limit state, AS 4100 also requires /TF bolts to be designed for the strength limit state. In this instance, the /TB strength design provisions are used with the strength limit state load factors, combination and capacity reduction factors.

The design value of the coefficient of friction depends on the surface preparation for the category 8.8/TF bolts. Table 8.6 has been compiled from various published sources, and it shows the importance of the surface preparation.

Table 8.6 Values of coefficient of friction, μ

Surface description	Coefficient of friction, μ Average	Minimum
Plain steel:		
- as rolled, no flaking rust	0.35	0.22
- flame-cleaned	0.48	0.35
- grit-blasted	0.53	0.40
Hot-dip galvanized:		
- as received	0.18	0.15
- lightly sandblasted	0.30	0.20
Hot-zinc sprayed:		
- as received	0.35	0.23
Painted:		
- ROZP primed	0.11	0.05
- inorganic zinc silicate	0.50	0.40
- other paint systems	Tests required	Tests required

8.2.6.3 Interference body bolts

Interference body bolts are manufactured for driving into well-aligned holes, facilitated by sharp protrusions or 'knurls' over the surface of the bolt shank. Because there is no gap between the bolt and the hole, it is possible to achieve a zero-slip joint.

The design of these bolts may be based on either friction-type (category 8.8/TF) or bearing-type joints (category 8.8/TB), whichever permits larger loads per bolt. This is because, for thinner plates, the friction-type design permits higher design capacities, while for thicker plates the bearing-type design will be more advantageous.

8.2.7 Installation and tightening of HS structural bolts

The first prerequisite is that the mating surfaces must be true, even without projections, so that they can be brought in close contact over the whole area to be bolted with a minimum of pressure.

If bolts are designed for friction mode (friction-type joint) the surfaces must be in the condition prescribed in the job specification or, if unspecified, in accordance with AS 4100. The job specification may require a bare steel mating surface complying with AS 4100 or may prescribe the type of surface preparation or finish that has been used as the basis of design. Sometimes galvanizing or zinc-rich paint is specified, but this must be checked against the design assumption of the slip factor value. The steelwork specification should note exactly which surface preparation is required by design.

When bolts are designed for bearing mode (bearing-type joint) the mating surfaces may be left primed or painted. In either case, any oil, dirt, loose rust, scale or other non-metallic matter lodged on the mating surfaces must be removed before assembly. When bolts are used in tension joints, no special preparation is required.

HS structural bolts may be provided with only one hardened washer under the nut or bolt head, whichever is turned during tightening. A tapered washer should always be placed against a sloping surface, in which case the turning end is on the opposite side to prevent rotation of the tapered washer. Three methods of tightening are in use:

- part-turn method (primary method)
- direct-tension indication method (primary method)
- torque control method (secondary method for inspection only).

The *part-turn method* (Clause 15.2.5.2 of AS 4100) relies on the known ratio of the bolt tensile force to the number of turns of the nut. This is described in AS 4100. For example, to achieve the minimum bolt tension in a 24 mm bolt, 150 mm long, it is necessary to rotate the nut or the head half a turn from the snug-tight condition. It is important to check that the other end does not rotate, which may occur as a result of seizure of the nut because of rusty threads or insufficient thread clearances. The latter can sometimes be experienced with galvanized bolts.

The *direct-tension indication method* (Clause 15.2.5.3 of AS 4100) makes use of a specially designed load-indicating device that are washers with protrusions on one face such that initially an air gap exists between the washer and the metal. The tightening of the bolt produces pressure on the protrusions, gradually squashing them. When the required bolt tension is reached, the gap diminishes to a specified minimum that can be ascertained by a blade gauge. This method is more reliable than the part-turn and torque control methods. See also Direct-tension indication (DTI) device definition in Section 8.2.2.

The *torque control method* (Appendix K of AS 4100) relies on the laboratory-established relationship between torque and bolt tension that is dependent on the coefficient of friction between the thread surfaces in contact. Oiled threads require considerably less torque than dry threads to achieve the minimum bolt tension (the torque ratio can be as much as 1:2). Rust on the threads would further increase the torque. Due to this variability, AS 4100 (and Australian practice) has relegated this tightening method to inspection purposes only. It is considered to be an independent method to assess the presence of gross under-tensioning.

The first stage in the bolt-tightening procedure requires the bolt to be brought into a snug-tight condition that is intended to draw the plies into firm contact, eliminating any air gaps over the entire joint area. If the surfaces cannot be drawn together, it is essential to eliminate the mismatch between the plates before the bolts are tightened to full tension. The final bolt tightening from snug-tight condition to the maximum prescribed bolt tension should proceed progressively from the central bolts to the peripheral bolts or, in another type of bolt disposition, from the bolts closest to the most rigid part of the joint towards the free edges.

Apart from the effect of the thread condition, there is a need to carefully calibrate power-operated wrenches, and to maintain these in calibrated condition. The 'stall' or cut-off torque of power-operated wrenches cannot be maintained for long at the predetermined level, and frequent calibrations are necessary.

Whichever method of tightening is employed, the minimum bolt tensions must comply with Clause 15.2.5 of AS 4100.

AS 4100 also notes that the reuse of HS structural bolts that have been fully tightened shall be avoided—as, for example, if a joint were required to be taken apart and reassembled. The reason for this is that the bolt material becomes strain-hardened when tensioned at or above the proof stress, and less than the original extension is available at the repeated tightening, with the consequent danger of fracture later in service. If re-used, HS structural bolts should only be placed in their original hole with the same grip. Retensioning of galvanized bolts is not permitted. The above bolt re-use provisions do not include "touching up" of previously tensioned bolts.

It is emphasised that tightening of HS structural bolts is not an exact science, and a lot of sound judgement and field experience is required if the design assumptions are to be realised on the job.

8.3 Design and verification of bolted connections

8.3.1 Capacity of a single bolt

The appropriate limit states for design of bolts are the strength limit state and the serviceability limit state. Durability also may be a factor for consideration.

Recent work on bolt strengths shows that the design criteria in the old AS 1250 (the predecessor to AS 4100) were too conservative. The bolt capacities specified in AS 4100 are substantially higher than the capacities back-calculated from the superseded AS 1250. Thus the number of bolts in a connection required by AS 4100 will be only about 60% of the number computed by the old code. It is now more important than ever to carry out an accurate design of bolted joints and take into account all design actions occurring at the joint. This should include prying action and the least favourable combination of loads.

8.3.1.1 Strength limit state

Clause 9.3.2 of AS 4100 considers the strength limit state design of bolts and applies to /S, /TB and /TF bolting categories.

(a) Bolts in shear

The nominal bolt shear capacity, V_f, is calculated as follows.

Where there are no shear planes in the threaded region and n_x shear planes in the unthreaded region:

$$V_f = 0.62\, k_r f_{uf} n_x A_o$$

Where there are n_n shear planes in the threaded region and no shear planes in the unthreaded region:

$$V_f = 0.62\, k_r f_{uf} n_n A_c$$

Where there are n_n shear planes in the threaded region and n_x shear planes in the unthreaded region:

$$V_f = 0.62\, k_r f_{uf} (n_n A_c + n_x A_o)$$

where

f_{uf} = minimum tensile strength of the bolt

A_c = core area (at the root of the threads)

A_o = bolt shank area

k_r = reduction factor for length of bolt line

= 1.0 for connections other than lap connections, otherwise

= 1.0 for lengths up to 300 mm, and 0.75 for lengths over 1300 mm (interpolation should be used in between).

It is worth noting that the capacities of bolts are based on the minimum tensile strength f_{uf} rather than the yield strength.

The overall inequality to be observed is:

$$V^* \leq \phi V_f$$

where V^* is the design shear force on the bolt/bolt group and the capacity reduction factor, ϕ, is 0.8 from Table 3.4 of AS 4100.

The normal commercial practice of thread lengths is based on the formula: length = two bolt diameter + 6 mm. The usual bolt projection allowance is roughly 1.25 diameter + 6 mm. This means that the length of thread projecting into the bolt hole is roughly 0.75 times the bolt diameter if the outer ply is thinner than 0.75 bolt diameters. This is quite common in building structures, where plate thicknesses are often less than 16 mm. In such instances it may be more economical to use shorter bolts and allow threads to project into the shear plane, provided the bolt capacity has been checked on this basis.

(b) Bolts in tension

The nominal capacity of a bolt in tension, N_{tf}, is calculated from:

$$N_{tf} = A_s f_{uf}$$

where f_{uf} is the tensile strength of bolt material, and A_s is the tensile stress area of the bolt (see AS 1275 or Table 8.4(a) or (b)). The tensile stress area is roughly 10% larger than the core area.

Tensile connections utilising end plates may be subject to an increase in tension because of the leverage or prying action. Section 8.3.2.4 discusses the measures required to minimise any prying action.

The overall inequality to be observed is:

$$N_{tf}^* \leq \phi N_{tf}$$

where N_{tf}^* is the bolt design tension force and the capacity reduction factor is 0.8.

(c) Bolts carrying shear and tension are required to satisfy the following interaction inequality:

$$\left(\frac{V_f^*}{(\phi V_f)}\right)^2 + \left(\frac{N_{tf}^*}{(\phi N_{tf})}\right)^2 \leq 1.0$$

where V_f is the nominal shear capacity, and N_{tf} is the nominal tension capacity with $\phi = 0.8$.

(d) Crushing capacity of the ply material from bolt bearing

The nominal ply crushing bearing capacity:

$$V_b = 3.2\, t_p d_f f_{up}$$

where t_p is the ply thickness, d_f is the bolt diameter, and f_{up} is the tensile strength of the ply.

The following condition must be satisfied: $V^* \leq \phi V_b$ with $\phi = 0.9$ (note the differing ϕ compared to bolt shear and tension). See Section 8.4.2 also.

(e) Bearing capacity of the ply material from bolt tearout
The nominal ply tearout capacity of the ply in contact with the bolt:

$$V_p = a_e t_p f_{up}$$

where a_e is the minimum distance from the ply edge to the hole centre in the direction of the bearing force, f_{up} is the tensile strength and t_p is the thickness of the ply.
The following condition must be satisfied: $V^* \leqslant \phi V_p$ with $\phi = 0.9$ (again, note the differing ϕ compared to bolt shear and tension). See Section 8.4.2 also.

8.3.1.2 Serviceability limit state

This limit state is relevant only for friction-type connections where connection slip is intended to be prevented at serviceability loads (i.e. for the /TF bolting category). From Clause 9.3.3.1 of AS 4100, the nominal shear capacity of a bolt for a friction-type connection, V_{sf}, is:

$$V_{sf} = \mu\, n_{ei} N_{ti} k_h$$

where

k_h = factor for hole type: 1.0 for standard holes, 0.85 for oversize holes and short slots, and 0.70 for long slotted holes;

μ = coefficient of friction between plies, which varies from 0.05 for surfaces painted with oil-based paints to 0.35 for grit-blasted bare steel. Tests are required for other finishes;

N_{ti} = minimum bolt tension imparted to the bolts during installation (see Table 8.5(c)), and

n_{ei} = number of shear planes.

The following inequality must be satisfied:

$$V^*_{sf} \leqslant \phi V_{sf}$$

where V^*_{sf} is the design shear force in the plane of the interfaces and $\phi = 0.7$.

8.3.2 Capacity of bolt groups

8.3.2.1 General

Connections in building structures use a minimum of two bolts and often more than eight bolts. The bolts used in a connection form a group. A bolt group may be acted on by loads and bending moments in the plane of the bolt group (in-plane) or at right angles to it (out-of-plane loading). See Figure 8.4.

Elastic analysis of bolt groups is permitted by AS 4100 and is subject to the following assumptions:

- Plate elements being bolted are rigid and all connectors fit perfectly.
- The overall *bolt group* design actions V^*_x, V^*_y and M^* may be superposed for simplicity.
- An 'instantaneous centre of rotation' (ICR) is evaluated for the bolt group. The ICR is the point at which the bolt group rotates about when subject to the overall bolt group design actions. This point may be the bolt group centroid or a point rationally determined from the joint's rotational behaviour (see below).

- The forces from the group design action effects acting on individual bolts are proportional to the distance from the bolt to the centre of rotation.
- The 'critical' bolt(s) is then considered to be the bolt(s) furthest from the ICR. The term 'critical' bolt(s) refers to the bolt (or bolts) subject to the greatest shear force arising from the combined effects from the overall bolt group design actions. The 'critical' bolt is then used for the design check of the overall bolt group.
- Conventional analysis uses the ICR concept in conjunction with superposition principles. Using a bolt group with in-plane design actions as an example, a pure moment acting only on the bolt group has the ICR positioned at the bolt group centroid. Whereas, when the same bolt group is subject to shear force only, the ICR is positioned at infinity. In-plane loadings are generally composed of moments and shear forces which can then be simply modelled by superposition of the above two individual action effects. That is, uniformly distributing shear forces to all bolts in the group whilst also assuming that bolt group rotation from moment effects occurs about the group centroid. Using this method the critical bolt is generally the furthest from the bolt group centroid.

The latter above-listed technique is the most commonly used method of analysis for bolt groups.

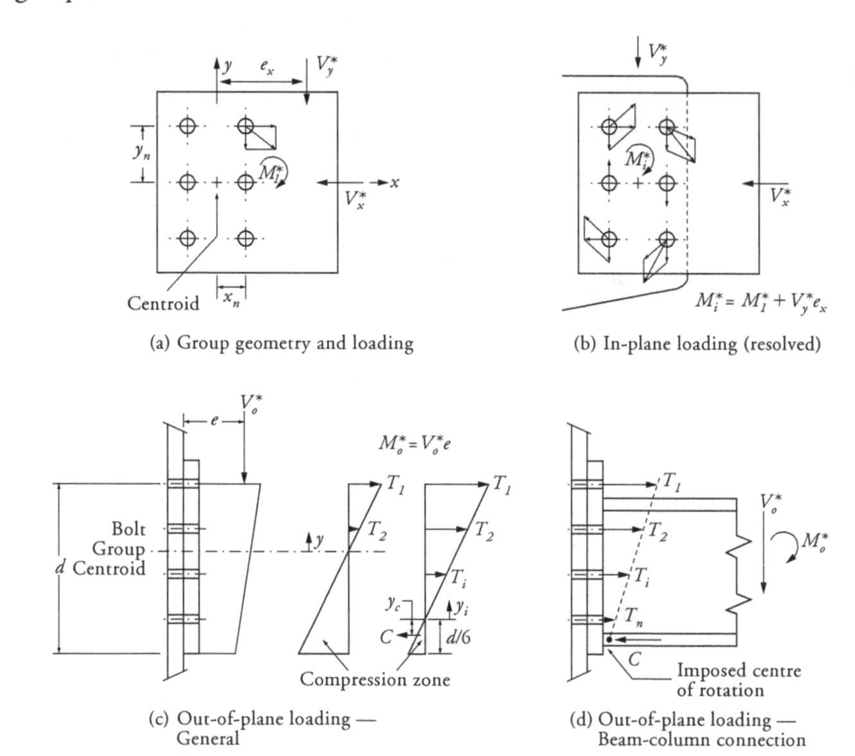

Figure 8.4 *Bolt groups*

The method of verifying the bolt group design capacity is well described in detail by Hogan & Munter [2007a]. A method of superposition will be used.

The first step is to determine the bolt group second moments of area (Figure 8.4(a)):

$$I_x = \Sigma x_n^2$$

$$I_y = \Sigma y_n^2$$

$$I_p = I_x + I_y = \Sigma(x_n^2 + y_n^2)$$

where I_x and I_y are second moments of area of bolts about the bolt group centroid axis, and I_p is the polar second moment of area of the bolts in the group, each bolt having a section area of unity (assuming all bolts are the same size in the bolt group).

8.3.2.2 Bolt groups subject to in-plane actions

Based on the above assumptions, and as noted in Figure 8.4(a) and (b), the resultant design bolt shear force in the bolt farthest away from the centre of bolt group is:

$$V_{res}^* = \sqrt{\left\{\left[\left(\frac{V_y^*}{n}\right) + \left(\frac{M_i^* x_{max}}{I_p}\right)\right]^2 + \left[\left(\frac{V_x^*}{n}\right) + \left(\frac{M_i^* y_{max}}{I_p}\right)\right]^2\right\}}$$

where n is the number of bolts in the bolt group, V_y^* and V_x^* are the applied forces in vertical and horizontal directions, M_i^* is the moment (applied and from eccentric shear forces) on the bolt group, and y_{max} and x_{max} are the distances from the centroid of the bolt group to the farthest corner bolt.

Generally, when the vector resultant loads are determined for the farthest bolt, V_{res}^* (which, in this instance, is considered to be the most critically loaded bolt), the following inequality can be used to check the adequacy of the whole bolt group:

$$V_{res}^* \leq \text{min.}\ [\phi V_f, \phi V_b, \phi V_p]$$

where, for the farthest bolt, ϕV_f is the design shear capacity, ϕV_b the ply crushing design bearing capacity and ϕV_p the ply tearout design bearing capacity (as noted in Sections 8.3.1.1(a), (d) and (e) respectively).

Hogan & Munter [2007a] provide further background, short-cuts and design aids on bolt groups loaded in this manner.

8.3.2.3 Bolt groups subject to out-of-plane actions

Figure 8.4(c) and (d) show typical bolt groups loaded out-of-plane. From force/moment equilibrium principles, Figure 8.4(c) notes that there are bolts which are not loaded as they are positioned in the bearing (compression) part of the connection and require no further consideration. The forces in the tension region bolts can be evaluated by assuming a linear distribution of force from the neutral axis to the farthest bolts—these latter bolts being the more critically loaded.

Due to the bolt, plate and support flexibility, a problem exists to determine where the neutral axis (NA) is placed. There is not much guidance available on precisely evaluating the NA position. However, some limited guidance is provided by such publications as AISC(US) [2011], CISC [2010] and Owens & Cheal [1989]. A conservative approach is to assume the NA is placed at the bolt group centroid line. A better approximation, which appears to be empirically based, is to assume the NA is positioned at $d/6$ from the bottom of the end plate which has a depth d (see Figure 8.4(c)). After the NA position has been assumed, the following can be undertaken to evaluate the tension load in the farthest (most critical) bolts.

Using equilibrium principles and Figure 8.4(c):

$$\Sigma T_i y_i + C y_c = M_o^* \qquad \text{and } \Sigma T_i = C$$

and the principle of proportioning from similar triangles provides:

$$T_i = T_1 \frac{y_i}{y_1}$$

the following can ascertained for the critically loaded farthest bolts:

$$T_1 = \frac{M_o y_1}{\Sigma[y_i(y_i + y_c)]}$$

The y_i terms can be determined from the geometry set by the NA placement. T_1 must be divided by the number of bolts in the top row, n_1, to evaluate the peak tension force on each critical bolt. Finally, the design shear force on each bolt, V^*, can be conservatively determined by uniformly distributing the bolt group shear force, V_o^*, to each bolt—i.e. $V^* = V_o^*/n$. The interaction equation of Section 8.3.1.1(c) is then used with $V_f^* = V^*$ and $N_{tf}^* = T_1/n_1$.

In specific joint configurations such as rigid beam-to-column connections (Figure 8.4(d)), some standard connection design models (Hogan & Van der Kreek [2009a,c,e], AISC [2010]) further assume that the two top rows of bolts about the top flange uniformly resist most—if not all—of the tension force from out-of-plane moments. The reasoning for this is due to the flange-web to end plate connection providing significant stiffness for tension forces to be drawn to it. Alternatively, some design models use an "imposed centre of rotation" placed at a "hard spot" through which all the compression force acts. However, these connection design models also commonly assume that the overall shear force is shared equally by all the bolts in the bolt group.

The above design models do not consider increased bolt tensions due to prying actions. This is considered in Section 8.3.2.4 below.

8.3.2.4 Prying action

As noted in Figure 8.5, prying action occurs in T-type or butt-type connections with bolts in tension. Bending of the end plate causes the edges of the end plate to bear hard on the mating surface, and the resulting reaction must be added to the bolt tension. AS 4100 has no specific provisions for determinating forces involved in prying action, but various authors have suggested methods of estimating the magnitude of prying forces (see Hogan & Thomas [1994] which suggests allowing for 20–33% increase in bolt tension force).

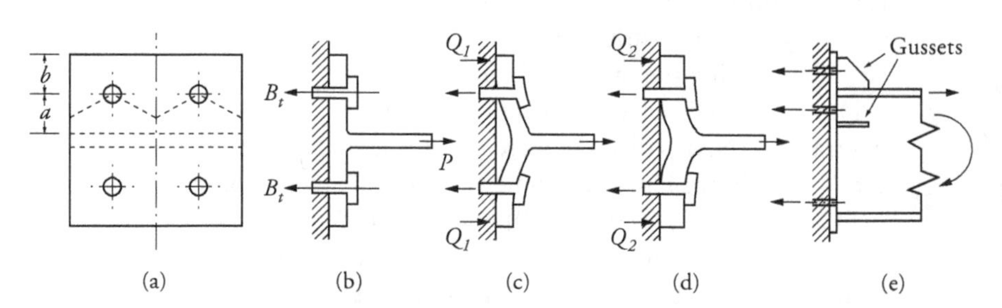

Figure 8.5 *Prying action: (a) end plate elevation; (b) rigid end plate; (c) medium thickness end plate; (d) thick end plate, and; (e) stiffened end plate. (Note: generally $Q_1 > Q_2$ where Q is the additional ply reaction to induce prying forces into the bolt).*

Based on the published results, prying action can be kept to a minimum by using the following measures:

- increasing the ratio *a*/*b* to at least 0.75 (see Figure 8.5(a))
- increasing the bolt spacing to at least 90 mm
- increasing the end plate thickness to at least 1.25 times the bolt size.

The other option is to stiffen the end plates, as shown in Figure 8.5(e). The latter option should be applied as a last resort, as welding of gussets will raise the cost of fabrication.

Figure 8.5 notes the instances when prying actions may occur. Relatively rigid end plates (b) unable to flex will separate from the support face rather than bend and no prying action occurs. Small thickness end plates (not shown) undergo pronounced double curvature bending and do not attract prying actions. Medium thickness end plates (c) undergo single curvature or limited double curvature bending causing the end plate edges to also contact the support surface, creating new reaction points thereby increasing the bolt tension loads. Thick plates (d) undergo single curvature bending under flexure and, as noted in (c), attract prying forces. Stiffened end plates (e) behave like (b) and prying forces are negligible if at all present.

8.3.2.5 Combined in-plane & out-of-plane actions

Occasionally, bolt groups are loaded both in-plane and out-of-plane. The procedure described in Hogan & Munter [2007a] combines the in-plane and out-of-plane forces using a general procedure.

8.4 Connected plate elements

8.4.1 Bolt holes and connection geometry

Bolt holes are usually made larger in diameter than the bolt shank for several practical reasons. First, the bolt shank diameter may vary by ±1%, while the hole diameter can also vary depending on the drill bits. Second, the fabrication of steel members or units can never be absolutely precise and is subject to position and fabrication tolerances.

As specified in AS 4100 and noted in Table 8.7, the bolt holes should be made larger than the nominal diameter by:

- 2 mm for M12, M16, M20 and M24 bolts
- 3 mm for bolts larger than M24.

One exception is that bolt holes in base plates are made larger by up to 6 mm, for all bolt sizes, to allow for the larger tolerances in holding-down bolt positions. Larger clearances are permitted only under the proviso that plate washers or hardened-steel washers are used, and that the possibility of significant connection slip has been examined.

Slotted holes are sometimes used to allow for temperature movements or to ease the problems of erection of complex units. Special provisions for slotted holes in AS 4100 are as in Table 8.7.

Short slots may be provided in all joined plies if plate washers or hardened washers are used. Long slots can be provided only in one ply of two-ply lap joints, or in alternate plies for multi-ply joints. In addition, the holes of long slots must be completely covered, including an overlap for joint movement, using plate washers 8 mm or thicker.

Table 8.7 Standard and slotted hole sizes as noted in Clause 14.3.5.2 of AS 4100

Bolt size	Nominal hole dia. mm	Base plate hole dia. mm	Oversize/slotted hole width/dia. mm	Short slotted hole length mm	Longslotted hole length mm
M12	14	18	20	22	30
M16	18	22	24	26	40
M20	22	26	28	30	50
M24	26	30	32	34	60
M30	33	36	38	40	75
M36	39	42	45	48	90

Circular bolt holes may be fabricated by drilling or by punching if special conditions of AS 4100 are met. For steel material of Grade 250/300 and static loading, the maximum thickness of plate that can be punched is 22.4/18.7 mm respectively. If the joint is dynamically loaded the maximum thickness for punching is reduced to 12 mm.

Slotted holes can be fabricated by machine flame-cutting, punching or milling. Hand flame-cutting would not comply with AS 4100. Figure 8.6 shows the spacing of bolts. Table 8.8 gives the edge distances.

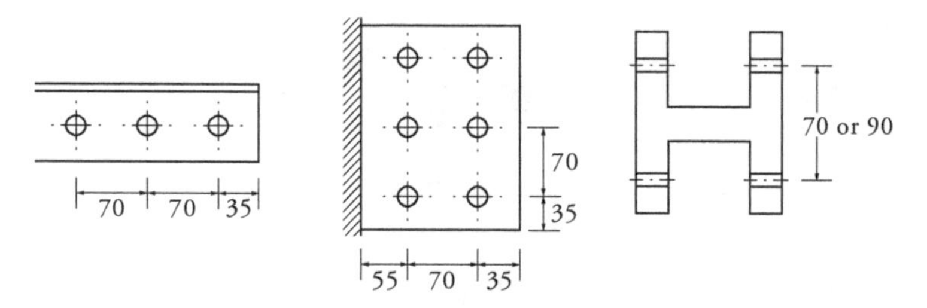

Figure 8.6 *Standard bolt gauges and pitches for M20 bolts (see ASI [2009a], Hogan & Munter [2007a–h], Hogan & Van der Kreek [2009a–e], Hogan [2011], Syam & Chapman [1996] for further information).*

Table 8.8 Minimum edge distances as specified in AS 4100

Condition of the plate element edge	Distance from centre of the hole
Sheared or hand flame-cut	1.75 d_f
Rolled plate or section or machine thermally cut, sawn, milled or planed	1.50 d_f
Rolled edge of hot-rolled flat bar or section	1.25 d_f
Edge distance will also be governed by bolt tearout failure on the ply (Clause 9.3.2.4 of AS 4100 or Section 8.3.1.1(e)).	

The minimum and the maximum pitch is also specified in Clause 9.6 of AS 4100:
Maximum edge distance: $12t_p \leq 150$ mm
Minimum pitch: $2.5d_f$
Maximum pitch: $15t_p \leq 200$ mm

8.4.2 Capacity of bolted elements

The capacity of the bolted element in a lap joint designed for bearing depends on the plate thickness, grade of steel and edge distance in the direction of force. The design must guard against bolt failure and the following types of connection failure:

- fracture across the connected element (Figure 8.7(a) and(e))
- bearing failure at bolt interface
- tearing failure.

The first noted failure mode, fracture across the connected element, is considered when designing the bolted element for tension (Section 7 of AS 4100 or Section 7.4.1 of this Handbook). In this instance, the check considers gross yielding and net fracture—the latter check takes into account the onset of fracture from reduced cross-section area from holes and non-uniform force distribution effects.

As noted in Section 8.3.1.1(d), the second of the above failure modes, bearing (or crushing) failure at the bolt-ply interface, is verified from:

$$V_b = 3.2 d_f t_p f_{up}$$

where f_{up} is the yield strength of the plate material.

The failure-bearing stress is thus 3.2 times the plate tensile strength because of the three-dimensional stress condition at the bolt-ply interface, whether the bolt threads are present at the bearing surface or not (see Figure 8.7(c)).

Tearing failure (see Section 8.3.1.1(e)) is usually more critical than bearing-type failure. The capacity of the connected element depends to a large degree on the end distance a_e (see Figure 8.7(d)). The tear-out capacity of the plate is verified by:

$$V_p = a_e t_p f_{up}$$

where a_e is the distance in the direction of force from the centre of the bolt (hole) to the edge of the member (note this could also be to the perimeter of an adjacent hole).

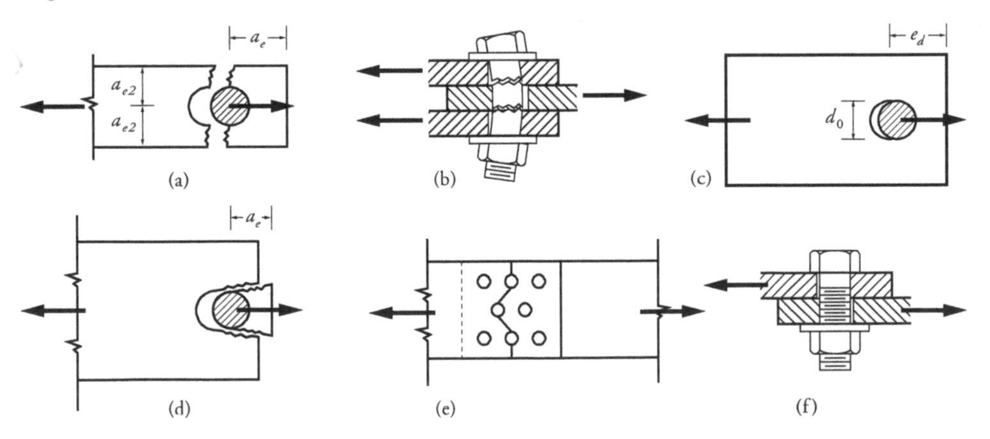

Figure 8.7 *Bolted shear connections and the potential modes of failure of joint (a) plate fracture (note: a_e and/or a_{e2} can contribute to this failure mode); (b) bolt failure; (c) crushing on ply to bolt shank interface; (d) plate tearing failure; (e) plate fracture where bolts are staggered; (f) bolt hole clearance leads to slippage*

It can be shown that tear-out failure will be more critical than bearing failure when $a_e < 3.2 d_f$, as is normally the case when standard end distances are used. The standard end distances are between 1.75 and 2 bolt diameters, simply to keep the connections as compact as possible. Using a thicker material is beneficial in raising the tear-out capacity. The last resort is in using extra bolts to compensate for the loss of end bearing capacity. Checks on standard spacing between bolt holes will see that these hole spacings are greater than $3.2d_f$

Reference should also be made to Section 8.1.5 for checks on Block Shear Failure of elements.

8.4.3 Pin connections

For pin connections refer to Section 8.10.2.

8.5 Welded connections

8.5.1 General

Electric metal arc welding has developed into a very efficient and versatile method for shop fabrication and construction of steelwork. The main areas of application of welding are:

(a) Fabrication
- compounding of sections—that is, joining of several plates or sections parallel to the long axis of the member (Figure 8.8)
- splicing of plates and sections to obtain optimal lengths for fabrication and transport to the site
- attachment of stiffeners and other details
- connection of members to one another
- attachment of the field connection hardware.

(b) Field work
- beam-to-column connections of the moment-resisting type
- column splices
- field splices for girders and trusses
- steel deck construction
- strengthening of existing steel structures
- jointing of plates in tank, silo, hopper and bunker construction

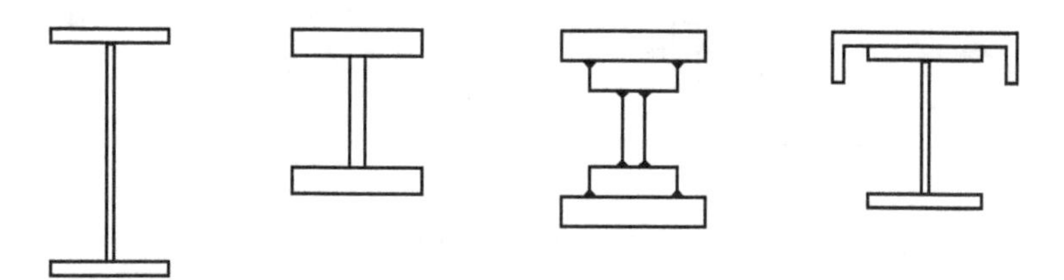

Figure 8.8 Compounding of sections

The principal use of welding is in the fabrication of steelwork, which can be regarded as transformation of plain rolled steel material into the constructional elements that can be erected with a minimum amount of site work. Welding is particularly useful for combining several plates or sections into built-up sections to produce large-capacity columns and girders, well over the capacity of the largest available rolled sections. Often it is required to produce built-up sections that are more compact than the standard rolled sections, and this can be done conveniently by welding.

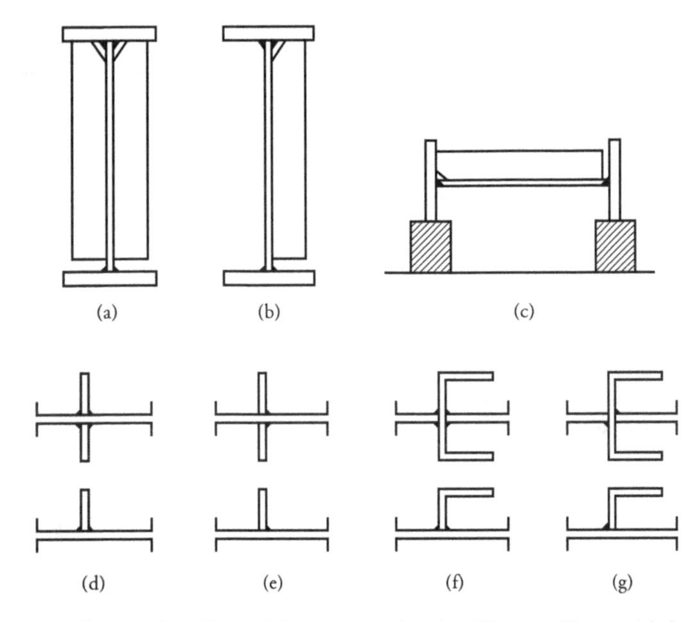

Figure 8.9 *Intermediate web stiffeners (a) symmetrical web stiffeners; (b) one-sided web stiffeners; (c) down-hand welding without turning is possible with one-sided stiffeners; (d) symmetrical fillet weld; (e) one-sided fillet welds may be adequate for stiffener attachment; (f) symmetrical fillets with angle/channel stiffeners; (g) same but with one-sided fillet welds*

The welds involved in compounding of sections are relatively long and uninterrupted, and permit the use of highly productive automatic welding machines. The welding of stiffeners, connection details and attachments usually requires relatively short lengths of weld runs and can be very labour-intensive. Reduction in the number of the individual pieces in such details can have a marked effect on lowering the costs of welded fabrication. Figure 8.8 shows typical members produced by welded fabrication, and Figure 8.9 shows the methods of attachment of intermediate stiffeners.

A large number of weld joint configurations are possible with welding. To improve communication between the design office and the welding shop, graphical symbols have been developed; some of the more frequently used symbols are shown in Tables 8.9 and 8.10 (see also AS 1101.3).

Table 8.9 Welding symbols

Location significance	Fillet	Plug or slot	Arc seam or arc spot	Butt welds				
				Square	V	Bevel	U	J
Arrow Side								
Other Side								
Both Sides		Not Used	Not Used					

Supplementary symbols

Weld all around	Field weld	Contour	
		Flush	Convex

Backing strip	Backing run

Table 8.10 Examples of use of welding symbols

No.	Symbols	Description of weld
1	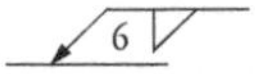 6	Continuous, one-sided fillet weld of 6 mm leg size along the length of the line indicated by arrow. Fillet weld is on the arrow side of the joint.
2	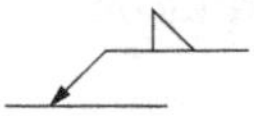	Same as for 1, but the weld is on the side opposite to where the arrow points.
3	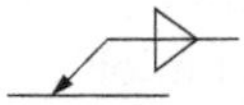	Continuous, double-sided fillet weld.
4	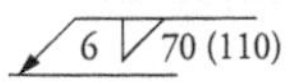 6 70 (110)	Intermittent 6 mm fillet weld having incremental lengths of 70 mm spaced at 180 mm. Arrow side only.
5	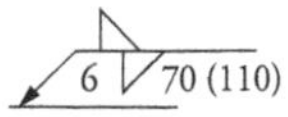 6 70 (110)	Staggered intermittent weld as for 4 (both sides).
6	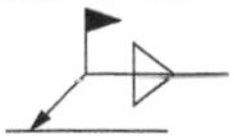	As for 3, but the flag indicates that this weld is to be done in the field.
7	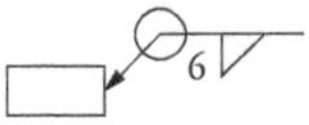 6	As for 1, but instead of being applied along a line, this weld is to be carried out all around, and this is indicated by a small circle.
8	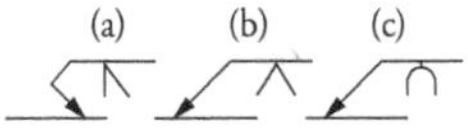 (a) (b) (c)	Butt welds: (a) single bevel; (b) single vee; (c) single U.
9	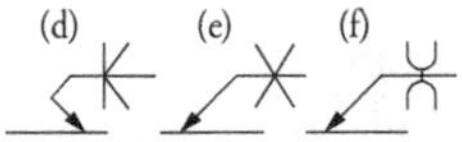 (d) (e) (f)	Butt welds: (d) double bevel; (e) double vee; (f) double U.
10	 9	Same as 9(d) but a special procedure is to be used as specified under item 9 of the procedure sheet.
11		Same as 8(b) but weld is to have a convex contour.
12		Same as 8(b) but weld is to have a flush contour obtained by grinding.
13		Same as 8(b) but a backing strip is to be used.
14	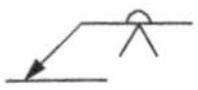	Same as 8(b) but the root of the weld is to be gouged and a backing weld run applied.
15	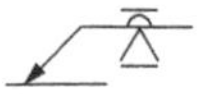	Same as 14, but both faces are to be ground flush.
16	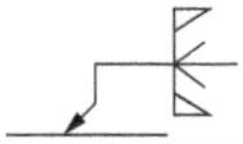	Double-bevel butt weld reinforced with fillet welds for a better stress dispersion.
17	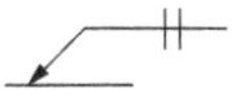	Square butt weld. No grooves are prepared for this weld (suitable only for thin plates).
18		Plug weld. Weld is on arrow side of the joint.

8.5.2 Definition of welding terms

Brittle fracture Sudden fracture of parts in tension without appreciable yield strain. The causes of brittle fracture are: first, a low notch toughness of the material at the particular service temperature or poor impact energy requirement leading to an inability to absorb energy inputs from impulse or dynamic loads; second, the presence of sharp notches in the form of cracks, crack-like inclusions, lack of fusion and incomplete penetration.

Butt weld A weld made by depositing weld metal into a groove at the join between the two elements being joined. The weld penetration may be through the entire thickness of the elements being welded (see Complete/Incomplete penetration butt welds also).

Complete penetration butt weld (CPBW) A butt weld completely filling the grooves and completely fusing all abutting faces.

Compound weld A hybrid weld defined as a fillet weld placed immediately adjacent to a butt weld.

Design throat thickness Applicable to fillet welds and incomplete penetration butt welds, the depth of weld metal for strength calculation purposes. For fillet welds it is the perpendicular distance from the unattached (non-fusion) weld face to the root (corner) of the weld. For incomplete penetration butt welds it is to the depth of the preparation (e.g. the bevel) and is dependent on the angle of bevel/Vee and the welding process used.

Effective length of weld Length of the full-size weld, excluding the end craters.

Effective throat thickness of butt weld For a complete penetration butt weld, this is the thickness of the thinner plate; for an incomplete penetration butt weld, the effective throat thickness is taken as the sum of the depths of fused weld metal. (See also Design throat thickness).

Electrode See Welding consumable.

Fillet weld Welds which generally have a triangular cross-section and are fused on two faces to the parent metal. Apart from the requirement of a clean surface, these welds typically require no edge or surface preparation.

Flux, welding A substance used during welding to help clean the fusion surfaces, to reduce oxidation, and to promote floating of slag and impurities to the surface of the weld pool.

Heat-affected zone (HAZ) A narrow zone of the parent metal adjacent to the weld metal; the changes in the grain size and absorption of gases, especially hydrogen, can promote brittleness in the HAZ.

Incomplete penetration butt weld (IPBW) A butt weld which, unlike complete penetration butt welds, has weld parent metal fusion occuring at less than the total depth of the joint.

Parent metal Metal to be joined by welding.

Penetration Depth of fusion of the weld into the parent metal.

Plug weld A weld deposited into a space provided by cutting or drilling a circular hole in a plate so that the overlapped plate can be fused.

Prequalified weld A term describing a weld procedure (including weld groove preparation) in accordance with AS/NZS 1554 known to be capable of producing sound welds, without the need for procedural tests, to behave in a manner as assumed in design.

Slag Fused, non-metallic crust formed over the exposed face of the weld that protects the deposited weld metal during cooling.

Slot weld Similar to a plug weld but slot-like in shape.

Weld category See Weld quality.

Weld metal The deposited metal from the electrode or wire (sometimes called weld consumable) which fuses with the parent metal components to be joined.

Weld preparation Preparation of the fusion faces for welding; such preparation may consist of removal of mill scale, grinding of groove faces, cutting of the bevels and aligning of the parts, to obtain correct root gap, etc.

Weld quality A measure of the permitted level of defects present on deposited welds. Weld qualities/categories can generally be either SP (structural purpose) or GP (general purpose) and possess a pre-determined capacity reduction factor, ϕ, for the weld in strength design calculations. See WTIA [2004] for further details on SP and GP welds.

Weldability Term used to describe the ease of producing crack-free welds under normal fabrication conditions. While all steels can be welded by observing the proper procedures and using the right amount of preheat, it should be realised that the inconvenience of using high-preheat temperatures and the constraints imposed by special procedures make certain types of steel unsuitable for building construction. Steels conforming with AS/NZS 1163 (all grades), AS/NZS 3678/3679 (Grades 200, 250, 300 and 350) are weldable without the use of preheat, subject to certain limits.

Welding consumable The weld metal in (covered) rod or wire form, prior to being melted and deposited as weld metal.

Wire/welding wire See Welding consumable.

8.5.3 Welding processes

A large number of welding processes are available for the joining of metals, but relatively few of these are in widespread use in steel fabrication. These are:

(a) Manual metal arc welding (MMAW)
    Welding consumable: stick electrode with flux coating
    Shielding medium: Gases and slag generated from flux coating
    Power source: generator, transformer or rectifier
    Deposition rate: low
    Suitability: extremely versatile, but low production rates increase the costs of welded fabrication.

(b) Semi-automatic metal arc welding
    (i) Gas metal arc welding (GMAW)
        Welding consumable: solid bare steel wire fed through gun
        Shielding medium: Gas-carbon dioxide (CO_2) or CO_2 mixed with Argon and Oxygen fed through gun

Power source: DC generator
Deposition rate: high
Suitability: most applications, except field welding.

(ii) Flux cored arc welding (FCAW)
Welding consumable: hollow steel tubular electrode, filled with flux, fed through gun
Shielding medium: may be used with or without inert shielding gas
Power source: AC transformer
Deposition rate: high
Suitability: most applications, including field welding, but good access is essential.

(c) Automatic metal arc welding: Submerged arc welding (SAW)
Welding consumable: solid steel wire electrode fed through gun
Shielding medium: granular flux fed through hopper at weld point
Power source: high-output AC transformer or generator
Deposition rate: very high
Suitability: best suited for long automated weld runs with excellent access.

(d) Stud welding (SW): special process for instant welding of steel studs (AS/NZS 1554.2).

(e) Electroslag welding: special process for welding thick plates and joints capable of depositing large volumes of weld metal in one automatic operation.

In terms of cost per kilogram of the deposited weld metal, the most costly welding process is MMAW, followed by GMAW and FCAW, while SAW is potentially the lowest-cost process. However, the choice of the welding process depends on other factors, among which are:

- accessibility of the weld runs, and the amount of turning and handling of components required to complete all welds in a member; the designer can at least partly influence the decision
- the inventory of the welding equipment held by the fabricator and the availability of skilled welders and operators
- the general standard established by the particular welding shop in achieving weld preparations and fitting tolerances
- the type of steel used for the structure and thickness of the plates
- the maximum defect tolerances permitted by specification.

The fabricator's welding engineer is the best-qualified person for choosing the optimal welding procedures to be used in order to produce welds of specified quality within other constraints. As far as the designer is concerned, it is the performance of the welds in the finished structure that is of primary concern, and the designer's safeguards are mainly in the inspection of welds.

8.5.4 Strength of welded joints

The term 'welded joint' embraces the weld metal and the parent metal adjoining the weld. A welded joint may fail in one of the following modes:

- ductile fracture at a nominal stress in the vicinity of the ultimate strength of the weld metal or the parent metal, whichever is the lower

- brittle fracture at a nominal stress lower than ultimate strength and sometimes lower than the working stress
- progressive fracturing by fatigue after a certain number of stress cycles
- other causes such as corrosion, corrosion fatigue, stress corrosion and creep, but these are relatively rare in steel structures.

The design objective of welded joints is to ensure that failure can occur only in a ductile mode and only after considerable yielding has taken place. This is particularly important, considering that yielding often occurs in many sections of a steel structure at, or below, the working loads as a result of the ever-present residual stresses and unintentional stress concentrations. Welded joints must be capable of undergoing a large amount of strain (of the order of 0.5%–1.0% preferably) without brittle fracture. To achieve this, the following factors must be controlled:

(a) The parent material must be ductile, or notch-tough, at the service temperature intended and for the thickness required

(b) The details of joints must be such that stress concentrations are minimised

(c) Reduction of ductility by triaxial stressing should be avoided at critical joints

(d) Weld defects should be below the specified maximum size

(e) Welded fabrication should not substantially alter material properties.

A welded structure made of Grade 300 steel having a Charpy V-notch impact energy value of at least 27 joules at the intended service temperature, and with a reasonable control over factors (b) and (e), will most likely behave in a ductile manner if loaded predominantly by static loads. Most building structures and industrial structures constructed during the past four decades have performed satisfactorily where they complied with these limitations.

Special-quality welding and special care in material selection and detailing are required for earthquake-resistant structures and structures subjected to fatigue and/or low service temperatures. Further information on these areas can be found in WTIA [2004].

The degree of care in preserving the ductile behaviour of the members and parts stressed in tension increases where the following factors occur: medium- and high-strength steels, thick plates and sections, complex welded joints, low temperatures, dynamic and impulse loads, metallurgical inclusions and welding defects. Impact testing of the materials to be used and procedural testing of the weldments are essential in safeguarding ductility.

8.5.5 Specification and validation of welded construction

The specification embraces the working drawings and written technical requirements containing complete instructions for welded fabrication and erection, including the permissible tolerances and defect sizes. The term 'validation' applies to a multitude of safeguards necessary to ensure that the intentions of the design (assumed to be fully specified) have been realised in the completed structure. The purpose of the specification is to communicate to the firm responsible for fabrication and erection all the geometrical and technical requirements for the particular project, and should include at least the following information.

On the working drawings:

(a) plans, sections and general details describing the whole structure

(b) the types and the sizes of all welds

(c) any special dressing of the exposed weld surfaces by grinding, etc.

(d) the minimum tensile strength of the welding electrodes, where more than one strength is used

(e) the class of welding: GP, SP or special-quality welding

(f) special welding sequences, and weld groove preparations where these are critical for design

(g) identification of the critical joints that will be subject to radiographic and/or ultrasonic inspection

(h) the relevant welding Standard (e.g. AS/NZS 1554.1, etc.).

In the technical requirements (specification):

(a) the chemical and the physical requirements for the steel sections and plates (this is generally done by specific reference to material Standards)

(b) the weld defect tolerances for each of the classes of the welds used

(c) alignment and straightness tolerances for welded joints

(d) special post-weld treatment of welds by peening, post-weld heat treatment, and the like, where required by design

(e) the nature, type and frequency of welding prequalification procedures and inspections

(f) whether the personnel involved in welding would be subjected to testing of any particular kind

(g) the type and frequency of non-destructive testing: visual, magnetic particle, radiographic, ultrasonic or other

(h) the type and frequency of material testing for physical validation of the welded joints

(i) other requirements, such as the use of low-hydrogen electrodes for manual metal arc welding, or the minimum preheating of the material prior to welding aimed at reducing the risk of cracking.

Items (b) to (i) in the technical specification may be covered by appropriate references to the welding Standard (e.g. AS/NZS 1554.1, etc.).

The specification has probably a greater impact on the economy of welded construction than all other considerations, mainly because of the various clauses covering the quality of workmanship and defect tolerances. The designer would ideally prefer the highest standards of workmanship to achieve a virtually defect-free weld. This may be technologically possible, but the cost of achieving such a goal would be prohibitive. The only practical solution is to specify weld defect tolerances and to introduce several weld-quality categories for the designer to choose from.

AS/NZS 1554.1 provides for two weld categories. Weld category SP is intended to be used for relatively high-stressed welds, and category GP for low-stressed and non-structural welds. Valuable guidance for correct choice may be found in Technical Note 11 (WTIA [2004]). From the designer's viewpoint, the validation of welds, or quality assurance, is particularly important so that the assumptions made in the calculations of strength of the structure can be verified. There are three stages in validation:

(a) Setting-up stage: selection of materials, welding electrodes and other welding consumables, procedural tests, testing of welders.

(b) Working stage: checking of the preparations for the weld grooves and fusion faces, preheat temperatures, slag removal, soundness of each weld run in a multi-run weld, the weld contour, and other defects capable of visual inspection.

(c) Post-weld stage: detection of defects that can reduce the strength of the structure below the acceptable limits using visual and other non-destructive techniques.

All three stages may be involved on critical projects where it is essential to set up an early warning system so that any problems detected in the shop can be resolved by prompt action without waiting for the final validation. On less critical projects, especially where static loads predominate and brittle fracture is not a serious threat, stage (c) may be sufficient.

The frequency of testing, or the percentage of the total length of weld to be examined, depends on many factors, such as:

- the nature of the load (static or dynamic) and stress level
- the consequences of the risk of failure
- the susceptibility of the weld type to welding defects (butt welds are more prone to cracking than fillet welds)
- the susceptibility of the specified steel to cracks
- the standard of workmanship in the particular welding shop.

8.5.6 Selection of weld type

Welded joints may be divided into butt splices, lap splices, T-joints, cruciform and corner joints. For each of these joints there is a choice of three main types of welds: butt, fillet or compound. Figure 8.10 notes some useful information on joint (butt splices and T-joints shown) and weld types (butt welds shown).

From purely strength considerations, butt joints are preferable. However, the care required in preparing the plates for welding plus carrying out the welding makes these welds relatively costly. Fillet welds, in contrast, require only minimal weld preparations and are more straightforward in execution, therefore less costly. Of course, from the structural point of view, fillet welds are inferior to butt welds because they substantially alter the flow of stress trajectories, and this becomes a serious drawback in welded joints subject to fatigue.

Compound welds consisting of a butt and a fillet weld are often used to provide a smoother transition. This is often done to reduce the stress concentrations at the corners. The choice of the weld groove type for butt welds is a matter of economy and reduction of distortion during welding. Butt welds of the double-vee, double-J and U type have considerably less weld metal than bevel butts and therefore require less labour and welding consumables for their execution.

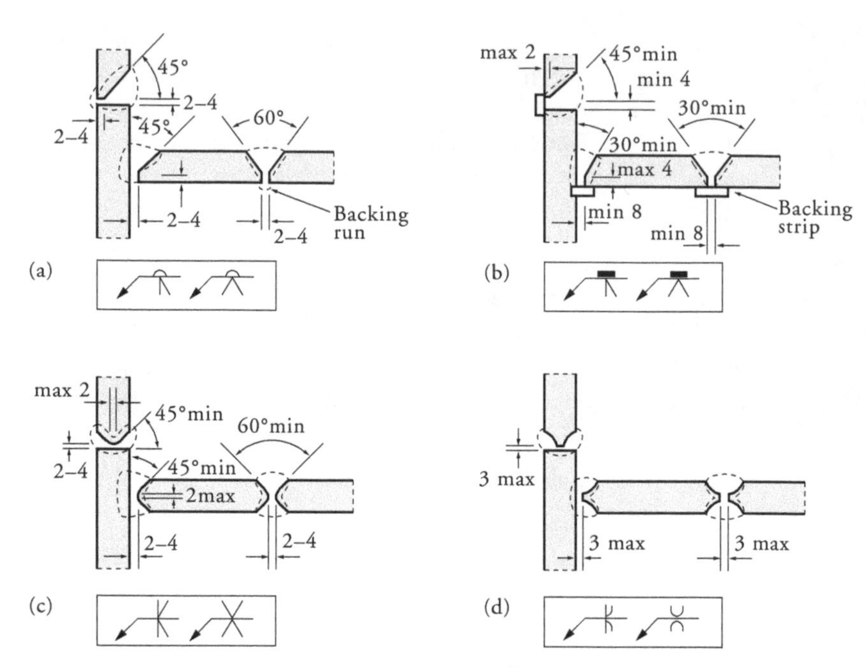

Figure 8.10 *Weld preparations and weld configurations for manually welded butt splices and T-joints: (a) single bevel and single vee with backing run; (b) single bevel and single vee with backing strip; (c) double bevel and double vee, and; (d) double J and double U*

8.5.7 Avoidance of lamellar tearing

Care is needed in detailing welded joints of the T and corner type, where restrained weld shrinkage forces are transmitted through the weld into a plate in the through-thickness direction, as shown in Figure 8.11. The tensile strength in the through-thickness direction of a relatively thick plate, say over 20 mm, is lower than the tensile strength in the plane of the plate. The reduction in strength has been attributed to non-metallic inclusions produced during the plate rolling process. Modern steels are manufactured with greater care and exhibit good through-thickness properties. Nevertheless, it is advisable when detailing welded joints in thick plates to avoid details that in the past have led to lamellar tearing. Figure 8.11 shows some acceptable and unacceptable joint types. Fabrication shops can play an important role in combating this problem by using suitable welding procedures. For critical joints involving plates more than 30 mm thick, the designer may consider specifying the use of through-thickness ultrasonically scanned plates. The scanning is carried out by the steelmaker (and a price surcharge applies).

8.5.8 Economy in detailing welded connections

The principal aim in detailing connections is to achieve the lowest-cost connection having an adequate strength and performance in service. The main cost components are material and labour, the latter being predominant. ASI [2009b] and Watson et al. [1996] provide some useful information and guidance in this regard.

Material savings can be achieved by:

- reduction in the number of parts making up a connection

- reduction in the volume of the deposited weld metal by choosing efficient weld types and weld groove shapes.

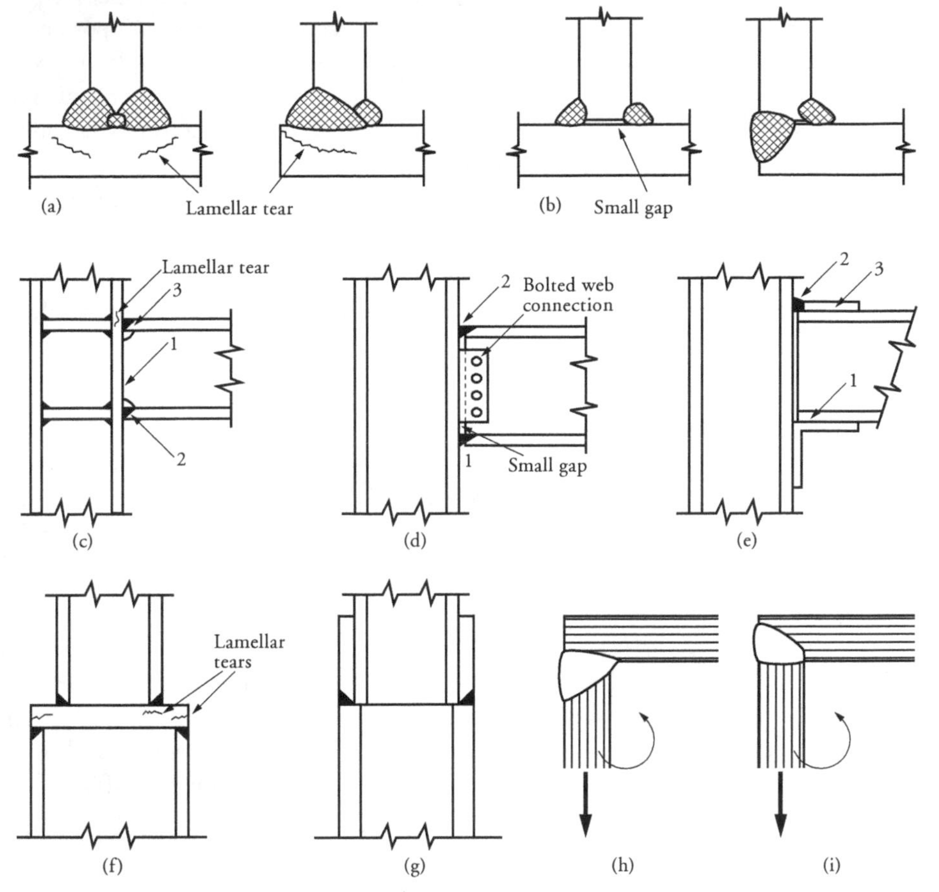

Figure 8.11 *Detailing of weldments to avoid lamellar tearing (a, h) should be avoided if at all possible; (b) better details; (c, f) sometimes susceptible to lamellar tearing; (d, e, g, i) safe detail. Note: For modern steels, the above may mainly apply for thick to very thick "plate" elements.*

Labour cost savings from these measures can be further enhanced by:

- standardisation of connections
- simplicity of the detail
- symmetrical arrangement of detail
- good access for welding and inspection
- realistic specification that matches the class of welding and inspection to the required performance
- using fillet welds in preference to butt welds, except where there are specific design reasons to the contrary
- avoiding the use of unnecessarily large sizes of fillet welds.

In statically loaded structures not subject to low service temperatures, fillet welds can be successfully used for most joints and details, except for flange and web splices without end plates. It is desirable to leave some freedom of choice for longitudinal welds (such as web-to-flange joints) to the fabricator, who may be well equipped to use deep-penetration fillets instead of normal fillets.

Detailing of joints subject to fatigue or other special circumstances requires a different approach, because strength considerations become much more important than economy in terms of the capital cost only. In such instances, fillet welds are avoided as far as possible because they cause stress concentrations and reduce the fatigue life of the joint. Detailing in general must be done with more care, so that abrupt changes in thickness or direction are completely avoided.

8.5.9 Weld metal consumables

As noted in Section 1.14(a), the amendment to AS 4100 (AS 4100 AMD 1—see Appendix D) has seen the following changes to Weld Metal Consumables:

(a) New process based Weld Metal Consumable designations. W40X and W50X still remain for Submerged arc weld metal consumables (to AS 1858.1) as that consumable Standard has not been converted to ISO. However, for Manual metal arc ("stick") processes (to AS/NZS 4855) the E4XX designation has been dropped totally and replaced by designations as A-E42, A-E46, B-E49XX, etc. The same occurs for Flux cored arc (to AS/NZS ISO 17632), Gas metal arc (to AS/NZS 2717.1 & ISO 14341) and Gas tungsten arc (to ISO 636) welding processes with differing designations.

(b) Prior to AS 4100 AMD 1, Table 9.7.3.10(1) of AS 4100 provided the nominal tensile strength of weld metals (f_{uw}) based on either the E41XX/W40X and E48XX/W50X weld metal consumable types. In this instance f_{uw} is 410 MPa and 480 MPa respectively. However, the AS 4100 AMD 1 (see Appendix D also) now has three weld metal consumable types with f_{uw} of 430, 490 and 550 MPa.

(c) AS 4100 AMD 1 (see Appendix D also) also introduces seven weld metal consumable types in Table 9.7.3.10(1) of AS 4100 for quenched and tempered steels which have f_{uw} of 430, 490, 550, 620, 690 760 and 830 MPa.

Industry practice till now has seen the use of either the lower strength E41XX/W40X or higher strength E48XX/W50X weld metal consumable type being used in engineering drawings and specifications with the latter designation being the standard used in structural steel design. However, there may be some confusion with the above noted amendment changes to Table 9.7.3.10(1) of AS 4100.

The Australian Steel Institute (ASI) has now published a Technical Note on the matter. ASI [2012b] provides useful background information on the changes to weld metal consumable designation and concludes that structural engineers will have to clearly identify on the structural drawings and in the specification the:

- weld size
- weld category
- nominal tensile strength of the weld metal (f_{uw}) as noted in Table 9.7.3.10(1) of AS 4100 AMD 1 (see Appendix D also)

Nominating f_{uw} in lieu of weld metal designation (e.g. A-G46, A-T46, W50X etc) is more workable from a design and specification perspective as the new designation method is dependent on welding process which is in the realm of the fabricator and determined after design. To perpetuate the standardised parameters previously noted by ASI, it may seem reasonable to standardise f_{uw} to 490 MPa which is somewhat conservative and reasonably close to previously standardised 480 MPa (i.e. E48XX/W50X).

8.6 Types of welded joints

8.6.1 Butt welds

The principal advantages of butt welds are the simplicity of the joint and the minimal change of the stress path. Static tension test results indicate that their average tensile strength is practically the same as that of the base metal, provided that the weld is free of significant imperfections and its contour is satisfactory. The disadvantage of butt welds is that they require expensive plate edge preparation and great care in following the correct welding procedures. Butt welds may be classed as complete penetration or incomplete penetration (see Section 8.5.2 for a definition of these weld types). The following applications are usually encountered:

(a) Butt welds subject to a static tensile force: Complete penetration butt welds (CPBW) with convex contour are the best choice. Single-bevel or V-joints prepared for down-hand welding keep the welding costs low, but distortion needs to be kept under control.

(b) Butt welds subject to a static compressive force: Because the stresses in compression members are usually reduced by buckling and bearing considerations, these welds can be incomplete penetration butt welds (IPBW) designed to carry the load.

(c) Butt welds subject to an alternating or fluctuating tensile stress with more than 0.5×10^4 repetitions during the design life (fatigue): CPBW with flush contours and a very low level of weld imperfections are essential in this application.

(d) As for (c) but carrying a predominantly compressive force: CPBW with convex contour and good-quality welding are sufficient.

(e) Butt welds subject to shear forces: These welds are less sensitive to the shape and form of the weld contour and to weld imperfections. Often, it is satisfactory to use IPBW.

The use of IPBW is subject to certain limitations, as specified in Clause 9.7.2 of AS 4100. Typical uses of IPBW are seen in:

(a) Longitudinal welds connecting several plates or sections to form a welded plate girder, a box girder or column and similar, especially when thick plates are used and the stresses in the welds are too low to require a complete-penetration weld. One exception is in structures or members subjected to fatigue, where these welds are not permitted, for example in bridges, crane bridges, and certain machinery support structures.

(b) Transverse welds to column splices or to column base plates, where axial forces are too low to warrant the use of a CPBW, or are milled for contact bearing so that the weld acts only as a positioning device, but providing that the columns are not subject to tension other than that due to wind loads.

The strength of butt welds can be affected by weld imperfections, causing stress concentrations and reduction in the cross-sectional area. The reduction in strength is particularly severe in welded joints subjected to fatigue. Figure 8.12 illustrates some of the weld imperfections: namely, undercut, over-reinforcement, notch effect caused by lack of gradual transition from the thicker to the thinner material, slag inclusions, gas pockets (porosity) and incomplete fusion. No cracks in the weld metal/HAZ can be tolerated.

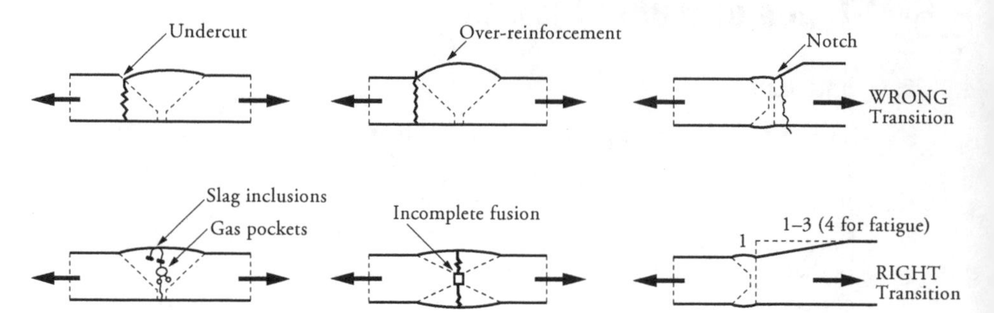

Figure 8.12 Weld defects in butt welds and how they influence (the position of) potential fracture

Figure 8.13 shows typical preparations of plates for manually deposited butt welds. The choice of preparation depends on economy and the need to control welding distortions. Double-V and U butt welds use less weld metal, and this can be a great advantage with thick plates.

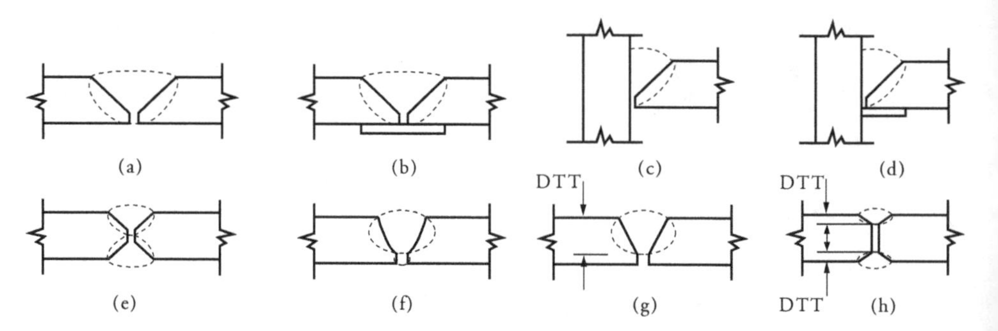

Figure 8.13 Plate edge preparations for complete penetration butt welds (CPBW)(except for (g)and (h)): (a) single vee butt; (b) single vee butt with backing bar; (c) bevel at T-joint; (d) bevel at T-joint with backing bar; (e) double vee butt; (f) U butt; (g)single vee preparation for incomplete penetration butt welds (IPBW), and (h) double vee preparation for IPBW. [Note: DTT = design throat thickness of IPBW].

IPBW produced by automated arc processes may possess deep penetration welds that can transmit loads that are higher than the more manual processes. Subject to the workshop demonstration of production welds, Clause 9.7.2.3(b)(iii) of AS 4100 permits the extra penetration to be added to the weld design throat thickness in this instance.

8.6.2 Fillet welds

The advantages and disadvantages of fillet welds can be stated in the following way:

(a) From the cost of fabrication perspective, fillet welds have an advantage over butt welds for the same force transmitted. This is mainly due to the absence of plate bevelling, which adds to the cost of weld preparation and fit-up required for butt joints. Also, the speed of welding, including all phases of preparation, is faster than for butt welds. For these reasons fillet welds are used to a much greater extent than butt welds for leg sizes smaller than 10 mm.

(b) As far as the distribution of stresses in welds are concerned, fillet welds are inferior to butt welds. The stress path through a side weld in a lap joint is not a direct one, and stress concentrations are always present; the same can be said for a fillet-welded T-joint or a cruciform joint. This is not a deterrent where the forces are predominantly static, as is the case with most building structures, as long as the design is carried out in accordance with established practice (see Figure 8.14).

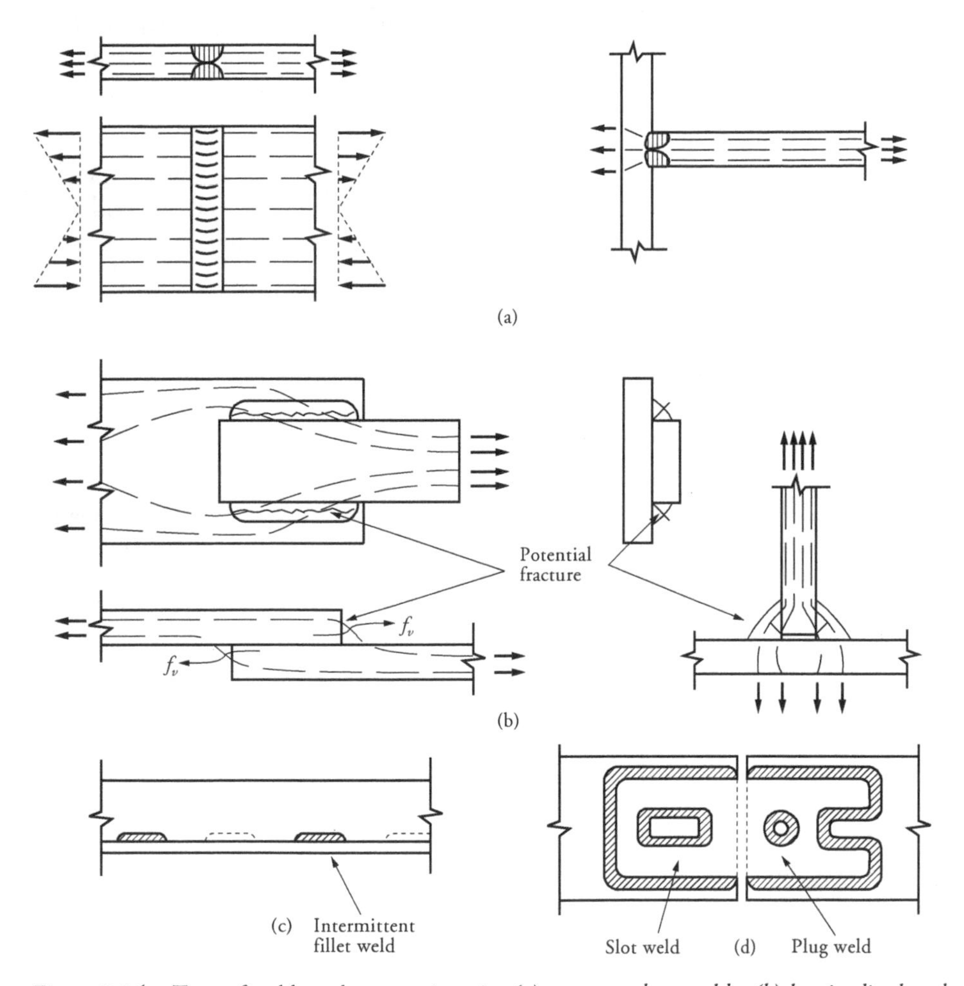

Figure 8.14 *Types of welds and stress trajectories (a) transverse butt welds; (b) longitudinal and transverse fillet welds; (c) intermittent fillet welds; (d) plug and slot welds*

(c) From the point of view of resistance to fatigue, fillet welds are inferior to butt welds. This is because there exists both an abrupt change in the direction of the stress trajectory and a notch-like effect at the root of the fillet weld. This results in stress concentrations and a triaxial stress state, and can lead to brittle fracture when the weldment is subjected to a large number of load cycles (fatigue failure).

The main uses of fillet welds are:

(a) For lap splices. The transfer of force from one plate to another is through shear in the weld. Fillet welds can be arranged to be parallel with the member axial force (longitudinal welds) or at right angles to it (transverse welds), or a combination of both.

(b) For T-joints. The two modes of transfer of forces are: compression or tension and shear through weld.

(c) For corner joints similar to T-joints.

(d) For structural plug and slot welds. Non-structural plug welds are permitted to be filled in flush with the surface of the plate, but such welds are rarely sound and they contain many cracks. Fillet welds run around the periphery of the hole can reliably be used to transmit the forces.

Where the forces transmitted by fillet welds are relatively small and the structure is not exposed to weather, it may be advantageous to use intermittent welds. Their benefits include using less filler metal and causing less distortion during welding. They may not show cost savings, however, because of frequent stop–start operations.

The inspection of fillet welds can usually be specified to include inspection during the preparation of material, fit-up and actual welding. Typical weld defects found in fillet welds are shown in Figure 8.15 as well as design concepts and terminology. Inspection must, of course, ascertain that the leg size and weld length specified in the design have been achieved.

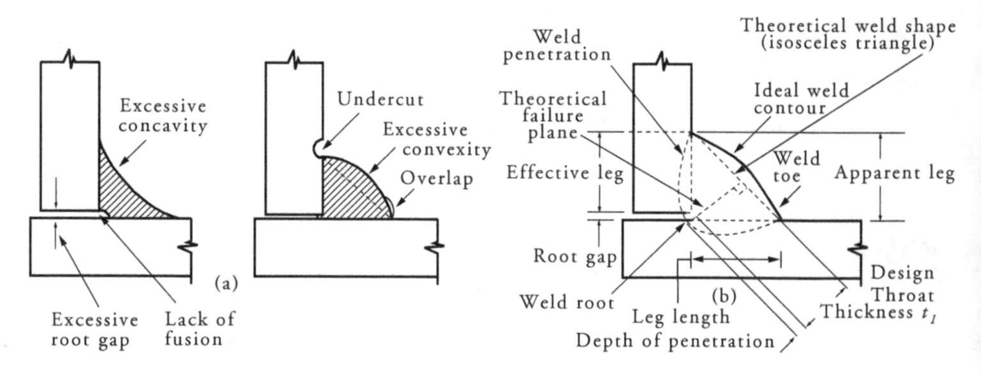

Figure 8.15 Fillet welds: (a) typical defects, and; (b) concepts and terminology.

The fact that the strength of fillet welds has a direct relationship to the nominal tensile strength of the weld consumable used leads to the necessity of specifying on the drawings not only the physical weld size but also the type of weld consumable to be used, especially when E48XX/W50X or higher-grade weld consumables have been assumed in the design.

Fillet welds produced by submerged arc welding will have a deep penetration into the root area, which is beneficial because a larger throat thickness is obtained; thus, for the same leg length, deep-penetration fillet welds will carry larger forces per unit length than manual welds. In order to increase the economy of welding, this type of deep penetration weld should be specified by the throat thickness rather than leg length, and the effective throat thickness calculated as a sum of 71% of the leg length plus 85% of the depth of penetration (Clause 9.7.3.4 of AS 4100). For this to occur, procedural tests are required to demonstrate that the specified weld dimensions have been achieved in the welding workshop.

8.6.3 Compound welds

A compound weld is considered to be a hybrid of a fillet and butt weld—i.e. by definition in AS 1101.3, the former weld type is superimposed onto the latter.

The design throat thickness (DTT) of a compound weld depends on whether there is a complete penetration butt weld (CPBW) or an incomplete penetration butt weld (IPBW) present. That is for a compound weld with:

- CPBW—the DTT is the size of the butt weld without reinforcement, and for;
- IPBW—the DTT is the shortest distance from the root of the IPBW to the face of the fillet weld.

Figure 9.7.5.2 of AS 4100 explains the compound weld configuration and the evaluation of the DTT.

8.7 Structural design of simple welds

8.7.1 Butt welds

8.7.1.1 General

Butt welds can be regarded as being integral to the parent metal, with the limiting stresses applicable to the parent metal also applying to the welds. As noted in Section 8.6.1, butt welds can be broadly split into two groups—complete penetration butt welds and incomplete penetration butt welds. This is not only due to the depth of weld fusion through the parent metal thickness but also in the methods used to assess their respective design capacities.

8.7.1.2 Complete penetration butt welds (CPBW)

Clause 9.7.2.7(a) of AS 4100 notes that the design capacity of a CPBW is equal to the nominal capacity of the weakest part being joined multiplied by a capacity reduction factor, ϕ, which is commensurate with the weld quality. From Table 3.4 of AS 4100, $\phi = 0.9$ for CPBW with SP quality and $\phi = 0.6$ for CPBW with GP quality. This applies to CBPW subject to transverse and shear loads.

Based on the above, for two similar plates joined by a CPBW with SP quality ($\phi = 0.9$) welded to AS/NZS 1554.1 or AS/NZS 1554.5, the AS 4100 definition notes that the weld is as strong as the joined plate elements and no further calculation is required (if the plates have been already sized for the design loads). If the lower quality GP category is used instead of the SP category for this connection type (i.e. with $\phi = 0.6$), the CPBW will have a lower design capacity than each of the two similar connected plates by a factor of (0.6/0.9=) 0.667.

8.7.1.3 Incomplete penetration butt welds (IPBW)

As the weld fusion in a IPBW does not cover the full depth of the joint, Clause 9.7.2.7(b) of AS 4100 states that IPBW are to be designed as fillet welds (see Section 8.7.2). The design throat thickness for IPBW are noted in Clause 9.7.2.3(b) of AS 4100, Section 8.6.1 of this Handbook and shown typically in Figure 8.13(g) and (h). The capacity reduction factor, ϕ, for IPBW is also the same as that for fillet welds.

8.7.2 Fillet welds

Stress distribution in a fillet weld is extremely complex, and certain simplifying assumptions are necessary to facilitate the design. The usual assumptions are:

(a) The failure plane intersects the root of the fillet and has an inclination such that it is at right angles to the hypotenuse of the theoretical weld shape of a 90-degree isosceles triangle (with the corner at the 90-degree angle being regarded as the weld root). See Figure 8.15(b).

(b) The stresses (normal and shear) on this failure plane are uniformly distributed.

The above assumptions become quite realistic at the ultimate limit state of the weld as plastic deformations take place. In general, the resultant forces acting on the failure plane may be composed of:
- shear force parallel to the weld longitudinal axis
- shear force perpendicular to the weld longitudinal axis and in the theoretical failure plane (Figure 8.15(b))
- normal force (compressive or tensile) to the theoretical plane (Figure 8.15(b)).

Clause 9.7.3.10 of AS 4100 provides a method for evaluating the design capacity of single fillet welds. The method is based on the premise that the capacity of a fillet weld is determined by the nominal shear capacity across the weld throat/failure plane (Figure 8.15(b)) such that:

v_w = nominal capacity of a fillet weld per unit length

$= 0.6 f_{uw} t_t k_r$

where

f_{uw} = nominal tensile strength of the weld metal

t_t = design throat thickness (see Figure 8.15(b), for equal leg fillet welds, t_t is equal to $t_w/\sqrt{2}$ where t_w = the fillet weld leg length)

k_r = reduction factor to account for welded lap connection length (l_w)

$= 1.0$ for $l_w \leq 1.7$ m

$= 1.10 - 0.06 l_w$ for $1.7 < l_w \leq 8.0$ m

$= 0.62$ for $l_w > 8.0$ m

The typical nominal tensile strengths, f_{uw}, used by the Australian/NZ steel construction industry are 410 MPa (i.e. E41XX/W40X) and 480 MPa (E48XX/W50X)—see Section 8.5.9 also. The design throat thickness of a equal-leg fillet weld is taken as $\sqrt{2}$ times the leg length (see Figure 8.15(b)).

The reduction factor, k_r , accounts for non-uniform shear flows that occur for long lengths of longitudinal welds in lap connections. Note this weld length must be greater than 1.7 m for the reduction factor to "kick-in"—hence k_r = 1.0 is generally used. The fillet weld is considered adequate if:

$$v^* \leqslant \phi v_w$$

where, from Table 8.12, ϕ is either 0.8 (for SP quality welds) or 0.6 (for GP quality welds) or 0.7 (for SP category longitudinal welds to RHS with $t < 3$mm) and v^* is the design force per unit length of weld. Note that, unlike other design Standards, this "force" is taken as the vector resultant of all the forces acting on the fillet weld and, hence, it is independent of direction of force. Values of ϕv_w for typical fillet weld sizes are listed in Table 8.11.

As noted in Section 8.7.1.3, the design capacity of an incomplete penetration butt weld is determined in the same manner as for fillet welds. In this instance, v_w is calculated with k_r = 1.0 and t_t can be evaluated by Clause 9.7.2.3(b) of AS 4100.

The minimum leg size of a fillet weld is governed by the thickness of the thinnest plate joined. The limitation is due to the difficulties in obtaining a sound weld if the plate thickness is much larger than the weld size.

It is not uncommon in the fabrication shop to set the plates slightly apart so as to obtain a small gap, which helps with control of weld shrinkage stresses. Gaps can also occur because of poor fit-up. When this occurs, the fillet leg size must be increased by the gap width, otherwise the effective throat thickness will be reduced.

Table 8.11 Design capacities of equal-leg fillet welds (in kN per 1 mm weld length)

Leg size	**SP welds**		**GP welds**		**Note**
t_w (mm)	E41	E48	E41	E48	
3	0.417	0.489	0.313	0.367	E
4	0.557	0.652	0.417	0.489	E
5	0.696	0.815	0.522	0.611	E
6	0.835	0.978	0.626	0.733	P
8	1.11	1.30	0.835	0.978	P
10	1.39	1.63	1.04	1.22	S
12	1.67	1.96	1.25	1.47	S

Notes: 1. E = economy size for welds carrying relatively small forces; P = preferred sizes, single-pass welds; S = special sizes for transmission of large forces where multi-pass welding is unavoidable.
2. E41 refers to E41XX/W40X (with f_{uw} = 410 MPa) and E48 refers to E48XX/W50X (with f_{uw} = 480 MPa) welding consumables. See Section 8.5.9 also.
3. For SP longitudinal fillet welds to RHS with $t < 3$ mm, multiply the listed SP design capacities by 0.875 (= 0.7/0.8) which is due to the differing capacity reduction factor for this type of parent material weld.

8.7.3 Compound welds

As noted in Clause 9.7.5.3 of AS 4100, the strength limit state design of compound welds shall satisfy the strength requirements of a butt weld (Section 8.7.1).

8.8 Analysis of weld groups

8.8.1 General

Analysis of weld groups is greatly simplified when the following assumptions are made:

(a) The welds are regarded as homogeneous, isotropic and elastic elements.

(b) The parts connected by welding are assumed to be rigid, but this assumption should not be made if there is doubt about the rigidity of adjoining plates.

(c) The effects of residual stresses, stress concentration and triaxial stress conditions are neglected on the assumption that the ultimate strength of weld groups is not significantly affected by these parameters.

Table 8.12 Capacity reduction factors, ϕ, for welds

Type of weld	Weld category SP	GP
Complete-penetration butt welds	0.90	0.60
Longitudinal fillet welds in RHS tubes (t < 3.0 mm)	0.70	NA
Other fillet welds, incomplete-penetration butt welds and weld groups	0.80	0.60

Note: NA stands for not applicable.

8.8.2 Weld groups subject to in-plane actions

The following assumptions are made.

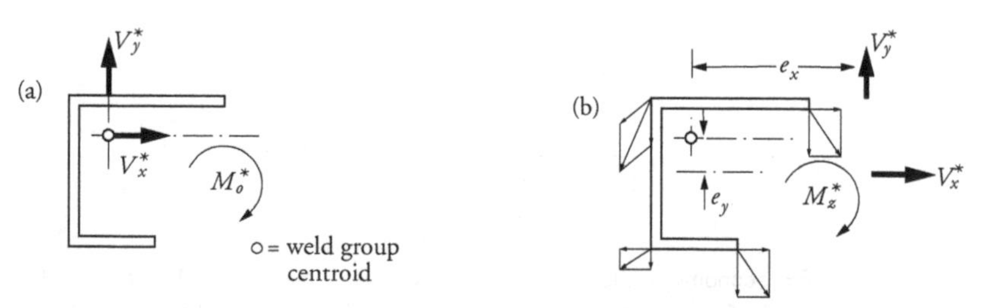

Figure 8.16 *Weld group loaded by in-plane actions: (a) Geometry and resolved actions about centroid, and; (b) initial in-plane actions.*

The plate elements being joined by fillet welds behave rigidly in the plane of the weld group. Design actions (V_x^*, V_y^*, M_z^*) applied away from the centroid of the weld group (Figure 8.16(b)) may be treated as being applied at the centroid plus moments (Figure 8.16(a)) with forces V_x^*, V_y^* and resolved moment (using the sign convention in Figure 8.16):

$$M_o^* = \Sigma(V_x^* e_y + V_y^* e_x) - M_z^*$$

The procedure for the analysis and design of weld groups subject to in-plane loadings is similar to that encountered for bolt groups loaded under the same conditions. The method follows the detailed proofs and outcomes from Hogan & Munter [2007a]. The resultant force per unit length, v^*_{res}, at the most critically loaded part of the weld group subject to in-plane forces and eccentric moment is:

$$v^*_{res} = \sqrt{[(v^*_x)^2 + (v^*_y)^2]}$$

where forces in the welds, per unit length, are:

$$v^*_x = \frac{V^*_x}{l_w} - \frac{M^*_o y_s}{I_{wp}}$$

$$v^*_y = \frac{V^*_y}{l_w} + \frac{M^*_o x_s}{I_{wp}}$$

where x_s and y_s are the for the weld segment farthest from the centroid of the weld group, l_w the total length of the weld in the weld group, and the polar second moment of area of the weld group is:

$$I_{wp} = \Sigma(x_s^2 d_s + y_s^2 d_s)$$

x_s and y_s are the coordinates of a weld segment, and d_s is its length, thus:

$$l_w = \Sigma d_s$$

8.8.3 Weld groups subject to out-of-plane actions

The analysis for out-of-plane actions on a weld group uses the same assumptions as adopted for in-plane actions. The results for rigid plate elements are similar to those of the in-plane actions. Should the transverse plate element be flexible, it will be necessary to neglect forces in welds in the flexible parts of the transverse plate element.

Allowing for coordinate axis changes from those in in-plane actions, the out-of-plane actions about the weld group centroid are the forces, V^*_y and V^*_z (see Figure 8.17), and the resolved moment:

$$M^*_o = \Sigma(V^*_y e_z + V^*_z e_y) - M^*_x$$

In many situations, V^*_z, e_y and M^*_x are taken as zero. Generally, the resultant force per unit length, v^*_{res}, at the most critically loaded part of the weld group subject to out-of-plane forces and eccentric moment is:

$$v^*_{res} = \sqrt{(v^*_y)^2 + (v^*_z)^2}$$

where forces in the welds, per unit length, are:

$$v^*_y = \frac{V^*_y}{l_w}$$

$$v^*_z = \frac{V^*_z}{l_w} + \frac{M^*_o y_s}{I_{wx}}$$

where y_s is for the weld segment farthest from the centroid of the weld group, l_w the total length of the weld in the weld group and the second moment of area about the x-axis of the weld group is:

$$I_{wx} = \Sigma y_s^2 d_s$$

y_s is the coordinate of a weld segment, and d_s is its length, thus:

$$l_w = \Sigma d_s$$

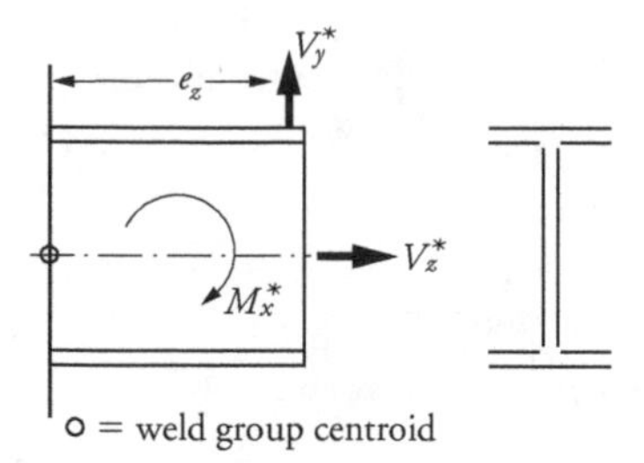

Figure 8.17 Weld group with out-of-plane actions.

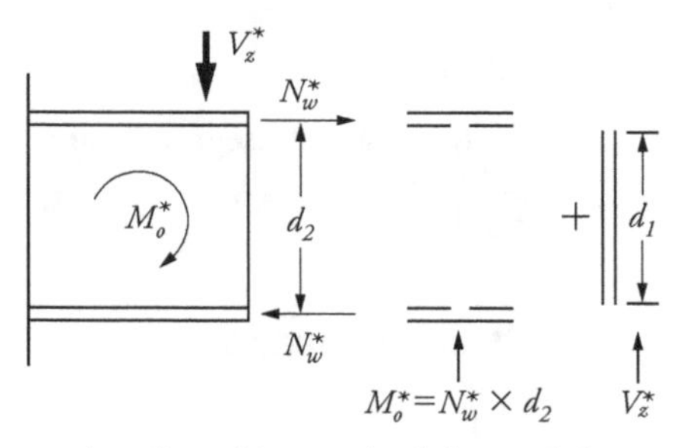

Figure 8.18 Alternative procedure for weld group loaded out-of-plane

A simpler alternative method to is break the weld group up into sub-groups based on the most significant form of loading seen by the sub-group. A case in point is where the weld group follows the perimeter of an I-section. The welds around the flanges are assumed to resist the full bending moment and the welds about the web resist the total shear force (see Figure 8.18). It is assumed that the fillet welds are ductile enough to allow some redistribution of internal forces. The method is executed as such:

N_w^* = flange forces (separated by a distance d_2 between flange centroids)

$$= \frac{M_o^*}{d_2}$$

The flange fillet welds then *each* resist the out-of-plane force N_w^* which is assumed to be uniformly distributed, i.e.

$$N_w^* \leqslant \phi v_{vf} l_{wf}$$

where

ϕv_{vf} = design capacity of flange fillet welds per unit weld length (Section 8.7.2 and Table 8.11)

l_{wf} = perimeter length of each flange fillet weld

The web fillet weld sub-group is assessed in the same manner by:

$$V_z^* \leqslant \phi v_{vw} l_{ww}$$

where

ϕv_{vw} = design capacity of web fillet welds per unit weld length (Section 8.7.2 and Table 8.11)

l_{ww} = perimeter length of web fillet welds

$= 2d_1$

Light truss webs are often composed of angles, being either single or double members. The balanced detailing of connections is important and, ideally, should be done so that the centroid of the connection is somewhat coincident with the centroidal line of the connected member (see Figure 8.19). However, as noted in Section 7.5.4, much research has been done to indicate that balanced connections are not required for statically (and quasi-statically) loaded structures as there is no significant decrease of connection capacity for small eccentricities. However, balanced connections are considered to be good detailing practice for connections in dynamically loaded applications subject to fatigue design.

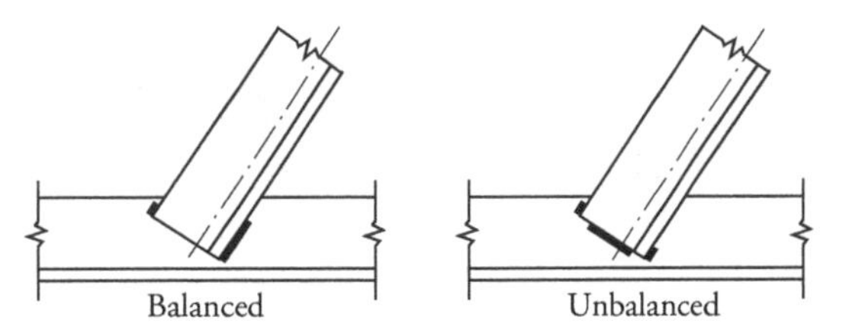

Figure 8.19 Balanced connections: Truss diagonal to chord connection

8.8.4 Combined in-plane & out-of-plane actions

Occasionally, welded connections are subjected to triaxial loadings. The general method described in Sections 8.8.2 and 8.8.3 can readily be adapted to deal with simultaneous application of in-plane and out-of-plane forces (see also Hogan & Munter [2007a]).

8.9 Design of connections as a whole

8.9.1 Design and detailing

8.9.1.1 General

Designing a connection as a whole means designing the part of the member being connected, the corresponding part of the other member or support, intermediate components as plates/gussets/brackets and the fasteners transferring the forces. Often other members connect to the same node, and so they have to be integrated into the node. The art of detailing connections and nodes is to use the simplicity and directness of force transfer. The tendency to excessive stiffening should be resisted in the interest of economy. Using slightly thicker material can produce an adequately strong connection at a lower cost.

The connection as a whole needs to be checked for strength and serviceability limit states. Careful attention should be paid to the trajectory of forces all the way from the connected member to the member being connected.

In detailing connections it is important to preserve the designer's intent with respect to the connection rigidity. If a pinned connection is intended, it should be detailed to rotate freely or at least to offer only minimal resistance to rotation. However, if a fixed connection was assumed in the analysis, then it should be detailed so as to offer adequate stiffness to resist joint rotation.

In general, structural steelwork connections can be considered to be composed of elements that work together much like the links in a chain. The weakest element ("link") controls the strength of the connection ("chain"). In the broadest sense, structural steelwork connections are composed of four types of elements:

- Bolts (or single-point type fasteners)
- Welds
- Components (i.e. connecting plates, gussets, cleats, brackets, etc.)
- (Supported/supporting) Member(s) at the joint

The ASI has published a valuable and significant suite of publications dealing with the above connection elements and overall steelwork connection models (Hogan & Munter [2007a-h], Hogan [2011], Hogan & Van der Kreek [2009a-e]). At the time of publication of this Handbook, these publications include:

(a) *Connection Handbook 1: Design of structural steel connections—Background and theory* (Hogan & Munter [2007a])

(b) *Connection Design Guide 1: Bolting in structural steel connections* (Hogan & Munter [2007b])

(c) *Connection Design Guide 2: Welding in structural steel connections* (Hogan & Munter [2007c])

(d) *Connection Design Guide 3: Simple Connections—Web side plate connections* (Hogan & Munter [2007d])

(e) *Connection Design Guide 4: Simple Connections—Flexible end plate connections* (Hogan & Munter [2007e])

(f) *Connection Design Guide 5: Simple Connections—Angle cleat connections* (Hogan & Munter [2007f])

(g) *Connection Design Guide 6: Simple Connections—Seated connections* (Hogan & Munter [2007g])

(h) *Design capacity tables for structural steel—Volume 3: Simple Connections—Open Sections: Web side plate, Flexible end plate, Angle cleat* (Hogan & Munter [2007h])

(i) *Connection Design Guide 7: Simple Connections—Pinned base plate connections for columns* (Hogan [2011])

(j) *Connection Design Guide 10: Rigid connections—Bolted moment end plates & Beam splice connections* (Hogan & Van der Kreek [2009a])

(k) *Connection Design Guide 11: Rigid connections—Welded beam to column moment connections* (Hogan & Van der Kreek [2009b])

(l) *Connection Design Guide 12: Rigid connections—Bolted end plate to column moment connections* (Hogan & Van der Kreek [2009c])

(m) *Connection Design Guide 13: Rigid connections—Splice connections* (Hogan & Van der Kreek [2009d]).

(n) *Design capacity tables for structural steel—Volume 4: Rigid Connections—Open Sections: Bolted moment end plates, Welded beam to column moment connection, Bolted and welded splices* (Hogan & Van der Kreek [2009e]).

(a) to (c) contain comprehensive background information and theory on the connection elements (bolts/bolt groups, welds/weld groups, components, and member(s) at a joint). (d) to (g) and (i) to (m) consider the design of the connection as a whole. (h) and (n) provide a consolidated summary of simple connection types noted in (d) to (g) and rigid connection types noted in (i) to (m). Design capacity tables for connections are provided in (d) to (n) and, in doing so, these publications also then provide standardized connections for the connections types considered. (See Section 8.9.2).

Apart from (d) and (i), the above ASI connection design models fundamentally consider structural steel Open Sections (e.g. I- or Channel type sections). Connection design models for structural steel hollow sections are considered in CIDECT [2008, 2009] and Syam & Chapman [1996]. At the time of publication of this Handbook, the ASI were working on a series of Tubular Steel Connection Design guides based on the above references and other key sources.

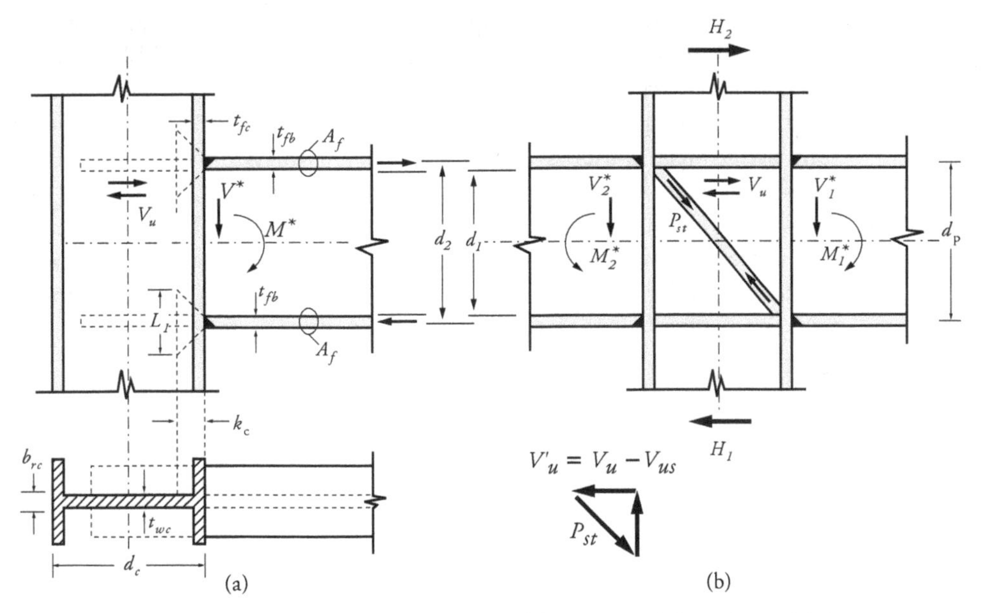

Figure 8.20 *Design of welded moment connections: (a) geometrical dimensions; (b) forces in column web and in stiffener plates (two beam connection shown but can be used for single beam connections)*

The design of a rigid welded connection (e.g. multi-storey beam to column connection or portal frame knee connection) is outlined here as an example. The connection detail shown in Figure 8.20 indicates the design action effects involved. Failure modes of the connection of this type (to name a few) are:

- weld failure, at beam flanges and web
- column web crushing failure (web yield)
- column web fracture (upper flange area)
- column web shear buckling failure
- column web compressive buckling.

The procedure for verifiying the connection capacity is as follows. A more comprehensive design model for this connection type can be found in Hogan & Van der Kreek [2009b].

8.9.1.2 Beam flange weld capacity

Using a simple procedure, it is assumed that the flange butt welds (CPBW) alone resist all of the bending moment (Figure 8.20(a)):

$$N_w^* = \frac{M^*}{d_2}$$

where d_2 is the mid-flange to mid-flange distance. As explained in Section 8.7.1.2 the nominal butt weld capacity is then:

$$N_w = f_{yf} A_f$$

where f_{yf} is the flange design yield stress and A_f is the area of one flange. For final check with $\phi = 0.9$ for SP quality welds:

$$N_w^* \leqslant \phi N_w$$

It should be noted that if this inequality is not satisfied then a total member check should be undertaken. Alternatively, if it is more than satisfied, then IPBW or continuous fillet welds should be considered.

Strictly speaking, no design checks are necessary for CPBW with SP weld category as long as the weld complies with AS 4100 and AS/NZS 1554.1/4/5. By definition in AS 4100, the nominal capacity of a CPBW is equal to the nominal capacity of the weakest part being joined. The design capacity of a CPBW is then calculated by multiplying the nominal member capacity by the capacity factor (ϕ), which for SP welds is 0.9. Hence, the design capacity of an SP CPBW is equal to the design capacity of the weakest member being joined.

8.9.1.3 Beam web weld capacity

Again using simple theory, the two web fillet welds carry all the shear force but no moment:

$$V_w = 2 v_{ww} d_1 t_t \quad = 2 \times (0.6 f_{uw} t_t k_r) d_1 t_t$$

where

V_w = nominal capacity of the web fillet weld group

v_{ww} = nominal capacity of the web fillet weld per unit length (Section 8.7.2)

f_{uw} = nominal tensile strength of the weld metal

t_t = design throat thickness of the fillet weld
k_r = lap length reduction factor (taken as 1.0 in this instance)
d_1 = clear web depth between flanges

Verify the web weld capacity with $\phi = 0.8$ for SP quality welds:

$$V^* \leqslant \phi V_w$$

8.9.1.4 Column web capacity in bearing (crushing)

Load from the beam flange is dissipated through the column flange a distance of 2.5 times the depth of dissipation, which is equal to the sum of the column flange thickness and the flange-web transition radius. This is distance k_c shown in Figure 8.20(a). The critical area of the column web is thus:

$$A_{cw} = (t_{fb} + 5k_c)\, t_{wc}$$

The bearing capacity of the web is thus:

$$R_{bc} = 1.25 f_{yc} A_{cw} \qquad \text{(see Section 5.8.5.2)}$$

The design beam flange force is conservatively:

$$R^*_{bf} = \frac{M^*}{d_2}$$

Verify capacity with $\phi = 0.9$:

$$R^*_{bf} \leqslant \phi R_{bc}$$

If the web capacity is insufficient, it will be necessary to stiffen the web. A web stiffener is in many ways similar to a beam bearing stiffener and should be designed to Clause 5.14 of AS 4100.

8.9.1.5 Column flange capacity at beam tension flange region

Several references (e.g. Hogan & Van der Kreek [2009b]) note that the column provides resistance to the beam tension flange pulling away by flexure of the column flanges and by the tension resistance of the "central rigid portion" connecting the column flanges to the column web. This overall resistance can be expressed as:

$$R_t = f_{yc}[(b_{sc} + 5k_c)t_{wc} + 6.25t_{fc}^2]$$

where f_{yc} is the design yield stress of the column, t_{wc} and t_{fc} are respectively the column web and flange thickness, b_{sc} is the stiff bearing dimension defined by t_{fc} (for butt welds) or $t_{fc} + 2t_w$ (for fillet welds with leg size t_w) and k_c is the distance on the column section from the outer flange face to inner termination of the root radius (Figure 8.20(a)). This capacity must then satisfy the following inequality with $\phi = 0.9$:

$$N^*_w \leqslant \phi R_t$$

8.9.1.6 Column web capacity in shear yielding and shear buckling

The column web panel bounded by the flanges is subjected to a design shear force of $(V^*_c + N^*_w)$ where V^*_c is the design shear force present in the column and N^*_w are the concentrated flange forces evaluated in Section 8.9.1.2.

The nominal column shear capacity, V_b, is determined in Sections 5.8.2 and 5.8.3:

$$V_b = 0.6\alpha_v f_{yc} d_1 t_w \leqslant 0.6 f_{yc} A_w$$

The capacity check now follows with $\phi = 0.9$:

$$(V_c^* + N_w^*) \leqslant \phi V_b$$

If the shear buckling capacity of the web is found to be inadequate, the capacity may be increased by a diagonal stiffener or by increasing the web thickness in the knee panel.

8.9.1.7 Column web capacity in compressive buckling

The unstiffened slender web may fail by compression buckling. The problem is similar to the beam web at a bearing support. Reference should be made to Chapter 5 (particularly Section 5.8.5) for a typical design method. In the event that web buckling occurs, it will be necessary to stiffen the web with a horizontal stiffener.

8.9.1.8 Other checks

Depending on the connection loadings, stiffness and other geometric conditions, subsequent investigations may be required for stiffener design for this connection type. Further information on this and the above connection design routines can be found in Hogan & Van der Kreek [2009b]. This reference is quite detailed and uses the above methodology of breaking the connection into components and investigates the component's strength and stiffness requirements. The above connection type is generally termed the "Welded Beam to Column Moment Connection".

8.9.2 Standardized connections

Standardized connections cover the most frequently used connection types. The advantage of using these connections is that they are rationalised to suit economical fabrication while providing the designer with predesigned details. An important feature of the standardized connections is that they are designed for bolting in the field and welding for fabrication. Experience over many years in Australia and New Zealand has shown that bolted field connections are more economical than the welded ones. The following field-bolted connection types are in widespread use:

- angle seat connection
- shear plate (bearing pad) connection
- flexible end plate connection
- angle cleat connection
- web side plate connection
- bolted moment end plate connection
- bolted splice connections.

Figure 8.21 shows examples of welded connections. See Section 8.9.1.1 for further information on standardized connections.

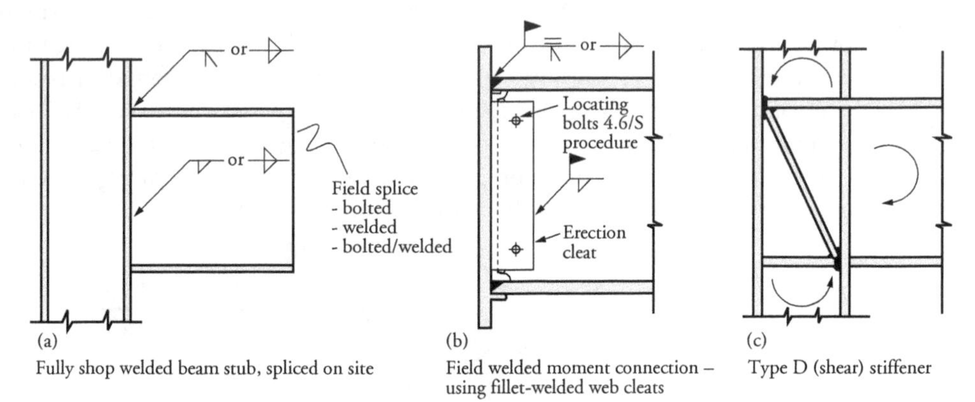

Figure 8.21 Typical welded connections

8.10 Miscellaneous connections

8.10.1 Hollow section connections

Structural hollow sections continue to gain favour with architects and engineers and are used in many structural forms. These include long-spanning applications, trusses, portal frame applications, portalised trusses, columns, bracing, etc. Their increased use comes from better engineering efficiencies and aesthetics for use in common and high profile applications.

Due to this there are now various connection types used for hollow sections. Many of these connection types are considered in CIDECT [1991, 1992, 2008, 2009], Syam & Chapman [1996], Packer & Henderson [1997], Packer et al. [2010] and Eurocode 3 (EC3). A key source document in this area is CIDECT [1991, 1992] which was then revised to CIDECT [2008, 2009] respectively. However, at the time of publication of this Handbook, CIDECT [2008, 2009] was not incorporated in EC 3 and all four CIDECT references may be used. A few of the aforementioned references have simple connection design models which can be developed from first principles via Standards as AS 4100. Other connection types require testing or other forms of analysis to determine behaviour. A few of the welded "tube-to-tube" connections—as noted in Figure 8.22—fall into this category.

In an N-type tubular lattice truss, for example, the web members connected to the chords transmit forces into the chord walls, creating a complex stress state. There are several modes of failure to be checked:

- punching through chord wall
- flexural failure of chord face
- tensile fracture of web members
- bearing failure of chord walls
- buckling of the chord side walls
- overall chord shear failure.

A large amount of research has been carried out in Europe, Canada, Japan and several other countries. Design rules have been formulated for easy use (as noted in the above mentioned references) to cover most of the connection types found in practice. At the time of publication of this Handbook, the ASI were working on a series of Tubular Steel Connection Design guides based on the above references and other key sources.

For more complex connections, recourse can be made to a finite element computer program using a material non-linearity option.

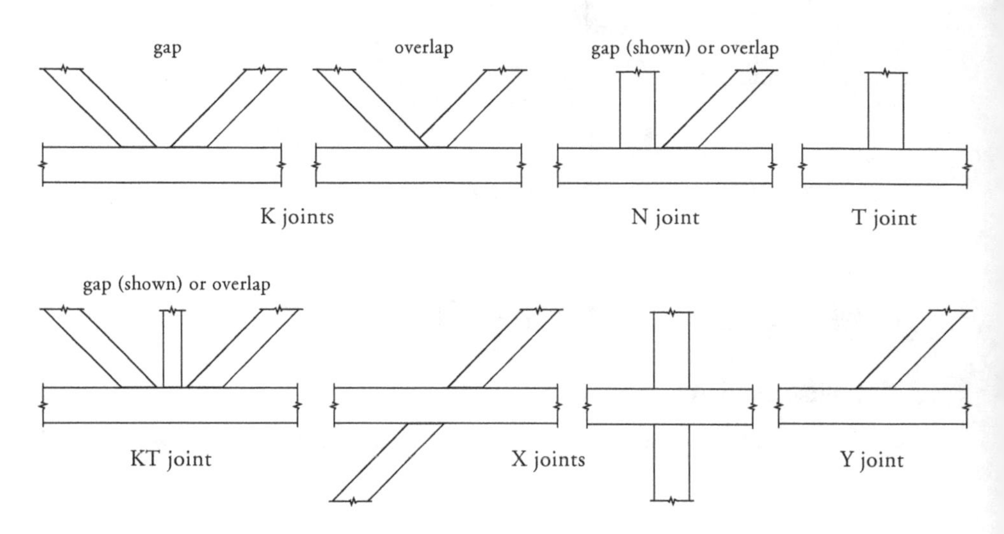

Figure 8.22 *Types of welded hollow section joints.*

8.10.2 Pin connections

8.10.2.1 General

Pin joints find their use where the connected parts must be free to rotate about one another during assembly and within the service life. The angle of rotation is usually very small—i.e. less than 1 degree. Often pins are used for the purpose of visually expressing the connection or to avoid the looks of an alternative large bolted joint. There are many precautions to be taken in the design of pins to ensure a satisfactory performance of these connections. Basically, the pin joint consists of eye plates, gussets, pin and pin caps. Optional radial spherical bearings may be needed to allow for the two-way rotations induced by lateral actions (Figure 8.23(e)).

The usual pin arrangement is a single "eye plate" (i.e. the plate element(s) connected to the *member*) fitted between two "gussets" (i.e. the plate element(s) at the support)—see Figure 8.23(a). Sometimes there are two member connected eye plates acting against a single gusset—Figure 8.23(b). Additional information on pin connections can be found in Riviezzi [1985].

The design of pin joints at the ends of tension members is covered by Clauses 7.5 and 9.5 of AS 4100. Clause 7.5 gives the required geometry for the connected plate element(s) and Clause 9.5 applies to the pins as fasteners with some reference to ply design. Clause 9.5 is also used for members subject to compression loads.

8.10.2.2 Ply design and detailing—Tension members

The two forms of pin joints used in practice are termed the 'dog-bone' and 'flush pin joint'.

(a) 'Dog-bone' form pin connection

The most materially efficient design, as shown in Figure 8.23(c), derives from the 'dog-bone' type shape. Clause 7.5 of AS 4100 is based on empirically derived provisions that were used in previous editions of the Australian/British Standards and successful past practice. These provisions are intended to prevent tear-out and plate "dishing" failures.

Clause 7.5(d) of AS 4100 implies the use of the more optimal requirement of constant plate thickness to avoid eccentricities of load paths within the connection—unless judicious placement of non-constant thickness plies are used. Applying the criterion of constant thickness leads to a design situation where the required width of the connected plate away from the connection, i.e. D_I as noted in Figure 8.23(c), is such that:

$$D_I \geqslant \frac{N^*}{\phi T_i}$$

where

N^* = design axial tensile force
ϕ = capacity factor = 0.9
T_i = min.$[f_{yi}t_i, 0.85k_{ti}f_{ui}t_i]$
f_{yi} = design yield stress of the eye plate/gusset element i
t_i = thickness of the eye plate/gusset element i
k_{ti} = correction factor for distribution of forces = 1.0 (away from the joint)
f_{ui} = design tensile strength of the eye plate/gusset element i

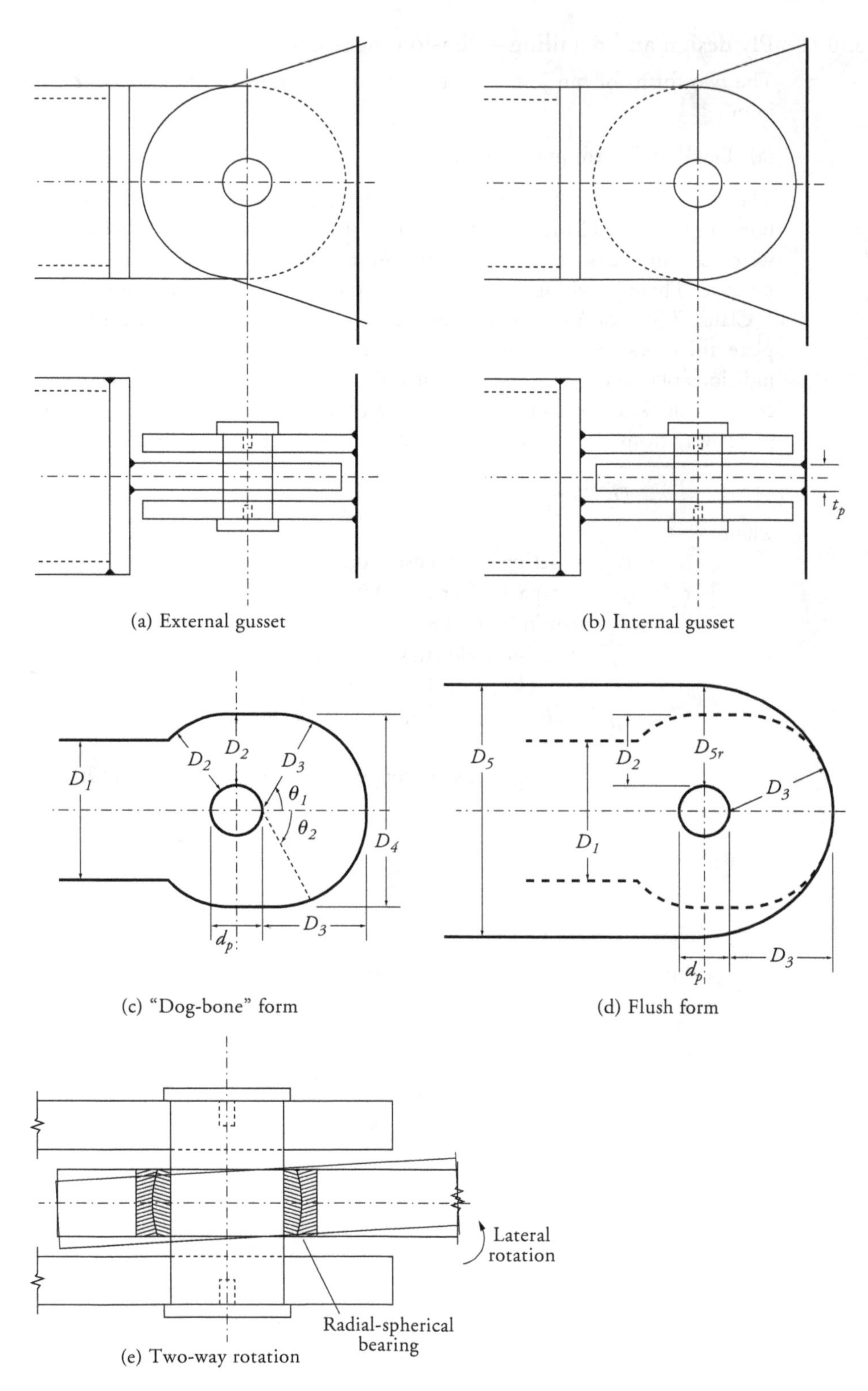

Figure 8.23 Pin connections

Clause 7.5(a) of AS 4100 requires t_i to satisfy the following for an unstiffened pin connection:

$$t_i \geqslant 0.25D_2$$

From Figure 8.23(c), Clause 7.5(b) of AS 4100 requires the following for D_3 on the eye plate/gusset with $\theta_1 = \theta_2 = 45$ degrees:

$$D_3 \geqslant D_1$$

This would seem generous according to research carried out at the University of Western Sydney and reported by Bridge [1999] which notes that $D_3 = 0.667D_1$ is more optimal. This research showed, at best, a very modest improvement in strength when using $D_3 = D_1$ in lieu of $D_3 = 0.667D_1$.

From Clause 7.5(c) of AS 4100, the overall width of the eye plate, D_4, is given by

$$D_4 \geqslant 1.33D_1 + d_p$$

where d_p is the pin hole/pin diameter.

It should be noted that the dog-bone design is seldom used in buildings due to the complex profile cutting of plates which produces additional effort and large amounts of off-cuts.

(b) Flush form pin connection

This type of connection (i.e flush edges with circular ends) is commonly used in tubular member to pin connections as depicted in Figure 8.23(d). The "design" required width of the connected plate, D_{5r}, is *conservatively* calculated in a similar manner to the dog-bone form (with k_t still taken as 1.0), i.e.

$$D_{5r} \geqslant \frac{N^*}{\phi T_i}$$

The overall width of the eye plate/gusset element is then:

$$D_5 = 2D_{5r} + d_p$$

This means that the distance from the edge of the eye plate/gusset to the edge of the pin hole is equal to the design required width of the connected plate, D_{5r} (= D_3 in Figure 8.23(d)).

In order to prevent local plate "dishing" under load, the eye plate/gusset thickness, t_i, is conservatively calculated to be not less than one-quarter of the effective outstand, i.e.:

$$t_i \geqslant 0.25D_{5r}$$

8.10.2.3 Ply design and detailing—Tension and/or compression members

Clause 9.5.4 of AS 4100 notes that the eye plate/gusset bearing and tearout provisions of Clause 9.3.2.4 for bolt design should also be observed for plies in pin connections, i.e.:

$$V_b^* \leqslant \phi V_b$$

where

V_b^* = design pin bearing force on ply element i

V_b = min.$[3.2 d_p t_i f_{ui}, a_e t_i f_{ui}]$

ϕ = capacity factor = 0.9

where d_p, t_i, f_{ui} are described above and the (directional) minimum edge distance from the edge of the hole to the edge of the ply, a_e, is explained in Section 8.4.2. Tension members with pin connections should observe these provisions as well as those listed in Section 8.10.2.2.

Bridge [1999], also recommends a further serviceability limit state check to safeguard excessive pin hole elongation:

$$t_p \geqslant \frac{N_s^*}{1.6\phi f_{yi} d_p}$$

where N_s^* is the serviceability limit state load and $\phi = 0.9$.

8.10.2.4 Design and detailing—Pins

(a) Pin in shear and bending

Pins are stressed in bending as well as shear due to transverse loads acting on the pin. For pins in double shear, and allowing for gaps between the plies and unavoidable eccentricity, the following calculation method is recommended. The strength limit state design actions on the pin are evaluated as:

$$V_f^* = \text{design shear force}$$
$$= \frac{1.2N^*}{2} = 0.6N^*$$

$$M^* = \text{design bending moment}$$
$$= \frac{N^*(2t_e + t_i + 2g)}{4}$$

where

N^* = member design axial load transmitted by the pin connection
t_e = external ply thickness
t_i = internal ply thickness
g = gap between plies required for ease of erection

The reason for the 1.2 shear force coefficient is that in practice there may be a less than perfect assembly of the eye plates/gussets with the possibility of lateral forces from wind and structural actions. The result is that the axial load could act eccentrically. Additionally, the pin must resist the whole load as a single element in a single load path situation which means the automatic analogy of the pin acting like a beam is not totally valid.

Clause 9.5.1 of AS 4100 notes that V_f^*, must satisfy:

$$V_f^* \leqslant \phi V_f$$

where

ϕ = capacity factor = 0.8
V_f = $0.62 f_{yp} n_s A_p$
f_{yp} = design yield stress of the pin
n_s = number of shear planes on the pin
= 2 for the design actions considered above
A_p = pin cross-section area

In terms of bending moments, Clause 9.5.3 of AS 4100 notes that M^* must satisfy:

$$M^* \leqslant \phi M_p$$

where

ϕ = capacity factor = 0.8

M_p = $f_{yp}S$

S = plastic section modulus of the pin

= $d_f^3/6$

d_f = pin diameter ($\approx d_p$)

(b) Pin in bearing

V_{bi}^* is the pin design bearing force from ply i, and the following is required from Clause 9.5.2 of AS 4100:

$$V_{bi}^* \leqslant \phi V_{bi}$$

where

ϕV_{bi} = design bearing capacity *on* the pin from ply i

ϕ = capacity factor = 0.8

V_{bi} = $1.4 f_{yp} d_f t_i k_p$

f_{yp} = design yield stress of the pin

d_f = pin diameter ($\approx d_p$)

t_i = thickness of ply i bearing on the pin

k_p = factor for pin rotation

= 1.0 for pins without rotation, or

= 0.5 for pins with rotation

Too high a contact pressure can produce "cold welding" of the pin to the pin hole surfaces, thus causing fretting of the surfaces subject to rotation. It is then further recommended that the pin diameter and material grade be chosen such that the contact pressure on the *pin* is limited to $0.8f_{yp}$ for the serviceability limit state check, i.e.:

$$V_{bs}^* \leqslant \phi 0.8 f_{yp} d_f t_i$$

where

V_{bs}^* = serviceability limit state bearing force on the pin, and

ϕ = capacity factor = 0.8

8.10.2.5 Design and detailing—Clearances and materials

Some clearance must exist between the hole and the pin so as to make assembly possible. The clearance should not exceed 0.05 times the pin diameter if the pin and the eye plates are not galvanized, and 1.0 mm larger if they are. Zinc coating tends to be uneven in the vicinity of holes and thus there is a need to increase the clearance and remove any blobs and runs left after the zinc bath.

Pins can be made of a variety of materials—e.g. mild steel, high strength steel and Grade 316 stainless steel. A lot of care needs to be given to the prevention of direct metallic contact between the contact surfaces to prevent galvanic corrosion.

The pin must be prevented from creeping out of the hole. One way to do this is by using shoulder bolts with crowned nuts for cotter pins, by retainer plates, or by cap plates screwed onto the ends of the pin. The latter solution can be quite pleasing in appearance.

8.10.2.6 Corrosion protection

The corrosion protection solutions used in practice for pins are a combination of:

- treating the pin with molybdenum disulfide in dry form or grease (Rocol)
- using Denco corrosion-inhibiting grease
- galvanizing the pin and the eye plates' surfaces
- applying a plasma-sprayed metallic protection coating.

The lubrication initially provided is meant to be renewed periodically, but in practice this rarely happens. Grease nipples, grease distribution channels and grooves should be considered for pins over 60 mm in diameter.

Corrosion protection of eye plate mating surfaces is equally important. The following points are relevant:

- Pin plates should be corrosion-protected on all surfaces.
- There is limited access for inspection in service, and long-lasting protection is required.
- Grease may not retain its quality for more than, say, 3 years, and thus the means for periodical regreasing should be built in.

8.11 Examples

8.11.1 Example 8.1

Step	Description and calculations	Result	Unit
	Determine the number of bolts and the geometry of a lap splice in the tension member shown. Use Grade 300 steel and M20 8.8/S bolts in double shear arrangement		

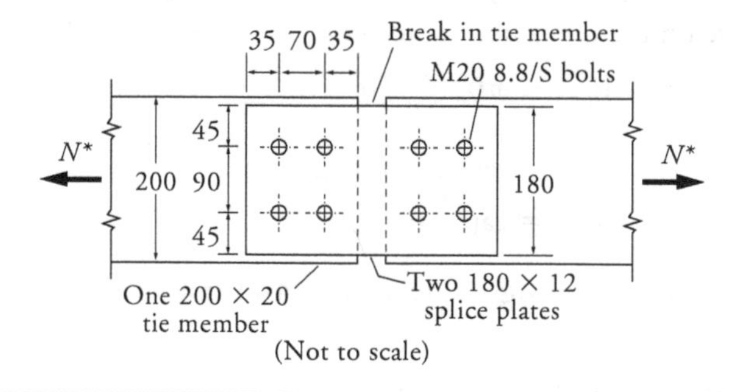

Step	Description and calculations	Result	Unit
	Data		
	Member design axial tension force N^* =	800	kN
	Member design bending moment M^* =	0	kN
	Tie member section: 200 × 20 flat (simple rectangular section)		
	Gross section area A_{gm} = 200 × 20 =	4000	mm^2
	Net section area A_{nm} = 4000 − (2 × 22 × 20) =	3120	mm^2
	Design yield stress f_{ym} ...Table 2.3 or AS 4100 Table 2.1 ...	280	MPa
	Design tensile strength f_{um} ...Table 2.3 or AS 4100 Table 2.1 ...	440	MPa

Splice plate section: 180 × 12 flat (simple rectangular section)

	Gross section area $A_{gs} = 180 \times 12 =$	2160	mm^2
	Net section area $A_{ns} = 2160 - (2 \times 22 \times 12) =$	1630	mm^2
	Design yield stress f_{ys} ...Table 2.3 or AS 4100 Table 2.1 ...	300	MPa
	Design tensile strength f_{us} ...Table 2.3 or AS 4100 Table 2.1 ...	440	MPa

---

1 Check tie member tension design capacity ϕN_{tm}

AS 4100 Clause 7.2 considers two possible failure modes:

1.1 Fracture

$$\phi N_{tfm} = \phi 0.85 k_t A_{nm} f_{um}$$

$$= \frac{0.9 \times 0.85 \times 1.0 \times 3120 \times 440}{1000} = \quad 1050 \text{ kN}$$

1.2 Gross yield

$$\phi N_{tym} = \phi A_{gm} f_{ym}$$

$$= \frac{0.9 \times 4000 \times 280}{1000} = \quad 1010 \text{ kN}$$

1.3 Member design capacity

$\phi N_{tm} = \min(\phi N_{tfm}, \phi N_{tym}) = \min(1050, 1010)$ — 1010 kN

$N^* \leqslant \phi N_{tm}$ is true as $800 \leqslant 1010$ is true ... satisfactory ... → OK

1.4 Block Shear Capacity check:

From Appendix E.2, ϕR_{bs} (Failure Mode B) controls

Block Shear Capacity with $\phi R_{bs} = 978$ kN.

$N^* \leqslant \phi R_{bs}$ is true as $800 \leqslant 978$ is true ... satisfactory ... → OK

---

2 Check minimum design actions on splice connection N^*_{cmin}

From AS 4100 Clause 9.1.4(b)(v)

$N^*_{cmin} = 0.3\phi N_{tm} = 0.3 \times 1010 =$ — 303 kN

As $N^* \geqslant N^*_{cmin}$ is true as $800 \geqslant 303$ is true ... then $N^* =$ — 800 kN

---

3 Check splice plate (from flats) tension design capacity ϕN_{ts}

Using the same method as in Step 1:

$$\phi N_{tfs} = \left(\frac{0.9 \times 0.85 \times 1.0 \times 1630 \times 440}{1000}\right) \times 2 = \quad 1100 \text{ kN}$$

$$\phi N_{tys} = \left(\frac{0.9 \times 2160 \times 300}{1000}\right) \times 2 = \quad 1170 \text{ kN}$$

$\phi N_{ts} = \min(1100, 1170)$ — 1100 kN

$N^* \leqslant \phi N_{ts}$ is true as $800 \leqslant 1100$ is true ... satisfactory ... → OK

From Appendix E.3, ϕR_{bs} (Failure Mode A) controls

Block Shear Capacity with $\phi R_{bs} = 1100$ kN.

$N^* \leqslant \phi R_{bs}$ is true as $800 \leqslant 1100$ is true ... satisfactory ... → OK

---

4 Total number of M20 8.8/S bolts required for each side of the splice $= N_b$

4.1 Capacity of an M20 8.8/S bolt in single shear with threads *excluded* from shear planes ϕV_{fx1}

Table 8.5(a) gives

$\phi V_{fx1} =$ ... for threads *excluded* from a *single* shear plane ... = — 129 kN/bolt

In *double* shear with threads *excluded* from shear planes ...

$\phi V_{fx2} = 2 \times \phi V_{fx1}$

$= 2 \times 129 =$ — 258 kN/bolt

Geometrical configuration requires two bolts/row in a transverse section, then ...

	$N_b = \dfrac{N^*}{\phi V_{fx2}} = \dfrac{800}{258} =$	3.10 bolts but ...	
4.2	It is rare to get a reasonable bolt length for the plies being joined that has threads *excluded* in both shear planes for this joint configuration unless there is a large "stickout" length ... try bolts with threads *included* in one shear plane ...		
	Table 8.5(a) gives		
	ϕV_{fn1} = ... for threads *included* in a *single* shear plane ... =	92.6	kN/bolt
	$N_b = \dfrac{N^*}{(\phi V_{fx1} + \phi V_{fn1})} = \dfrac{800}{(129 + 92.6)} = 3.61$ = say	4	bolts
	Use **4** M20 8.8/S bolts in each end of member. Threads may be included in one shear plane but not two.	**Answer**	
5	Check the above against AS 4100 Clause 9.3.2.1, Tables 9.3.2.1 and 3.4		
5.1	Shear capacity ϕV_f of a bolt		
	Distance from first to last bolt centreline on half the joint is 70 < 300 mm, then		
	Reduction factor for splice length, k_r =	1.0	
	$\phi V_f = \phi 0.62\, f_{uf}\, k_r\, (n_n A_c + n_x A_o)$		
	... n = no. of shear planes for A_c and A_o = 1		
	$= 0.8 \times 0.62 \times 830 \times 1.0 \times \dfrac{(1 \times 225 + 1 \times 314)}{1000} =$	222	kN
	where f_{uf} is from Table 8.3		
	and A_c and A_o are from AISC [1999a] Table 10.3-2 Page 10-10		
	Four bolts have a capacity = 4 ϕV_f = 4 × 222 =	888	kN
	$N^* \leqslant 4\,\phi V_f \rightarrow 800 \leqslant 888 \rightarrow$ true, satisfied ϕV_f is ...	OK	
5.2	Bolt bearing capacity ϕV_b of a bolt AS 4100 Clause 9.3.2.4		
5.2.1	Bolt bearing design capacity based on *edge distance*, ϕV_{bed}		
	a_e = centre of hole to edge or adjacent hole in direction of N^*		
	= min (35, 70) =	35	mm
	t_p = min (20, 2 × 12) = min (20, 24) =	20	mm
	f_{up} is the same for the member and splice plates		
	$\phi V_{bed} = \phi a_e t_p f_{up}$		
	$= \dfrac{0.9 \times 35 \times 20 \times 440}{1000} =$	277	kN
	Four bolts have a capacity = 4 ϕV_{bed} = 4 × 277 =	1110	kN
	$N^* \leqslant 4\,\phi V_{bed} \rightarrow 800 \leqslant 1110 \rightarrow$ true, satisfied for a_e = 35	OK ...	
	... ϕV_{bed} for edge distance (tearout) is satisfactory.		
5.2.2	ϕV_{bcon} of a bolt in local bearing failure (crushing) of plate due to *bolt contact*		
	$\phi V_{bcon} = \phi\, 3.2\, d_f\, t_p\, f_{up}$		
	$= \dfrac{0.9 \times 3.2 \times 20 \times 20 \times 440}{1000} =$	507	kN
	Four bolts have a capacity = 4 ϕV_{bcon} = 4 × 507 =	2030	kN
	$N^* \leqslant 4\,\phi V_{bcon} \rightarrow 800 \leqslant 2030 \rightarrow$ true, satisfied in local bearing	OK ...	
	... ϕV_{bcon} for bolt bearing/failure (crushing) is satisfactory		
5.3	Conclusion on bolted splice with four M20 8.8/S bolts on each end with minimum edge distance 35 is adequate. Threads as in Step 4.2.	Check is OK	

8.11.2 Example 8.2

Step	Description and calculations	Result	Unit
	Determine the adequacy of the bolts on the bracket shown. Use Grade 300 steel, and M20 8.8/S bolts in single shear.		
	This is an eccentric connection with the action/load in the plane of the fasteners/bolts. Action is eccentric to the centroid of the bolts.		
	Solution uses ASI [2009a].		

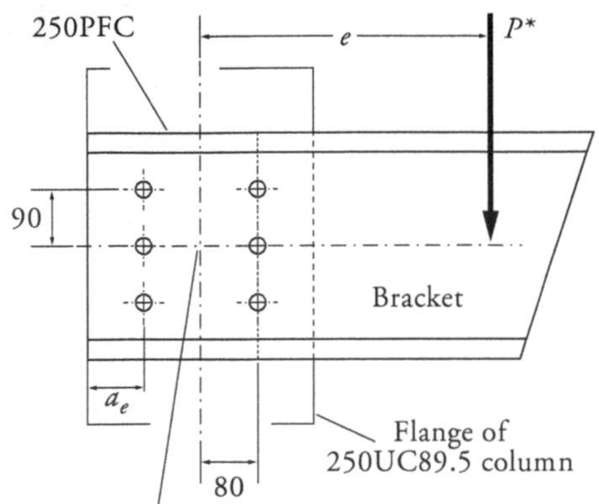

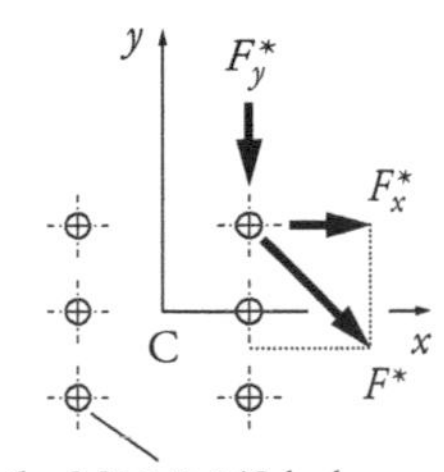

Step	Description and calculations	Result	Unit
	Data		
	Design force on bracket, P^* =	120	kN
	Eccentricity of P^* from bolt group centroid = e =	300	mm
	Data for plate tear-out and bearing:		
	Bracket is 250PFC Grade 300		
	ASI [2009a] Table 3.1-7 (A) page 3–18		
	Thickness of bracket web = t_w =	8.0	mm
	Edge distance, centre of bolt to end of bracket = a_e = (see a_e below for UC) =	48	mm
	Tensile strength = f_u =	440	MPa
	Column is 250UC89.5 Grade 300		
	ASI [2009a] Table 3.1-4 (A) page 3–12:		
	Thickness of flange = t_f =	17.3	mm
	Width of flange = b_f =	256	mm
	Edge distance to edge of flange = $a_e = \frac{[b_f - (2 \times 80)]}{2} = \frac{[256 - 160]}{2} =$	48	mm
	Tensile strength = f_u =	440	MPa
1	Design action effects on group of bolts, M^*, V^*		
	$M^* = P^*e$		
	$= 120 \times 0.300 =$	36.0	kNm
	$V^* = P^* =$	120	kN
	Note: This is a sample exercise on bolt/ply adequacy and no checks on minimum design actions are undertaken (AS 4100 Clause 9.1.4). See Examples 8.1, 8.4 to 8.6 for such examples.		

Step	Description	Value	Unit
2	Polar moment of inertia/second moment of area I_p about C		
	Bolt size is chosen for simplicity initially with shank area =	1	mm^2
	... disregards level of stress ... set when size is selected/specified.		
	$I_p = \Sigma(x_i^2 + y_i^2)$ where $i = 1$ to n_b		
	$= 6 \times 80^2 + 4 \times 90^2 =$	70 800	mm^4/mm^2
3	Consider design action effect F^* on the top right *corner* bolt as the critical bolt.		
	Using the simplified method noted in Sections 8.3.2.1 and 8.3.2.2.		
	F_x^* = horizontal/x-component of F^*		
	$= \dfrac{M^* y_c}{I_p}$ ... with y_c to critical bolt		
	$= \dfrac{36.0 \times 10^6 \times 90}{70800} =$	45 800	N
	F_y^* = vertical/y-component of F^*		
	$= \dfrac{M^* x_c}{I_p} + \dfrac{V^*}{n_b}$ ... with x_c to critical bolt		
	$= \dfrac{36.0 \times 10^6 \times 80}{70800} + \dfrac{120 \times 10^3}{6} =$	60 700	N
	F^* = resultant force on *corner* bolt(s)		
	$= \sqrt{(F_x^{*2} + F_y^{*2})}$		
	$= \sqrt{(45800^2 + 60700^2)} =$	76 000	N
	= design capacity of bolt required =	76.0	kN
	... Table 8.5(a) or ASI [2009a] Table T9.3 page 9-6		
	1- M20 8.8/S bolt in single shear, threads in shear plane has		
	design capacity ϕV_{fn} =	92.6	kN
	... and $F^* \leq \phi V_{fn} \rightarrow 76.0 \leq 92.6 \rightarrow$ true, satisfied	OK	
	6- M20 8.8/S threads in shear plane are adequate for bolt shear capacity	**Answer**	
4	Plate tear-out ... edge distance bearing		
	f_u = ... for both UC and PFC bracket...	440	MPa
	$t_p = \min(t_w, t_f)$		
	$= \min(8.0, 17.3) =$	8.0	mm
	a_e = edge distance =	48	mm
	$\phi V_{bed} = \phi a_e t_p f_{up} = 0.9 \times 48 \times 8.0 \times 440/10^3 =$	152	kN
	As $F^* \leq \phi V_{bed} \rightarrow 76.0 \leq 152$ is true, satisfied ... Plate tear-out is ...	OK	
5	Bearing ... bolt contact bearing on ply with $t_w = t_p = 8.0$ ...		
	i.e. the thinner ply with the same f_{up}		
	$\phi V_{bcon} = \phi 3.2 d_f t_p f_{up} = 0.9 \times 3.2 \times 20 \times 8.0 \times 440/10^3 =$	203	kN
	Note from ASI [2009a] Table T9.3 page 9-6, ϕV_{bcon} =	203	kN
	As $F^* \leq \phi V_{bcon} \rightarrow 76.0 \leq 203$ is true, satisfied ... Bearing on bracket/web ...	OK	
6	Bolt, bracket, bearing capacity and edge distance are adequate.		

8.11.3 Example 8.3

Step	Description and calculations	Result	Unit
	Check the capacity of the bolts in the rigid end plate connection in which the plate is welded to the end of the beam shown. The bolts are specified as M20 8.8/TB. The plate material is Grade 300 steel.		
	Note this is an eccentric connection in which P^* is out-of-plane of the bolts at distance e, perpendicularly.		

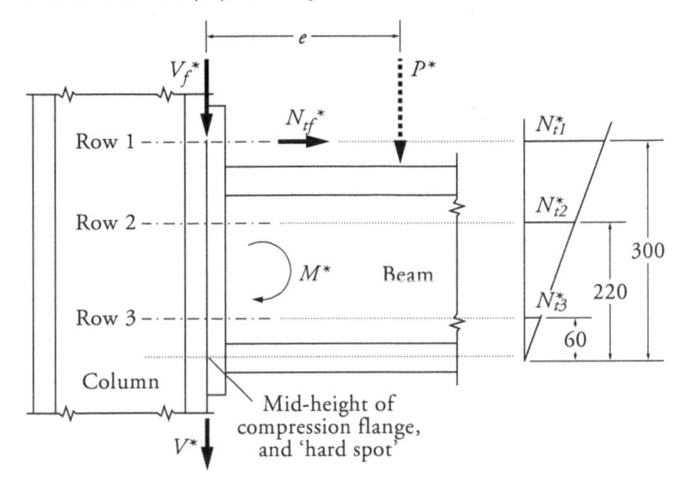

Step	Description and calculations	Result	Unit
	Data		
	Action/load effects on connection		
	Design moment M^* = ...from analysis or P^*e ... =	120	kNm
	Design shear force V^* =	110	kN
	Number of bolts in the bolt group n_b =	6	
1	Assumptions:		
1.1	Connection is regarded as a beam end.		
1.2	At ultimate load the beam rotates about 'hard-spot' having a centroid at the mid-height of the compression flange through which the compressive bearing load acts. See Section 8.3.2.3 for further information.		
1.3	Bolts behave perfectly elastic.		
1.4	Minimum design actions (AS 4100 Clause 9.1.4) is not considered in this instance as the beam size is not known and the capacity of the bolts and associated plies are being investigated. For typical calculations on minimum design actions see Examples 8.1, 8.4 to 8.6		
2	I_x = Second moment of area of the bolt group composed of 3 rows each with 2 bolts.		
2.1	For simplest calculations, initially assume bolts of a size with shank area of 1 mm^2. Level of stress in bolt is ignored until a choice of bolt size is made later to keep within the specified bolt capacity ...		
2.2	$I_x = \Sigma(n_{bi}\, y_i^2)$ ... where n_{bi} = no of bolts in row i (= row number 1, 2, and 3) $= 2 \times 300^2 + 2 \times 220^2 + 2 \times 60^2 =$	284 000	mm^4/mm^2

3	Design actions N_{tf}^* and V_f^* in *one* fastener/bolt in *top* row ... axial tensile strain/force is greatest here in row 1 with 2 bolts:		
3.1	$N_{tf}^* = N_{t1}^*$ = horizontal/*tension* force on fastener		
	$= \dfrac{M^* y_1}{I_x}$		
	$= \dfrac{120 \times 10^6 \times 300}{284000} =$	127 000	N
	$=$	127	kN
3.2	$V_f^* = \dfrac{V^*}{n_b}$ = vertical/*shear* force uniformly distributed to each fastener		
	$= \dfrac{110000}{6} =$	18 300	N
	$=$	18.3	kN
3.3	Capacities of a M20 8.8/TB threads excluded from shear plane ... Table 8.5(b) gives:		
	$\phi V_f =$	129	kN
	$\phi N_{tf} =$	163	kN
3.4	Combined shear and tension ... AS 4100 Clause 9.3.2.3 ...		
	$\left[\dfrac{V_f^*}{\phi V_f}\right]^2 + \left[\dfrac{N_{tf}^*}{\phi N_{tf}}\right]^2 = \left[\dfrac{18.3}{129}\right]^2 + \left[\dfrac{127}{163}\right]^2 =$	0.627	
	$\leqslant 1.0$	OK	
3.5	6- M20 8.8/TB bolts threads excluded from shear plane is ...	OK	
Note:	if threads were included in the shear plane, the bolts would still be adequate as the above interaction equation would equal 0.646.		
4	Prying action affecting the 2 top bolts/fasteners Measures to counteract prying are:		
4.1	Make end plate thickness at least 1.2 d_f = 1.2 × 20 =	24	mm
4.2	Increase the top bolt design tension force by 20% (see Section 8.3.2.4): $N_{tf}^* = 127 \times 1.20 =$	152	kN
	Recalculate the combined force check		
	$\left[\dfrac{18.3}{129}\right]^2 + \left[\dfrac{152}{163}\right]^2 =$	0.890	
	$\leqslant 1.0$	OK	
4.3	End plate connection is adequate for prying action		
Comments:	The simple analysis used in the above example is considered adequate for the connection considered – i.e. an I-section welded to an end plate which is then bolted to a support. The other simple analysis technique noted in Section 8.3.2.3 can basically give the same result. As the beam section or flange proportions were not known, the simpler Hogan & Munter [2007a] model could not be used in this instance.		

8.11.4 Example 8.4

Step	Description and calculations	Result	Unit
	Determine the size of the fillet weld to connect one end of the diagonal tension member to the joint in the truss shown.		
	Member size is 1- 100 × 100 × 8 EA. Grade 300 steel.		
	Electrode is E48XX/W50X. Weld is SP category.		

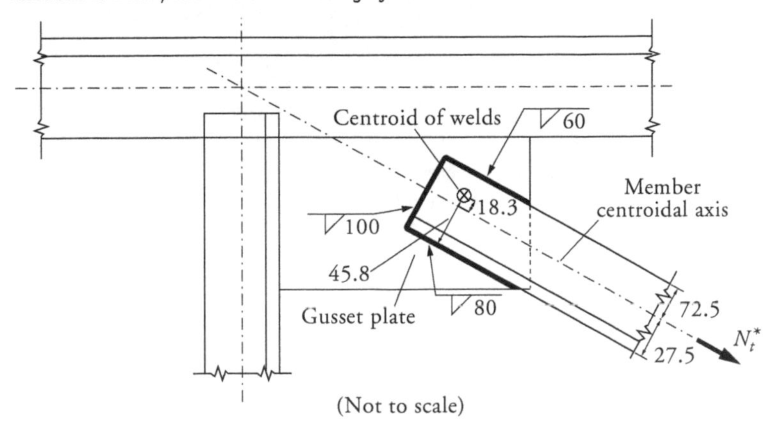

(Not to scale)

Step	Description and calculations	Result	Unit
	Data		
	Member size: 100 × 100 × 8 EA Grade 300 steel		
	A_g =	1500	mm^2
	Design axial tensile action/force, N_t^* =	190	kN
	Design action effect/bending moment =	0	kNm
1	Static/quasi-static action/load effects on connection		
1.1	Design data		
	Design axial tension force, N_t^* =	190	kN
	Design bending moment, M^* =	0	kNm
1.2	Minimum actions on the connection		
1.2.1	Tension members ... AS 4100 Clause 9.1.4(b)(iii) ... requires $0.3\phi N_t$ ASI [2009a] Tables 7-20(2) page 7-15 gives the following for a 100x100x8.0EA		
	ϕN_t = ... for welded no holes, eccentric connection =	430	kN
	$0.3\phi N_t = 0.3 \times 430$ =	129	kN
	$N_t^* > 0.3\phi N_t$ → 190 > 129 → true, satisfied → N_t^* =	190	kN
1.3	Design actions for calculations		
	N_t^* =	190	kN
2	Try fillet weld with leg size, t =	6	mm
	... made with weld electrode of Grade and category ...	E48XX/W50X, SP	
3	Strength of fillet weld ϕV_v		
	... AS 4100 Clause 9.7.3.10 and Table 3.4		
	$\phi V_v = \phi 0.6\, f_{uw}\, t_t\, k_r$		
	$= 0.8 \times 0.6 \times 480 \times \frac{6}{\sqrt{2}} \times 1.0$ =	978	N/mm

	Alternatively, ϕV_v =	0.978	kN/mm
	=	978	N/mm
	... from ASI [2009a] Table 9.8 page 9-13		
4	(Total) length, l_w, of fillet weld required to resist N_t^*		
	$l_w = \frac{N_t^*}{\phi V_v} = \frac{190000}{978} =$	194	mm
	Available length = 100 + 60 + 80 =	240	mm
	Length of 6 mm equal leg fillet weld	OK	
5	Position of weld group centroid Centroid should coincide as closely as practicable with the centroidal axis of the connected member to minimise the eccentricity e of N_t^*, and therefore M^*. Though not important for statically (and quasi-statically) loaded connections (see Section 8.8.3 and Figure 8.19) it may become noteworthy for fatigue applications.		
6	6 mm equal leg fillet weld (see Step 2) is satisfactory.		

8.11.5 Example 8.5

Step	Description and calculations	Result	Unit
	Check the welded connection between a continuously laterally restrained 410UB53.7 beam and a 310UC96.8 column using E48XX/W50X electrodes and SP category welding procedure/quality for structural purpose. Steel is Grade 300. Note this is an eccentric connection with out-of-plane weld loading due to M^*		

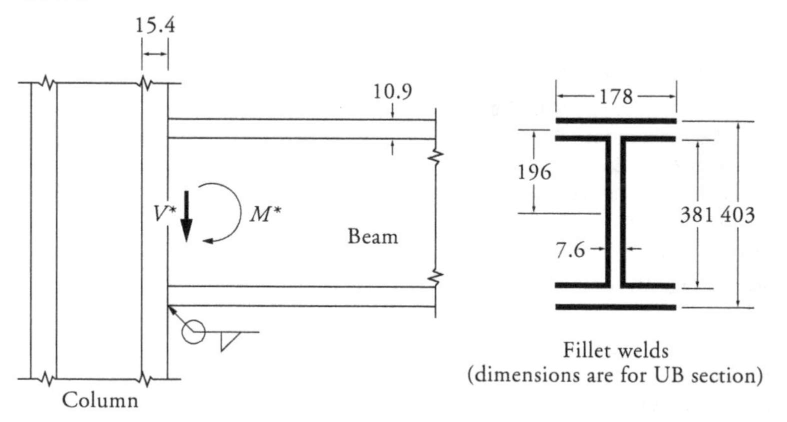

1	Action/load effects on connection		
1.1	Design data		
	Design bending moment, M^* =	166	kNm
	Design shear force, V^* =	36.0	kN

1.2	Minimum actions on the connection		
1.2.1	Moment ... AS 4100 Clause 9.1.4(b)(i) ... requires $0.5\phi M_{bx}$ Beam is continuously laterally restrained ASI [2009a] Table 3.1-3(B) page 3-11 gives the following for a 410UB53.7		
	Z_{ex} = effective section modulus about the x-axis =	1060×10^3	mm^3
	f_{yf} =	320	MPa
	then $\phi M_{sx} = \phi M_{bx} = \dfrac{0.9 \times 320 \times 1060 \times 10^3}{10^6}$ =	305	kNm
	$0.5\phi M_{bx} = 0.5 \times 305$ =	153	kNm
	$M^* > 0.5\phi M_{bx} \rightarrow 166 > 153 \rightarrow$ true, satisfied $\rightarrow M^*$ =	166	kNm
1.2.2	Shear force ... AS 4100 Clause 9.1.4(b)(ii) ... not applicable as the Clause only specifies this for simple construction.		
1.3	Design actions for calculations		
	M^* =	166	kNm
	V^* =	36.0	kN
2	Section properties ...		
2.1	Column		
	310UC96.8 Grade 300 ... ASI [2009a] Table 3.1-4 (A) page 3-12		
	t_f =	15.4	mm
2.2	Beam		
	410UB53.7 Grade 300: ASI [2009a] Tables 3.1-3(A) page 3-10 gives		
	$d = 403$ $b_f = 178$ $d_1 = 381$ $t_f = 10.9$ $t_w = 7.6$		mm
2.3	E48XX/W50X electrodes ... Table 8.11 or AS 4100 Table 9.7.3.10(1) ...		
	f_{uw} =	480	MPa
(a)	**Simple solution—proportioning method**		
3	Method is as follows		
3.1	Flange forces N_f^* due to M^* as a couple are resisted by flange welds alone		
	d_f = distance between flange centres		
	$= \dfrac{(381 + 403)}{2}$ =	392	mm
	=	0.392	m
	$N_f^* = \dfrac{M^*}{d_f}$		
	$= \dfrac{166}{0.392}$ =	423	kN
	l_{ff} = length of two fillet welds on each side of a flange (assume continuous through web) $= 2 \times 178$ ... disregard the weld returns at the flange ends ...		
	=	356	mm
	V_{ff}^* = design shear force on fillet welds on flange		
	$= \dfrac{N_f^*}{l_{ff}}$		
	$= \dfrac{423}{356}$ =	1.19	kN/mm

3.1.1	Minimum size fillet weld $t_{w\,min}$ AS 4100 Table 9.7.3.2		
	$t = \max(t_{f\,col}, t_{f\,beam}) = \max(15.4, 10.9) =$	15.4	mm
	$t_{w\,min} =$	6 min	mm
3.1.2	Maximum size of fillet weld $t_{w\,max}$ AS 4100 Clause 9.7.3.3		
	$t_{w\,max}$ = Not applicable … no fillet weld along a free edge …	-	
3.1.3	Select size of fillet weld t_{wf} for flanges using weld capacity ϕv_w in Table 8.11 for Category SP welds and E48XX/W50X electrodes gives $\phi v_{wf} = 1.30$ kN/mm for $t_w = 8$mm $V^*_{ff} \leqslant \phi v_{wf} \rightarrow 1.19 < 1.30$ then		
	$t_{wf} =$ … flange fillet welds …	8	mm
	$t_{w\,min}$ and $t_{w\,max}$ are satisfied	**Answer flange**	
3.2	Web shear forces are wholly resisted by web fillet welds		
3.2.1	l_{fw} = length of two fillet welds to each side of web $= 2 \times 381 =$	762	mm
3.2.2	V^*_{fw} = design shear force on fillet welds on web $= \dfrac{V^*}{l_{fw}}$ $= \dfrac{36.0}{762} =$	0.0472	kN/mm
3.2.3	Minimum size fillet weld $t_{w\,min}$ AS 4100 Table 9.7.3.2		
	$t = \max(t_{f\,col}, t_{w\,beam}) = \max(15.4, 7.6) =$	15.4	mm
	$t_{w\,min} =$	6	mm
3.2.4	Maximum size of fillet weld $t_{w\,max}$ AS 4100 Clause 9.7.3.3		
	$t_{w\,max}$ = Not applicable … no fillet weld along a free edge …	-	
3.2.5	Select size of fillet weld t_{ww} for web using weld capacity ϕv_w in Table 8.11 for Category SP welds and E48XX/W50X electrodes …		
	$t_w = 3$ with $\phi v_w = 0.489 > 0.0472$ … web fillet welds …	3 but …	mm
	in which $t_w = 3$ for web does not satisfy $t_{w\,min} = 6$ Use $t_w = t_{w\,min} =$ … for web fillet welds …	6	mm
		Answer web	
3.2.6	Use **8 mm fillet welds in flanges** and **6 mm fillet welds** along **web. E48XX/W50X** and **SP** quality.	**Answer(a)**	

(b)	**Alternate solution—using rational elastic analysis**		
4	Second moment of area of weld group I_{wx} … welds with throat thickness 1 mm		
	$I_{wx} = 4 \times 178 \times 196^2 + 2 \times 381^3/12 =$ … approx. …	36.6×10^6	mm^4/mm
	A_w = area of weld group with throat thickness 1 mm $= 4 \times 178 + 2 \times 381 =$ … approx. …	1470	mm^2/mm
4.1	Forces in N/mm^2 at extremity of weld group with throat thickness 1 mm are:		
4.1.1	Flange force $f^*_x = \dfrac{M^* y_c}{I_{xw}}$		
	$= 166 \times 10^6 \times 202/(36.6 \times 10^6) =$	916	N/mm
4.1.2	Web force $f^*_y = \dfrac{V^*}{A_w}$		
	$= \dfrac{36.0 \times 10^3}{1470} =$	24.5	N/mm

4.1.3	Resultant force $f_r^* = \sqrt{(f_x^{*2} + f_y^{*2})}$		
	$= \sqrt{(916^2 + 24.5^2)}$ ... at most critically loaded point ...		
	$=$	916	N/mm
4.1.4	Throat size of fillet weld t_t required		
	$t_t = \dfrac{f_r^*}{(\phi 0.6\, f_{uw})}$ ... AS 4100 Clause 9.7.3.10 ...		
	$= \dfrac{916}{(0.8 \times 0.6 \times 480)} =$ ... throat size ...	3.98	mm
	Size of fillet weld t_w		
	$t_w = \dfrac{t_t}{0.707} = \dfrac{3.98}{0.707} =$	5.63	mm
4.1.5	Minimum size of fillet weld shown in (a) solution is	6	mm
4.1.6	Use **6 mm fillet welds** along flanges and web. **E48XX/W50X** and **SP.**	**Answer (b)**	

Comment Solution (a) ignores any moment capacity in the fillet welds on the web, necessitating larger-size fillet welds in the flange. Moreover, minimum fillet weld requirements impose larger fillet welds in the web to satisfy thermal (in terms of welding heat input) rather than strength demands, leading overall to less economical size welds in this instance.

8.11.6 Example 8.6

Step	Description and calculations	Result	Unit
	Determine the weld size for the 250PFC Grade 300 bracket shown. Welding consumables are E48XX/W50X and welding quality is SP, structural purpose.		

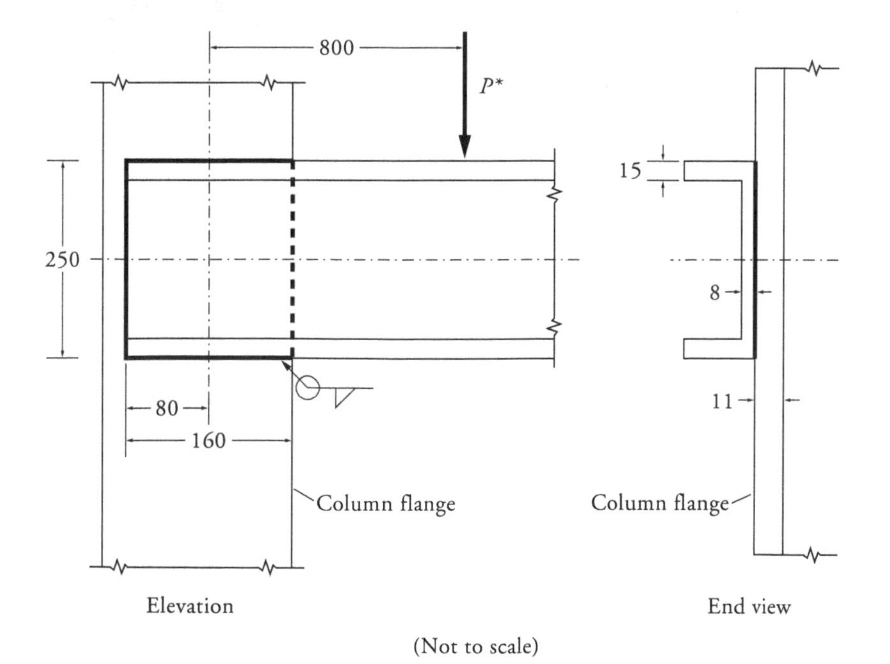

1	Action/load effects on connection		
1.1	Design data		
	Design action/force, $P^* =$	80.0	kN
	Design shear force, $V^* =$	80.0	kN
	Design bending moment, $M^* = P^*e = 80.0 \times 0.800 =$	64.0	kNm
1.2	Minimum actions on the connection		
1.2.1	Moment ... AS 4100 Clause 9.1.4(b)(i) ... requires $0.5\phi M_{bx}$ Bracket is a *cantilever* with load acting *down* on *top* flange as critical flange. Assuming no other loads or restraints after this point, the effective length to load point is (see Table 5.2.2 also) $l_e = k_t k_l k_r l$ $= 1.1 \times 2.0 \times 1.0 \times 0.800 =$	1.76	m
	Moment modification factor... AS 4100 Table 5.6.2 ... $\alpha_m = 1.25$ ASI [2009a] Table 5.3-9, page 5-56 gives the following for a 250PFC		
	$\phi M_{bx1} =$ design member moment capacity with $\alpha_m = 1.0 =$	93.7	kNm
	$\phi M_{bx} = 93.7 \times 1.25 =$	117	kNm
	$0.5\phi M_{bx} = 0.5 \times 117 =$	58.5	kNm
	$M^* > 0.5\phi M_{bx} \rightarrow 64 > 58.5 \rightarrow$ true, satisfied $\rightarrow M^* =$	64.0	kNm
1.2.2	Shear force ... AS 4100 Clause 9.1.4(b)(ii) ... not applicable as the Clause only specifies this for simple construction.		
1.3	Design actions for calculations		
	$V^* =$	80.0	kN
	$M^* =$	64.0	kNm
2	Second moment of area of weld group I_{wx} ... welds with throat thickness 1 mm		
	$I_{wx} = 2 \times 160 \times 125^2 + \frac{2 \times 250^3}{12} =$	7.60×10^6	mm^4/mm
	$I_{wy} = 2 \times 250 \times 80^2 + \frac{2 \times 160^3}{12} =$	3.88×10^6	mm^4/mm
	$I_{wp} = I_{wx} + I_{wy}$ $= 7.60 \times 10^6 + 3.88 \times 10^6 =$	11.5×10^6	mm^4/mm
	$A_w =$ area of weld group with throat thickness 1 mm $= 2 \times 250 + 2 \times 160 =$	820	mm^2/mm
3	Forces f^* in corner of two welds. Throat thickness 1 mm. Coordinates of centroid C of weld group: $x_c = 80$ and $y_c = 125$ from bottom left corner in elevation view		mm
	$f_x^* =$ force in x-direction $= \frac{M^* y_c}{I_{wp}}$ $= \frac{64.0 \times 10^6 \times 125}{11.5 \times 10^6} =$	696	N/mm
	$f_y^* =$ force in y-direction $= \frac{M^* x_c}{I_{wp}} + \frac{V^*}{A_w}$ $= \frac{64.0 \times 10^6 \times 80}{11.5 \times 10^6} + \frac{80.0 \times 10^3}{820} =$	543	N/mm
	$f_r^* =$ resultant force on unit weld @ critical point $= \sqrt{(f_x^{*2} + f_y^{*2})}$ $= \sqrt{(696^2 + 543^2)} =$	883	N/mm

4	Size of fillet weld, t_w required ... with SP category, then $\phi = 0.80$		
4.1	Throat size $t_t = \dfrac{f_r^*}{(\phi 0.6\ f_{uw})}$		
	$= \dfrac{883}{(0.8 \times 0.6 \times 480)} =$	3.83	mm
4.2	Fillet weld size t_w ... size of leg, taken as an equal leg length fillet weld		
	$t_w = \sqrt{2} t_t$		
	$= \sqrt{2} \times 3.83 =$	5.42	mm
4.3	Minimum size of fillet weld to minimise weld cracking from too rapid cooling ... AS 4100 Table 9.7.3.2 ...		
4.3.1	PFC flange to column flange: t = max (15, 11) =	15	mm
	minimum fillet weld on PFC flange =	5	mm
4.3.2	PFC web to column flange: t = max (8, 11) =	11	mm
	minimum fillet weld on PFC web =	5	mm
4.4	Maximum size of fillet weld that can be accommodated ... for PFC web to column flange fillet weld (see end view) AS 4100 Figure 9.7.3.3(b) gives $t_w \leqslant t - 1 = t_{PFC\ web} - 1 = 8 - 1 =$	7	mm
4.5	Use **6 mm fillet weld.** E48XX/W50X SP category	**Answer**	

Comment: On step 3—Corner welds at the top right and bottom right have the highest shear stress, being the furthest from the 'modelled' Instantaneous Centre of Rotation (ICR).

8.12 Further reading

- For additional worked examples see Chapter 9 of Bradford, et al. [1997], Hogan & Munter [2007a–h], Hogan [2011], Hogan & Van der Kreek [2009a–e] and Syam & Chapman [1996].
- For typical structural steel connections also see ASI [2009b], AISC [2001], Hogan & Munter [2007a–h], Hogan [2011], Hogan & Van der Kreek [2009a–e] and Syam & Chapman [1996].
- Welding symbols are further explained in AS 1103.1.
- Two good practical references on welding processes and related activities are Taylor [2001, 2003].
- A practical designer's guide to welding can be found in Technical Note 11 of the WTIA [2004].
- An internationally respected reference on bolts and other single point fasteners is Kulak, et al. [1987].
- Material Standards for bolts and screws include AS 1275, AS 1110, AS 1111, AS 1112, AS/NZS 1252, AS/NZS 1559 and AS 4291.1. For bolts, nuts and washers see also Hogan & Munter [2007a,b].
- Some very good information on structural steelwork connections from a New Zealand perspective can be found in SCNZ [2007].
- From an Australian and New Zealand perspective, some excellent structural steel connection design software include Limcon (a stand-alone program) and also the packages found in Microstran and Space Gass (see Appendix A.4)

chapter 9

Plastic Design

9.1 Basic concepts

The plastic method of frame analysis is concerned with predicting the ultimate load-carrying capacity of steel structures. In this method of analysis, the structure is considered to be at an ultimate (i.e. strength) limit state. Any further increase in load would cause the framework to collapse, rather like a mechanism. One of the prerequisites of plastic frame analysis is that the structure will not fail by rupture (brittle failure) but by deformations, which may progressively grow without any increase in the load. Another prerequisite is that no buckling or frame instability will occur prior to the formation of a collapse mechanism.

The simple plastic design method implicit in AS 4100 is based on the following assumptions:

(a) The material has the capacity to undergo considerable plastic deformation without danger of fracture.

(b) Ductility of steel (that is, a long yield plateau) is important for the development of plastic zones (plastic hinges).

(c) Rigid connections must be proportioned for full continuity and must be able to transmit the calculated plastic moment (the moment attained at plastic hinge).

(d) No instability (buckling) must occur prior to the formation of a sufficient number of plastic hinges, to transform the structure into a mechanism.

(e) The ratio between the magnitudes of different loads remains constant from the formation of the first plastic hinge to the attainment of the mechanism.

(f) Frame deformations are small enough to be neglected in the analysis.

In summary, the plastic design method is a (ultimate) limit-state design procedure for the derivation of 'plastic moments' for given design loads, followed by the selection of steel sections matching these moments.

9.2 Plastic analysis

Plastic analysis of structures differs fundamentally from elastic analysis, as can be seen from the following comparison:

(a) Elastic analysis of indeterminate two-dimensional structures is based on the conditions of equilibrium and continuity.
   (i) *Equilibrium.* The three conditions of equilibrium must be maintained, that is:

$$\Sigma F_x = 0, \Sigma F_y = 0 \text{ and } \Sigma M_z = 0$$

   (ii) *Continuity.* The shape of the elastically deformed structure must show no discontinuities at the rigid joints between members or at the supports or along the members, unless hinges or sliding supports are incorporated in the design.

(b) Plastic analysis of indeterminate structures is based on the conditions of equilibrium, mechanism and plastic moment limit:
   (i) *Equilibrium.* As for elastic design.
   (ii) *Mechanism condition.* Discontinuities in the deflected shape of the structure form at points where plastic deformations lead to formation of plastic hinges, allowing the structure to deform as a mechanism.
   (iii) *Plastic moment limit.* No bending moment in the structure may 'exceed' the maximum moment capacity (plastic moment—i.e. the strength limit state design section moment capacity).

It must not be overlooked that the principle of superposition of load cases does not apply to plastic design but only to elastically analysed structures free of second-order effects (see Beedle [1958]). The two principal methods for plastic analysis are:

(a) *Mechanism method.* This is essentially an upper bound procedure: that is, the computed maximum (ultimate) load corresponding to an assumed mechanism will always be greater or at best equal to the theoretical maximum load.

(b) *Statical method.* This is a lower bound solution using the principles of statics in finding the location of plastic hinges at a sufficient number of sections to precipitate transformation of the structure into a mechanism. Equlibrium equations are only used in this instance.

Only a brief description of the mechanism method is given here. (See Beedle [1958] for a more detailed outline of the design procedure.)

The first step consists of determining the types of mechanisms that can possibly form under the given load pattern. The position of the plastic hinges can only be set provisionally, and the final positions are determined during the process of analysis, which is by necessity a trial-and-error procedure. The second step involves setting the virtual work equations and their solution for the plastic moment, M_p. The third step consists of a moment diagram check to see that there is no location between the plastic hinges where the bending moment is larger than the plastic moment, because by definition the plastic moment is the largest moment that the section can resist. Furthermore, the location of the plastic hinge must coincide with the point where the plastic moment occurs.

The idealised stress–strain diagram used in plastic design is composed of two straight lines: a rising line representing the elastic response up to the yield level, and a horizontal line at the level of the yield plateau representing the plastic response. The strain hardening portion of the stress–strain diagram (see Figure 9.1) is disregarded in simple plastic theory. The plastic strain at the end of the yield plateau should be numerically equal to at least six times the elastic strain at the onset of yielding. Steels exhibiting plastic strains of this magnitude are suitable for structures designed by plastic theory, because they assure a ductile behaviour. (See also Section 9.4).

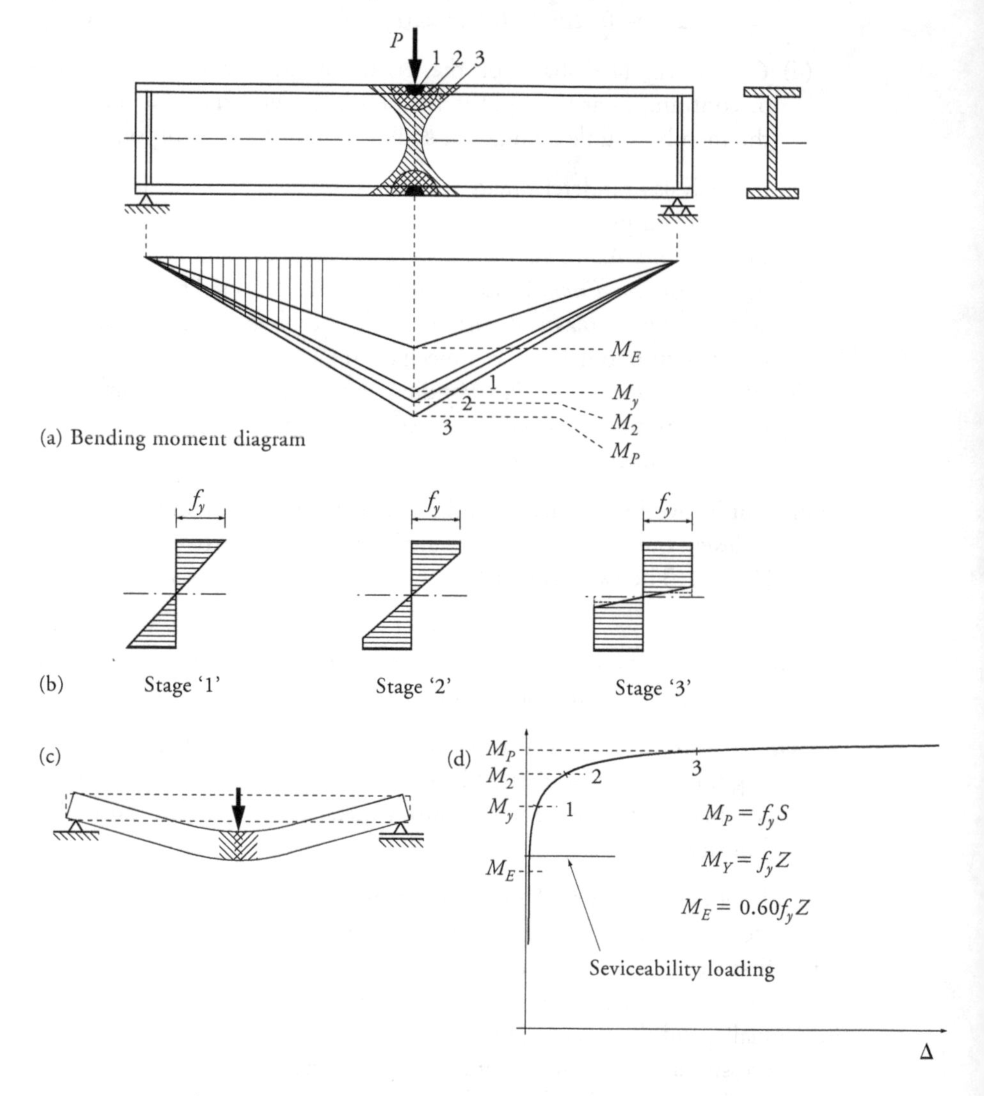

Figure 9.1 *Development of plastic hinge in a simple supported beam (a) bending moment diagram for gradually increasing load; (b) flexural stress diagrams corresponding to the various stages shown in (a); (c) deflected shape of a plastified beam; (d) load versus deflection diagram*

9.3 Member design

9.3.1 Plastic modulus of the section

The neutral axis used in elastic design has no real counterpart in plastic design. The change of stress from compression to tension occurs at an axis which, for want of a better term, will be called the 'equal area' axis. This axis divides the section into two equal areas, A_h, each being equal to one-half of the cross-sectional area, A. For equilibrium, $A_h = 0.5A$, as $f_{yc} = f_{yt}$ where f_{yc} and f_{yt} are the design yield stresses for the compression and tension regions respectively. The plastic modulus for a symmetrical section is thus:

$$S = 2(A_h\, y_h) = Ay_h$$

where y_h is the distance (lever arm) from the 'equal area' axis to the centroid of A_h.
For an unsymmetrical section:

$$S = A_{h1} \times y_{h1} + A_{h2} \times y_{h2} = A_h\,(y_{h1} + y_{h2})$$

where the terms are defined in Figure 9.2. The modulus used in plastic design is the effective modulus, Z_e, which is equal to S, the plastic section modulus, but must not exceed the value of $1.5Z$ (where Z is the elastic section modulus). It is a requirement of AS 4100 that only compact sections be used at the locations of plastic hinges. Table 5.4 lists the comparisons between S and Z values for several common sections.

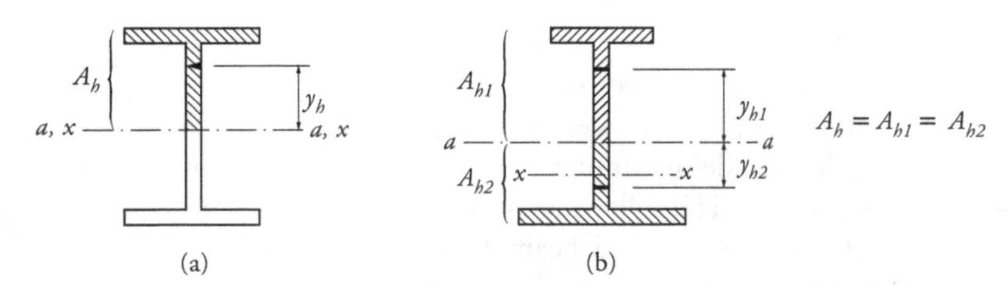

Figure 9.2 *Computation of plastic modulus (a) for a doubly symmetrical section; (b) for a section with one axis of symmetry only—a-a is the 'equal area' axis and x-x is the neutral axis*

From Table 5.4, it can be seen from the comparison of the values of S_x and Z_x the plastic section modulus is numerically larger by 10%–18% for the standard doubly symmetric I-sections (UB, UC, WB, WC) sections bending about the *x*-axis. Thus, if a compact beam is designed by the plastic method, the design load on the beam would be at least equal to the value derived by elastic design. This is particularly so for statically *determinate* (non-redundant) beam members and flexural elements. For other cases, particularly in statically *indeterminate* (redundant) members, plastic design gives values of the design load which can be up to 33% higher than the one derived by elastic design (Trahair & Bradford [1998]). See also Section 9.3.4. However, these benefits are negated if deflections control the design (Section 9.6).

9.3.2 Plastic moment capacity

The design moment capacity of a section, M_p (the fully plastic moment), is the maximum value of the bending moment that the section can resist in the fully yielded condition: that is, all the steel in the cross-section is stressed to the yield stress. No further increase of moment is assumed to be possible in simple plastic theory, and a segment of a beam would, under this condition, deform progressively without offering the slightest increase of resistance. The maximum moment capacity of a fully restrained beam, in the absence of axial load, is given by:

$$M_p = \phi Z_e f_y$$

where Z_e is the effective section modulus computed in accordance with Section 5.5.1.2 or 9.3.1.

The presence of axial force has the effect of reducing the value of M_p (see Clause 8.4.3 of AS 4100).

9.3.3 Plastic hinge

'Plastic hinge' is the term applying to the localised zone of yielding where the moment capacity, M_p, is reached. The length of the yielded zone depends on the member geometry and distribution of the transverse loads, but for the purposes of the simple plastic theory it is assumed that this length is very small and can be likened to a hinge. Unlike a hinge in the usual sense, a plastic hinge will allow rotation to take place only when the moment at the hinge has reached the value of the plastic moment, M_p.

9.3.4 Collapse mechanism

Collapse mechanism or 'mechanism' is a term applying to the state of the structure approaching total collapse. The mechanism condition is reached when a sufficient number of plastic hinges have developed and, even though no further load is or can be applied, the deformation of the structure progressively increases until the structure ceases to be stable. The following examples illustrate the mechanism condition.

A simply supported beam needs only one plastic hinge to form a mechanism, as illustrated in Figure 9.3. The response of the beam remains linear and elastic almost to the stage when the stress in the extreme fibre has reached the value of the yield stress. From there on further increases of the load induce progressive yielding of the cross-section until the whole section has yielded and the plastic hinge has fully developed. The slightest increase of load beyond this point will trigger the plastic hinge into rotation, that will continue over a short space of time and lead to very large deformations, ending in failure. No fracture is expected to occur before the hinge location becomes quite large, because structures designed by simple plastic theory are required to be made of a ductile steel and designed to remain ductile. (For further reading, consult Neal [1977]).

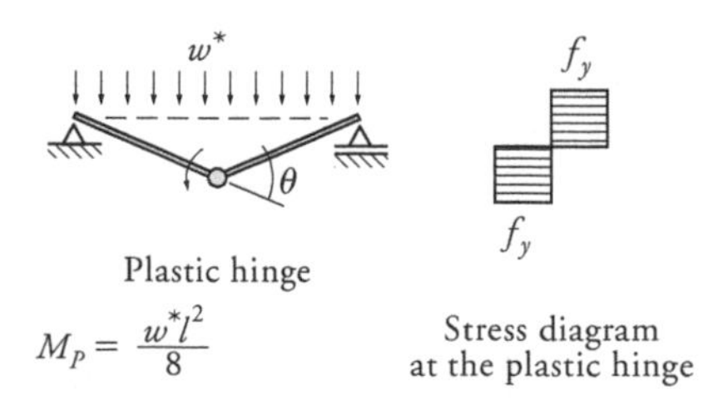

Figure 9.3 Collapse mechanism of a simply supported beam

One noteworthy observation is that statically determinate systems such as single-span simply-supported beams will give the same result for elastic and plastic analysis/design in terms of ultimate beam loading. This is due to the system "failing" when the elastic peak moment is reached which, in terms of plastic analysis, sees the coincidental formation of a hinge and the structure collapses from becoming a mechanism.

In the case of a fixed-end beam, three plastic hinges are required to transform it into a mechanism. The first plastic hinges to form as the load increases are those at the fixed ends, as in the earlier elastic stage the maximum bending moments develop at these positions. After the formation of plastic hinges at the ends, further increase of the load will not produce any further increase of the moment at the fixed ends (which are at the plastic moment capacity). However, moment redistribution will occur from any load increase and the moment at mid-span will then increase until the maximum moment capacity is reached, accompanied by the formation of the third plastic hinge, as shown in Figure 9.4. Thereafter, collapse occurs as the beam/structure becomes a mechanism. As can be seen, the plastic moment (design moment) is given by:

$$M_p = \frac{w^* l^2}{16} \qquad \left(\text{i.e. } w^*_{max} = \frac{16M_p}{l^2}\right)$$

This compares with the limiting negative moment over the supports calculated by elastic theory:

$$M^* = -\frac{w^* l^2}{12} \qquad \left(\text{i.e. } w^*_{max} = \frac{12M_p}{l^2} \text{ assuming that } M_p \text{ is the maximum moment in this instance}\right).$$

The elastic (non-critical) moment at mid-span (not relevant for design) is one-half of the elastic support moment. From the above, it can be seen that plastic analysis and design permits a ($\frac{16}{12} \times 100 =$) 33% increase in beam load carrying capacity over elastic analysis methods. This translates to a 25% reduction in required moment capacity for plastic design and is particularly true when the strength limit state governs.

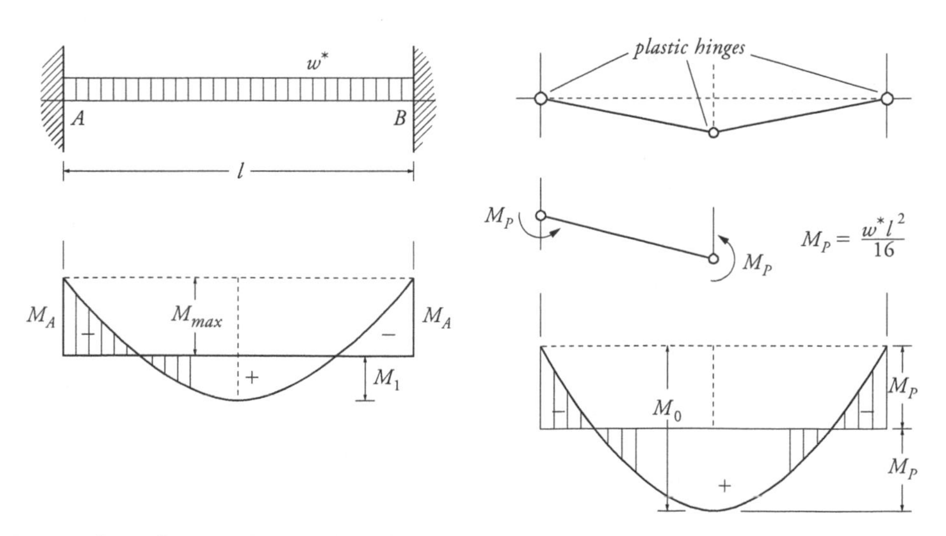

Figure 9.4 *Collapse mechanism of fixed-end beam*

Continuous beams and frames also benefit from plastic design. Three cases should be considered for equal span continuous beams:

- External supports hinged: $M_p = \dfrac{w^* l^2}{12}$
- External supports fixed: $M_p = \dfrac{w^* l^2}{16}$
- External supports elastically fixed into columns: intermediate of the above two values

Plastic (design) moments are thus approximately 25% less than the moments computed by elastic theory. Material savings are realisable even when the loads are arranged in the least favourable pattern (checkerboard loading).

From the above, it can be surmised that distinct advantages can be had from bending moment distributions/redistributions which can only occur in statically *indeterminate* structures. However, statically *determinate* structures cannot display moment redistribution characteristics and, hence, there is no advantage in using plastic design for such structural systems. Additionally, as noted in Section 9.3.1, any advantage obtained from plastic design may be offset by serviceability constraints.

9.4 Beams

The applicability of plastic analysis and design are listed in Clause 4.5 of AS 4100 which notes specific limitations to be observed (see below) and also the requirement for equilibrium and boundary conditions to be satisfied.

Due to the products tested, the current AS 4100 limitations restrict plastic analysis/design to the following member types:

- hot-formed, compact, doubly symmetric I-sections
- minimum yield stress shall not exceed 450 MPa
- to ensure adequate moment redistribution, the steel's stress-strain characteristics shall not be significantly different to those of the AS/NZS 3678 & 3679.1 steels (unless the steel has a yield stress plateau extending at least six times the yield strain; $f_u / f_y \geqslant 1.2$; the AS 1391 tensile test elongation is not less than 15%, and; the steel displays strain-hardenability)
- not subject to impact loading or fluctuating loading requiring fatigue assessment, and
- with connections that are either "full strength" to cope with plastic hinge formation at the joint or "partial strength" that do not suppress the generation of plastic hinges in the structural system

The above must be observed unless adequate structural ductility and member/connection rotational capacities can be demonstrated.

Embracing the above, there is no subsequent distinction made in AS 4100 between section capacities for beam (only) actions designed by elastic analysis and those designed by plastic analysis. Consequently, for plastically analysed and designed beams, Clause 5.1 of AS 4100 notes the beam nominal section capacity, M_s, to be:

$$M_s = Z_e f_y$$

where

$$Z_e = \text{effective section modulus}$$
$$= \text{plastic section modulus, } S\text{, for compact doubly-symmetric I-sections}$$
$$f_y = \text{yield stress used in design } (\leqslant 450 \text{ MPa})$$

and, somewhat like the elastic analysis method (which requires $M^* \leqslant \phi M_s$), the following must be satisfied:

$$M^* \approx \phi M_s$$

$$\text{i.e. } Z_e = S \approx \frac{M^*}{\phi f_y}$$

Any significant difference between M^* and ϕM_s will invalidate the analysis or design—i.e. if M^* is much lower than ϕM_s then the plastic mechanism(s) cannot occur and if M^* is much higher than ϕM_s the member is inadequate. References such as Pikusa & Bradford [1992] and Trahair & Bradford [1998] note (via worked examples) that due to the approximate nature of load estimation and analysis, the actual member S should only be a few percent less (say up to 5%–7%) than the above calculated S. Design checks in terms of ultimate/collapse loads or moments may also be used instead of the above method using S.

The above provisos for plastic analysis/design fundamentally ensure the development of plastic collapse mechanisms. This is explicitly seen with the suppression of local buckling effects with the requirement for compact sections.

Though not explicitly stated in AS 4100, but implied by its basic aims, the possibility of flexural-torsional buckling must also be suppressed so that plastic hinges can be generated where required. In essence this means that for plastic analysis/design, restraint spacing on beams must be such that the member moment capacity, M_b, equals the section moment capacity, M_s (i.e there is no need to calculate α_m, α_s, etc unless these parameters are used to establish the fundamental criterion of the section moment capacity being fully mobilised).

Other checks required for plastic design of beams include shear capacity (based on the fully plastic shear capacity of the web—see Section 5.8.2), bending and shear interaction (see Section 5.8.4) and bearing (Sections 5.8.5 and 5.8.6). For shear ductility, web stiffeners are also required at or near plastic hinges where the factored shear force exceeds the section shear capacity by 10% (Clause 5.10.6 of AS 4100).

9.5 Beam-columns

9.5.1 General

Members subject to combined bending and axial load are termed 'beam-columns'. All practical columns belong to this category because the eccentricity of the load is always present, no matter how small, or from flexural loads via elastic connections to beams. Each beam-column should be verified for section capacity at critical sections and for member capacity of the beam-column as a whole. When required, Clause 4.5.4. of AS 4100 requires plastically analysed beam-columns to consider second-order effects from the interaction of bending and compression (see also Section 9.5.6).

9.5.2 Section capacities—uniaxial bending with axial loads

Since its first release, AS 4100 only considers the plastic design of single-plane beams, beam-columns and frames. The reasoning for this was due to the complexity of assessing the plastic interaction behaviour of biaxial bending effects and translating this into practical design provisions. Consequently, Clause 8.4.3 of AS 4100 only considers one type of combined action check for plastic analysis/design, that of in-plane capacity of a member subject to uniaxial bending and axial load. Though complex, the reader is directed to Trahair & Bradford [1998] for further information on the plastic design of members subject to biaxial bending.

Combined bending and axial load result in a reduced section moment capacity, which is determined by the same interaction equations as for elastically designed structures (see Chapter 5).

For a compact, doubly symmetrical I-section member subject to uniaxial bending and tension or compression, the reduced plastic moment capacity (ϕM_{prx} or ϕM_{pry}) shall be calculated as follows:

(a) For doubly symmetrical section members bent about the major principal axis:

$$\phi M_{prx} = 1.18\phi M_{sx}\left(1 - \frac{N^*}{\phi N_s}\right) \leq \phi M_{sx}$$

(b) For members bent about the minor principal axis:

$$\phi M_{pry} = 1.19\phi M_{sy}\left(1 - \left(\frac{N^*}{\phi N_s}\right)^2\right) \leq \phi M_{sy}$$

where ϕM_{sx} and ϕM_{sy} are the design section moment capacities (see Section 5.3), N^* the design axial load and ϕN_s the axial design section capacity (see Section 6.2.7(a) for axial compression or Section 7.4.1.1 for axial tension with $\phi N_s = \phi N_t$). The peak design moment, M^*, must then satisfy either:

$$M_x^* \leq \phi M_{prx} \quad \text{or}$$

$$M_y^* \leq \phi M_{pry}$$

There may be some further iterations to this check as the limiting moment is reduced when axial load is present and the analysis may need to be redone. Clause C8.4.3.4 of the AS 4100 Commentary provides further guidance on this design routine. Note that for beam-columns subject to axial compression, there are additional limits on member and web slenderness—see Section 9.5.4 and 9.5.5 below. For further details, refer to Clause 8.4.3.4 of AS 4100.

9.5.3 Biaxial bending

Not considered in AS 4100. See Section 9.5.2.

9.5.4 Member slenderness limits

From Clause 8.4.3.2 of AS 4100 the member slenderness for members subject to axial compression and bent in the plane of the frame containing a plastic hinge shall satisfy:

$$\frac{N^*}{\phi N_s} \leqslant \left[\frac{(0.6 + 0.4\beta_m)}{\left(\frac{N_s}{N_{ol}}\right)^{0.5}}\right]^2 \qquad \text{when } \frac{N^*}{\phi N_s} \leqslant 0.15 \text{ and}$$

$$\frac{N^*}{\phi N_s} \leqslant \frac{\left[1 + \beta_m - \left(\frac{N_s}{N_{ol}}\right)^{0.5}\right]}{\left[1 + \beta_m + \left(\frac{N_s}{N_{ol}}\right)^{0.5}\right]} \qquad \text{when } \frac{N^*}{\phi N_s} > 0.15$$

It should be noted that the member may not have plastic hinges and should be designed elastically when:

$$\frac{N^*}{\phi N_s} > \frac{\left[1 + \beta_m - \left(\frac{N_s}{N_{ol}}\right)^{0.5}\right]}{\left[1 + \beta_m + \left(\frac{N_s}{N_{ol}}\right)^{0.5}\right]} \qquad \text{and } \frac{N^*}{\phi N_s} > 0.15$$

where β_m = the ratio of the smaller to the larger end bending moments, and

$$N_{ol} = \frac{\pi^2 EI}{l^2}$$

9.5.5 Web slenderness limits

In members containing plastic hinges, the design axial compression force shall satisfy:

$$SR = \frac{N^*}{\phi N_s}, \quad d_n = \left(\frac{d_1}{t}\right)\left(\frac{f_y}{250}\right)^{0.5}, \text{ and Table 9.1}$$

Table 9.1 Web slenderness limits as noted in Clause 8.4.3.3 of AS 4100.

Inequality to be satisfied	Range
$SR \leqslant 0.60 - \frac{d_n}{137}$	$45 \leqslant d_n \leqslant 82$
$SR \leqslant 1.91 - \frac{d_n}{27.4} \leqslant 1.0$	$25 < d_n < 45$
$SR \leqslant 1.0$	$0 \leqslant d_n \leqslant 25$

Where web slenderness exceeds 82, the member must not contain any plastic hinges: that is, the member must be designed elastically in the plastically analysed structure or the frame should be redesigned.

9.5.6 Second-order effects

When required, Clause 4.5.4 of AS 4100 notes that second-order effects (Section 4.3) need to be evaluated for beam-column members and/or frames analysed by first-order plastic analysis methods. The rational evaluation of second-order effects may be neglected when:

- $10 \leq \lambda_c$
- $5 \leq \lambda_c < 10$ provided the design load effects are amplified by a factor δ_p

$$\text{where } \delta_p = \frac{0.9}{1 - \left(\frac{1}{\lambda_c}\right)}$$

A second-order plastic analysis must be undertaken when the elastic buckling load factor, λ_c, is less than 5. For the member/frame, λ_c is the ratio of the elastic buckling load set to the design load set. See Clause 4.7 of AS 4100 or Section 4.4.2.2 for methods of evaluating λ_c. Woolcock et al. [1999] also notes that most practical portal frames satisfy $\lambda_c > 5$ and that second-order plastic analysis is generally not required.

9.6 Deflections

Until the formation of the first plastic hinge, the frame behaves elastically and the deflections can be determined by linear elastic theory. The formation of the first plastic hinges nearly always occurs after reaching the serviceability load. Further loading induces formation of further plastic hinges and the behaviour becomes markedly non-linear. The deflections at serviceability loads can be computed by the elastic method (Beedle [1958], Neal [1977]). Some deviations occur because the connections are not perfectly rigid, but this is mostly offset by the likelihood of the partial fixity of column bases. However, deflections in plastically designed frames are usually larger than with frames that have been elastically designed.

If deflections are needed to be computed for ultimate loads, the computations become rather involved. Beedle [1958] and Neal [1977] give methods for the evaluation of plastic deflections. For gable frames, Melchers [1980] and Parsenajad [1993] offer relatively simple procedures, the latter with useful charts and tables for rafters.

Interestingly, from an Australian perspective, plastic analysis and design is not commonly used. Some general reasons reported include the complexity of analysis and extra fabrication requirements to ensure no onset of instabilities (local buckling, web buckling, flexural-torsional buckling, etc). In terms of analysis, where superposition principles cannot be used, it is said that Australian loading conditions on structures are unsymmetrical—i.e. typically wind loads govern. Whereas in Europe, where gravity loads govern (especially snow loading, etc), such symmetrical loading types can be better handled by plastic analysis.

However, one of the main reasons offered for the low popularity of plastic design in Australia is that serviceability limits generally govern flexural/sway designs. Consequently, any savings offered by plastic analysis will be more than offset by compliance with serviceability requirements.

9.7 Portal frame analysis

Example 9.1 illustrates an application of the mechanism analysis method of fixed-end beams. For a preliminary design of portal frames with horizontal rafters, Table 9.2 gives the necessary coefficients.

Table 9.2 Formulae for plastic moments of portal frames; $I_{col} = I_{beam}$

Vertical loads only

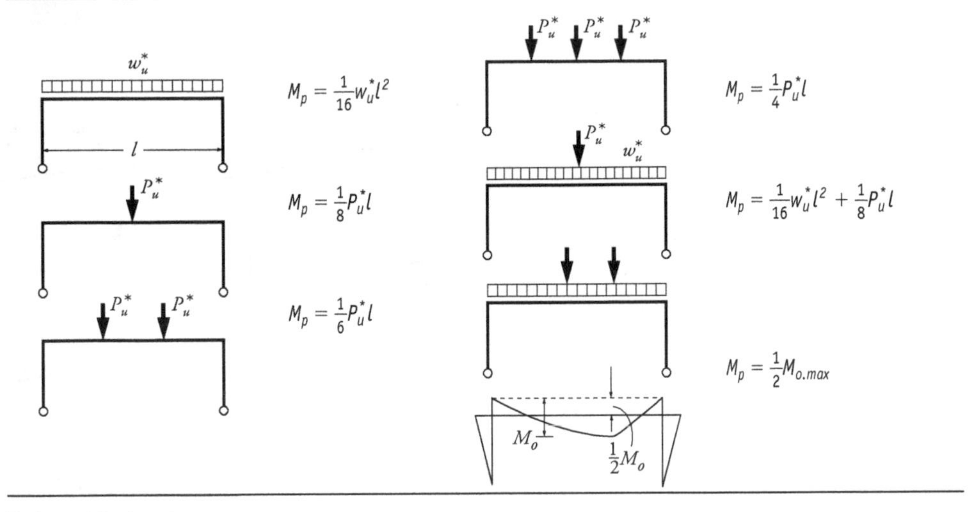

Horizontal loads only

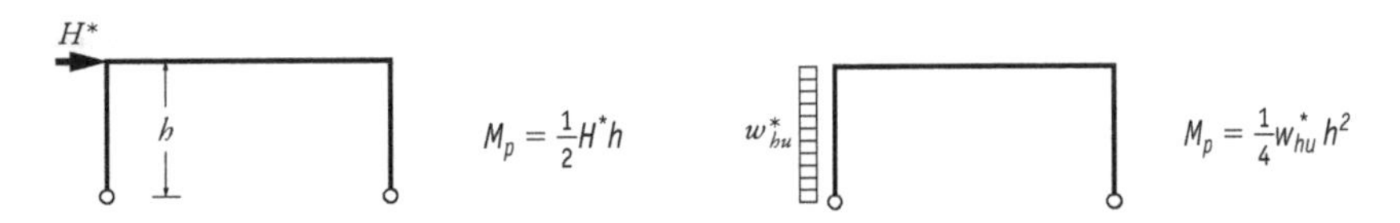

Mixed loads

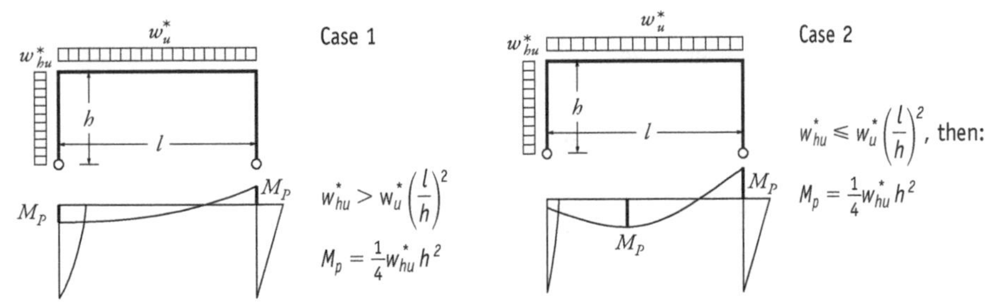

Note: The principle of superposition of loads does not hold in plastic analysis; all loads must therefore be considered simultaneously, case after case.

Gable frames with a small ratio of apex rise to half spans can be designed as if the rafters were straight, because there is a negligible effect of a small kink such as occurs in rafters when metal cladding (4 degree slope, or 1 in 14) is used.

Plastic moments of a single bay portal frame with pinned bases and a rafter slope not exceeding 6 degrees can be determined using coefficient K_p given in Figure 9.5:

$$M_p = K_p w^* l^2$$

See Section 9.5.6 for information on second-order effects and the above. Pikusa and Bradford [1992] and Woolcock et al. (1999) should be consulted for further study of plastic analysis of portal frames. However, Woolcock et al. [2011] note that, unlike the detailed consideration of plastic analysis in previous editions of their popular publication, this form of analysis is not covered in their 2011 edition for portal frame buildings. The main reasons are due to plastic design being labour intensive, not widely used nor software being readily available and eventually elastic analysis has to be undertaken for deflection checks.

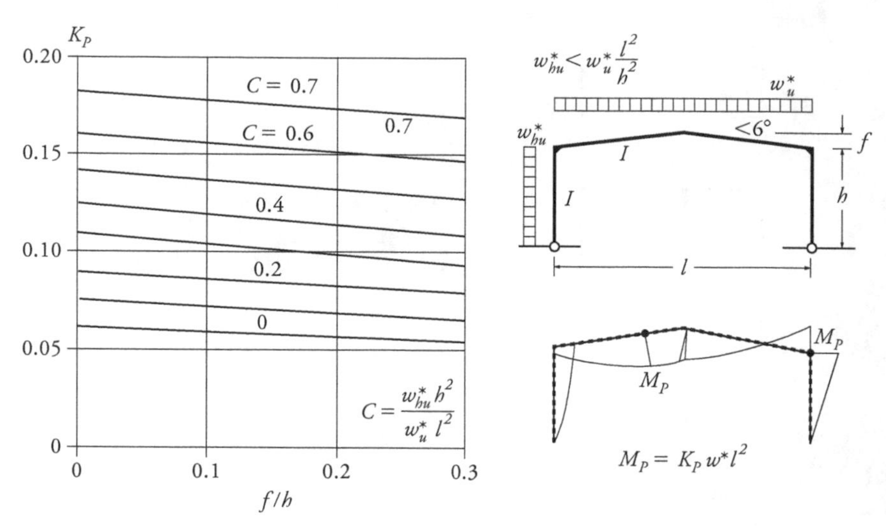

Figure 9.5 *Chart for plastic moment coefficients for pinned base, gabled portal frames*

9.8 Examples

9.8.1 Example 9.1

Step	Description and computation	Result	Unit
	Verify the size of the beam shown using plastic design method. The beam has adequate restraints to ensure no flexural-torsional buckling occurs.		

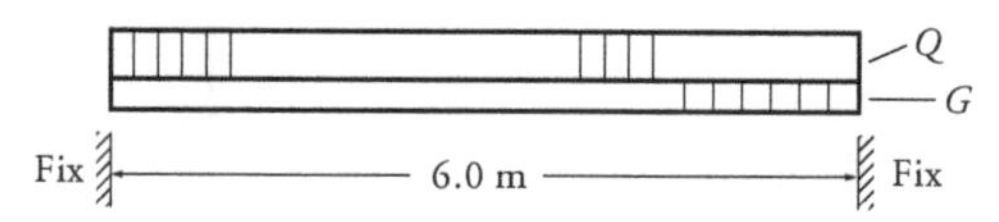

Step	Description and computation	Result	Unit
	Data:		
	Trial section: 410UB53.7 Grade 300		
	... from ASI [2009a] Table 3.1-3(B), page 3-11...		
	Section compactness for plastic design...	Compact-OK	
	Loads:		
	Uniformly distributed permanent action/dead load, G	27.0	kN/m
	Uniformly distributed imposed action/live load, Q	45.0	kN/m
1	Simple span bending moment:		
	$M_o = \left(\frac{1}{8}\right) \times (1.2G + 1.5Q) \times 6.0^2$		
	$= \left(\frac{1}{8}\right) \times (1.2 \times 27.0 + 1.5 \times 45.0) \times 6.0^2 =$	450	kNm

Step	Calculation	Value	Unit
2	Draw the bending moment diagram and move the closing line so as to obtain equal negative and positive bending moments (see Figure 9.4 also):		
	$M_A^* = -0.5M_o$		
	$= \left(\frac{1}{2}\right) \times 450$ ... neglecting sign =	225	kNm
	$M_1^* = 0.5\ M_o =$	225	kNm
3	For the plastic analysis/design of beams:		
	$M_{bx} = M_{sx}$ ... i.e. no flexural-torsional buckling ...		
	From ASI [2009a] Table 5.2-5, page 5-38 for a 410UB53.7 Grade 300		
	$\phi M_{sx} =$	304	kNm
	$> M_A^*$ and M_1^*		
	The difference between ϕM_{sx} and M_A^* or M_1^* is excessive and a lighter section is sought. Try a 360UB44.7 Grade 300 with		
	$\phi M_{sx} =$	222	kNm
4.	Check adequacy of 360UB44.7 Grade 300 for plastic moment:		
	$\phi M_s \leqslant M_A^*$ and $M_A^* \approx \phi M_s$	True	
	$\phi M_s \leqslant M_1^*$ and $M_1^* \approx \phi M_s$	True	
	NOTE: See Section 9.4 on satisfying these inequalities and the relative magnitudes of M^* to ϕM_s.		
5.	Verify adequacy of 360UB44.7 Grade 300 for other design checks		
5.1	Additional data		
	$V^* = \left(\frac{1}{2}\right) \times (1.2G + 1.5Q)l$ ... at the end supports		
	$= \frac{1}{2} \times (1.2 \times 27 + 1.5 \times 4.5) \times 6.0 =$	300	kN
	From ASI [2009a] Tables 3.1–3(A) and (B), pages 3–10 and 3–11		
	d = section depth =	352	mm
	t_w = web thickness =	6.90	mm
	f_{yw} = design yield strength of the web =	320	MPa
	End connections are full depth welded end plates		
5.2	Plastic web shear capacity, ϕV_u (Section 5.8.2)		
	$\phi V_u = \phi 0.6 f_{yw} A_w = \phi 0.6 f_{yw} d t_w$		
	$= 0.9 \times 0.6 \times 320 \times (352 \times 6.90)/10^3 =$	420	kN
	Check with ASI [2009a] Table 5.2-5, page 5–38		
	$\phi V_u = \phi V_v =$	420	kN
	and indicates the web is compact and confirms the above calculation		
	$V^* \leqslant \phi V_u$ is true as $300 \leqslant 420$ →	OK	
5.3	Shear-bending interaction (Section 5.8.4)		
	$0.6\phi V_u = 0.6 \times 420 =$	252	kN
	$V^* \leqslant 0.6\phi V_u$ is not true as $300 \leqslant 252$ is false	Not OK	
	and the reduced design shear capacity must be evaluated		
	ϕV_{vm} = design web capacity in the presence of bending moment		
	$= \phi \alpha_{vm} V_v$		
	$= \left[2.2 - \frac{1.6M^*}{\phi M_s}\right]\phi V_u$ .... as $0.75\phi M_s \leqslant M^* \leqslant \phi M_s$		

	$= \left[2.2 - \frac{1.6 \times 225}{222}\right] \times 420 =$	243	kN
	$V^* \leqslant \phi V_{vm}$ is not true either as 300 ≤ 243 is false	Web stiffeners required	
	The webs require shear stiffeners in the vicinity of the plastic hinges at the end supports (no further calculations done here—see Section 5.8.7 and aim for $V^* \leqslant 0.6\phi(R_{sb} + V_b)$ as a suggested minimum) with no re-analysis required. Also, thickening the web (by plate reinforcement) at the end supports is an option, however, this changes the ultimate load for plastic collapse and re-analysis is then required.		
5.4	Load bearing		
	Bearing capacity calculations at reaction points do not need to be evaluated as fully welded end plate connections are used. These "full strength" connections are required as plastic hinges occur at the reaction points. Clause 5.4.3(a) of AS 4100 requires the full strength connection to have a moment capacity not less than the connected member. Additionally, some constraints on rotation capacity are noted—this is typically complied with by testing or using industry accepted rigid connections (Chapter 8).		
6.	The second trial can be adopted for the design.		
7.	Comparison with the elastic design: The (peak) bending moment at the fixed supports would then be:		
	$M_A^* = \left(\frac{1}{12}\right) \times (1.2 \times 27.0 + 1.5 \times 45.0) \times 6.0^2 =$	300	kNm
	As 300 > 222, the second trial section would be inadequate. Thus a larger section would be required if elastic design were used (i.e. 410UB53.7 with ϕM_{sx} = 304 kNm which is heavier by 20%.)		

9.9 Further reading

- For additional worked examples see Chapters 4 and 8 of Bradford et al. [1997], Chapter 8 of Woolcock et al. [1999] and Trahair & Bradford [1998].
- Well-known texts on plastic analysis and design include Beedle [1958], Baker & Heyman [1969], Morris & Randall [1975], Neal [1977], Horne [1978] and Trahair & Bradford [1998].
- Further references on the application of plastic analysis and design include Heyman [1971], Massonnet [1979], Manolis & Beskos [1979], Horne & Morris [1981] and Woolcock et al. [1999].
- The fundamental aim of plastic analysis and design is to ensure the adequate development of plastic collapse mechanisms. Unless otherwise demonstrated, this may be attained by the member having full lateral restraint or ensuring that discrete restraints (F, P, L) are placed so that segment lengths cannot undergo flexural-torsional buckling—i.e. segments are of sufficient length that they are considered to be fully laterally restrained (Clause 5.3.2.4 of AS 4100). These restraints and their spacings can be determined by referring to Chapter 5 and Trahair et al. [1993c].

chapter 10

Structural Framing

10.1 Introduction

Structural framing comprises all members and connections required for the integrity of a structure. Building structures may vary in size from a single-storey dwelling framing to a large mill building or high-rise framing, but the design principles involved are much the same. This section deals with practical design aspects and miscellaneous design matters. It starts with the selection of the form of structural framing and continues with the practical design of structural components (Table 10.1).

Table 10.1 Contents

Subject	Subsection
Mill-type buildings	10.2
Roof trusses	10.3
Portal frames	10.4
Steel frames for low-rise structures	10.5
Purlins and girts	10.6
Floor systems for industrial buildings	10.7
Crane runway girders	10.8
Deflection limits	10.9
Fire resistance	10.10
Fatigue	10.11
Painted and galvanized steelwork	10.12

Structures serve many functions, stemming from the overall building design:

- providing the support for the building envelope (walls, roof, fenestrations)
- resisting the environmental forces acting on the building envelope
- supporting the floors, machinery and service (live) loads/actions
- supporting their own weight.

The structure can be designed to resist horizontal loads by:

- columns cantilevered from the footings (beam and post frames)
- shear walls (concrete or brick walls)
- rigid frames (rectangular frames and portal frames)
- bracing systems or stayed cables
- combinations of the above methods.

Solely from the point of view of economy, braced frames (one-way or two-way) would be most appropriate. However, bracing panels often interfere with the functional layout of the building. The use of rigid frames overcomes this drawback. In workshop buildings and warehouses, the framing system often employed consists of portal frames in the transverse direction and bracing panels in the longitudinal direction. For maximum flexibility in layout it is sometimes necessary to use a two-way rigid frame system, but there is a cost penalty involved in this. Typical industrial building framing systems are shown in Figure 10.1.

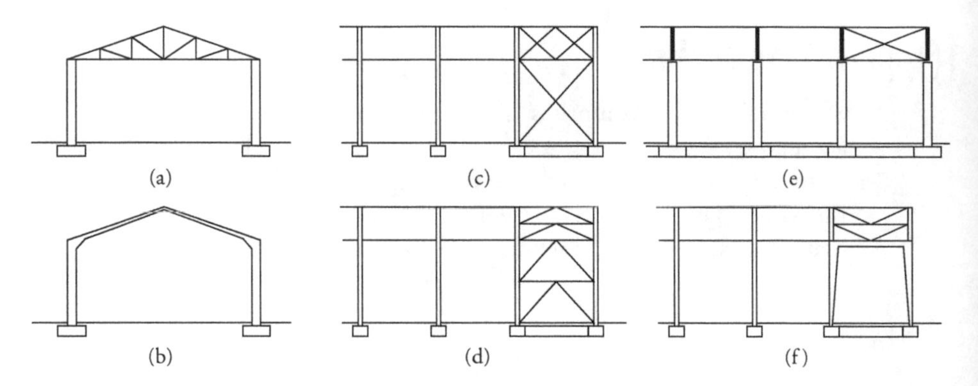

Figure 10.1 *Industrial building framing systems: (a) truss and cantilever column bent; (b) portal frame; (c) longitudinal X-bracing; (d) longitudinal K-bracing; (e) longitudinal cantilever columns; (f) portal-type bracing*

Before proceeding with the selection of an appropriate system, the designer should gather as much data as possible concerning the site of construction, foundation materials and various constraints on structural solution. The following checklist, by no means exhaustive, should provide a starting point for information gathering:

- site topography, location of adjacent buildings, services, other constraints
- foundation material profile, bearing capacity, suitable types of footings
- ground water level, drainage
- terrain features for wind loading assessment
- access for construction equipment
- building layout, column grids
- minimum headroom, maximum depth of beams and trusses
- possible locations of braced bays
- provisions for future extensions
- loads imposed on the structure: live loads, machinery and crane loads
- materials used in the building envelope and floors
- type and extent of building services, ventilation and air conditioning
- fire rating and method of fire protection

- means of egress and other statutory requirements.

Some additional variables influencing the selection of the structural form are as follows:
- roof slope (minimum, maximum)
- type of sheeting and purlins/girts
- spacing of column grids
- height of columns between the floors
- feasibility of using fixed column bases
- architectural preferences for the overall form of the building.

The building framework consists of a variety of members and connections, and the individual elements can number tens of thousands in a high-rise building. The tendency, nowadays, is to reduce the number of framing elements to a minimum and optimise the weight for reasons of economy. The designer of steel-framed buildings should develop a mastery in marshalling relevant facts and developing a satisfactory framing solution.

This section is intended to give practical hints on the design of framing for low-rise buildings, connections and miscellaneous other considerations. Further reading on the topic can be found in ASI [2009b].

10.2 Mill-type buildings

10.2.1 General arrangement—in-plane of frame

The structural framing for 'mill-type' (i.e. medium–heavy type industrial) buildings usually consists of a series of bents, arranged in parallel. These bents (portal-type frames) resist all vertical and lateral loads acting in their plane. Forces acting in the longitudinal directions—that is, perpendicularly to the bents—are resisted by bracing panels arranged at intervals. The following framing systems are commonly used:
- cantilevered columns with simply supported roof trusses
- rigid-jointed frames with fixed or pinned bases.

Figure 10.2 illustrates these framing types.

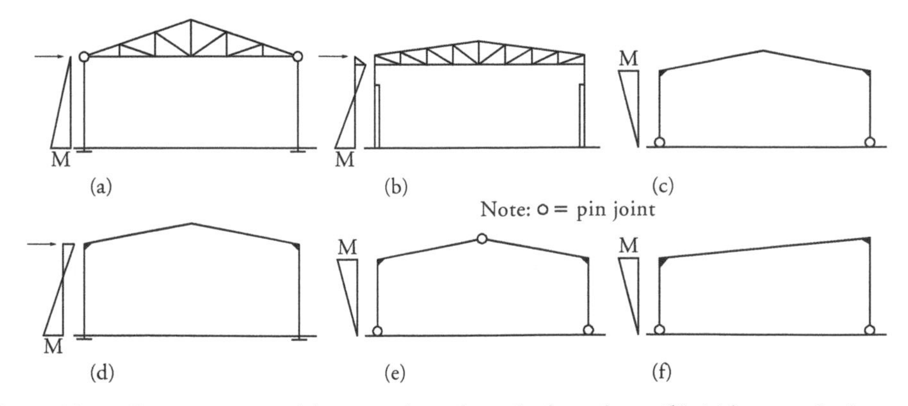

Figure 10.2 *Framing systems: (a) truss and cantilevered column bent; (b) rigid truss and column bent; (c) two-pin portal; (d) fixed-base portal; (e) three-pin portal; (f) two-pin mono-slope rafter portal*

10.2.1.1 Cantilevered columns with simply supported roof trusses

The main advantage of this type of framing is that it is not too sensitive to the foundation movement. Another advantage is in its relatively easy erection. Because there is only a pin connection between the columns and the roof trusses, the wind forces and lateral forces from cranes are resisted solely by the columns, cantilevering from the footings.

10.2.1.2 Rigid frames

With regard to the rigidity of column-to-footing connections, rigid frames can have fixed or pinned bases. The roof member can be a rolled or welded plate section, or alternatively a truss. This may also apply to the columns or both roof and column members to act like a portalised truss. Frames designed for base fixity rely on the rigidity of the foundation, particularly on the rotational rigidity. Fixed-base frames are structurally very efficient, and therefore very economical, but their footings tend to be costly (if feasible).

Pinned-base frames—that is, frames with flexible connections to the footings—derive their resistance to lateral forces from their rigid knee action. This type of framing is commonly used for buildings of relatively low span-to-column height. The main advantage is in reduced footing costs, as the footings need not resist very large bending moments.

10.2.2 Longitudinal bracing

Columns of mill-type buildings are oriented such that their strong axis is parallel to the longitudinal axis of the building. This means that, without bracing in the longitudinal direction, the building would be too flexible—if not unstable. Figure 10.3 shows typical longitudinal bracing systems. The location of the braced bays should be carefully planned to avoid interfering with the operational requirements and to obtain the right structural solution. Wherever possible, the bracing should be situated close to the position where the first frames will be erected, to facilitate the overall building erection.

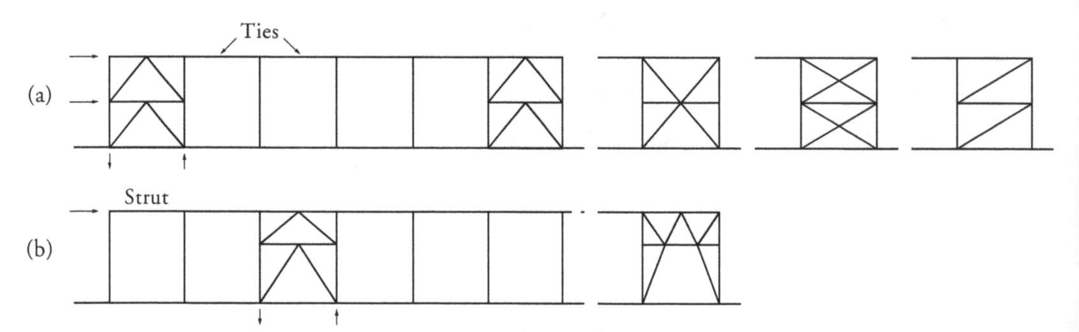

Figure 10.3 *Longitudinal bracing systems (alternate configurations shown at end): (a) wall panels between frames with minimal openings, and; (b) wall panels with significant openings*

Thermal expansion joints may be required in buildings longer than 80 metres, unless it can be proven by analysis that the principal members will not be overstressed if expansion joints are placed farther apart. The columns and rafters are usually doubled at the expansion joint to achieve complete separation of the structure within a small gap.

Bracing elements must be designed for all forces from wind and crane operation (acceleration, braking, surge and impact against buffers). Where high-capacity cranes are used, it is best to provide double bracing systems, one for wind bracing and another in the plane of the crane runway girders.

Bracing in the plane of the roof has the function of transferring wind loads acting on the end walls to the vertical bracing. The roof bracing is usually designed in the form of a horizontal truss or 'wind girder'. Two examples are shown (see Figure 10.4). See also Section 10.3.1 and Figure 10.8.

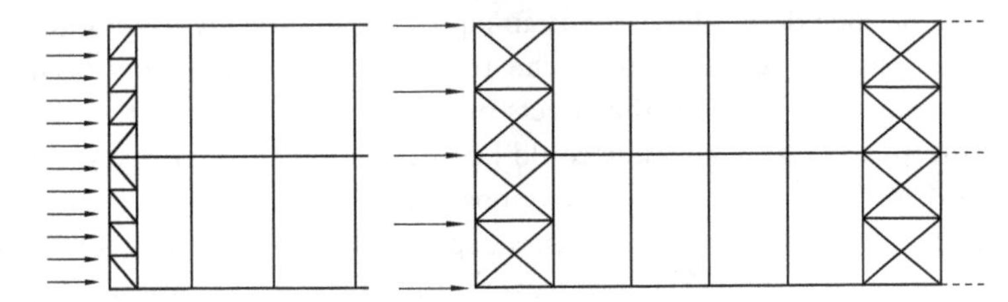

Figure 10.4 *Wind bracing in the roof plane*

10.3 Roof trusses

10.3.1 General

Roof rafters spanning more than 20 m can often be designed, quite economically, in the form of trusses. The saving in weight stems from the fact that truss web members use less steel than the solid webs of UB sections or plate girders. The fabrication cost of trusses is marginally higher, but this is offset by the saving in steel.

The usual span-to-depth ratio of steep roof trusses is 7.5 to 12, depending on the magnitude of loads carried. The truss can be designed with sloping top chords and a horizontal bottom chord, but this could make the mid-span web members too long. Constant depth (near-flat roof) design is preferable as far as the fabrication is concerned. The Warren and Pratt types of trusses are used extensively because they have structural advantages and a good appearance. The following design rules should be observed:

- the panel width should be constant
- even number of panels avoids cross-braces
- diagonal web members should be in tension under worst-case loading (unless hollow sections are used)

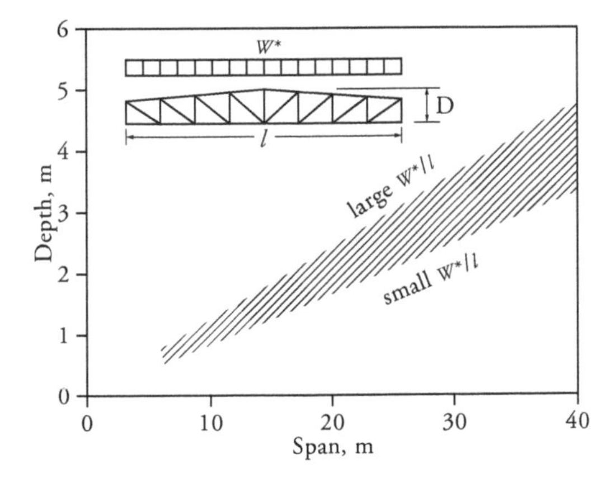

Figure 10.5 *The usual range of depths of roof trusses*

- the inclination angle of the diagonals should be between 35° and 50°
- if at all possible, the purlins and verticals should closely coincide.

Figure 10.5 gives the range of depths for various spans between simple supports of a Warren or Pratt truss in which the parameter W^*/l at each end of the indicative range has the total design load W^* as an equivalent uniformly distributed load along the span. The length l is the span of the truss between simple supports. The parameter is a combined indicator of the effects of varying the span and the intensity of loading. For example, a large/high value could be due to either a small span or high concentration of loads or both. Conversely, the opposite is true. It is seen that the magnitude of W^* and span influences the required truss depth and thus upper and lower limits are indicated.

Figure 10.6 shows the roof truss framing types and Figure 10.7 gives some data on the 'older' type of roof truss designs.

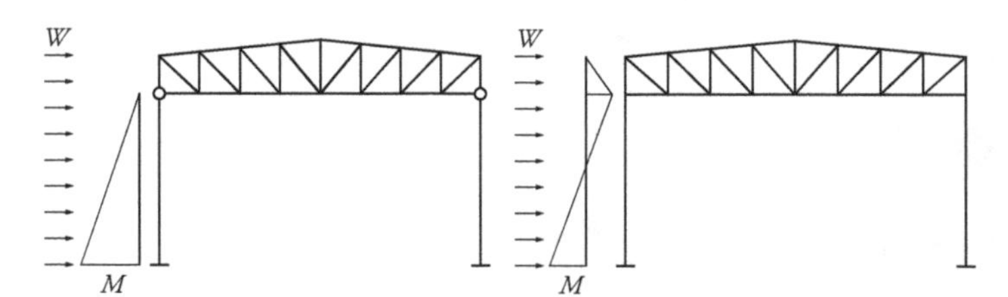

Figure 10.6 *Framing systems incorporating roof trusses of the Pratt type: (a) cantilevered column and pin connected trusses; (b) portal-type truss*

Diagram	Span, m	Mass, kg	kN
	5 – 7	160 – 210	1.6 – 2.1
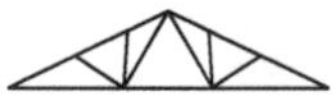	8 – 11	210 – 270	2.1 – 2.7
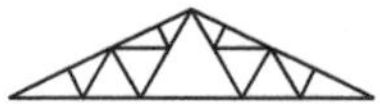	12 – 14	270 – 400	2.7 – 4.0
	15 – 16	400 – 620	4.0 – 6.2
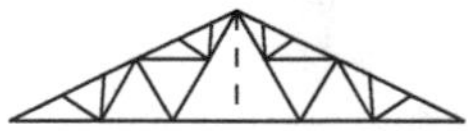	17 – 20	620 – 950	6.2 – 9.5
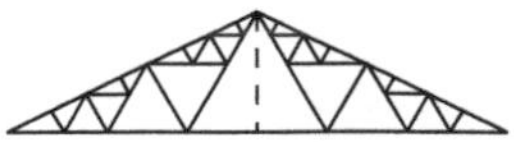	22 – 30	950 – 1100	9.5 – 11.0

Figure 10.7 *Approximate mass for roof trusses of high pitch, steep roof (Fink type)*

As an aid to calculating the self-weight of trusses, Table 10.2 gives some rough weight estimates.

Table 10.2 Approximate self-weight of trusses.

Span (m)	Self-weight (kN/m^2), over span × truss spacing
10	0.11
20	0.12
30	0.16
40	0.22

As noted above, portal frame/mill-type buildings resist in-plane actions (e.g. cross wind-loads, etc) by in-plane flexural stiffness. Also, Section 10.2.2 notes the role of longitudinal bracing to stabilise the primary structural frame in the out-of-plane (building longitudinal) direction. The loads transmitted by longitudinal bracing include those arising from:

- wind loads acting on the upper half of the end walls
- frictional drag effects on the roof, and
- accumulated "lateral" bracing system restraint forces (e.g. from purlins and fly-braces).

These longitudinal loads are subsequently transmitted to the roof bracing (Figure 10.4) and then down the wall bracing to the foundations. In terms of load eccentricities, it is advantageous to place the roof bracing in close proximity to the purlins as the latter elements may also form part of the bracing system as well as providing restraint to the critical compression elements of primary roof members.

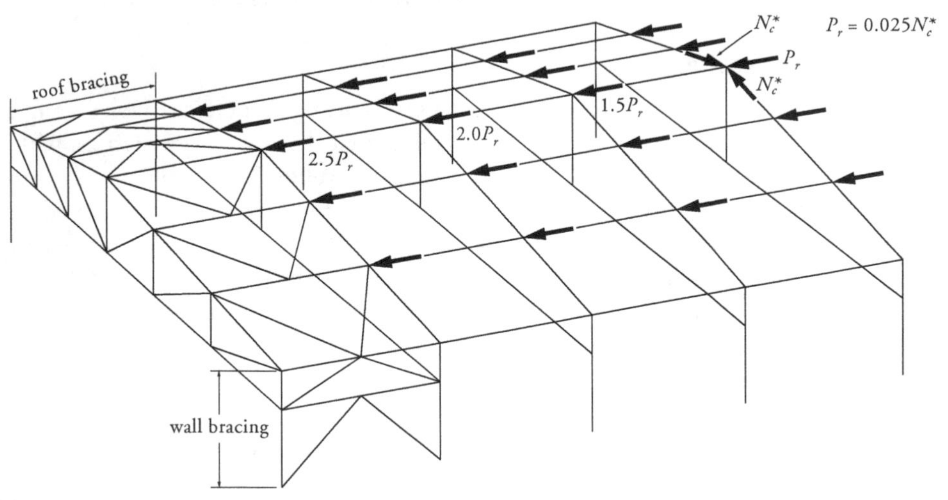

Notes:

(1) Unless special studies are made, lateral forces from rafters should be accumulated at the braced bay.

(2) N_c^* = compression force in top truss chord, which for illustrative purposes, is assumed to be the same for all roof trusses in the figure.

Figure 10.8 *Forces in the longitudinal bracing system in the plane of the compression chords*

Clauses 6.6.1 to 6.6.3 of AS 4100 should be consulted for the evaluation of accumulated restraint forces. Woolcock et al. [2011] argue that the AS 4100 approach is conservative and not as rational as Eurocode EC 3 (2005). The latter approach sets initial realistic geometric imperfections in a member and then undertakes a second-order analysis of the bracing system and permits load sharing between the restraints along the length of the member. Woolcock et al. [2011] also note that as the AS 4100 provisions (Clauses 6.6.1 to 6.6.3) are "unrealistic and unmanageable" as they tend to be typically disregarded in such designs. They conclude that the Eurocode approach and/or engineering judgement should be used for this situation.

10.3.2 Truss node connections

The term 'node' applies to the juncture of two or more members. Truss node connections can be designed as:

- direct connections
- gusseted connections
- pin connections.

Direct node connections occur where members are welded directly to one another, without the need for gussets or other elements (e.g. tubular joints). Where the chords are made from large angles or tee-sections, it is possible to connect angle web members directly to the chords.

Gusseted node connections used to be predominant at the time when rivets were used as fasteners, and later when bolting was introduced. Their main disadvantage is that the transfer of forces is indirect and that they are not aesthetically pleasing. Their advantage is that it is easier to make all members intersect at the theoretical node point—in contrast to direct connections, where some eccentricity is unavoidable.

Pin connections are generally used when aesthetics are important.

10.3.3 Open sections

A truss design popular with designers is using double-channel or angle chords and double-angle web members. The result is that the truss is symmetrical with respect to its own plane, and that no torsional stresses develop. The sections used with this type of design are shown in Figures 10.10(g) to (i). Trusses of small to intermediate span can be built from single angles, as shown in Figure 10.9. The designer may be tempted to place the verticals and the diagonals on different sides (as in Figure 10.9(a)), but research results show that this practice produces twisting of the chords and bending in the web members. See Clause 8.4.6 of AS 4100 for design provisions on this topic. It is recommended that the verticals and the diagonals be placed on the same side of the chord members, as shown in Figure 10.9(b). The same figure, to its right, indicates how to start distributing bending moments induced by the eccentricity of the web member with respect to the node by the moment distribution method.

The typical sections used in trusses are shown in Figure 10.10. Large-span trusses and trusses carrying heavy loading are often composed of rolled steel sections, as shown in Figures 10.11 and 10.12, though with the ready availability of larger and thicker CHS, RHS & SHS, hollow sections are also used in such applications.

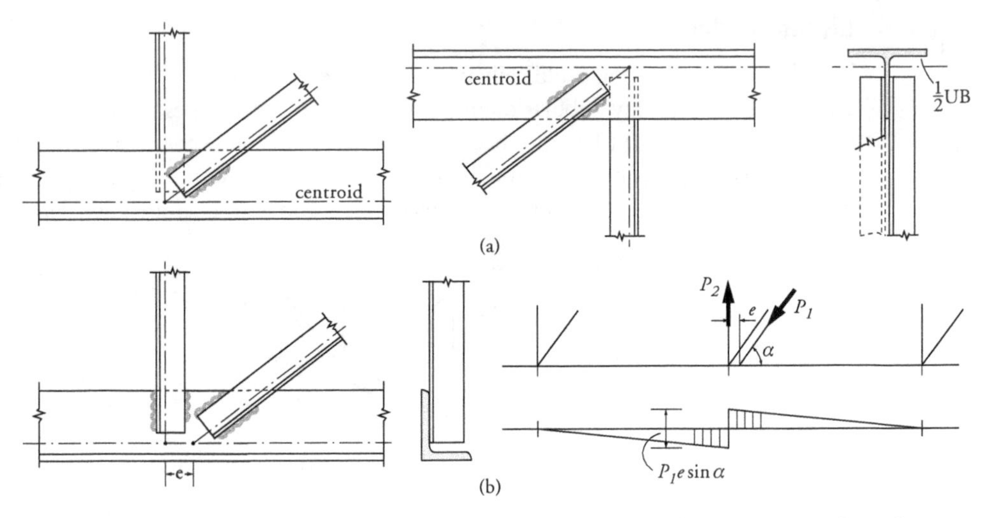

Figure 10.9 *Gusset-free connections for trusses (a) centre of gravity lines intersect at the node; (b) eccentric connection can be a practical way of detailing but additional bending stresses are induced*

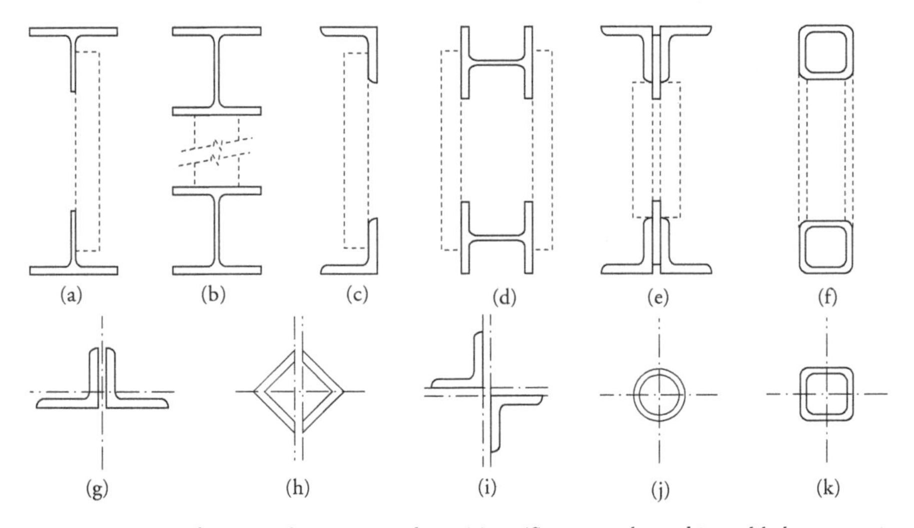

Figure 10.10 *Typical sections for truss members: (a) to (f) commonly used in welded construction (though (a), (c), (d) and (e) may be bolted), and; (g) to (k) common sections used for chord and web/diagonal members*

10.3.4 Closed sections

Tubular trusses are being prolifically used because of their structural efficiency and inherent clean lines. Structurally, tubular members offer superior capacities, because the steel used is Grade C350 or C450 (the latter becoming more popular nowadays) and, except for RHS, their radius of gyration is the same in all directions. Tubes need less paint per linear metre, which is particularly important when upkeep costs and aesthetics are considered.

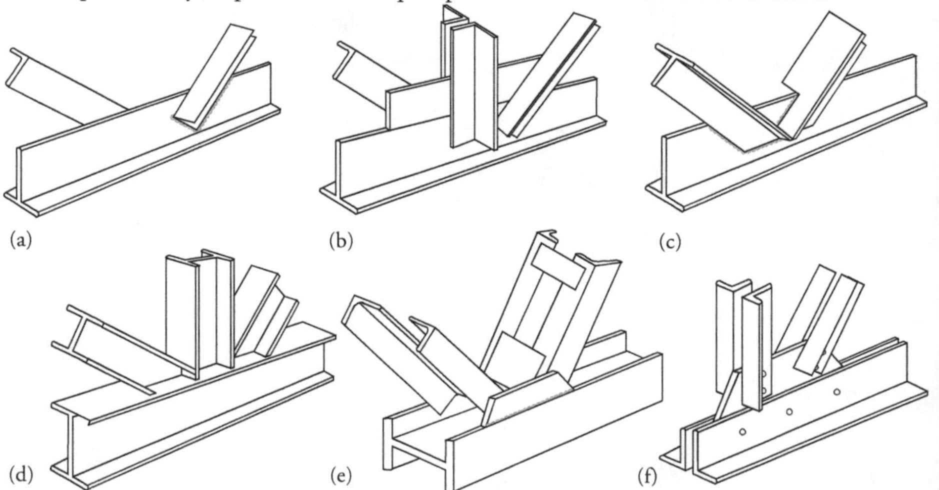

Figure 10.11 *Typical node connections for trusses composed of rolled sections: (a) gussetless construction using Tee-chords; (b) gussets are required where diagonals carry large forces; (c) Tee-diagonals and chords, gussetless; (d) and (e) node detail for heavy trusswork, and (f) riveted/bolted nodes*

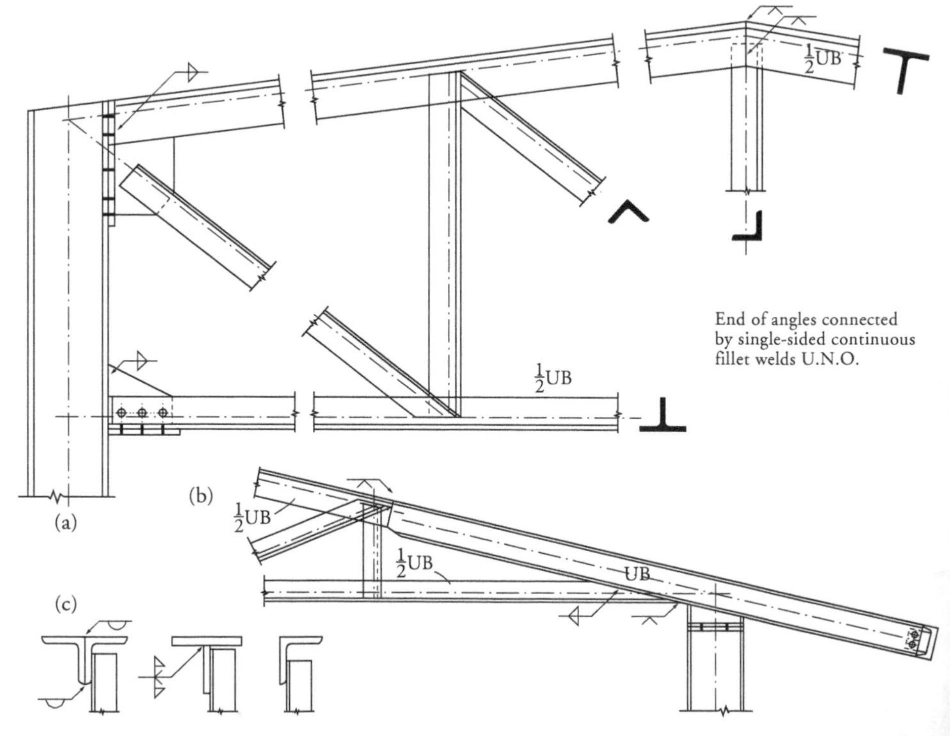

Figure 10.12 *Typical connections for roof trusses composed of rolled-steel sections: (a) portal-type Pratt truss; (b) Fink truss with large eaves overhang; (c) alternative chord cross-sections*

Tubular node connections can be of a direct or gusseted (i.e. plate face reinforcement) type. The latter is used only where large loads are being transmitted through the node.

Web members can be connected with no gap for higher strength (Figure 10.14(a) and (b)), although a positive gap makes fabrication easier and is more commonly used (Figure 10.14(c) without a chord face reinforcement plate). Some possible tubular splices and node connections are shown in Figures 10.13 and 10.14, respectively. Further reading on behaviour, design, detailing and fabrication of hollow section trusses can be found in Packer & Henderson [1997], CIDECT [1991, 1992, 2008, 2009] and Syam & Chapman [1996]. At the time of publication of this Handbook, the ASI were working on a series of Tubular Steel Connection Design guides based on the above references and other key sources.

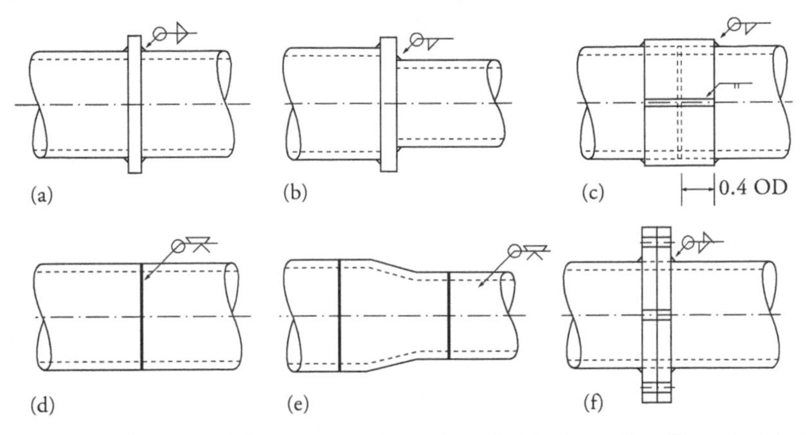

Figure 10.13 *Splices for tubular truss members: (a) sandwich plate splice; (b) sandwich plate splice at chord reduction; (c) jacket splice; (d) welded butt splice; (e) welded butt splice with reducer, and; (f) flange splice. (Note: CHS shown but (a), (b), (d) and (f) also apply to RHS/SHS members).*

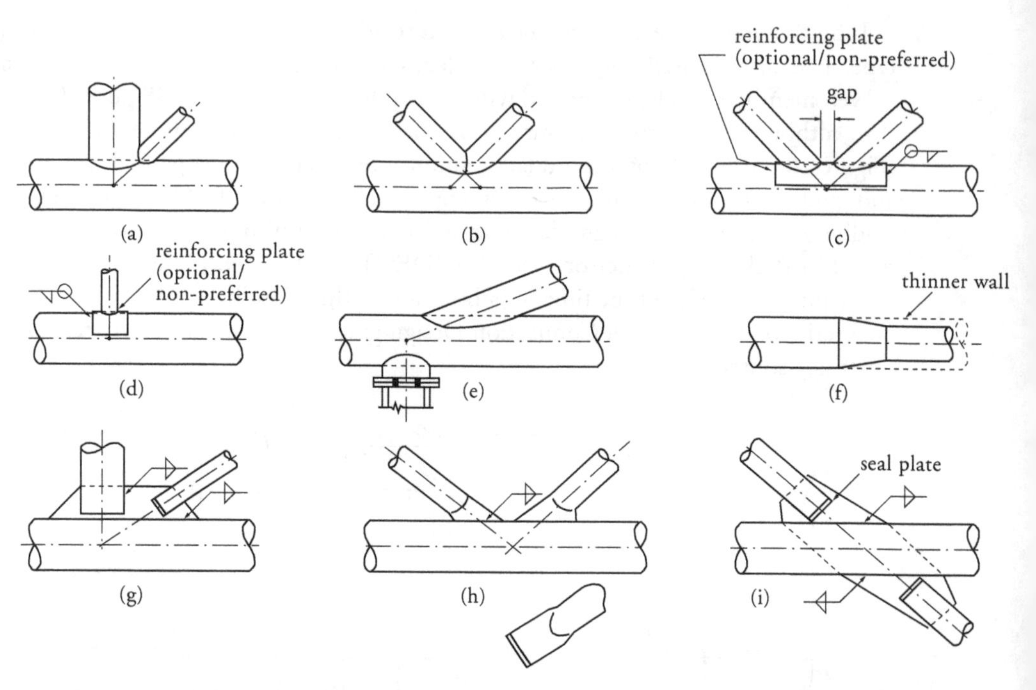

Figure 10.14 *Some typical connections for roof trusses composed of tubular sections: (a) direct contact overlap connection without eccentricity; (b) direct contact overlap connection with eccentricity; (c) direct contact gap connection with/without eccentricity (with chord face reinforcing plate shown—without reinforcing plate is very common); (d) T-joint with chord face reinforcing plate (for very heavy loads—otherwise no reinforcing plate is also popular); (e) connection detail at support (note vertical stub portion with flange splice for lifting onto support); (f) concentric reducer where chord section is stepped down (alternatively, if the overall section is not stepped down then the wall thickness is reduced—the latter applies for RHS/SHS); (g) slotted-gusset connections; (h) flattened end connections, and; (i) slit tube connections. (Note: CHS shown but all connections, except (f) with reducer and flattened tube in (h), readily apply for RHS/SHS. See Figure 8.22 for other tubular truss connection configurations).*

10.4 Portal frames

10.4.1 General

Portal frames are used extensively for the framing of single-storey buildings. Portal frames offer cost advantages over other framing systems for short to medium spans. Other advantages they offer over truss systems are low structural depth, clean appearance and ease of coating maintenance.

Portal frames derive their resistance to vertical and lateral loads through frame action. When fixed bases are used, the structural action is enhanced because all members are then fully utilised, but there is a cost penalty for larger footings that often precludes their use.

The design of the knee joint is of prime importance, as this is one of the critical elements in the frame. Figure 10.15 shows some typical knee connections. Figure 10.16 gives typical depths and self-weights for portal frames that can be used in the preliminary design. Frame forms for single-bay and multi-bay portal frames are shown in Figure 10.17.

An excellent publication on the analysis and design of portal frame buildings is Woolcock et al. [2011].

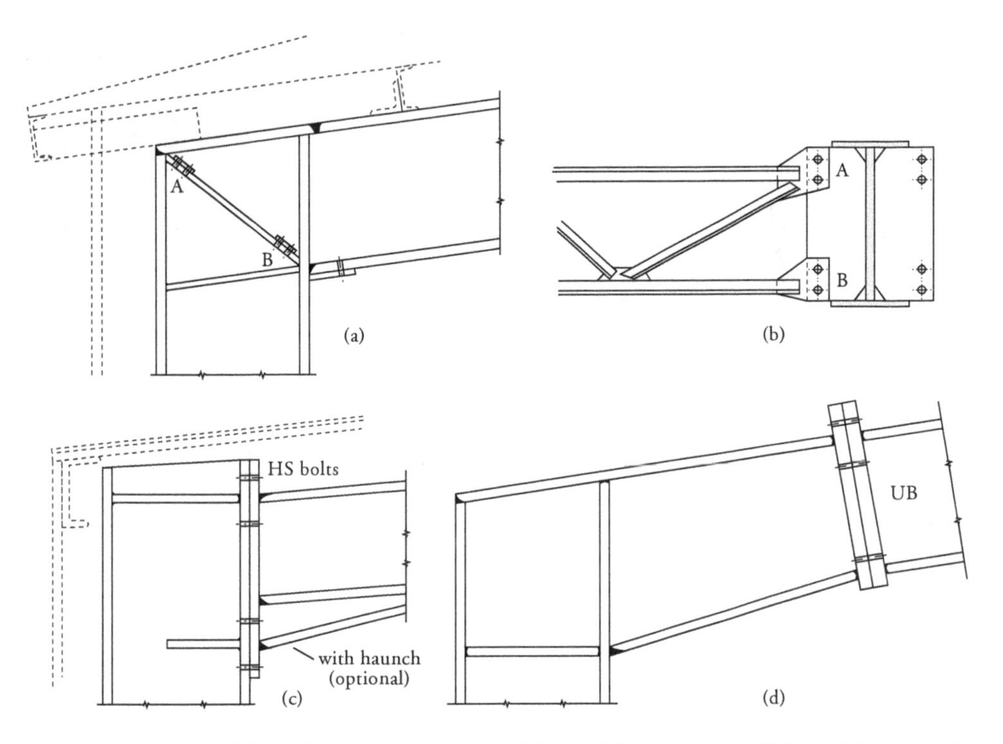

Figure 10.15 *Portal frame knee connections: (a) field welded connections; (b) lateral bracing A-B (see (a)—note a single tube member placed midway between A and B may also be used); (c) bolted moment end plate connection; (d) stub connection. (See Hogan & Van der Kreek [2009a-e] for details and design models for the above connections and Hogan & Syam [1997] for haunching of portal frame knee joints).*

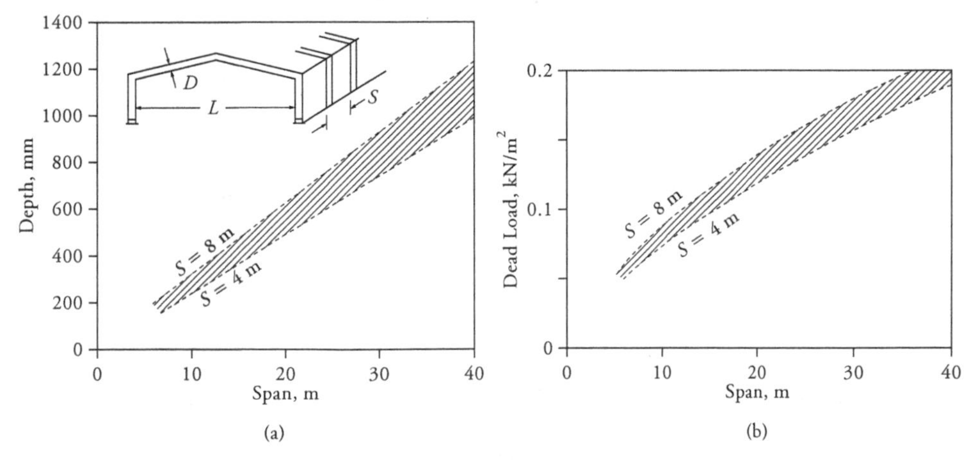

Figure 10.16 *Data for preliminary sizing of hinged base frames: (a) usual range of rafter depths; (b) dead load of rafters expressed as kN/m²; design height to eaves is 5 m, wind Region A.*

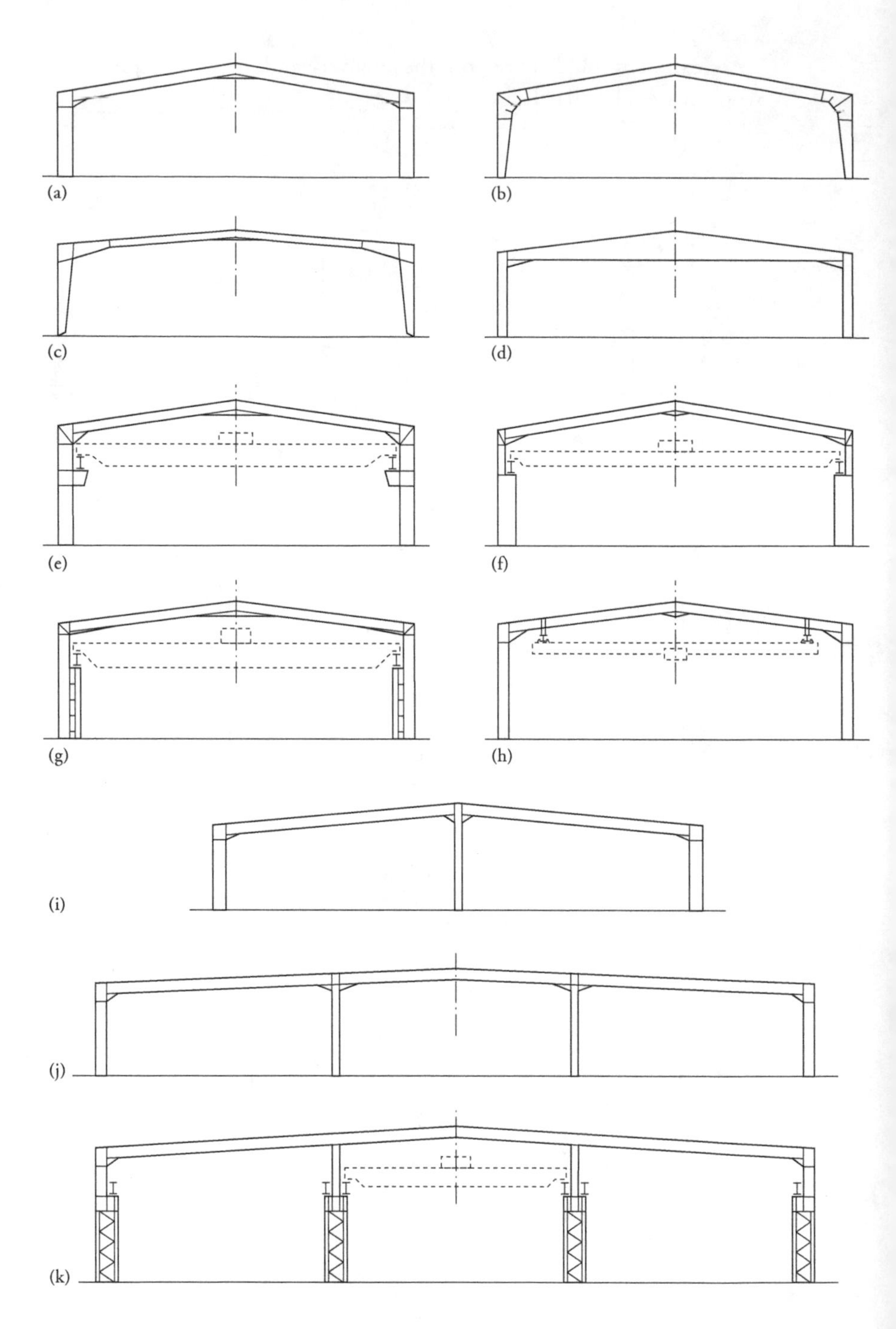

Figure 10.17 *Types of portal frames for industrial buildings: (a) constant (sometimes called prismatic) cross-section; (b) to (d) tapered members; (e) portal with column crane runway brackets; (f) stepped column portal; (g) separate crane post; (h) rafter hung crane runways; (i) to (k) multiple bays*

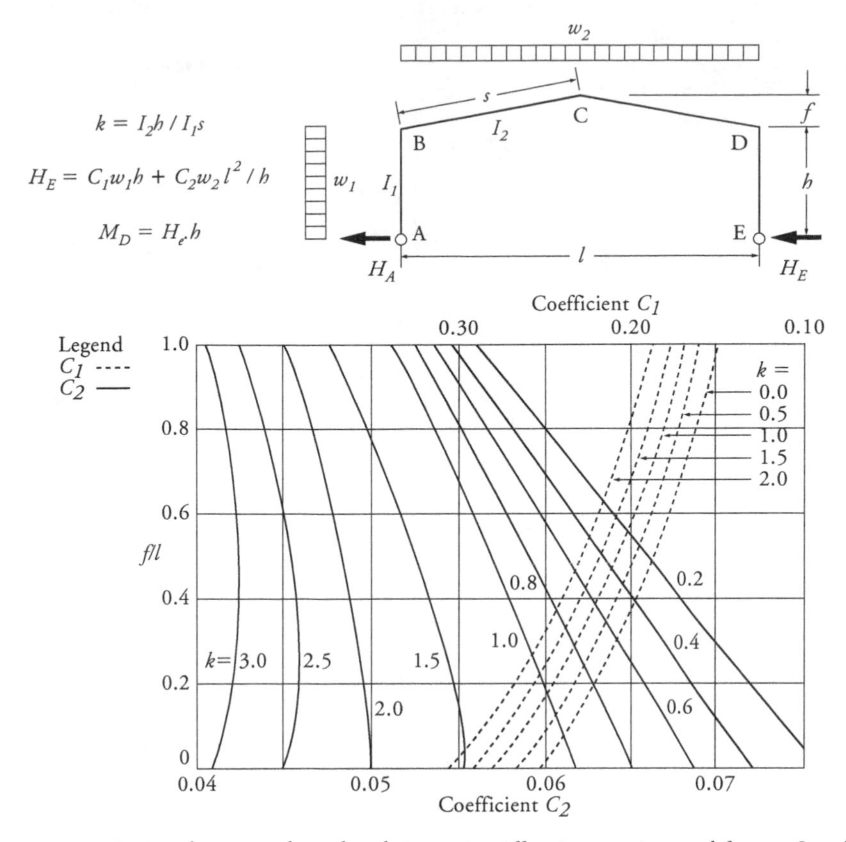

Figure 10.18 *Design chart to evaluate key design action effects in two-pin portal frames. See also Section 4.3 for calculation of second-order effects.*

10.4.2 Structural analysis

Structural analysis of portal frames is not covered in this Handbook (Woolcock et al. [1999, 2011] is very useful). As an aid to preliminary design, use can be made of the plot in Figure 10.18 and of the design formulae in Table 10.3. For a more accurate hand analysis, the designer should refer to Kleinlogel [1973] for a classic text on the subject. Table 10.3 only considers first-order elastic analysis. However, AS 4100 requires such frames to include second-order effects where relevant. Consequently, reference should also be made to Section 4.3 for the evaluation of second-order effects for elastic analysis of portal frames. In lieu of 'manual' techniques, sophisticated structural analysis software packages are readily available [Microstran, Multiframe, SpaceGass and Strand7] which can rapidly handle linear/non-linear analysis with design along with many other functions.

As considered in Section 10.3.1, Woolcock et al. [2011] note that plastic analysis is not popular let alone effective for structural analysis of portal frames. Additionally, with the ready availability of computer software packages on second-order elastic analysis, the most effective way to undertake structural analysis for portal frames is to use these packages and not manual methods to model second-order effects (e.g. moment amplification, etc.).

Table 10.3 Formulae for forces/bending moments in portal frames—first-order elastic analysis

Applied loads	Forces and moments
Notation	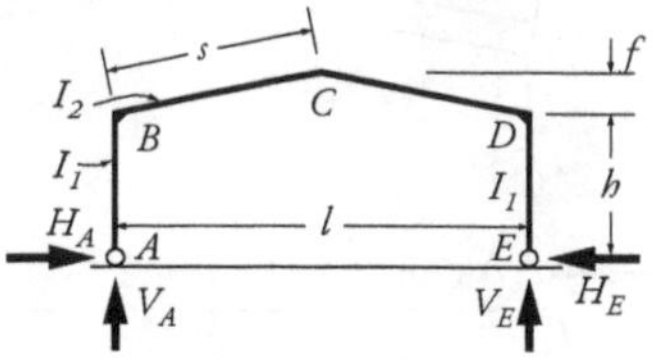 $b_1 = \frac{h}{l}$; $f_1 = \frac{f}{h}$ $k_1 = \frac{I_2 h}{I_1 s}$; $k_2 = \frac{l}{h}$; $k_3 = f_1^2 + 3f_1 + k_1 + 3$

Applied loads	Forces and moments
1	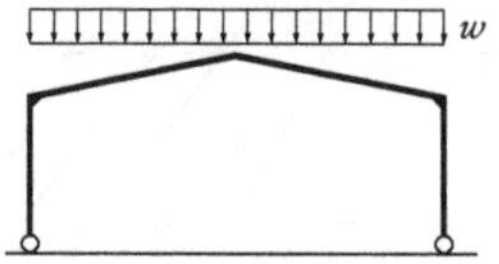 $H_A = H_E = -\frac{wl^2(1 + 0.625f_1)}{4hk_3}$ $V_A = V_E = 0.5wl$ $M_B = M_D = -H_A h$
2	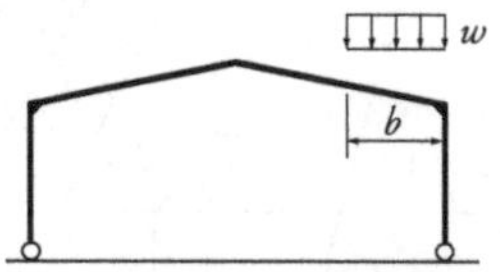 $H_A = -H_E = \frac{wb^2(6 + 3f_1 - 4b_1 - 2f_1 b_1^2)}{8hk_3}$ $V_A = \frac{wb^2}{2l}$ $M_B = M_D = -H_A h$
3	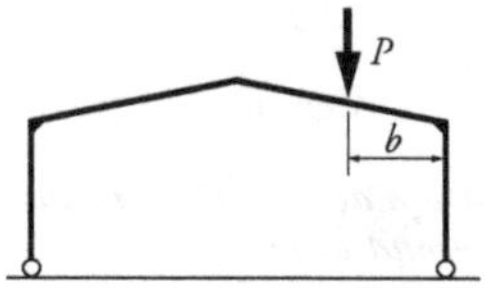 $H_A = -H_E = \frac{Pb(6 - 6b_1 + 4f_1 b_1 - 3f_1)}{4k_3}$ $V_A = \frac{Pb}{l}$ $M_B = M_D = -H_A h$
4	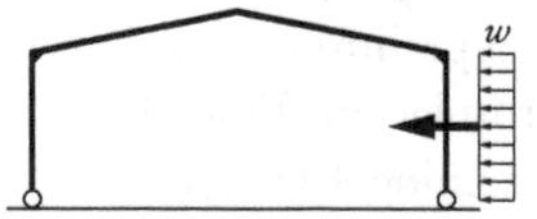 $H_A = \frac{wh(5k_1 + 6f_1 + 12)}{6k_3}$ $H_E = H_A - wh$ $V_A = -V_E = \frac{wh^2}{2l}$ $M_B = H_A h$ $M_D = H_E h - \frac{(wh^2)}{2}$
5	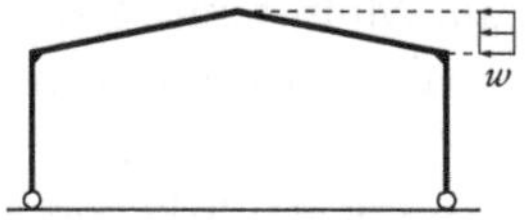 $H_A = \frac{wf(3 + k_1 + 2.5f_1 + 0.625f_1^2)}{2k_3}$ $H_E = H_A - wf$ $V_A = -V_E = \frac{wf(2h + f)}{2l}$ $M_B = -H_A h$ $M_D = H_E h$

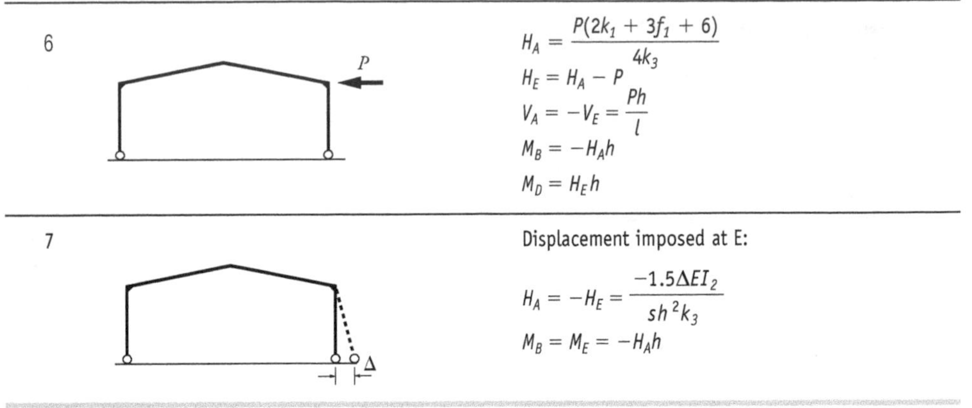

6		$H_A = \frac{P(2k_1 + 3f_1 + 6)}{4k_3}$ $H_E = H_A - P$ $V_A = -V_E = \frac{Ph}{l}$ $M_B = -H_A h$ $M_D = H_E h$
7		Displacement imposed at E: $H_A = -H_E = \frac{-1.5\Delta EI_2}{sh^2 k_3}$ $M_B = M_E = -H_A h$

Notes: See Section 4.3 for the evaluation of second-order effects for the above structural and loading configurations.

10.5 Steel frames for low-rise buildings

Framing systems for low-rise buildings can take many forms. The framing systems commonly used are:

- two-way braced
- core braced, using concrete or steel core with steel gravity frame and simple connections
- one-way braced, using rigid frame action in the other direction
- two-way rigid frames with no bracing.

The frames of the two-way braced and the core braced types are often used for their low unit cost and simple construction. The beam-to-column connections for such frames can be in the form of flexible end plates or web-side plate connections.

Rigid frames featuring two-way rigidity may require more steel material but permit less restrictrions in layout and functions and, like braced frames, are sometimes employed in severe earthquake zones because they offer higher ductility.

The framing connections for low-rise buildings are usually of the field bolted type. Welding in-situ is not favoured by erectors because welding could not be properly carried out without staging and weather shields. Site welding can cause delays in construction because of the need for stringent inspection and consequent frequent remedying of defects.

Figures 10.19 to 10.24 show some often used connection details.

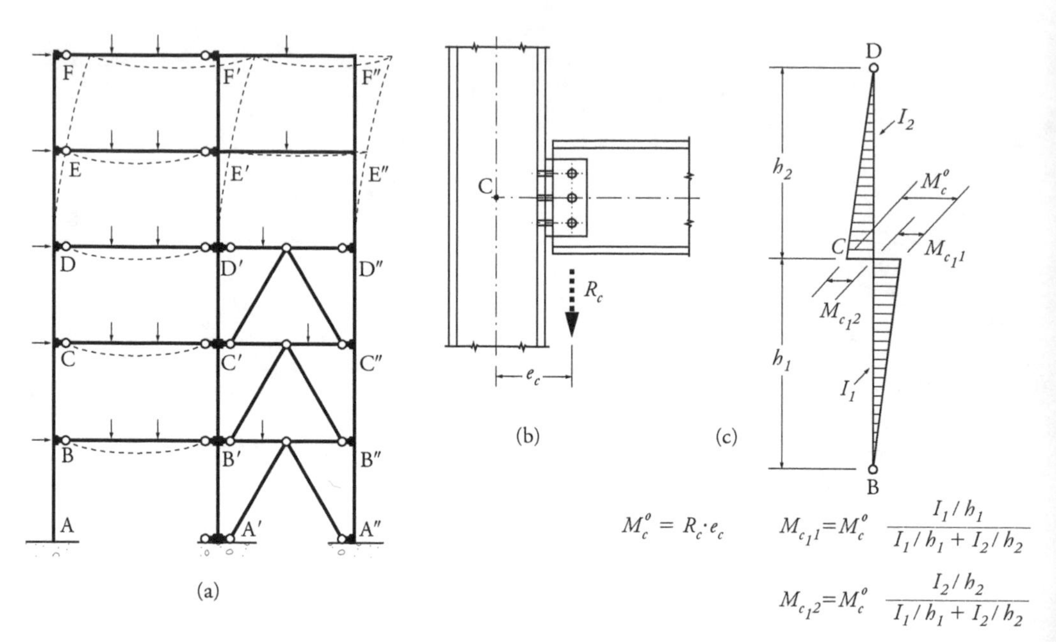

Figure 10.19 *Simple connection design method illustrated: (a) typical braced multi-storey frame; (b) assumed connection eccentricity; (c) bending moment distribution on upper and lower column shafts*

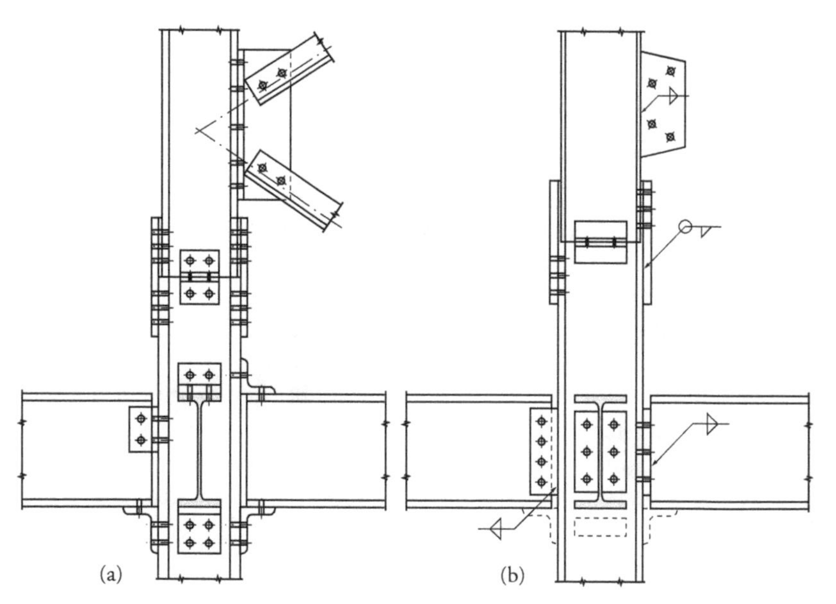

Figure 10.20 *Bolted connections for structures designed by simple method. See Hogan & Munter [2007d-h], Hogan & Van der Kreek [2009d] and ASI [2009b] for further information on these connections.*

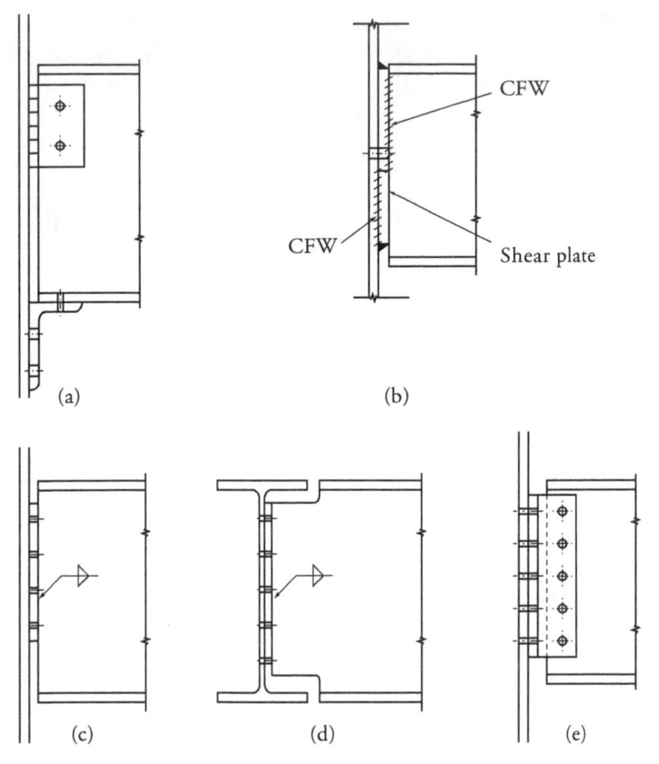

Figure 10.21 *Flexible bolted connections: (a) flexible angle seat; (b) bearing pad connection; (c) flexible end plate; (d) coped flexible end plate; (e) angle cleat connection. See Hogan & Munter [2007d-h] for futher information on these connections.*

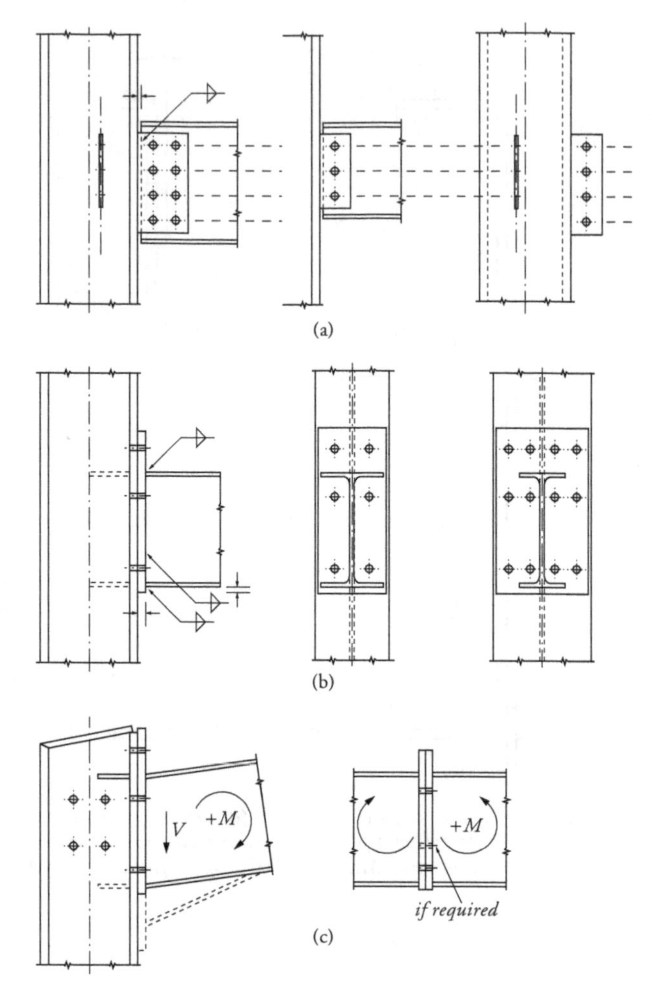

Figure 10.22 *Bolted beam-to-column connections: (a) web side plate (flexible) connection; (b) bolted moment (rigid) end plate connection; (c) other applications of bolted moment (rigid) end plate. See Hogan & Munter [2007d-h] and Hogan & Van der Kreek [2009a-e] for further information on these connections.*

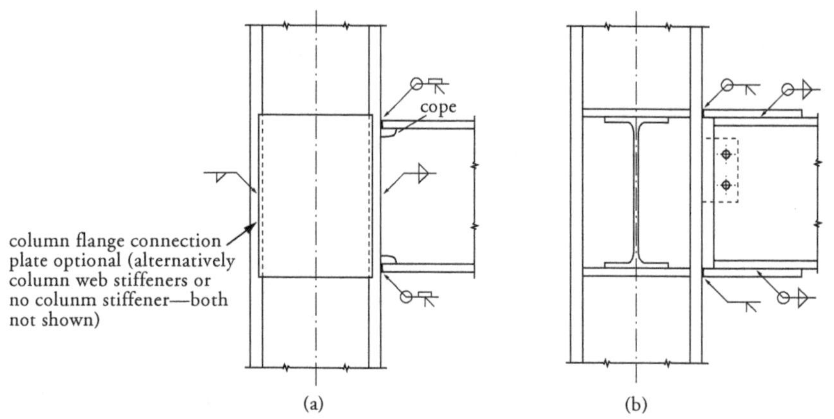

Figure 10.23 *Welded rigid beam-to-column connections: (a) directly welded connection (note beam may not be coped and is typically connected by fillet and butt welds along all perimeter to column face—not shown); (b) fully welded connection using moment plates. See Hogan & Van der Kreek [2009b,e] for further information on these connections. (Note: (b) is not considered in the aforementioned reference and is considered to be less commonly used.)*

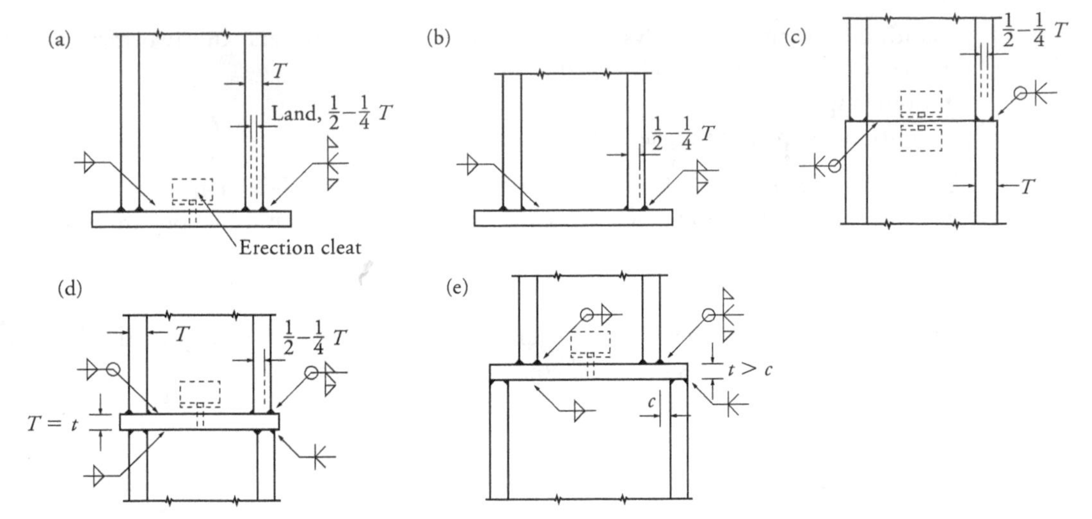

Figure 10.24 *Typical welded column-to-base connections and column splices with partial-penetration butt welds. See Hogan [2011] for pinned base plate connections and Hogan & Thomas [1994] for column splice connections.*

10.6 Purlins and girts

The main function of purlins and girts (shown in Figure 10.27) is to support the metal cladding. In addition, purlins and girts are used to provide lateral restraint to the rafters and columns. Both elements use the same sections and materials, and so in further discussion girts can be omitted. Purlins are made by cold-forming operations using high-strength steel strip. Some manufacturers offer yield strengths of 550 MPa for thinner sections, say up to 2.0 mm. Cold-formed purlins are therefore very cost-effective, and have almost completely displaced hot-rolled sections, timber and other related materials.

The usual cross-sectional shapes are (Cee) C- and (Zed) Z-sections in sizes up to 300 mm, and larger if specially ordered. The section thicknesses range from 1.0 to 3.0 mm depending on the purlin size. The strip is pre-galvanized and usually requires no further corrosion protection.

Owing to the small torsion constant of purlins, special care is required to avoid lateral and torsional instability. All purlin sections have flanges stiffened with downturned lips to increase their local and flexural-torsional buckling resistance. The rules relevant for design of cold-formed steel purlins are contained in AS/NZS 4600.

To enhance the flexural-torsional behaviour of purlins, it is necessary to approach the design as follows:

- For inward loading (i.e. for dead, live and wind loads on purlins or wind pressure on girts), it is assumed that the metal cladding provides enough lateral resistance to achieve an effective length factor of 1.0.
- For outward loading (negative wind pressure), the inner flanges are assumed not to be restrained by the cladding, and purlin bridging is used where necessary. The bridging provides the necessary lateral and torsional resistance and can be counted on as the means of division into beam segments.

Continuous spans are an advantage in reducing the purlin sizes, and two methods are available for that purpose (Figure 10.26):

- double spans
- continuous spans, using lapped splices in purlins.

An important secondary benefit of continuity is that the deflections are significantly reduced, in contrast with the simple spans. The typical deflection reductions are as follows:

- the first outer span: 20%
- an interior span: 50%.

Figure 10.25 shows the relevant purlin details, and Figure 10.26 gives the bending moment coefficients for simple and continuous purlins.

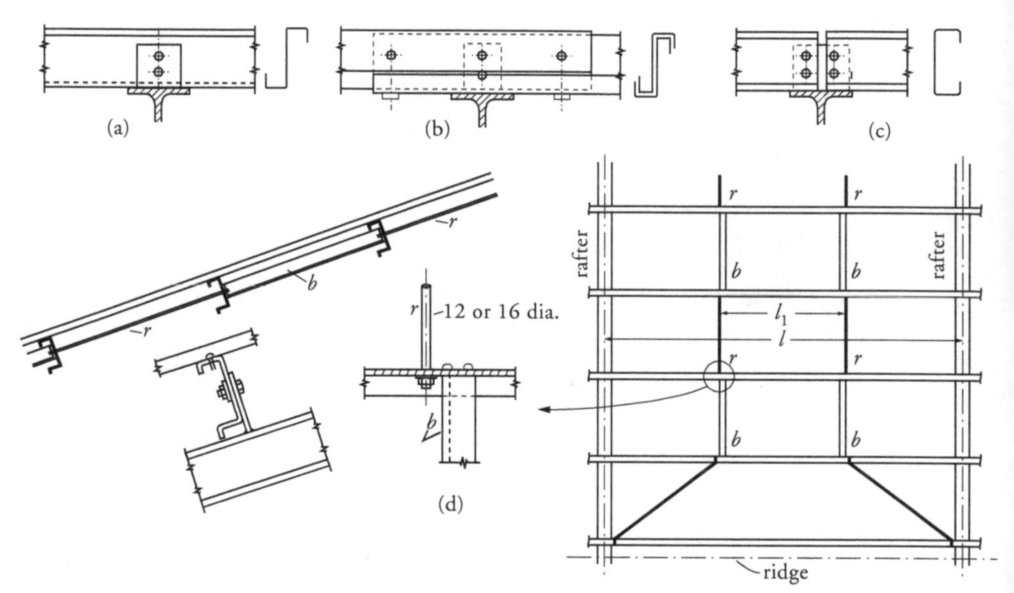

Figure 10.25 *Purlins and purlin bracing ("bridging"): (a) Zed (or Cee) purlin over end/intermediate support rafter bolted (by generally 2 bolts) to a purlin cleat welded to the rafter top flange; (b) Lapped Zed (not applicable for Cee) purlin over intermediate support rafter bolted (by generally 2 bolts to a purlin cleat, 2 bolts in the lapped Zed bottom flange and 2 bolts in the lapped Zed upper web) with the purlin cleat welded to the rafter top flange; (c) Simply supported Cee purlins over intermediate support rafter bolted (by generally 4 bolts) to a purlin cleat welded to the rafter top flange; (d) plan of roof purlin layout between two adjacent rafters showing bridging "b" and tie rod "r" to control purlin flexural-torsional buckling and sag. Bridging may be proprietary systems from certain manufacturers for quick installation. Tie rods may be replaced by adjustable bridging to control sag. Also shown to the left are Cee purlins with down-turned lip to reduce dust, etc build up.*

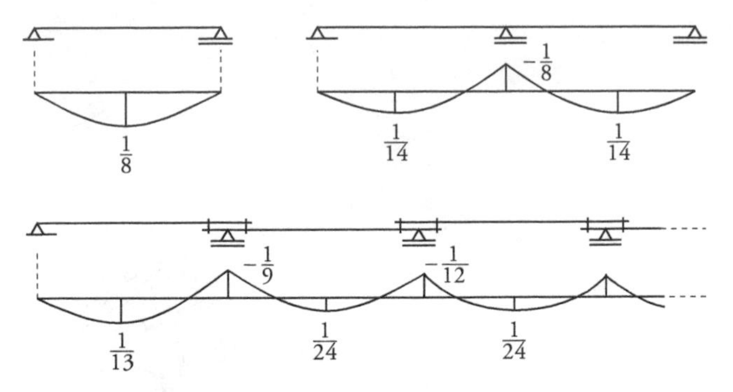

Figure 10.26 *Bending moment coefficients for simple, double-span and continuous purlins*

Due to increased wind loads in eaves strips and gable bays, it is necessary to rationalise the purlin sizes such that uniform purlin spacing and sizes can be used. There are three ways in which this can be achieved:

- Design all purlins for the worst location loads (expensive).
- Use smaller spacings for purlins in the eaves strip so that the same purlin size can be used, and in gable bays use the same size but a heavier purlin section.
- Reduce the gable bay span so that the same-size purlin can be used throughout (the building owner may object to this).

The spacing of purlins is not only a function of purlin capacity, it also depends on the ability of the roof sheeting to span between the purlins. In moderate wind zones, the purlin spacing ranges between 1200 and 1700 mm, depending on the depth of the sheeting profile and thickness of the metal. The same comments apply to the girts and wall sheeting. Advice should be sought from the purlin and sheeting manufacturer on the optimum spacing in various wind regions.

Most manufacturers provide commonly used design tables and instructions to make the purlin selection task simpler. The larger manufacturers produce some high quality publications and load tables for purlin sections, bridging and connection systems (e.g. Fielders [2011], Bluescope Lysaght [2008], Stramit [2010]). Manuals on cladding are also available from these manufacturers. To understand its use and outcomes, the designer should verify the use of the tables by carrying out a number of test calculations. Designers should also note that due to the thin gauge of purlin sections, any slight variation in dimensions (i.e. depth, width, thickness, lip stiffener height, corner radius) and steel grade can affect the actual purlin performance which may significantly depart from that noted in the respective tables.

To illustrate the geometry and interaction of the primary structure (portal frame, roof and wall bracing), secondary structural elements (purlins, girts, bridging, rods) and the slab/foundation, Figure 10.27 shows typical details for such elements of portal frame building construction.

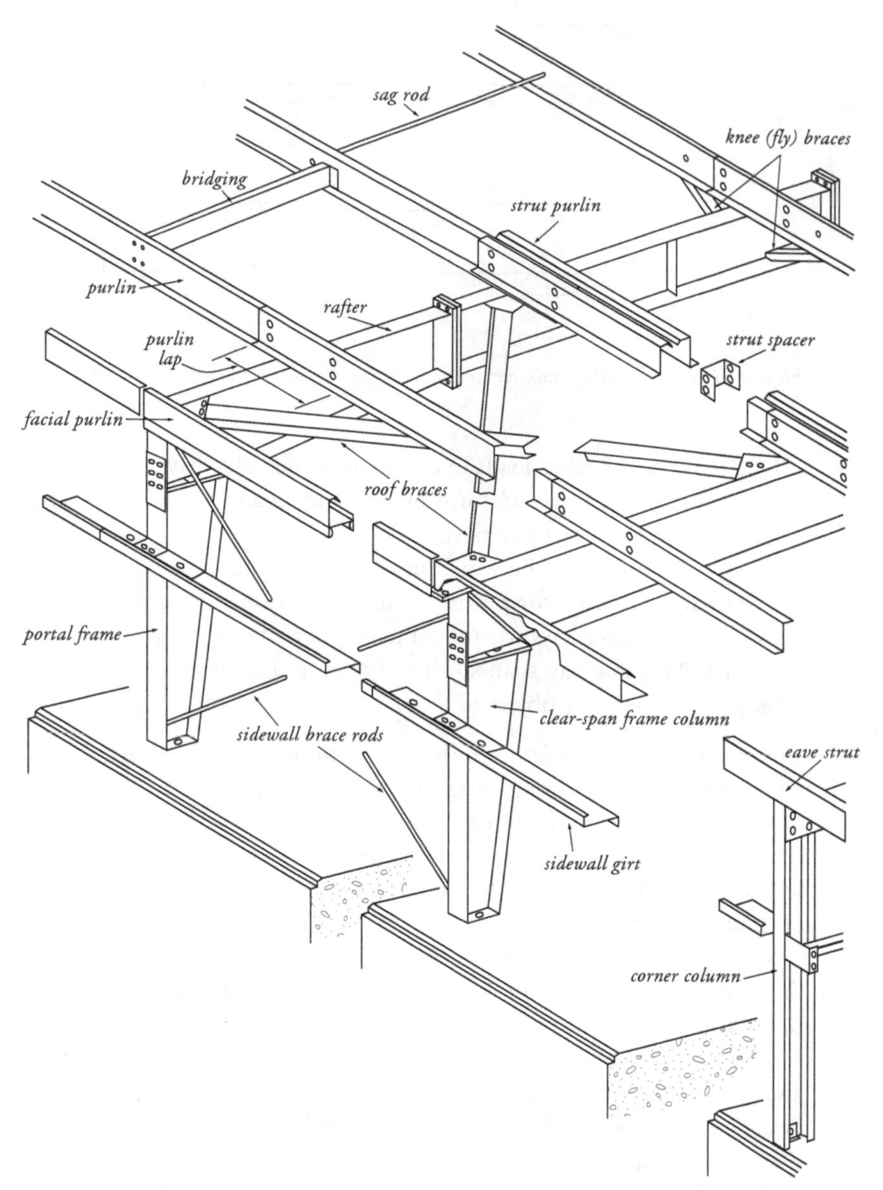

Figure 10.27 *Isometric view of typical single-storey framework showing pinned-base portal frames, purlins, girts and other stabilising elements*

10.7 Floor systems for industrial buildings

10.7.1 Types of construction

The selection of the type of industrial floor depends on the floor use and the load intensity and application. Consideration should be given to wear and skid resistance, and also the need to change layout, plant and services.

The following types of floor are in frequent use:

(a) *Steel floor plate (chequer plate) supported on closely spaced floor joists* Advantages are: rapid construction, ability to give lateral restraint to the beams and to provide floor bracing. Disadvantages are: high cost and high noise transmission.

(b) *Steel grid flooring over steel joists* Main advantages are: self-cleaning, low weight, rapid placing and ability to alter layout. Disadvantages are: low resistance to point loads, need for a separate floor bracing system and difficulty of upkeep of its coating system.

(c) *Reinforced concrete floor slabs* Advantages are: good load-carrying resistance, increased stiffness, suppression of noise and even surface for rolling loads. Friction between the slab and the beam may provide full lateral restraint. Disadvantages are: high unit weight, difficulties in making alterations and relatively high cost.

(d) *Composite reinforced concrete slab* Same as (c), with an additional advantage of economy offered by the composite action with steel beams (via shear studs) and the facility of providing full lateral restraint to the beams.

10.7.2 Steel floor plate

Steel floor plate or 'checker plate' is rolled with an angular pattern to improve skid resistance. It is available in thicknesses ranging from 5 to 12 mm and is manufactured from Grade 250 steel. The best corrosion protection for the floor plates is by hot-dip galvanizing, but care must be exercised to prevent distortion.

The highest resistance to floor loads is obtained with all four edges supported, but the ratio of sides should not exceed 1:4. This requires floor beams and joists at relatively close spaces of between 900 and 1200 mm. The fixing to the supporting frame is best provided in the form of slot welds (20 × 40 mm) at roughly 1000 mm spacing. Welding is also used to join the individual panels.

The plate tables in ASI [2009a] are very useful for design. Deflections are usually limited to $l/100$ under localised loads.

10.7.3 Steel grid flooring/grating

A popular type of steel grid flooring consists of vertical flats (20–65 mm high and 3–6 mm wide) spaced at 30–60 mm, and cross-connected at 50–200 mm for stability and load sharing. The load capacities of grid flooring are usually based on a computational model derived from tests, because there are too many variables to consider in designing grids from first principles. The manufacturer's data sheets are usually employed in the design (Weldlok [2010], Webforge [2010]).

The fixing of the steel flooring is by proprietary clips at approximately 1.0 m spacing. It is important that clips be installed as soon as the grid flooring is laid, as a precaution against fall-through accidents.

10.8 Crane runway girders

10.8.1 General

Lifting and transporting heavy loads and bulk materials is almost entirely done by mechanical appliances. Overhead travelling cranes and monorails are commonly used in industrial buildings. Cranes travel on rails supported by runway girders. Monorails are runway beams that carry hoists on units travelling on their bottom flanges.

The design of crane runway beams differs from the design of floor beams in the following ways:

- the loads are determined in accordance with AS 1418.1, the Crane Code
- the loads are moving
- lateral loading is usually involved
- localised stresses occur in the web at the top flange junction
- lateral buckling with twisting needs to be considered
- fatigue assessment may be required because of repetitive load cycling.

In recognition of these special aspects of crane runway design, a new part of the crane code has been published as AS 1418.18. For an introduction to the design of crane runway girders, see Gorenc [2003] and an authoritative treatment of the topic is given in Woolcock et al. [2011].

10.8.2 Loads and load combinations

Because the operation of a crane is not a steady-state operation, there are significant dynamic effects to consider. This is done in practice by applying dynamic load multipliers to the loads computed for a static system.

These multipliers take into account the travelling load fluctuations, the hoisting impacts, and lateral inertial and tracking loads. AS 1418.18 provides a method for the evaluation of special loads occurring in operation and the dynamic multipliers to be used with these loads.

Several types of lateral loads occur with cranes:

- lateral loads caused by the acceleration/braking of the crane trolley
- lateral loads caused by the oblique travelling tendency of the cranes
- lateral loads caused by longitudinal acceleration/braking of the crane bridge.

The load combination table of AS 1418.18 is much more elaborate than the one used for building design. It specifies, for each load combination, the value of multipliers for the dynamic effects and for the likelihood of more than two load types occurring at once.

10.8.3 Bending moments and shear forces

The maximum design bending moment is a function of the load position. The load position at which the maximum moment occurs must be found by trial and error or by a suitable routine. Any linear elastic computer program can be used by incrementing the position of the leading load. For a two-axle crane, it is possible to arrive at the direct evaluation of design moments using the methods shown in Table 10.4 and Figure 10.28.

Table 10.4 Load effects for moving loads

<table>
<tr><th>Load disposition</th><th>Load effects</th></tr>
<tr><td>1

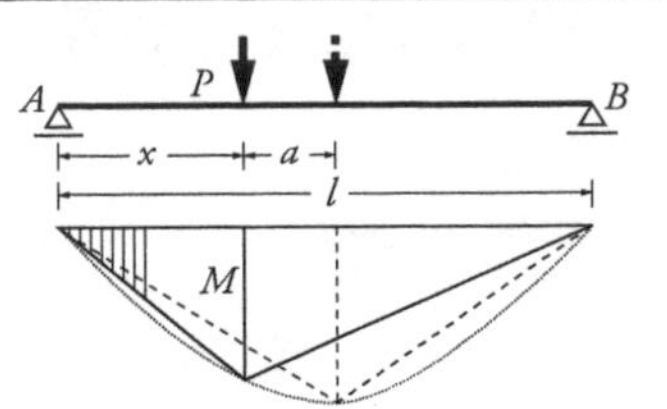

</td><td>One concentrated moving load:

$(R_A)_{max} = P$, when $x = 0$

$M_1 = \dfrac{Px(L - x)}{L}$, at $x = 0.5L - 0.25a$

$M_{max} = \dfrac{PL}{4}$, when $x = \dfrac{L}{2}$
</td></tr>
<tr><td>2

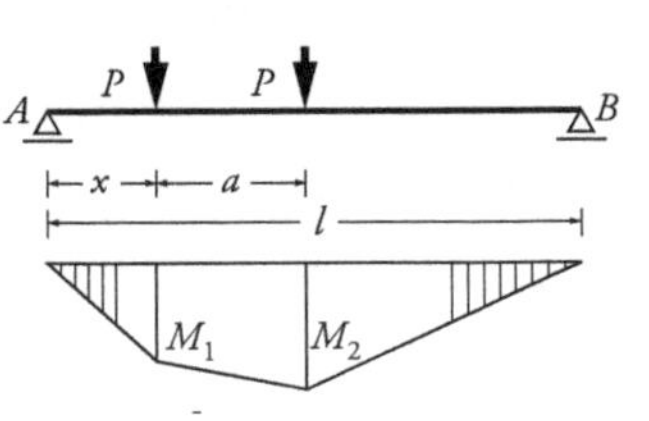

</td><td>Two equal moving loads:

Case 1: $a \leqslant 0.586L$

$(R_A)_{max} = P\left[1 + \dfrac{(L - a)}{L}\right]$, when $x = 0$

$R_A = \dfrac{P(2L - 2x - a)}{L}$

$M_1 = R_A x$

$M_2 = R_A(x + a) - Pa$

$M_{max} = \dfrac{P(L - 0.5a)^2}{2L}$, at $x = 0.5L - 0.25a$
</td></tr>
<tr><td>3

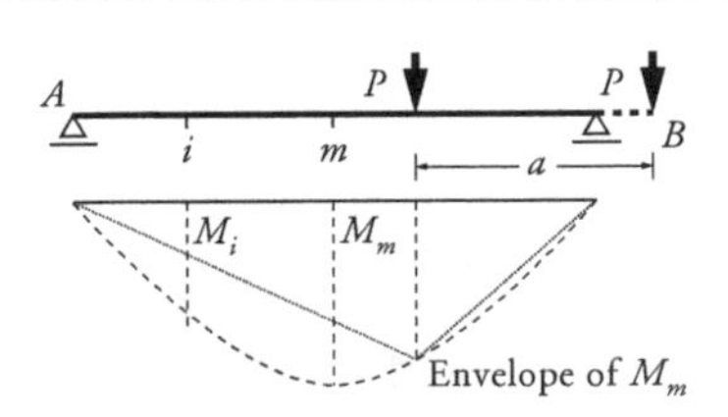

</td><td>Two equal moving loads:

Case 2: $a > 0.586L$

The maximum bending moment occurs when only one load is on the span:

$M_m = \dfrac{PL}{4}$, when $x = \dfrac{L}{2}$
</td></tr>
<tr><td>4

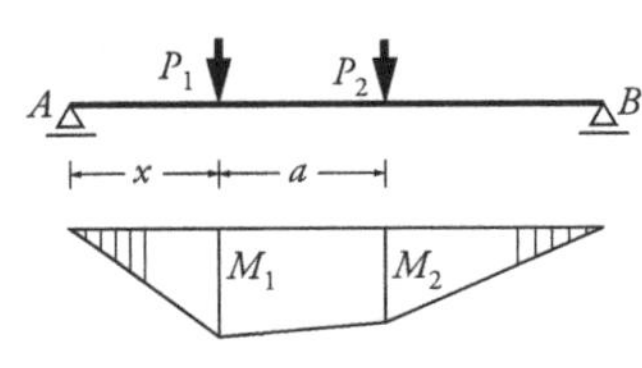

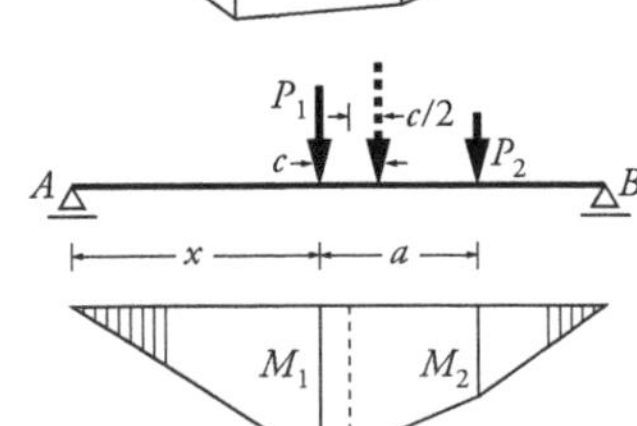

4(a)

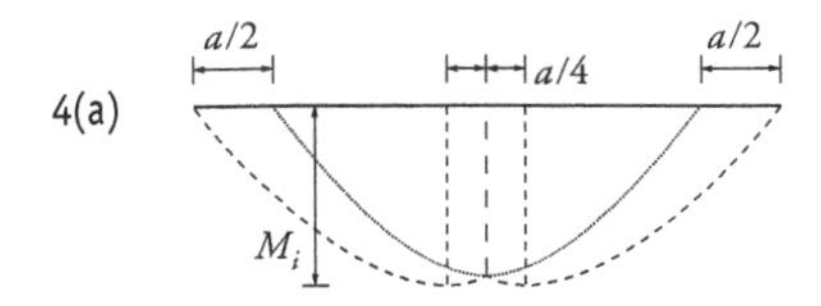

</td><td>Two unequal moving loads:

Case 3: $a \leqslant \left(0.5 + \dfrac{0.086P_2}{P_1}\right)L$

$(R_A)_{max} = P_1 + \dfrac{P_2(L - a)}{L}$, when $x = 0$

$R_A = \dfrac{P_1(L - x)}{L} + \dfrac{P_2(L - x - a)}{L}$

$M_1 = R_A x$

$M_2 = M_1 + (R_A - P_1)a$

The maximum bending moment occurs at:

$x_m = 0.5\left[L - \dfrac{P_2 a}{(P_1 + P_2)}\right]$

$M_{max} = \dfrac{(P_1 + P_2){x_m}^2}{L}$
</td></tr>
</table>

continued

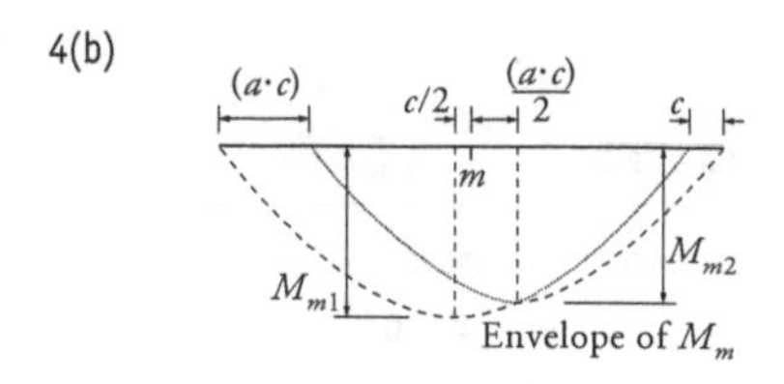

Two unequal moving loads:

Case 4: $a > \left(0.5 + \dfrac{0.086P_2}{P_1}\right)L$

Only the larger load is relevant; the other is off the span:

$$M_m = \frac{P_1 L}{4}, \text{ when } x = \frac{L}{2}$$

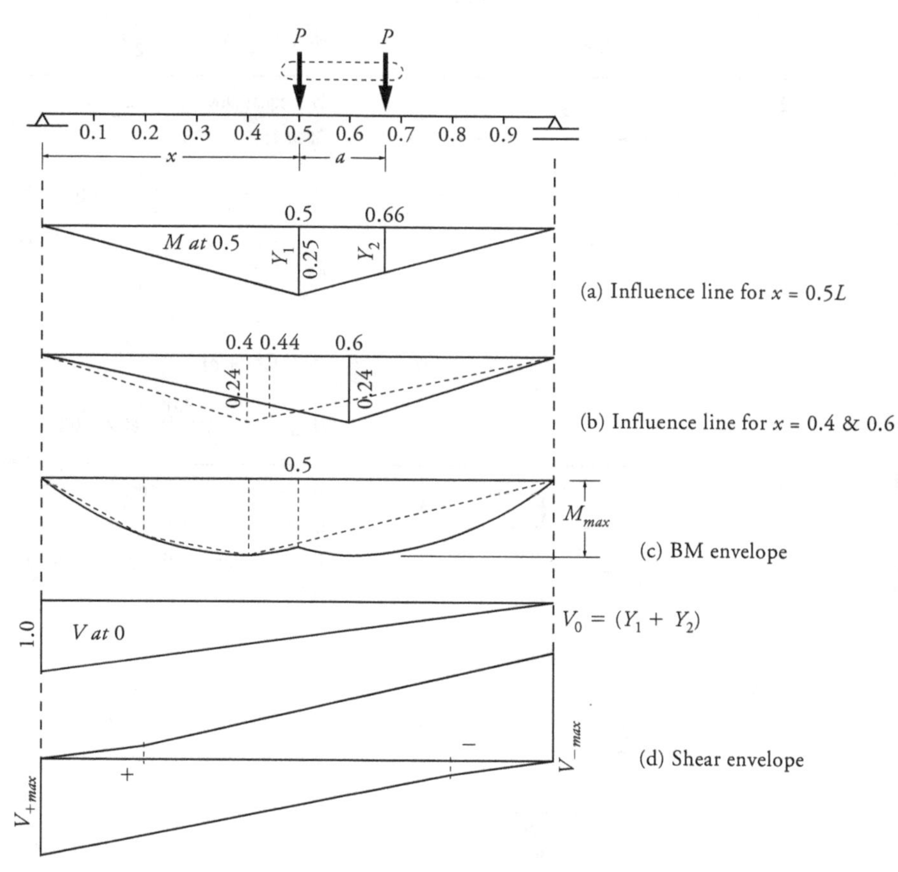

Figure 10.28 *Influence lines for moments and shears and absolute maximum envelopes for moments and shears*

10.9 Deflection limits

10.9.1 General

In contrast to the stringent code requirements for strength design, the deflection limits are largely left to the designer to decide on. One reason for the lack of detailed rules on deformation limits is that the subject is too variable and complex; the other is that designers are in a better position to set realistic limits appropriate to the particular project.

In AS 4100, the deflection limits come under the serviceability limit state. A structure may become unserviceable for a number of reasons:

- inability to support masonry walls without inducing cracking
- possibility of ponding of rainwater because of excessive sag
- floor slopes and sags interfering with the operations performed on the floor (bowling alley, forklift truck operations)
- damage to machinery that may be sensitive to floor movement
- mal-operation of cranes when runway beam deformations are excessive
- floor vibrations that may be felt by users as disturbing
- deflections and tilting that may cause damage or malfunction to the non-structural building components (walls, doors, fenestration)
- large differential deflection between adjacent frames that may cause damage to cladding or masonry wall.

The difficulty in stipulating appropriate deflection limits lies in the fact that there are many different types of buildings, occupancies and floor uses. The only practical way out is to determine, at an early design stage, all the constraints on deflections of critical building elements. Clause 3.5.3 of AS 4100 states that deflection limits should be appropriate to the structure, its intended use, the nature of loading and the elements supported by it. The same Clause gives very general instructions for determining the deflection limits. Appendix B of AS 4100 suggests some deflection limits which have now been somewhat superseded by AS/NZS 1170.0. Refer to a summary of deflection limits in Sections 1.8 and 5.10.

The following additional drift (sway) limits are also suggested:

- $\frac{H}{500}$ where masonry partition walls are built between the frames
- $\frac{H}{300}$ where reinforced concrete walls abut the frames
- $\frac{H}{200}$ to $\frac{H}{300}$ where lightweight partitions are in contact with frames
- $\frac{H}{100}$ to $\frac{H}{200}$ where the operation of doors and windows could be impaired

where H is the floor-to-floor/ceiling/roof height. The results of a survey into portal frame deflection limits applied in practice are reported by Woolcock et al. [1986, 2011]. As noted in Section 1.14(a), the amendment to AS 4100 (AS 4100 AMD—see Appendix D also) also has revised the suggested relative horizontal deflection limits between frames of industrial (portal) buildings to that noted in Woolcock et al. [2011].

10.9.2 Deflection calculations

Load combinations for the serviceability limit state are noted in Section 1.8. For example, for the dead plus live load case, the load combination is:

$$1.0G + \psi_s Q$$

where ψ_s ranges from 0.7 to 1.0. Other load combinations are given in AS/NZS 1170.0.

Thus the structural analysis has to be organised in such a way that strength limit state load cases are separated from the serviceability load cases—the latter case only requiring first-order elastic analysis (i.e. without second-order effects). It is also important for the modelling to be such that the maximum deflections can readily be obtained.

For simple structures it is sufficiently accurate to carry out a manual method of deflection calculation. For example, the UDL on a simply supported beam produces the following maximum deflection:

$$y_m = \frac{5M_s^* l^2}{48EI}$$

where M_s^* is the serviceability design bending moment obtained from an elastic analysis. For other load distributions and end moments, it is simple to use a correction coefficient, K_1, from Table C.2.3 in Appendix C of this Handbook. The deflection equation is then:

$$y_m = \frac{521 \times 10^3 M_s l^2}{I} K_1$$

where K_1 depends on the type and distribution of loads. For continuous beams, the effect of end bending moments can be taken into account at mid-span by subtracting the deflection induced by these moments. A useful reference for manual calculations of deflections is Syam [1992] and Appendix C of this Handbook.

10.10 Fire resistance

The fire resistance levels are specified in terms of endurance (in minutes) of the structural framing when it is subjected to a notional fire event. The notional fire event is defined by means of a standard time–temperature relationship given in the Building Code of Australia National Construction Code Series (NCC[2011]). The required levels of fire resistance are given in the NCC[2011] on the basis of standard fire tests performed to date. The member being tested is deemed to reach its period of structural adequacy (PSA) when the deflection of the member exceeds the specified limits—that is, the limit state of fire endurance.

Now that AS 4100 includes a section on fire resistance, it is no longer necessary to subject the steel structure to a standard fire test in order to determine its PSA. Section 12 of AS 4100 contains the rules for design verification of structural steel members in a fire event. Both unprotected and fire-protected members are covered. The difference between the two types of members is as follows:

- Unprotected members can resist the fire for a period that is a function of the ratio of mass to exposed surface area.
- Protected members rely on the thickness and thermal properties of protective material to endure for a specified period of fire exposure (PSA). (Note: these provisions do not consider concrete encasement or concrete filling.)

The properties of the protective material are given in Bennetts et al. [1990].

Typical examples of unprotected steelwork are single-storey industrial structures, parking stations and hangar roofs. Fire-protected members are usually employed in medium- and high-rise building structures.

One of the consequences of the high temperatures that develop in a fire event is that the modulus of elasticity and the yield stress of steel members reduce very significantly (see Section 12 of AS 4100). A secondary effect is that the coefficient of temperature expansion increases, causing the members affected to expand, resulting in additional action effects in the steelwork.

Because the incidence of fire is a relatively rare event, the applicable load combination, given in AS/NZS 1170.0, is:

$$1.1\,G + \psi_l\,Q$$

where ψ_l is the long-term combination factor which is between 0.4 to 1.0. This may be typically taken as 0.4 for UDLs on general floors (see Table 4.1 of AS/NZS 1170.0). It is not required to include wind or earthquake forces in any load combinations when this limit state is considered.

In lieu of the above, the NCC[2011] also permits a risk-based approach to the fire engineering of steel structures. Verification of the structural adequacy against fire is a very complex design task that requires considerable training and experience beyond the knowledge of the requirements of the NCC[2011] and AS 4100. Thomas et al. [1992] and O'Meagher et al. [1992], provide a good introduction to the methods used. Rakic [2008] also provides very useful information on the topic, particularly on fire protection materials and some case studies. Obtaining professional advice from relevant experts could prove invaluable.

10.11 Fatigue

10.11.1 Introduction

Machinery support elements in industrial structures are often subject to fluctuating loading with a large number of load cycles. The resulting stresses in structural elements vary cyclically and the difference between the upper and lower stresses is termed the 'stress range'. Repeated cycling can induce fatigue damage to a steel element and can lead to fatigue fracture. Distinction should be made between the high-cycle/low-stress fatigue, where stresses rarely reach yield, and the low-cycle/high-strain fatigue. The latter type of fatigue is characterised by repeated excursions into yield and strain hardening regions, as for example in an earthquake event. The rest of this Section is concerned with high-cycle/low-stress fatigue.

The fatigue damage potential increases with:

- magnitude of the stress range experienced in service and the number of stress cycles
- existence of notches, stress risers and discontinuities, weld imperfections and injuries to material
- thickness of the plate element if it exceeds 25 mm.

10.11.2 Stress range concept

Exhaustive research effort conducted in Europe and the USA in the past three decades has confirmed that stress range is the major parameter in fatigue assessment. AS 4100 reflects these findings. Fatigue damage will occur even if stress fluctuations are entirely in compression. The stress range, f, at a point in the structure or element is expressed as:

$$f = f_{max} - f_{min}$$

Prior to the European fatigue research findings, there was a school of thought that said compression stress cycles pose little risk of fatigue damage. The true situation with welded structures is that welds and the material adjacent to welds are in a state of high residual tension due to weld shrinkage forces. Thus the effective tensile stress in these areas fluctuates from tension yield to compression.

The second fact borne out by the European research is that the steel grade is largely irrelevant in welded steelwork. On the other hand, the higher-strength steels machined and free of severe notches do have a higher fatigue strength/endurance. High-strength steel elements fabricated by welding are not treated differently from Grades 250 or 300 steels because the imperfections produced by product manufacturing, welding and fabrication are the main problem. AS 4100 does not give guidance here, as machined structural elements are rare in building structures.

Stress risers are geometric features at which stress concentrations develop, usually because of local peaks of stress where the calculated nominal stress locally increases several-fold. Stress risers such as those that occur at weld imperfections, and others can be found in:

- small weld cracks (micro-cracks) along the toes of butt and fillet welds (large weld cracks are not permitted at all)
- porosity (gas bubbles) and slag inclusions
- lack of fusion between the parent plate and the weld in butt welds
- undercut in fillet and butt welds
- misalignment of adjoining plates
- excessive weld reinforcement
- rough weld surface contours
- weld craters at ends of weld runs
- rapid geometric or configuration changes in a localised sense.

Other stress risers occur at thermally cut surfaces that are excessively rough, and around punched holes. All these are generally associated with the notch-like defects or tiny cracks inherent in these processes. Cyclic stressing at such defects gives rise to stress concentrations and subsequently leads to lower fatigue strength.

Reliable welding inspection and good structural detailing is very important and can detect and/or mitigate the effects of most of these defects. The welding code, AS/NZS 1554.1, gives the permissible weld defect tolerances for each of the two 'weld categories', GP and SP. Weld category GP is for general-purpose welding and SP for structural (special-purpose) welding. Weld category GP should not be used in structures subject to fatigue. The third, very special category, which may be termed 'SX', is specified in AS/NZS 1554.5. It entails higher weld quality and thus tighter defect tolerances, as required for structures subject to high cycle fatigue (e.g. bridges).

10.11.3 Detail category

Some details are more prone to fatigue damage than others. It is thus necessary to categorise the welded joints and other details by the severity of the expected stress concentration. The geometry of the detail is an important factor in categorisation.

The term 'detail category' (DC) is used to describe the severity of stress concentrations and triaxial stressing of details, and for normal stresses, this ranges from DC 36 to 180. DC 180 is about the best that can be achieved with as-rolled steel plate or shapes, with no welding, holing and notching. DC 36, on the other hand, applies to details with considerable discontinuities. For example, a typical welded plate web girder flange is given a DC 112 if web stiffeners are not used, but if the web stiffeners are used the DC drops to 71. Table 11.5.1 of AS 4100 gives the DC numbers for a great variety of situations. For detail categories above 112 it is necessary to apply more stringent weld inspection and tolerance limits, such as specifying weld category SX from AS 1554.5, applicable to fatigue-loaded structures.

As noted in the fatigue strength versus number of cycles (S–N) curves, the DC numbers for normal stresses in AS 4100 coincide with the stress ranges at 2 million cycles, as can be seen in Figure 10.29. The slope of the lines representing the DC numbers is 1 in 3 for less than 5 million cycles and 1 in 5 beyond 5 million cycles. The S–N curves for shear stresses are also shown in Figure 10.29.

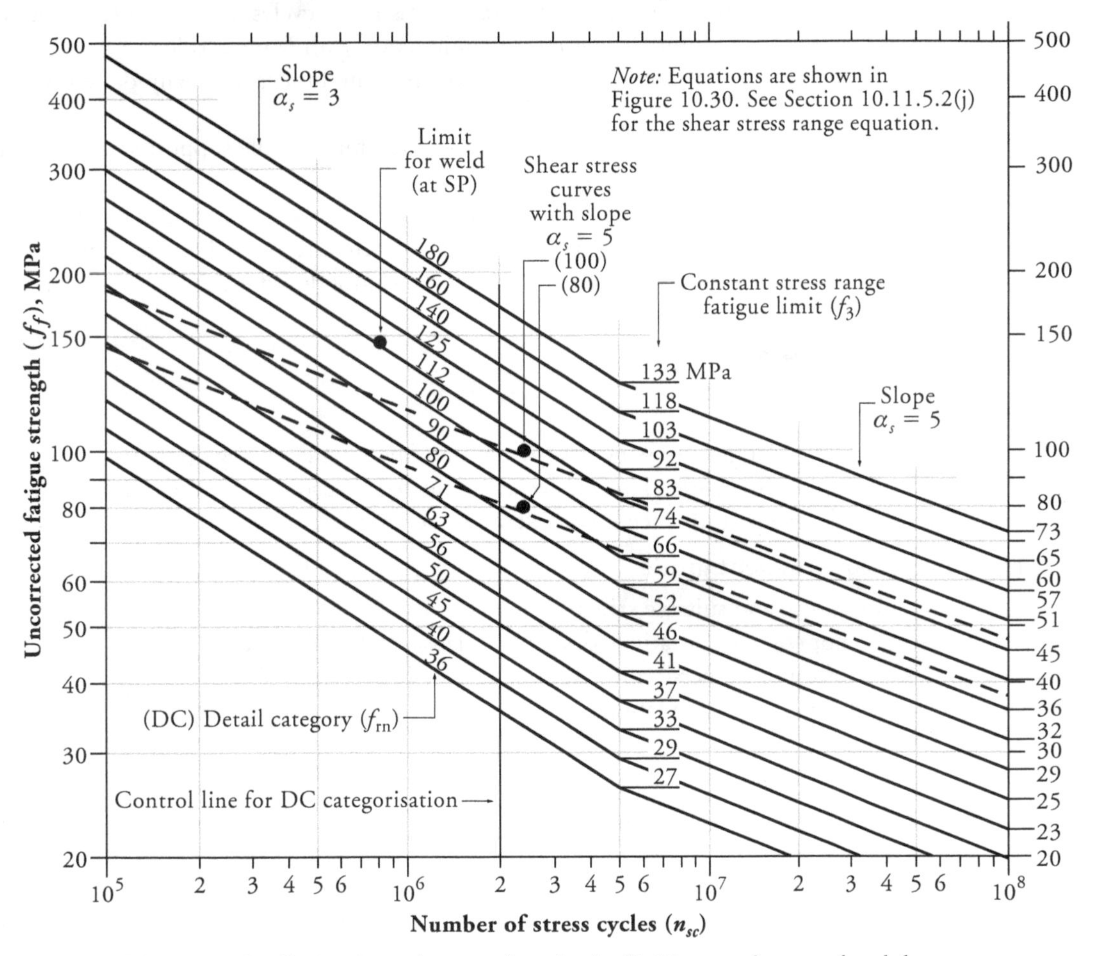

Figure 10.29 *Fatigue strength vs number of cycles (S–N) curves for normal and shear stresses (see Figure 11.6.1 and 11.6.2 of AS 4100 for further information)*

10.11.4 Number of cycles

The number of stress cycles over the life of the structure has a predominant influence on the fatigue resistance of a member or detail. Once the fatigue crack has started growing it will propagate with every stress cycle. A crack of say 1 mm in length may take years to reach 10 mm in length, gradually accelerating. In general, the fatigue life has three stages: crack initiation, crack propagation and fracture. The fatigue life is usually the period measured from the commissioning of the structure to the time when fatigue crack growth becomes a safety and/or maintenance problem. It is not possible to determine that point in time with any precision except by documenting the inspection and monitoring the yearly cost of weld repairs.

The design life is usually (at least) 30 years for buildings and bridges. It is accepted by most building (asset) owners that periodical inspection and repairs will be necessary in the later part of the structure's life.

Fluctuations in loads may be uniform or variable. Uniform fluctuations produce constant stress range cycles. Variable fluctuations produce variable stress range cycles. Variable cycling is first converted to constant stress range cycles by various techniques, such as the 'rainflow' counting method in conjunction with Miner's rule.

The evaluation of the number of cycles and stress amplitudes is usually done on one of the following bases:

- characteristic parameters of the vibrator (machine), for example support steelwork for a pump or vibrating screen—constant ranges
- prescribed number of cycles, as in crane and bridge Standards—constant ranges
- time and motion analysis, as in a container crane structure—variable stress ranges
- Wind tunnel tests, where stress cycles result from wind-induced oscillations—stochastic.

10.11.5 Fatigue assessment

10.11.5.1 Introduction

The purpose of fatigue assessment is to estimate the fatigue damage potential over the design life of the structure. The stresses are evaluated by using unfactored loads because structural fatigue performance is considered to be a serviceability limit state.

10.11.5.2 Fatigue assessment guide

(a) Restrictions, and suitable structures:

Areas not covered for fatigue design by Clause 11.1 of AS 4100 are:

- corrosion or immersion in reducing fatigue life
- high stress-low cycle fatigue
- thermal fatigue
- stress corrosion cracking.

Other structures that are suitable for the application of Section 11 of AS 4100 must satisfy its other requirements for their design and construction. The examples of the structures that *may* be suitable are: rail and highway bridges, crane runway girders, machinery support structures, cranes and the like. For existing structures only, the fatigue loading is to embrace the actual service loading for a design life that includes the cumulated fatigue damage from previous service and its planned future use.

Limits on yield stress, yielding in structures, stress range and weld quality:

- although the S–N curves in Clause 11.6 of AS 4100 may be applied to structural steels with $f_y \leq 700$ MPa, the lower limit in Clause 1.1.1(b) of AS 4100 governs with $f_y \leq 450$ MPa. The curves can be used for bolts with $f_y \leq 1000$ MPa.
- prohibits the application of Section 11 of AS 4100 fatigue design to structures which are designed to yield, or if the stress range exceeds $1.5 f_y$.
- weld category SP is mandatory in Clause 11.1.5 of AS 4100 for up to and including DC 112. Above this, weld quality is to AS/NZS 1554.5. Further details are given in Sections 10.11.2 and 10.11.3.

(b) Fatigue loads to Clause 11.2 of AS 4100.
They are based upon emulating the actual service loads (including dynamic effects, impact actions, etc.) as:
- loading should imitate as much as possible the actual service loading anticipated throughout the design life of the structure. The design life must include its accumulated fatigue damage.
- various crane loads are given in AS 1418.1, 3 and 5.
- do include the loads from perturbations and resonance, dynamic effects from machinery, and induced oscillations (e.g. structural response of lightly damped structures as in some cranes, and in masts, poles, towers, vents, and chimneys by wind).
- the effect of impact may be *very* important.

(c) A nominal event is the loading sequence on the structure, (connection or detail):
- An event may cause one or more sequence of stress cycles.

(d) Stresses at a point due to loads:
- A cyclic sequence of varying loads causes stresses to repeat at different points in the structure. The type and intensity of stress depends on the location of the points being investigated, type of structure, materials used, the quality of the construction, fabrication defects, design limitations, transportation, erection and the type and arrangement of the loads at each instant. Normal (perpendicular to plane) stresses, and shear (parallel to plane) stresses are required in the evaluation of the fatigue damage.

(e) Design spectrum:
An elastic analysis is performed to obtain the design stresses. An alternative is to deduce the stress history from strain measurements. See Clause 11.3.1 of AS 4100.
- holes, cut-outs and re-entrant corners (details of which are not a characteristic of the DC) have additional effects taken into account separately, and are included by using the appropriate stress concentration factors. See for example, Table 10.5 for hollow sections.
- joint eccentricity, deformations, secondary bending moments, or partial joint stiffness should also have their effects determined and included.
- compile and sum the spectra of stress ranges (f) versus number of stress cycles (n_{sc}) for each of the loading cases to give the design spectrum for the fatigue assessment.
- do not use the constant stress fatigue limit (f_3) unless it is certain that other stress ranges, which may occur during fabrication, transportation, erection or service of the structure, will not exceed it. Caution is required in using f_3 (see Figure 10.29).
- be conservative (with less risk), and adopt the premise that compressive stress ranges are as damaging as tensile stress ranges (unless proven to the contrary).

(f) Detail Category (DC):
Match, select and assign a DC to the structure from Tables 11.5.1(1) to (4) of AS 4100. See f_{rn} in Figure 10.30 and Section 10.11.2 for further information on DC. Variations in f_{rn} for some DCs are noted in Clause C11.5.1 of the AS 4100 Commentary.
- it is the responsibility of the *design* engineer to produce complete documented details to which all subsequent work done by others, do not vary in any respect,

including those of a seemingly trivial nature, such as making good temporary cut-outs, and attachments, during fabrication, transportation or erection. The *design* engineer of the structure must be kept fully cognizant of all matters affecting the detail at all times to commissioning and during operation.

(g) Stress evaluation:

- The stress analysis (at a point) on a plane, having normal and shear stresses, for simple load cases, is accomplished with principal stresses. Using principal stresses is only acceptable if there is a *complete* coincidence of point, plane, time, and reasonable cycle regularity. In the absence of a complete coincidence, the fatigue damage from the normal and shear stress ranges are added using Miner's rule. If this does not occur, Clause C11.3.1(b) of the AS 4100 Commentary notes the following to be used:

$$\left[\frac{f_n^*}{\phi f_{rn}}\right]^3 + \left[\frac{f_s^*}{\phi f_{rs}}\right]^5 \leqslant 1.0$$

where f_n^* and f_s^* are the design normal and shear stress ranges respectively.

f_{rn} and f_{rs} are the reference fatigue strengths (= DC number) for the normal and the shear stress ranges respectively (see Figure 10.29).

ϕ is the capacity reduction factor. See Section 10.11.6.

Should the simultaneous presence of the normal and shear stresses on a plane lead to the formation of fatigue cracks at two distinct locations, no combinations need be considered because they indicate more than one load path and structural redundancy.

(h) Design stress ranges f^*.

- The difference of the extremes of the stresses at each point gives a stress range, f, for the point i.e. $f = f_{max} - f_{min}$. Then the design stress range, f^*, is the maximum of f, i.e. $f^* = \max(f)$ where f is from the *subset* of all the points common to being in the same cycles band for the incident load (active in the load sequence). The *subset* varies with the incident loads. This f^* is used to assess the fatigue damage to the structure for the same cycles band and repeated for the other cycles band. Figure 10.30 also emphasises the fact that the fatigue strength is evaluated differently in each cycles band. Repeat for all the points.
- In *variable* stress range evaluations, both the constant stress range fatigue limit f_3, (as ϕf_{3c}), and the design stress ranges f^*, (as f_i^* and f_j^*), are simultaneously dependent on exponent 3 when $\phi f_{3c} \leqslant f_i^*$ or n_{sc} is within the cycles band 3, and dependent on exponent 5 when $\phi f_{5c} \leqslant f_i^* < \phi f_{3c}$ or n_{sc} is within the cycles band 5, *provided* $\phi f_{5c} \leqslant f^*$ is satisfied. f_{5c} is the cut-off limit. See Figure 10.30, and the i and j ranges in Clause 11.8.2(a) of AS 4100. f^* is dependent on the (sub)set of relevant points (and their disposition) in the structure during the incident load, and being within the same cycles band.

(i) Fatigue strengths and limits (Normal-*constant* stress range):

- The characteristics of a typical of S–N curve (two lines at different slopes of α_s) are summarised in Figure 10.30. The line expresses the relationship between the (uncorrected) fatigue strength, f_f, and the number of stress cycles, n_{sc}. The corresponding equations are shown at the top. Grundy [1985] describes the development of S–N curves.

There are two cycle bands, and from Clause 11.6.2 of AS 4100, each curve is bilinear and is defined by:

$$f_f^3 = \frac{f_{rn}^3 \times 2 \times 10^6}{n_{sc}} \quad \text{when } n_{sc} \leqslant 5 \times 10^6\text{, and}$$

$$f_f^5 = \frac{f_5^5 \times 10^8}{n_{sc}} \quad \text{when } 5 \times 10^6 < n_{sc} \leqslant 10^8$$

where f_{rn} is the reference fatigue strength (uncorrected) at $n_r = 2 \times 10^6$ cycles. Also see Figure 10.30.

(j) Fatigue strength (Shear-*constant* stress range)
- Two S–N curves, DC 80 and 100, are shown in Figure 10.29. There is only one cycles band, and from Clause 11.6.2 of AS 4100, each curve appears as a single straight line defined by:

$$f_f^5 = \frac{f_{rs}^5 \times 2 \times 10^6}{n_{sc}} \quad \text{when } n_{sc} \leqslant 10^8$$

where f_{rs} is the reference fatigue strength (uncorrected) at $n_r = 2 \times 10^6$ cycles.

(k) Thickness effect:
- A thicker material (plate) creates a three-dimensional distress effect to reduce its fatigue strength. Clause 11.1.7 of AS 4100 provides a thickness correction factor, β_{tf}:

$$\beta_{tf} = \left(\frac{25}{t_p}\right)^{0.25}$$ for a transverse fillet/butt welded connection involving a plate thickness, t_p, greater than 25 mm, or otherwise

$$= 1.0$$

- The corrected fatigue strength, f_c, is given by:

$$f_c = \beta_{tf} f_f$$

f_f has the variant forms f_{rn}, f_{rs}, f_3 and f_5 in the same clause.

(l) Compliance with the fatigue strength (*Constant* stress range *only*):
- the structure complies with Clause 11.8.1 of AS 4100 if it satisfies *at all points*, the inequality (it is not explicit whether it is normal or shear stress, or both):

$$\frac{f^*}{\phi f_c} \leqslant 1.0$$

where: f^* is the design *constant* stress range and ϕf_c is the corrected fatigue strength.

(m) Exemption from fatigue assessment is available (*Constant* stress range for both normal, and shear stress ranges), if Clause 11.4 of AS 4100 is satisfied:

$$f^* < \phi \times 27 \text{ MPa}$$

$$n_{sc} < 2 \times 10^6 \left(\frac{\phi \times 36}{f^*}\right)^3$$ (Note: worst DC = 36 = f_{rn})

where f^* is the design stress range (for each of normal and shear stress), n_{sc} is the number of stress cycles and ϕ is noted in Section 10.11.6.

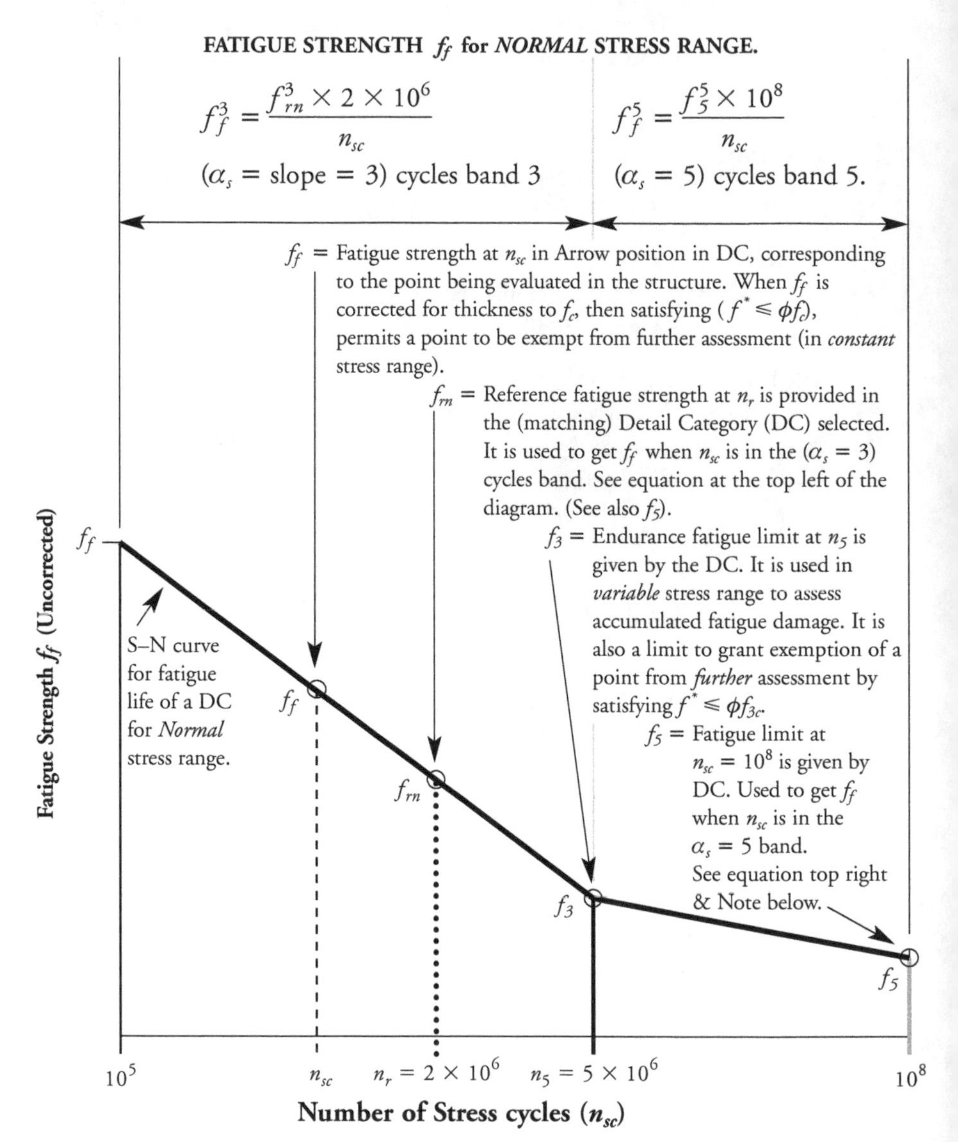

Note: The *constant* stress range f_f depends on f_{rn} when n_{sc} is in the cycles band 3, and on f_5 in the cycles band 5. In *variable* stress range f_f is f_3 used with index 3 (n_{sc} in band 3) or f_5 (band 5) and f^* relative to ϕf_{3c}. Use the correction factor, β_{tf}, given in Paragraph (k) to get $f_c = \beta_{tf} f_f$. At n_5, the highest f_c (i.e. $f_c \leq f_{3c}$), at which, if there are any cracks, they are not expected to grow. At $n_{sc} = 10^8$, the highest f_c (i.e. $f_c \leq f_{5c}$), at which cracks are not considered to occur. See the Glossary in Paragraph (r) to clarify the descriptions of f_3, and amplitude. There is more on the *variable* stress range in Paragraphs (h) and (o).

Figure 10.30 *An S–N curve for normal constant stress range. The significance of reference value f_{rn}, and fatigue limits f_3 and f_5 are noted.*

(n) Exemption from further assessment at a point (normal *constant* stress range) is available if Clause 11.7 of AS 4100 is satisfied:

$$f^* < \phi f_{3c}$$

where f_{3c} is the fatigue limit f_3 corrected for thickness effect. See Figure 10.30, and Paragraph (k).

(o) Compliance with the fatigue strength (*Variable* stress range):

The basis for evaluating the fatigue damage in *one* design stress range (and its number of cycles in the incident load sequence) is shown in Figure 10.30, and in *two* design stress ranges (albeit involving normal and shear stresses) in Paragraph (g) as Miner's rule. As an extension, it is used here in a similar manner, to *sum* the fatigue damage cumulatively (as constant stress ranges). See Paragraphs (h) and (s) for more details:

- normal stress range: See Clause 11.8.2(a) of AS 4100 which requires compliance with the inequality:

$$\frac{\Sigma_i n_i (f_i^*)^3}{5 \times 10^6 (\phi f_{3c})^3} + \frac{\Sigma_j n_j (f_j^*)^5}{5 \times 10^6 (\phi f_{3c})^5} \leqslant 1.0$$

- shear stress range: See Clause 11.8.2(b) of AS 4100. It is necessary to satisfy the inequality:

$$\frac{\Sigma_k n_k (f_k^*)^5}{2 \times 10^6 (\phi f_{rsc})^5} \leqslant 1.0$$

where:

- Σ_i is for i design stress ranges (f_i^*) for which $\phi f_{3c} \leqslant f_i^*$
- Σ_j is for j design stress ranges (f_j^*) for which $\phi f_{5c} \leqslant f_j^* < \phi f_{3c}$
- Σ_k is for k design stress ranges (f_k^*) for which $\phi f_{5c} \leqslant f_k^*$
- n_i, n_j and n_k are the respective numbers of cycles of nominal loading event producing f^*. See Notation in Clause 11.1.3 of AS 4100.
- f_{3c}, see Paragraph (n).
- f_{rsc} is the DC reference fatigue strength at n_r for shear stress, corrected for the thickness of material. See Figure 10.30.
- ϕ is the capacity reduction factor, (1.0 or $\leqslant$ 0.7). See Section 10.11.6.

(p) Punching limitation on plate thickness:
Punched holes are only permitted in plates less than or equal to 12.0 mm. See Clause 11.9 of AS 4100.

(q) Notes:

- The evaluation of the history of fatigue loads, and the culling of the stress cycles in the spectrum (of stress range versus the number of stress cycles) for counting are very complex. See Grundy [2004] who cites fatigue damage to civil engineering structures (e.g. bridges) as predominantly being due to *variable* stress range cycling. Examples of this, and other aspects, such as the pairing of stress maxima and minima (from a multitude of stresses) for amplitude, are given.
- Also see the Flow chart for fatigue assessment in Figure C11 of the AS 4100 Commentary.

(r) Glossary:

Amplitude, and the limits, f_3 and f_5, have multiple/differing descriptions:

- Amplitude (depending on context used) is: (1) the vertical distance between the peak and zero of a stress range cycle, (2) the stress range, and (3) the *maximum* of the stress ranges (in the number of cycles of the nominal loading event in either of the cycles band. See Figure 10.30).
- CAFL = Constant Amplitude Fatigue limit. See Grundy [2004].
- f_3 = DC Fatigue *strength* = DC Fatigue *limit* = CAFL, (*all*, at $n_5 = 5 \times 10^6$ cycles). See Clause 11.1.3 of AS 4100. (Definition 1).
- f_3 = DC Constant stress range Fatigue *limit* f_3 (at $n_5 = 5 \times 10^6$ cycles). See Figure 11.6.1 of AS 4100. (Definition 2).
- f_5 = DC Fatigue *strength* = DC Fatigue *limit*, set at $n_{sc} = 10^8$ cycles where n_{sc} has been cut-off (to position the limit f_5).

(s) Comment on *variable stress range* fatigue damage assessment:

In the area of *variable stress range* cycling, use is made of the uncertain (statistical) quantum of reserve capacity (in the region across the n_{sc} band) *above* the fatigue limit f_3, to accommodate the damage that results from any number of cycles of loading event, ($n_i \leqslant 5 \times 10^6$ cycles). See Figure 10.30, and consider for example, in a given incident load (or loading event), the more the design stress range f_i^*, exceeds the fatigue limit f_3, the less the number of cycles, n_i, is available for the other stress ranges. The design stress range f^* has an irregularity, at times being above, and at other times below f_3. The cumulated damage for all the design stress ranges has already been addressed in the inequalities in Paragraph (o). It turns out, that in the longer term, under some circumstances, those earlier stress ranges of f^* below f_3 also contribute ultimately to the fatigue damage. For more details see Grundy [2004]. It indicates that, the design S–N curve is two standard deviations below the mean life curve. Even more uncertain capacity is suggested because of the wide scattering of test data.

10.11.6 Dispensation and capacity reduction factor ϕ

AS 4100 provides for some exemptions and relaxations on fatigue assessments. These include:

- exemptions noted in Section 10.11.5.2 Steps (m) and (n), and
- a two "tiered" capacity reduction factor, ϕ, which is dependent on the compliance of the 'reference design' condition.

The 'reference design' condition given in Clause 11.1.6 of AS 4100 is described as follows:

- The detail is on a redundant load path: that is, the member failure will not lead to overall collapse of the structure.
- The stress history is determined by conventional means.
- The load-in each cycle are not highly irregular.
- Access for weld inspection is available.
- Regular inspection of welds will be the part of the owner's overall maintenance plan.

The value of the design capacity factor is taken as follows:

- $\phi = 1.0$ for elements complying with the 'reference design' condition
- $\phi \leq 0.7$ when the element in question does not comply with the reference design condition. For instance, the element is on a non-redundant load path, e.g. where the failure of the member being considered is not kept in check by an adjoining structure or other fail-safe feature (see Clause 11.1.6 of AS 4100).

Welded tubular elements may be subject to high local stress concentrations, which are not of great consequence to the design of ordinary building structures but are important when stresses are fluctuating. The AS 4100 method is to increase the stress range using a multiplier (Table 10.5) instead of considering the effects of connection stiffness and eccentricities. Clause 11.3.1 of AS 4100 provides information on the use and limitations of this method—e.g. the design throat thickness of a fillet weld must be greater than the hollow section thickness. Alternatively, CIDECT [2000] can be used for hollow sections subject to high levels of fatigue.

Table 10.5 Stress range multipliers for node connections of tubular trusses

Description	Type of detail	Chords	Verticals	Diagonals
Gap joints	K-type	1.5 (1.5)	1.0 (1.0)	1.3 (1.5)
	N-type	1.5 (1.5)	1.8 (2.2)	1.4 (1.6)
Overlap joints	K-type	1.5 (1.5)	1.0 (1.0)	1.2 (1.3)
	N-type	1.5 (1.5)	1.65 (2.0)	1.25 (1.4)

Note: Values in parenthesis apply to RHS/SHS and values outside parenthesis apply to CHS.

10.11.7 Improvement of fatigue life

Worthwhile improvements in the fatigue strength of welded components can be achieved by such weld improvement techniques as weld toe peening, toe grinding and toe remelting. It is important to obtain advice from an experienced welding engineer before embarking on these refinements. This is because small cracks often exist at toes (micro-cracks) which act as fatigue initiators in butt and fillet welds. Fillet welds are particularly prone to such defects, as shown by their lower DC numbers. Hammer peening is particularly effective and easy to apply; it consists of hitting the weld toes along the whole length of weld with a hammer having small round point. The benefit is seen in reversing the tensile stress arising from weld shrinkage at the toe of the weld, into a compressive stress. The other two techniques aim to remove the weld metal along the weld toes where micro-cracks are found by grinding away the burr or by arc remelting. It must be stressed that experienced operators supervised by a welding engineer should be employed. Butt welds can also be given a 'beneficiation' treatment by improving the weld contours through grinding and peening the weld toes. Other specialised methods of reducing residual stresses in the welds and in parent material are:

- stress relieving, normalising—a process for reducing the residual stresses in the welds using heat (500°–620°C) followed by (forced) air cooling.
- warm stressing: a process of subjecting the welded structure to its service loads when the temperature of the steel is above, say, 20°C.

Advice from and supervision by a welding engineer is essential when contemplating the use of these processes.

10.12 Corrosion protection

10.12.1 Methods of protection

Structural steels are susceptible to corrosion, and some method of corrosion protection is necessary for steelwork exposed to a corrosive environment. However, corrosion protection can be omitted where weather-resistant steels are used in a mildly corrosive environment, or where standard steel grades are shielded from a corrosive environment as, for example, within the ceiling space or within the clad envelope of an air-conditioned building.

The principal methods of corrosion protection are painting and galvanizing. However, the whole subject of corrosion protection is too wide to cover in this Handbook, and other references should be consulted. AS/NZS 2312 is a good source of information.

It should be noted that AS/NZS 2312 considers paint and metallic (e.g. galvanized) coatings and provides guidance on coating systems for the protection of steelwork against corrosion. The guidance entails selecting systems based on expected service life (typically to first maintenance) for various corrosivity environments around Australia and New Zealand. This entails considering regional/macro as well as micro/site/local effects in corrosivity zone influence, as well as planning, design and detailing matters. Additional information is also provided on surface preparation, characteristics of coating types, factors influencing coating selection, coating application methods, other systems (e.g. powder coatings, wrapping tapes), maintenance, inspection & testing, specifications and other useful background on the topic.

AS/NZS 2312 was published in 2002 with an amendment in 2004 and, for various reasons, the metallic (specifically zinc-based) coatings industry subsequently published AS 2309 in 2008 to update their respective coating durability information in the area of AS/NZS 2312 but more so to create a Standard for the classification of metallic coatings. In this regard, AS 2309 goes further than AS/NZS 2312 in classification and durability assessment of metallic coatings but is dependant on the balance of background information of AS/NZS 2312. Also, AS 2309 specifically refers to AS 4312 instead of AS/NZS 2312 for atmospheric corrosivity classification. See Francis [2011b] for a current review of AS 4312.

So, at the time of publication of this Handbook, the paint coating systems industry still use AS/NZS 2312 as a fundamental reference/guide, whereas the metallic coatings industry use AS 2309 and AS 4312 (with supplementary background information in AS/NZS 2312). In late 2011, Standards Australia has indicated its intent to revise AS/NZS 2312 with work on the Standard beginning in early February 2012.

See also Section 10.13 on further reading.

10.12.2 Painting steelwork

Painting is specified for the majority of steelwork, and a variety of coating systems are available. The usual coating system consists of a corrosion-inhibiting primer and several topcoats. Traditional painting systems relied on anti-corrosive agents incorporated in the primer coat, and on impermeability of the coating system. Modern coating systems use high-performance anti-corrosive and/or 'sacrificial' ingredients, and a thinner coat of paint that require meticulous surface preparation. An example of such a paint system consists of inorganic zinc primer applied over a surface that has been grit-blasted to bright metal.

The importance of good surface preparation cannot be overemphasised. Steel, as manufactured, is covered with a tightly adherent layer of mill scale, light rust, oil and impurities. No paint can adhere to such a surface and stay on without blistering and peeling. Brush cleaning used to be a traditional method of surface preparation, but the durability of the coating over such a surface was not satisfactory. See also Section 10.13 on further reading.

10.12.3 Galvanizing steelwork

Galvanizing provides effective protection against the corrosion of steelwork in most operating environments. The process known as hot-dip galvanizing consists of immersing the specially prepared steel elements in a bath of molten zinc, under conditions of carefully controlled temperature (about 450°C) and duration. The resulting protective zinc layer is metallurgically and mechanically bonded to the steel, to provide reliable barrier and sacrificial protection.

The preparation of steel surfaces must be thorough in order to achieve effective bonding. All loose scale, rust and contaminants such as grease, dirt, wax-crayon, paint marks, welding slag and flux must be removed; this is often done by caustic cleaning, followed by pickling in hydrochloric acid. The last step, before immersion in the zinc bath, is fluxing. This prevents the formation of an oxide layer and promotes complete fusing of the zinc coat.

A well-executed galvanizing treatment provides a thickness of protection of up to 100+ microns (depending on parent metal thickness amongst other factors)—if not more. The coating layer is basically split into two sub-layers: the lower layer, consisting of an alloy of zinc and iron; and an upper layer of pure zinc. Steel composition is important in this respect because an excess content of silicon or phosphorus can result in a zinc/iron layer of excessive thickness, featuring unsightly dark-grey spots and/or a brittle coating. Alternatively, too little ('trace') or no silicon levels may produce a shortfall in zinc/iron layer thickness. When ordering steel sections and plates, the steel manufacturer/supplier must be made aware that galvanizing will be applied.

The steel elements to be galvanized should be designed with galvanizing in mind. The design aspects of particular importance for well-executed galvanized protection are as follows:

- the disposition of long-run welds; unbalanced weld shrinkage forces. Also, asymmetrical compound sections can increase the bow distortion in the member after galvanizing
- venting and drainage of hollow sections: without venting there is a danger of bursting during galvanizing and inadequate drainage will see unnecessary accumulation of zinc
- detail design: full drainage of the molten zinc during withdrawal from the zinc bath must be assured, for example full-length stiffeners should be snipped at the corners to allow drainage.
- for further information on this area see GAA [1999] or the GAA website (www.gaa.com.au).

Fabricators should also be careful when sequencing the welded fabrication so as to minimise distortion. This is of particular importance when the welding elements consist of a mixture of thin and heavy plates and sections.

It is sometimes necessary to straighten members exhibiting relatively large asymmetrical patterns of residual stresses induced during fabrication. It is preferable to identify those prior to galvanizing so that hot straightening can be carried out prior to galvanizing. Cold straightening is less desirable because new residual stress patterns can be introduced, with the risk that distortion may not be completely avoided. Straightening after the galvanizing is often unavoidable, but care should be exercised to avoid damage to the galvanized surface through physical injury, cracking and peeling of the zinc coating.

One way of avoiding problems with distortion is to pre-galvanize the sections, then carry out welded splices, and protect the galvanized coat damaged by welding with a zinc-rich protective coating applied over grit-blasted bare steel.

In general, galvanizing contractors have developed expertise in dealing with the above problems and good-quality work can be achieved with proper planning and consultation.

As far as the economy of galvanizing is concerned, the decision should be based on a life-cycle costing that takes into account the initial cost of corrosion protection together with the cost of coating maintenance. The cost analysis should be prepared with particular attention to severity of exposure to corrosive media. In a benign environment it is unlikely that galvanizing would be more economical than some other low-cost protective treatment, while in a mild or aggressive (e.g. industrial) environment galvanizing is likely to be very cost-effective.

There are now many galvanized coating systems available. In its broadest sense galvanizing can be considered to be either the:

- traditional "post-fabrication" (or "batch") galvanizing which provides relatively thick, robust and reliable zinc coatings (these are provided to AS/NZS 4680)
- automated (or "in-line") galvanizing in which the zinc layer is controlled during a mechanised process (supplied to such Standards AS/NZS 4791 and AS/NZS 4792).

Additionally, for the automated processes, galvanizing can be further broken down into:

- thermal immersion (i.e. "hot-dip" type)—see above
- electro-galvanized (non-thermal)—as say, AS 4750
- zinc sprayed.

Publications such as GAA [1999] and the current (at the time of this Handbook's publication) GAA CD "After Fabrication Hot Dip Galvanizing" provide very useful information on galvanizing to AS/NZS 4680 as well as its properties, design, detailing, specification, inspection and durability. AS/NZS 2312 in conjunction with AS 2309 and AS 4312 are also good references on the durability of the above galvanizing systems for particular corrosion environments. The Galvanizers Association of Australia (GAA) should be contacted (see Appendix A.6) for expert advice on all the above galvanizing systems—specifically inherent properties and suitable applications.

10.13 Further reading

- As noted earlier, ASI [2009b] and Woolcock, et al. [2011] provide some good information on structural framing systems and sub-systems. The former reference offers general advice on all systems and the latter considers more detailed information on portal frame buildings. Though slightly dated, and in imperial measurements, AISE [1979] could be consulted as a guide on the design and construction of mill buildings.
- More recent practical references on multi-storey steel-framed buildings include Ng & Yum [2008] and Durack & Kilmister [2007], whereas for portal-framed buildings, ASI [2010] should be consulted.
- Interestingly, another good reference to consider is Ogg [1987]. Though written for architects, (the author is a qualified engineer and architect) the text and figures in terms of history and technical detail is of high quality and makes for noteworthy reading of structural steelwork and cladding systems.
- For fire design, AS 4100 and references as Bennetts, et al. [1987,1990], O'Meagher, et al. [1992] and Thomas, et al. [1992] consider various aspects of the standard fire test as noted in AS 1530.4. A recent reference on fire protection materials is Rakic [2008]. Alternatively, a risk-based assessment could be undertaken to the satisfaction of the regulators.
- For fatigue design, the internationally renowned fundamental text by Gurney [1979] should be consulted for steel and other metals. Alternatively, some very good information can be obtained from Grundy [1985,2004] which reflects on local Standards. The AS 4100 Commentary (Figure C11) provides a flow-chart on the use of the fatigue design provisions in AS 4100.
- The situation with corrosion protection coatings is not static with many galvanized coating systems now present and sophisticated paint coating systems continually evolving. Some excellent references to consult are AS/NZS 2312, AS 2309 and AS 4312 which describe these systems and provide guidance on system durability when the corrosion environment is assessed—they also assist in assessing the latter item as well. Account must be taken of "macro" and "micro" environmental effects which are noted by the Standards (though, understandably, it is termed a "Guide"). AS/NZS 2311 should also be consulted for painting of buildings. Francis [1996] provides a good summary on the mechanisms for steelwork corrosion and an update of the systems available. Readers should also consult Francis [2011a] for an up-to-date (at the time of this Handbook) reference on the design and details for the corrosion protection of structural steelwork.
- There are various galvanizing Standards such as AS/NZS 4680, AS/NZS 4792, etc. Of these, AS/NZS 4680 (which replaced the now withdrawn AS 1650) considers traditional after-fabrication galvanizing, which provides the thickest deposition of zinc and is commonly used in medium to aggressive environments. The other galvanizing Standards consider automated and semi-automated processes for the deposition of zinc with controlled thickness/coverage. GAA [1999] and the Galvanizers Association of Australia (GAA) can provide guidance in this area (see Appendix A.6).
- Paints for steel structures are noted in AS/NZS 2312 and AS/NZS 3750. Of these, zinc-rich paints are popular for structural steelwork (e.g. inorganic zinc silicates) which not only provide corrosion protection by barrier action but also by sacrificial action. However, in many instances, structural steelwork within building envelopes may only require a primer-coat (say red-oxide zinc phosphate)—if there is to be any coating at all.

Appendix

Bibliography

A.1 Contents

Appendix A contains the following sub-sections, references and other information:

A.2 Standard and codes

Readers should note that the list of Standards in this section is not exhaustive and is limited to the most essential Standards. The year of revision (coupled with amendments (not listed) current at the time of publication of this Handbook) can change at any time, and the reader should make sure that only the current revisions/amendments are used.

Australian Building Codes Board (ABCB) – *www.abcb.gov.au*

NCC, 2011, *Building Code of Australia (BCA) – National code of construction series*

Committé Européen de Normalisation (CEN) [European Committee for Standardization] – *www.cenorm.be*

EC 3, *EN 1993-1-1:1992(E) Eurocode 3, Design of steel structures – Part 1-1: General rules and rules for buildings*, European Committee for Standardization (CEN), 1992.

EC 3 (2005), *EN 1993-1-1:2005(E) Eurocode 3, Design of steel structures – Part 1-1: General rules and rules for buildings*, European Committee for Standardization (CEN), 2005.

EC 4, *EN 1994-1-1:1994(E) Eurocode 4, Design of composite steel and concrete structures – Part 1-1: General – Common rules and rules for buildings*, European Committee for Standardization (CEN), 1994.

EC 4 (2004), *EN 1994-1-1:2004(E) Eurocode 4, Design of composite steel and concrete structures – Part 1-1: General – General rules and rules for buildings*, European Committee for Standardization (CEN), 2004.

Standards Australia – *www.standards.com.au*

AS 1085.1, *Railway track material – Part 1: Steel rails*, 2002

AS 1101.3, *Graphical symbols for general engineering – Part 3: Welding and non-destructive examination*, 2005.

AS 1110, *ISO metric hexagon bolts and screws.*

AS 1110.1, *Product grades A & B – Bolts*, 2000.

AS 1110.2, *Product grades A & B – Screws*, 2000.

AS 1111, *ISO metric hexagon bolts and screws.*

AS 1111.1, *Product grades C – Bolts*, 2000.

AS 1111.2, *Product grades C – Screws*, 2000.

AS 1112, *ISO metric hexagon nuts.*

AS 1112.1, *Style 1 – Product grades A and B*, 2000.

AS 1112.2, *Style 2 – Product grades A and B*, 2000.

AS 1112.3, *Product grade C*, 2000.

AS 1170.1, *Minimum design loads on structures – Part 1: Dead and live loads and load combinations*, 1989.

AS 1170.2, *Minimum design loads on structures – Part 2: Wind loads*, 1989.

AS 1170.3, *Minimum design loads on structures – Part 3: Snow loads*, 1990 (plus supplementary Commentary).

AS 1170.4, *Minimum design loads on structures – Part 4: Earthquake loads*, 1993 (plus supplementary Commentary).

AS 1170.4, *Structural design actions – Part 4: Earthquake actions in Australia*, 2007.

AS 1210, *Pressure vessels*, 2010.

AS 1250, *SAA steel structures code*, 1981 (superseded & withdrawn).

AS 1275, *Metric screw threads for fasteners*, 1985.

AS 1391, *Methods for tensile testing of metals*, 2007.

AS 1397, *Continuous hot-dip metallic coated steel sheet and strip – Coatings of zinc and zinc alloyed with aluminium and magnesium*, 2011.

AS 1418.1, *Cranes, hoists and winches – Part 1: General requirements*, 2002.

AS 1418.3, *Cranes, hoists and winches – Part 3: Bridge, gantry, portal cranes (including container cranes) and jib cranes*, 1997.

AS 1418.5, *Cranes, hoists and winches – Part 5: Mobile cranes*, 2002.

AS 1418.18, *Cranes hoists and winches – Part 18: Crane runways and monorails*, 2001.

AS 1530.4, *Methods for fire tests on building materials, components and structures – Part 4: Fire-resistance tests of elements of building construction*, 2005.

AS 1548, *Steel plates for pressure equipment*, 2008.

AS 1650, *Galvanized coatings*, 1989 (superseded).

AS 1657, *Fixed platforms, walkways, stairways and ladders – Design, construction and installation*, 1992.

AS 1858.1, *Electrodes and fluxes for submerged-arc welding – Carbon steels and carbon-manganese steels*, 2003.

AS 2074, *Cast steels*, 2003.

AS 2309, *Durability of galvanized and electrogalvanized zinc coatings for the protection of steel in structural application – Atmospheric*, 2008.
AS 2327.1, *Composite structures – Part 1: Simply supported beams*, 2003.
AS 3597, *Structural and pressure vessel steel – Quenched and tempered plate*, 2008.
AS 3600, *Concrete structures*, 2009.
AS 3774, *Loads on bulk solids containers*, 1996 (plus supplementary Commentary).
AS 3828, *Guidelines for the erection of building steelwork*, 1998.
AS 3990, *Mechanical equipment – Steelwork*, 1993.
AS 3995, *Design of steel lattice towers and masts*, 1994.
AS 4100, *Steel structures*, 1998.
AS 4100 Commentary, *AS 4100 Supplement 1, Steel Structures – Commentary*, 1999.
AS 4100 AMD 1, *Amendment No. 1 to AS 4100 – 1998 Steel Structures*, 29 February 2012 (see also Appendix D of this Handbook).
AS 4291.1, *Mechanical properties of fasteners made of carbon steel and alloy steel – Bolts, screws and studs*, 2000.
AS 4312, *Atmospheric corrosivity zones in Australia*, 2008.
AS 4750, *Electrogalvanized (zinc) coatings on ferrous hollow and open sections*, 2003.
AS 5100, *Bridge design*, (numerous parts), 2004, plus supplementary Commentaries.
AS 5100.6, *Bridge design – Part 6: Steel and composite construction*, 2004.

Standards Australia & Standards New Zealand

AS/NZS 1163, *Cold-formed structural steel hollow sections*, 2009.
AS/NZS 1170.0, *Structural design actions – Part 0: General principles, 2002* (plus supplementary Commentary).
AS/NZS 1170.1, *Structural design actions – Part 1: Permanent, imposed and other actions*, 2002 (plus supplementary Commentary).
AS/NZS 1170.2, *Structural design actions – Part 2: Wind actions*, 2011.
AS/NZS 1170.3, *Structural design actions – Part 3: Snow and ice actions*, 2003 (plus supplementary Commentary).
AS/NZS 1252, *High strength steel bolts with associated nuts and washers for structural engineering*, 1996.
AS/NZS 1554.1, *Structural steel welding – Part 1: Welding of steel structures*, 2011.
AS/NZS 1554.2, *Structural steel welding – Part 2: Stud welding (Steel studs to steel)*, 2003.
AS/NZS 1554.3, *Structural steel welding – Part 3: Welding of reinforcing* steel, 2008.
AS/NZS 1554.4, *Structural steel welding – Part 4: Welding of high strength quench and tempered steels*, 2010.
AS/NZS 1554.5, *Structural steel welding – Part 5: Welding of steel structures subject to high levels of fatigue loading*, 2011.
AS/NZS 1554.6, *Structural steel welding – Part 6: Welding stainless steels for structural purposes*, 2012.
AS/NZS 1554.7, *Structural steel welding – Welding of sheet steel structures*, 2006.
AS/NZS 1559, *Hot-dip galvanized steel bolts with associated nuts and washers for tower construction*, 1997.
AS/NZS 1594, *Hot-rolled steel flat products*, 2002.
AS/NZS 1595, *Cold rolled, unalloyed steel sheet and strip*, 1998.
AS/NZS 2311, *Guide to painting of buildings*, 2009.

AS/NZS 2312, *Guide to the protection of structural steel against atmospheric corrosion by the use of protective coatings*, 2002.

AS/NZS 2717.1, *Welding - Electrodes – Gas metal arc – Ferritic steel electrodes*, 1996.

AS/NZS 3678, *Structural steel – Hot-rolled plates, floorplates and slabs*, 2011.

AS/NZS 3679.1, *Structural steel – Hot-rolled bars and sections*, 2010.

AS/NZS 3679.2, *Structural steel – Welded I sections*, 2010.

AS/NZS 3750, *Paints for steel structures*, (numerous parts and years).

AS/NZS 4600, *Cold-formed steel structures*, 2005.

AS/NZS 4600 Supplement 1, *Cold-formed steel structures – Commentary*, 1998.

AS/NZS 4671, *Steel reinforcing materials*, 2001.

AS/NZS 4680, *Hot-dip galvanized (zinc) coatings on fabricated ferrous articles*, 2006.

AS/NZS 4791, *Hot-dip galvanized (zinc) coatings on ferrous open sections, applied by an in-line process*, 2006

AS/NZS 4792, *Hot-dip galvanized (zinc) coatings on ferrous hollow sections, applied by a continuous or specialized* process, 2006.

AS/NZS 4855, *Welding consumables – Covered electrodes for manual metal arc welding of non-alloy and fine grain steels – Classification*, 2007.

Standards New Zealand – *www.standards.co.nz*

NZS 1170.5, *Structural design actions – Part 5: Earthquake actions – New Zealand*, 2004.

NZS 1170.5, *Supplement 1, Structural design actions – Earthquake actions – New Zealand Commentary*, 2004.

NZS 3404.1, *Part 1: Steel structures standard*, 1997.

NZS 3404.1, *Part 1: Steel structures standard – Materials, fabrication and construction*, 2009.

NZS 3404.2, *Part 2: Commentary to the steel structures standard*, 1997.

Standards Australia, Standards New Zealand and International Organization for Standardization

AS/NZS ISO 17632, *Welding consumables – Tubular cored electrodes for gas shielded and non-gas shielded metal arc welding of non-alloy and fine grain steel – Classification (ISO 17632:2004, MOD)*, 2006.

International Organization for Standardization (ISO) – *www.iso.org*

ISO 636, *Welding consumables – Rods, wires and deposits for tungsten inert gas welding of non-alloy and fine-grain steels – Classification*, 2004.

ISO 14341, *Welding consumables – Wire electrodes and weld depositis for gas shielded metal arc welding of non-alloy and fine grain steels – Classification*, 2010.

A.3 References

Notes:

1) AISC/ASI: In 2002, the Australian Institute of Steel Construction (AISC) merged with the Steel Institute of Australia (SIA) to become the Australian Steel Institute (ASI).

2) BHP Steel Long Products/Tubemakers became OneSteel in 2000.

3) BHP Steel/Bluescope: BHP Steel Flat Products became Bluescope Steel in 2003.

4)PTM/SSTM: Palmer Tube Mills became Smorgon Steel Tube Mills in 2003.

5)OST P&T/SSTM: OneSteel Pipe & Tube merged with SSTM as part of the overall OneSteel Limited and Smorgon Steel Group Ltd merger in 2007.

AISC, 1987, *Safe load tables for structural steel*, sixth ed., Australian Institute of Steel Construction.

AISC, 2001, *Australian Steel Detailers' Handbook*, Syam, A.A. (editor), first edition, second printing, Australian Institute of Steel Construction.

AISC(US), 1993, *Manual of Steel Construction – Volume II Connections: ASD (9'th ed.)/LRFD (1'st ed.)*, American Institute of Steel Construction.

AISC(US), 2002, *Designing with structural steel – A guide for Architects*, second ed., American Institute of Steel Construction.

AISC(US), 2011, *Steel construction manual*, American Institute of Steel Construction.

AISE, 1979, *Guide for the design and construction of mill buildings*, AISE Technical Report No. 13, American Society of Iron and Steel Engineers.

ASI, 2003, *Australian Steel Detailers' Handbook*, Syam, A.A. (editor), first edition, third printing, Australian Steel Institute.

ASI, 2004, *Design capacity tables for structural steel, Vol 2: Hollow sections*, second ed., Syam, A.A. & Narayan, K. (editors), Australian Steel Institute.

ASI, 2009a, *Design capacity tables for structural steel, Vol 1: Open sections*, fourth ed., Hogan, T.J. & Syam, A.A. (editors), Australian Steel Institute.

ASI, 2009b, *Economical structural steelwork – Design of cost effective steel structures*, fifth ed., Gardner, J.R. (editor), Australian Steel Institute.

ASI, 2010, *Design guide – Portal frame steel sheds and garages*, first ed. – revised 2010, (compiled by the ASI Steel Shed Group), Australian Steel Institute.

ASI, 2012a, *Documentation of structural steel*, 2012, ASI Technical Note TN009 V1 (April 2012, Author: T J Hogan & P Key), Australian Steel Institute.

ASI, 2012b, *Welding consumables and design of welds in AS 4100–1998 with Amendment 1, 2012,* ASI Technical Note TN008 V1 (February 2012, Author: T J Hogan), Australian Steel Institute.

Baker, J.F. & Heyman, J., 1969, *Plastic design of frames – 1. Fundamentals*, Cambridge University Press.

Beedle L.S., 1958, *Plastic design of steel frames*, John Wiley.

Bennetts, I.D., Thomas, I.R. & Hogan, T.J., 1986, *Design of statically loaded tension members*, Civil Engineering Transactions, Vol. CE28, No. 4, Institution of Engineers Australia.

Bennetts, I.D., Proe, D.J. & Thomas I.R., 1987, *Guidelines for assessment of fire resistance of structural steel members*, Australian Institute Steel Construction.

Bennetts, I.D., Thomas, I.R., Proe, D.J. & Szeto, W.T., 1990, *Handbook of fire protection materials for structural steel*, Australian Institute of Steel Construction.

Bisalloy, 1998, *Bisplate range of grades*, Bisalloy Steel.

Bisalloy Steels, 2011, *Bisplate technical guide*, Bisalloy Steels.

Bleich, F., 1952, *Buckling strength of metal structures*, McGraw-Hill.

Bluescope Lysaght, 2008, *Zeds and cees, Users guide, Purlins and girts structural sections*, Bluescope Steel.

Bluescope Steel, 2010, *Xlerplate Steel Product Information*, Bluescope Steel.

Bradford, M.A., Bridge, R.Q. & Trahair, N.S., 1997, *Worked examples for steel structures*, third ed., Australian Institute of Steel Construction.

Bridge, R.Q. & Trahair, N.S., 1981, *Thin walled beams*, Steel Construction, Vol. 15, No.1, Australian Institute of Steel Construction.

Bridge, R.Q., 1994, *Introduction to methods of analysis in AS 4100-1990,* Steel Construction, Vol. 28, No. 3, Paper 1, Syam, A.A. (editor), Australian Institute of Steel Construction.

Bridge, R.Q., 1999, *The design of pins in steel structures*, Proceedings of the 2'nd International Conference on Advances in Steel Structures, Hong Kong.

Bridon, undated, *Structural systems – Structures Brochure*, Bridon International Ltd (www.bridon.com).

Bruno, A., Bollinger, K., Davies, J.M., Feldmann, M., Grohmann, M., Mazzolani, F.M., O'Sullivan, G., Rambert, F., Reichel, A. and van Wyk, L., (co-authors Berlanda, T., Cardosa, F., Engel, P., Schnell, G., Turcaud, S.), 2009, *Featuring steel – resources, architecture, reflections*, ArcelorMittall.

Burns, P.W., 1999, *Information technology in the Australian steel construction industry*, Vol. 33, No. 3, Syam, A.A. (editor), Australian Institute of Steel Construction.

CASE, 1993, Pi, Y.L., & Trahair, N.S., *Inelastic bending and torsion of steel I- beams*, Research Report No. R683, November 1993, Centre for Advanced Structural Engineering, The University of Sydney.

CASE, 1994, Pi, Y.L., & Trahair, N.S., *Plastic collapse analysis of torsion*, Research Report No. R685, March 1994, Centre for Advanced Structural Engineering, The University of Sydney.

CASE, 1994, Pi, Y.L., & Trahair, N.S., *Torsion and bending design of steel members*, Research Report No. R686, March 1994, Centre for Advanced Structural Engineering, The University of Sydney.

CIDECT, 1991, Wardenier, J., Kurobane, Y., Packer, J.A., Dutta, D. & Yeomans, N., *Design guide for circular hollow section (CHS) joints under predominantly static loading*, Verlag TUV Rheinland.

CIDECT, 1992, Packer, J.A., Wardenier, J., Kurobane, Y., Dutta, D. & Yeomans, N., *Design guide for rectangular hollow section (RHS) joints under predominantly static loading*, Verlag TUV Rheinland.

CIDECT, 1994, Twilt, L., Hass, R., Klingsch, W., Edwards, M. & Dutta, D., *Design guide for structural hollow section columns exposed to fire*, Verlag TUV Rheinland.

CIDECT, 1998, Bergmann, R., Matsui, C., Meinsma, C. & Dutta, D., *Design guide for concrete filled hollow section columns under static and seismic loading*, Verlag TUV Rheinland.

CIDECT, 2000, Zhao, X.L., Herion, S., Packer, J.A., Puthli, R.S., Sedlacek, G., Wardenier, J., Weynand, K., van Wingerde, A.M. and Yeomans, N.F., *Design guide for circular and rectangular hollow section welded joints under fatigue loading*, Verlag TUV Rheinland.

CIDECT, 2008, *Design Guide for Circular Hollow Section (CHS) Joints Under Predominantly Static Loading*, Wardenier, J., Kurobane, Y., Packer, J.A., van der Vegte, G.J. and Zhao, X.-L., 2nd Edition, CIDECT.

CIDECT, 2009, *Design Guide for Rectangular Hollow Section (RHS) Joints Under Predominantly Static Loading*, Packer, J.A., Wardenier, J., Zhao, X.-L., van der Vegte, G.J. and Kurobane, Y., 2nd Edition, CIDECT.

CISC, 2010, *Handbook of Steel Construction*, tenth ed., Canadian Institute of Steel Construction.

CRCJ, 1971, *Handbook of structural stability*, Column Research Committee of Japan, Corona.

Colombo, C., 2006, *New technology delivers fabrication throughput efficiencies*, Steel Construction, Vol. 40, No.1, Munter, S.A. (editor), Australian Steel Institute.

Crandall, S.H., Dahl, N.C. & Lardner, T.J., 1978, *An introduction to the mechanics of solids*, second ed. with SI units, McGraw-Hill.

Durack, J.M. & Kilmister, M.B., 2007, *Composite steel design*, Australian Steel Institute.

EHA, 1986, *Engineers Handbook Aluminium – Design Data*, second ed., The Aluminium Development Council of Australia (Limited).

Fielders, 2011, *Purlin & girt design manual*, Fielders Australia Pty Ltd.

Firkins, A. & Hogan, T.J., 1990, *Bolting of steel structures*, Syam, A.A. (editor), Australian Institute of Steel Construction.

Francis, R., 2011a, *Design and details for the corrosion protection of structural steelwork*, Half-day seminar notes (c. March 2011), Australian Steel Institute.

Francis, R., 2011b, *AS 4312: An Australian atmospheric corrosivity standard*, Steel Construction, Vol. 45, No.1, Australian Steel Institute.

Francis, R.F., 1996, *An update on the corrosion process and protection of structural steelwork*, Steel Construction, Vol. 30, No.3, Syam, A.A. (editor), Australian Institute of Steel Construction.

GAA, 1999, *After-fabrication hot dip galvanizing*, fifteenth edition, Galvanizers Association of Australia. (Alternatively, refer to www.gaa.com.au if the publication or CD is not readily available.)

Gorenc, B.E. & Tinyou, R., 1984, *Steel designers' handbook (working stress design)*, fifth ed., University of New South Wales Press.

Gorenc, B.E., 2003, *Design of crane runway girders*, second edition, Australian Institute of Steel Construction.

Grundy, P., 1985, *Fatigue limit state for steel structures*, Civil Engineering Transactions, Vol. CE27, No.1, The Institution of Engineers Australia.

Grundy P, 2004, *Fatigue design of steel structures*, Steel Construction, Vol. 38 No. 10, Australian Steel Institute.

Gurney, T.R., 1979, *Fatigue of welded structures*, second ed., Cambridge University Press.

Hall, A.S., 1984, *An introduction to the mechanics of solids*, second ed., John Wiley.

Hancock, G.J., 1994a, *Elastic method of analysis of rigid jointed frames including second order effects*, Steel Construction, Vol. 28, No. 3, Paper 2, Syam, A.A. (editor), Australian Institute of Steel Construction.

Hancock, G.J., 1994b, *Second order elastic analysis solution technique*, Steel Construction, Vol. 28, No. 3, Paper 3, Syam, A.A. (editor), Australian Institute of Steel Construction.

Hancock, G.J., 2007, *Design of cold-formed steel structures*, fourth ed., Australian Steel Institute.

Harrison, H.B., 1990, *Structural analysis and design*, Parts 1 and 2, Pergamon Press.

Hens, C.P. & Seaberg, P.A., 1983, *Torsional analysis of steel members*, American Institute of Steel Construction.

Heyman, J., 1971, *Plastic design of frames – 2. Fundamentals*, Cambridge University Press.

Hogan, T.J., 2009, *An updated overview of design aids for structural steelwork*, Steel Construction, Vol. 43, No. 1, Australian Steel Institute.

Hogan, T.J., 2011, *Connection Design Guide 7 – Simple Connections, Design Guide 7: Pinned base plate connections for columns*, Munter, S.A. (editor), Australian Steel Institute.

Hogan, T.J. & Munter, S.A., 2007a, *Connection Handbook 1 – Background and theory, Handbook 1: Design of structural steel connections*, Munter, S.A. (editor), Australian Steel Institute.

Hogan, T.J. & Munter, S.A., 2007b, *Connection Design Guide 1 – Bolting, Design Guide 1: Bolting in structural steel connections*, Munter, S.A. (editor), Australian Steel Institute.

Hogan, T.J. & Munter, S.A., 2007c, *Connection Design Guide 2 – Welding, Design Guide 2: Welding in structural steel connections*, Munter, S.A. (editor), Australian Steel Institute.

Hogan, T.J. & Munter, S.A., 2007d, *Connection Design Guide 3 – Simple Connections, Design Guide 3: Web side plate connections*, Munter, S.A. (editor), Australian Steel Institute.

Hogan, T.J. & Munter, S.A., 2007e, *Connection Design Guide 4 – Simple Connections, Design Guide 4: Flexible end plate connections*, Munter, S.A. (editor), Australian Steel Institute.

Hogan, T.J. & Munter, S.A., 2007f, *Connection Design Guide 5 – Simple Connections, Design Guide 5: Angle cleat connections*, Munter, S.A. (editor), Australian Steel Institute.

Hogan, T.J. & Munter, S.A., 2007g, *Connection Design Guide 6 – Simple Connections, Design Guide 6: Seated connections*, Munter, S.A. (editor), Australian Steel Institute.

Hogan, T.J. & Munter, S.A., 2007h, *Structural steel – Simple connections: Web side plate, Flexible end plate, Angle cleat; Design capacity tables for structural steel, Volume 3: Simple Connections – Open Sections*, Munter, S.A. (editor), Australian Steel Institute.

Hogan, T.J. & Syam, A.A., 1997, *Design of tapered haunched universal section members in portal frame rafters*, Steel Construction, Vol. 31, No. 3, Syam, A.A. (editor), Australian Institute of Steel Construction.

Hogan, T.J. & Van der Kreek, N., 2009a, *Connection Design Guide 10 – Rigid connections; Design Guide 10: Bolted moment end plates & Beam splice connections*, Australian Steel Institute.

Hogan, T.J. & Van der Kreek, N., 2009b, *Connection Design Guide 11 – Rigid connections; Design Guide 11: Welded beam to column moment connections*, Australian Steel Institute.

Hogan, T.J. & Van der Kreek, N., 2009c, *Connection Design Guide 12 – Rigid connections; Design Guide 12: Bolted end plate to column moment connections*, Australian Steel Institute.

Hogan, T.J. & Van der Kreek, N., 2009d, *Connection Design Guide 13 – Rigid connections; Design Guide 13: Splice connections*, Australian Steel Institute.

Hogan, T.J. & Van der Kreek, N., 2009e, *Structural steel – Rigid connections: Bolted moment end plates, Welded beam to column moment connection, Bolted and welded splices; Design capacity tables for structural steel, Volume 4: Rigid Connections – Open Sections*, Australian Steel Institute.

Horne, M.R., 1978, *Plastic theory of structures*, second edition, Pergamon Press.

Horne, M.R. & Morris, L.J., 1981, *Plastic design of low-rise frames*, Granada.

Hutchinson, G.L., Pham, L. & Wilson, J.L., 1994, *Earthquake resistant design of steel structures – An introduction for practicing engineers*, Steel Construction, Vol.28, No.2, Syam, A.A. (editor), Australian Institute of Steel Construction.

Johnston, B.G., Lin, F.J. & Galambos, T.V., 1986, *Basic steel design*, Prentice Hall, Engelwood, NJ.

Keays, R., 1999, *Steel stocked in Australia – A summary for designers of heavy steelwork*, Steel Construction Vol. 33, No. 2, Syam, A.A. (editor), Australian Institute of Steel Construction.

Kitipornchai, S. & Woolcock, S.T., 1985, *Design of diagonal roof bracing rods and tubes*, Journal of Structural Engineering, Vol. 115, No.5, American Society of Civil Engineers.

Kleinlogel, A., 1973, *Rigid frame formulae*, Frederick Unger, UK.

Kneen, P., 2001, *An overview of design aids for structural steelwork*, Steel Construction, Vol. 35, No. 2, Australian Institute of Steel Construction.

Kotwal, S., 1999a, *The evolution of Australian material Standards for structural steel*, Steel Construction, Vol. 33, No. 2, Syam, A.A. (editor), Australian Institute of Steel Construction.

Kotwal, S., 1999b, *The evolution of Australian material Standards for pressure vessel plate*, Steel Construction, Vol. 33, No. 2, Syam, A.A. (editor), Australian Institute of Steel Construction.

Kulak, G.L., Fisher, J.W. & Struik, J.H.A., 1987, *Guide to design criteria for bolted and riveted joints*, second edition, John Wiley and Sons.

Lambert, M.J., 1996, *The life of mast stay ropes*, Journal of the International Association of Shell and Spatial Structures, Vol.37, No.121.

Lay, M.G., 1982a, *Structural Steel Fundamentals*, Australian Road Research Board.

Lay, M.G., 1982b, *Source book for the Australian steel structures code AS 1250*, Australian Institute of Steel Construction.

Lee, G.C., Morrell, M.L. & Ketter, R.L., 1972, *Design of tapered members*, Bulletin No. 173, Welding Research Council, Australia.

McBean, P.C., 1997, *Australia's first seismic resistant eccentrically braced frame*, Steel Construction, Vol.31, No.1, Syam, A.A. (editor), Australian Institute of Steel Construction.

Manolis, G.D. & Beskos, D.E., 1979, *Plastic design aids for pinned-base gabled frames*, (reprint from American Institute of Steel Construction), Steel Construction, Vol. 13, No. 4, Australian Institute of Steel Construction.

Marks, T., 2010, *Evaluation of footfall vibration in commercial buildings*, Steel Construction, Vol. 44, No.1, Australian Steel Institute.

Massonnet, C., 1979, *European recommendations (ECCS) for the plastic design of steel frames*, (reprint from Acier-Stahl-Steel), Steel Construction, Vol. 13, No. 4, Australian Institute of Steel Construction.

Melchers, R.E., 1980, *Service load deflections in plastic structural design*, Proceedings, Part 2, Vol. 69, Institution of Civil Engineers, UK.

Meyer-Boake, 2011, *Understanding Steel Design – An Architectural Design Manual*, Birkhäuser GmbH, Basel.

Morris, L.J. & Randall, A.L., 1975, *Plastic design*, Constrado, London, UK.

Munter, S.A., 2006, *World best practice in steel construction: UK and international markets, JIT fabrication and issues*, Steel Construction, Vol. 40, No.1, Munter, S.A. (editor), Australian Steel Institute.

Murray, T.M., 1990, *Floor vibration in buildings: Design methods*, Australian Institute of Steel Construction.

Neal, B.G., 1977, *The plastic methods of structural analysis*, third ed., Chapman and Hall.

Ng, A. & Yum, G., 2005, *Floor Vibrations in Composite Steel Office Buildings*, Steel Construction, Vol. 39, No.1, Australian Steel Institute.

Ng, A. & Yum, G., 2008, *Design aspects for construction – Composite steel framed structures*, Australian Steel Institute.

O'Meagher, A.J., Bennetts, I.D., Dayawansa, P.H. & Thomas, I.R., 1992, *Design of single storey industrial buildings for fire resistance*, Steel Construction Vol. 26, No. 2, Syam, A.A. (editor), Australian Institute of Steel Construction.

Oehlers, D.J. & Bradford, M.A., 1995, *Composite steel and concrete structural members – Fundamental behaviour*, first edition, Pergamon.

Oehlers, D.J. & Bradford, M.A., 1999, *Elementary behaviour of composite steel and concrete structural members*, Butterworth-Heinemann, Oxford.

Ogg, A., 1987, *Architecture in steel*, Royal Australian Institute of Architects (RAIA).

OneSteel, 2011, *Hot rolled and structural steel products catalogue*, fifth ed., OneSteel.

OneSteel, 2012a, *Product manual*, Syam, A.A. (editor), OneSteel Australian Tube Mills.

OneSteel, 2012b, *Design capacity tables for structural steel hollow sections*, Syam, A.A. (editor), OneSteel Australian Tube Mills.

Owens, G.W., & Cheal, B.D., 1989, *Structural Steelwork Connections*, Butterworths & Co.

Packer, J.A. & Henderson, J.E., 1997, *Hollow structural section connections and trusses – A design guide*, Canadian Institute of Steel Construction.

Packer, J., Sherman, D., and Lecce, M., 2010, *Hollow Structural Section Connections*, Steel Design Guide 24, American Institute of Steel Construction (AISC).

Parsenajad, S., 1993, *Deflections in pinned-base haunched gable frames*, Steel Construction, Vol. 27, No. 3, Syam, A.A. (editor), Australian Institute of Steel Construction.

Petrolito, J. & Legge, K.A., 1995, *Benchmarks for non-linear elastic frame analysis*, Steel Construction, Vol. 29, No. 1, Syam, A.A. (editor), Australian Institute of Steel Construction.

Pi, Y.L. & Trahair, N.S., 1994, *Inelastic bending and torsion of steel I- beams*, Journal of Structural Engineering, Vol. 120, No. 12, American Society of Civil Engineers.

Pikusa, S.P. & Bradford, M.A., 1992, *An approximate simple plastic analysis of portal frame structures*, Steel Construction, Vol. 26, No. 4, Syam, A.A. (editor), Australian Institute of Steel Construction.

Popov, E.P., 1978, *Introduction to the mechanics of solids*, second edition, Prentice-Hall, Englewood Cliffs, NJ.

Rakic, J., 2008, *Structural steel fire guide – Guide to the use of fire protection materials*, Steel Construction, Vol. 42, No. 1, Australian Steel Institute.

Riviezzi, G., 1984, *Curving structural steel*, Steel Construction, Vol. 18, No. 3, Australian Institute of Steel Construction.

Riviezzi, G., 1985, *Pin connections*, Steel Construction, Vol. 19, No. 2, Australian Institute of Steel Construction.

SCNZ, *Structural steelwork connections guide – Design procedures*, SCNZ Report 14.1:2007, Authors/editors: Hyland, C., Cowie, K., Clifton, C., Steel Construction New Zealand, February 2008.

Schmith, F.A., Thomas, F.M. & Smith, J.O., 1970, *Torsion analysis of heavy box beams in structures*, Journal of Structural Division, Vol. 96, No. ST3, American Society of Civil Engineers.

Stramit, 2010, *Stramit purlins, girts and bridging*, Stramit Building Products.

Syam, A.A., 1992, *Beam formulae*, Steel Construction, Vol. 26, No.1, Syam, A.A. (editor), Australian Institute of Steel Construction.

Syam, A.A. & Hogan, T.J., 1993, *Design capacity tables for structural steel*, Steel Construction, Vol. 27, No.4, Syam, A.A. (editor), Australian Institute of Steel Construction.

Syam, A.A., 1995, *A guide to the requirements for engineering drawings of structural steelwork*, Steel Construction, Vol. 29, No. 3, Syam, A.A. (editor), Australian Institute of Steel Construction.

Syam, A.A. & Chapman, B.G., 1996, *Design of structural steel hollow section connections, Vol. 1: Design models*, Australian Institute of Steel Construction.

Taylor, J., 2001, *An engineer's guide to fabricating steel structures – Vol. 1: Fabrication methods*, Australian Institute of Steel Construction.

Taylor, J., 2003, *An engineer's guide to fabricating steel structures – Vol. 2: Successful welding of steel structures*, Australian Steel Institute.

Terrington, J.S., 1970, *Combined bending and torsion of beams and girders*, Journal, No. 31, Parts 1 and 2, Constructional Steelwork Association, UK.

Thomas, I.R., Bennetts, I.D. & Proe, D.J., 1992, *Design of steel structures for fire resistance in accordance with AS 4100*, Steel Construction, Vol. 26, No. 3, Syam, A.A. (editor), Australian Institute of Steel Construction.

Tilley, P.A., 1998, *Design and documentation deficiency and its impact on steel construction*, Steel Construction, Vol. 32, No. 1, Australian Institute of Steel Construction.

Timoshenko, S.P., 1941, *Strength of materials*, Vol.2, Van Nostrand Co.

Timoshenko, S.P. & Gere, J.M., 1961, *Theory of elastic stability*, second ed., McGraw-Hill.

Trahair, N.S., 1992a, *Steel structures: Lower tier analysis*, Limit States Data Sheet DS02, Australian Institute of Steel Construction/Standards Australia. (Also reprinted in the AS 4100 Commentary).

Trahair, N.S., 1992b, *Steel structures: Moment amplification of first-order elastic analysis*, Limit States Data Sheet DS03, Australian Institute of Steel Construction/Standards Australia. (Also reprinted in the AS 4100 Commentary).

Trahair, N.S., 1992c, *Steel structures: Elastic in-plane buckling of pitched roof portal frames*, Limit States Data Sheet DS04, Australian Institute of Steel Construction/Standards Australia. (Also reprinted in the AS 4100 Commentary).

Trahair, N.S., 1993a, *Steel structures: Second-order analysis of compression members*, Limit States Data Sheet DS05, Australian Institute of Steel Construction/Standards Australia. (Also reprinted in the AS 4100 Commentary).

Trahair, N.S., 1993b, *Flexural-torsional buckling of structures*, E. & F.N. Spon.

Trahair, N.S., Hogan, T.J. & Syam, A.A., 1993c, *Design of unbraced beams*, Steel Construction, Vol. 27, No. 1, Syam, A.A. (editor), Australian Institute of Steel Construction.

Trahair, N.S., 1993d, *Design of unbraced cantilevers*, Steel Construction, Vol. 27, No. 3, Syam, A.A. (editor), Australian Institute of Steel Construction.

Trahair, N.S. & Bradford, M.A., 1998, *The behaviour and design of steel structures to AS 4100*, third ed. – Australian, E. & F. N. Spon.

Trahair, N.S. & Pi, Y.L., 1996, *Simplified torsion design of compact I-beams*, Steel Construction, Vol. 30, No. 1, Syam, A.A. (editor), Australian Institute of Steel Construction.

Watson, K.B., Dallas, S., van der Kreek, N. & Main, T., 1996, *Costing of steelwork from feasibility through to completion*, Steel Construction, Vol. 30, No. 2, Syam, A.A. (editor), Australian Institute of Steel Construction.
Webforge, 2010, *Access products Division*, Webforge.
Weldlok, 2010, *Weldlok Steel Grating*, Weldlok Industries (a division of Graham/Nepean Group).
Woodside, J.W., 1994, *Background to the new loading code – Minimum design loads on structures, AS 1170 Part 4: Earthquake loads*, Steel Construction, Vol. 28, No. 2, Syam, A.A. (editor), Australian Institute of Steel Construction.
Woolcock, S.T. & Kitipornchai, S., 1985, *Tension bracing*, Steel Construction, Vol. 19, No. 1, Australian Institute of Steel Construction.
Woolcock, S.T. & Kitipornchai, S., 1986, *Portal frame deflections*, Steel Construction, Vol. 20, No. 3, Australian Institute of Steel Construction.
Woolcock, S.T., Kitipornchai, S. & Bradford, M.A., 1999, *Design of portal frame buildings*, third ed., Australian Institute of Steel Construction.
Woolcock, S.T., Kitipornchai, S., Bradford, M.A. & Haddad, G.A., 2011, *Design of portal frame buildings – including crane runway beams and monorails*, fourth ed., Australian Steel Institute.
WTIA, 2004, *Technical Note 11 – Commentary on the Standard AS/NZS 1554 Structural steel welding*, Cannon, B. (editor) & Syam, A.A. (editor in part), Welding Technology Institute of Australia/Australian Steel Institute.
Young, W.C., Budynas, R.G. & Sadegh, A., 2012, *Roark's formulas for stress and strain*, eighth ed., McGraw-Hill.

A.4 Computer software

Structural analysis and design software referred to in this Handbook:

- Coldes, Limcon and Microstran—by Engineering Systems (www.microstran.com)
- Multiframe—Formation Design (www.formsys.com/multiframe)
- SpaceGass—by Integrated Technical Software (www.spacegass.com)
- Strand7—by Strand7 Pty Ltd (www.strand7.com)

Structural steel software developers/suppliers providing other types of structural steel software:

- AceCad software (www.acecadsoftware.com)—for structural steel project lifecycle software.
- ProSteel (www.bentley.com/en-AU/Products/ProSteel)—steel detailing, 3D modelling, Building Information Modelling.
- Revit (www.usa.autodesk.com/revit)—steel detailing, 3D modelling, Building Information Modelling.
- Strucad (refer to Tekla and strucad.support@tekla.com)—steel detailing, 3D modelling. See also AceCad software.
- Tekla (previously XSteel) (www.tekla.com)—steel detailing, 3D modelling, Building Information Modelling.

A.5 Steel manufacturer/supplier websites

Though the listing is not exhaustive, some major steel manufacturer/supplier websites of relevant interest include :

- Ajax Fasteners (www.ajaxfast.com.au)
- Bisalloy Steel (www.bisalloy.com.au)
- Bluescope Steel (www.bluescopesteel.com)
- Build with Standards (www.buildwithstandards.com.au)
- Emrails (www.emrails.com.au)
- Fielders (www.fielders.com.au)
- Hobson Engineering (www.hobson.com.au)
- ILB [includes the Industrial Light Beam] (www.ilbsteel.com.au)
- Lincoln Electric (www.lincolnelectric.com and www.lincolnelectric.com.au)
- Lysaght (www.lysaght.com)
- OneSteel (www.onesteel.com)
- OneSteel Australian Tube Mills (www.austubemills.com)
- Stramit (www.stramit.com.au)
- Webforge (www.webforge.com.au)
- Weldlok/Graham Group (www.weldlok.com.au/www.grahamgroup.com.au)

A.6 Steel industry association websites

Though the listing is not exhaustive, some major steel manufacturer/supplier websites of relevant interest include:

- American Institute of Steel Construction [AISC(US)] – www.aisc.org
- Australasian Corrosion Association [ACA] – www.corrosion.com.au
- Australian Stainless Steel Development Association [ASSDA] – www.assda.asn.au
- Australian Steel Institute [ASI – previously the Australian Institute of Steel Construction (AISC) and Steel Institute of Australia (SIA)] – www.steel.org.au
- British Constructional Steelwork Association [BCSA] – www.steelconstruction.org
- European Convention for Constructional Steelwork [ECCS] – www.steelconstruct.com
- Galvanizers Association of Australia [GAA] – www.gaa.com.au
- Galvanizing Association of New Zealand [GANZ] – www.galvanizing.org.nz
- Heavy Engineering Research Association [HERA] – www.hera.org.nz
- National Asssociation of Steel Framed Housing [NASH] – www.nash.asn.au
- National Association of Testing Authorities [NATA] Australia – www.nata.asn.au
- NATSPEC – www.natspec.com.au
- Steel Construction Institute (U.K.) [SCI] – www.steel-sci.org
- Steel Construction New Zealand [SCNZ] – www.scnz.org
- Surface Coatings Association Australia [SCAA] – www.scaa.asn.au
- Welding Technology Institute of Australia [WTIA] – www.wtia.com.au
- Worldsteel Association (previously the International Iron and Steel Institute [IISI]) – www.worldsteel.org

Appendix B

Elastic Design Method

B.1 Contents

Appendix B contains the following sub-sections:

B.2 Introduction

The elastic design method or "permissible stress" method is occasionally used in design for the following applications:

- elastic section properties for deflection calculations
- triaxial and biaxial stresses (e.g. as used in structural/mechanical components modelled by finite element (F.E.) methods using elements with more than one dimension—plates, bricks, etc)
- stress range calculations for fatigue assessment
- stresses in connection elements
- elastic torsion analysis
- other permissible stress-based design procedures (e.g. in mechanical, pressure vessel, vehicular, etc applications).

Additionally, past designs based on the permissible stress method may need to be appraised to give the designer some knowledge of pre-limit states design. Readers should note that the permissible stress method of design is also known as "working stress" or "allowable stress"—the latter term being more commonly used in the USA. The purpose of this section is to outline material pertaining to the elastic method of design.

Due to the nature of loading and differing "limiting" design conditions, mechanical and process engineers prefer—if not require—to stay with permissible stress methods to evaluate the "working" capacity of steel members and connections. AS 3990 (which is essentially a "rebadged" version of AS 1250—the predecessor to AS 4100) is used quite commonly in these instances. Another good reference on the topic is a previous edition of this Handbook (Gorenc & Tinyou [1984]).

Lastly, as is evidenced in the body of the Handbook and this Appendix, many of the parameters used in elastic design methods are also used in evaluating the various limiting conditions for the limit states design of steel structures.

B.3 Elastic section properties

B.3.1 Cross-section area

It is necessary to distinguish the "gross" cross-section area and the "effective" (or "net") cross-section area. The effective cross-section area is determined by deducting bolt holes and other ineffective areas of the cross-section which have excessive slenderness (see Section 4 of AS 3990). The cross-section may then be considered to be a "compound" area—i.e. composed of a series of rectangular and other shape elements which form the gross cross-section.

B.3.2 Centroids and first moments of area

After establishing the compound area, the centroid of the section about a specific axis is then determined. The centroid is sometimes called the centre of area (or mass or gravity) and is further explained in numerous fundamental texts on mathematical, structural and mechanical engineering. Essentially, the coordinate of the centroid position (x_c, y_c) from a datum point for the compound area made up of n elements about a specific axis can be evaluated by:

$$x_c = \frac{\sum_1^n (A_i x_i)}{\sum_1^n A_i}$$

$$y_c = \frac{\sum_1^n (A_i y_i)}{\sum_1^n A_i}$$

where A_i, x_i and y_i are respectively the area and centroidal coordinates from the datum point/axis for area element i. The terms $A_i x_i$ and $A_i y_i$ are called the first moments of area.

B.3.3 Second moments of area

The second moment of area (I_x and I_y) of a compound section is given by:

$$I_x = \sum_1^n \left[I_{xi} + A_i (y_i - y_c)^2 \right]$$

$$I_y = \sum_1^n \left[I_{yi} + A_i (x_i - x_c)^2 \right]$$

If the datum point is shifted to the centroid then x_c and y_c are taken as zero in the above equations for I. Values of I for standard hot-rolled open sections, welded I-sections and structural steel hollow sections are listed in AISC [1987] (given with other permissible stress design data), ASI [2004, 2009a] and OneSteel [2011, 2012a-b].

B.3.4 Elastic section modulus

The elastic section modulus (Z) is used in stress calculations and is determined from the second moment of area in the following manner:

$$Z_x = \frac{I_x}{y_e} \quad \text{and} \quad Z_y = \frac{I_y}{x_e}$$

where y_e and x_e are the distances from the neutral axis to the extreme point(s) of the section. Z has two values for unsymmetrical sections about the axis under consideration, the more critical being the lower Z value which has the larger distance from the neutral axis.

B.3.5 Sample calculation of elastic section properties

The calculation of the above section properties can be done in the following manner with the trial section noted in Figure B.1. Assume initially that all plate elements have f_y = 250 MPa and use Table 5.2 of AS 4100 to ascertain the slenderness limits of the flange outstand and web.

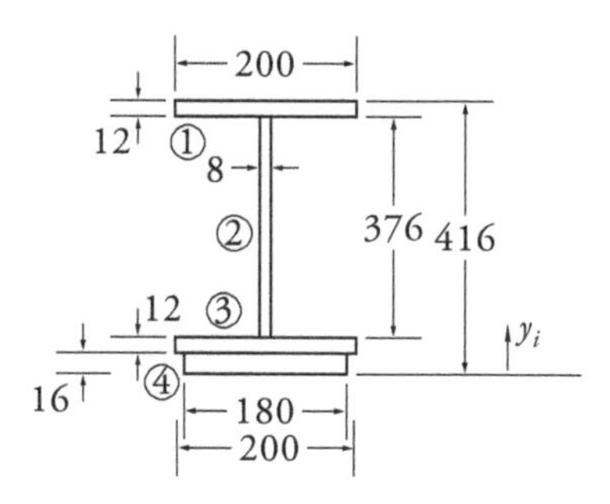

Figure B.1 *Trial section*

(a) Compactness of flanges: The top flange is the more critical of the flanges as it is the thinnest for the equal flange width section. The element slenderness of the top flange outstand, λ_{ef}, is:

$$\lambda_{ef} = \frac{b}{t}\sqrt{\frac{f_y}{250}} = \frac{(200-8)}{2 \times 12} \times \sqrt{\frac{250}{250}}$$

$$= 8.00 \leqslant \lambda_{ep} \text{ (= 8 from Table 5.2 of AS 4100 for LW plate)}$$

The flange outstand is compact and fully effective though λ_{ef} is at its limit. If $\lambda_{ef} > \lambda_{ep}$ then only the effective part of the outstand—i.e. b_{eff} (= $\lambda_{ep}t$) is considered in the section calculation and the balance of the outstand is disregarded.

(b) Compactness of web:

$$\lambda_{ew} = \frac{b}{t}\sqrt{\frac{f_y}{250}} = \frac{376}{8} \times \sqrt{\frac{250}{250}}$$

$$= 47.0 \leqslant \lambda_{ep} \text{ (= 82 from Table 5.2 of AS 4100)}$$

The web is compact and therefore fully effective.

The total area and first moment of area are listed in Table B.1 with centre of gravity subsequently calculated below.

Table B.1 Area and first moment of area for trial section

Section part, i	Size mm × mm	Areas, A_{ei} mm^2	Distance, y_i mm	Product, $A_{ei}y_i$ mm^3
1 top flange	200 ×12	2400	410	984×10^3
2 web	376 ×8	3010	216	650×10^3
3 bottom flange	200 ×12	2400	22	52.8×10^3
4 btm. flange plate	180 ×16	2880	8	23.0×10^3
Sum totals:		10 700 mm^2		1710×10^3 mm^3

Distance from base to neutral axis (i.e. centre of graxity x-axis):

$$y_c = \frac{\Sigma A_{ei}y_i}{\Sigma A_{ei}} = \frac{1710 \times 10^3}{10700} = 160 \text{ mm}$$

The intermediate values required for the evaluation of the second moment of area are listed in Table B.2.

Table B.2 Second moment of area, I_x

Section part, i	Areas, A_{ei} mm^2	Distance, e_i mm	Product, $A_{ei}e_i^2$ mm^4	I_{xi} mm^4
1 top flange	2400	250	150×10^6	0 (negligible)
2 web	3010	56.0	9.44×10^6	35.4×10^6
3 bottom flange	2400	-138	45.7×10^6	0 (negligible)
4 btm. flange plate	2880	-152	66.5×10^6	0 (negligible)

Note: $e_i = y_i - y_c$ and $I_{xi} = ab^3/12$ where a and b are the relevant plate dimensions.

The trial section second moment of area about the x-axis is:

$$I_x = \sum_1^4\left[I_{xi} + A_i(y_i - y_c)^2\right]$$

$$= (35.4 + 150 + 9.44 + 45.7 + 66.5) \times 10^6 = 307 \times 10^6 \text{ mm}^4$$

The trial section elastic section moduli about the x-axis is:

$$Z_{xT} = \frac{I_x}{y_{eT}} = \frac{I_x}{(D - y_c)} = \frac{307 \times 10^6}{(416 - 160)} = 1200 \times 10^3 \text{ mm}^3$$
(to top flange outer fibre)

$$Z_{xB} = \frac{I_x}{y_{eB}} = \frac{I_x}{y_c} = \frac{307 \times 10^6}{160} = 1920 \times 10^3 \text{ mm}^3$$
(to bottom flange outer fibre)

B.3.6 Calculated normal stresses

The term 'calculated' stress is used to denote a nominal stress determined by simple assumptions and not a peak stress that occurs in reality. Stresses occurring in reality would need to include the residual stresses induced by rolling and thermal cutting, stress concentrations around holes and at notches, and triaxial stresses. This can be done experimentally or by carrying out a finite element analysis but is a more complex exercise.

The normal or "axial" stress on a tension member or tie, σ_a, with an effective cross-section area, A_e, subject to an axial force, N, is:

$$\sigma_a = \frac{N}{A_e} \leqslant 0.6\, f_y$$

The same applies to a stocky compression member (not prone to buckling).

A longer strut fails by buckling, and the following check is necessary:

$$\sigma_{ac} = \frac{N}{A_e} \leqslant \text{permissible stress for buckling}$$

The permissible stress for buckling depends on the strut slenderness ratio, L/r, where:

$$\frac{L}{r} = \frac{L_{eff}}{r} \text{ where } L_{eff} \text{ is the effective buckling length and } r = \sqrt{\left(\frac{I}{A}\right)}$$

For further details, see Section 6 of AS 3990.

Bending stresses, σ_b, on a 'stocky' beam with an elastic section modulus, Z, subject to bending moment, M, are given by:

$$\sigma_b = \frac{M}{Z} \leqslant 0.66 f_y$$

For a slender beam, that is, a beam prone to flexural-torsional buckling:

$$\frac{M_x}{Z_x} \leqslant \text{permissible stress for flexural-torsional buckling}$$

See Section 5 of AS 3990 for the method of determining permissible bending stresses.

For bending combined with axial tension or bending with axial compression on a short/stocky member:

$$\frac{\sigma_a}{0.6f_y} + \frac{\sigma_b}{0.66f_y} \leqslant 1.0$$

Bending and axial compression of a slender member must be verified by the method of calculation given in Section 8 of AS 3990.

B.3.7 Calculated shear stresses

In general, for a flanged section subject to shear down the web, the transverse shear stress is:

$$\tau = \frac{VQ}{I_x t_w}$$

where V is the shear force, Q is the first moment of area of the part of the section above the point where the shear stress is being calculated, I_x is the second moment of area about the x-axis and t_w is the web thickness. For an equal-flanged I-beam, the average shear stress is:

$$\tau_{av} = \frac{V}{t_w d}$$

where d is the depth of the I-section. As the maximum shear stress at the neutral axis are only slightly higher than the shear stresses at the flanges, the average shear stress is taken as the maximum shear stress for such sections.

Conversely, a plate standing up vertically has a maximum shear stress at the neutral axis of:

$$\tau_{\max} = \frac{1.5V}{td}$$

which is 50% more than the average shear stress.

The maximum permissible shear stress is generally taken as:

$$\tau_{vm} \leqslant 0.45f_y$$

For I- sections, channels, plate girders, box and hollow sections, the average shear stress should satisfy:

$$\tau_{av} \leqslant 0.37f_y$$

this would reduce for slender webs, or be different for stiffened webs as noted in Clause 5.10.2 of AS 3990.

B.4 Biaxial and triaxial stresses

The design provisions noted in many typical structural steel design Standards (AS 3390, AS 4100, NZS 3404 and other similar national Standards) consider stress states which are generally uniaxial (single direction) in nature. These assumptions readily lend themselves to structural elements with:

elemental thicknesses (t) < member depth (d) << member length (L)

However, for structural elements in which the stress state is more complex with normal stresses in three mutually orthogonal directions and associated shear stresses, other methods of assessing the "failure" condition are required. Such applications include plates with double curvature bending and through-thickness shearing, pressure/cylindrical vessel design, stressed skin construction, induced stresses in thick/very thick elements, connection stresses from welding, etc. In these instances, an assessment of the triaxial stress state would have to be undertaken. These methods lend themselves to permissible design principles.

Indeed, many of the structures and elements considered in AS 3990, AS 4100, etc could be analysed for the triaxial stress state condition—e.g. in structural connections such as the beam flange welded to column flange connections with/without column flange/web stiffeners. The stress state is more than uniaxial and complex. However, to assess these and other situations by triaxial stress state methods for "stick" type structures requires much effort and in many instances can be avoided by simplifying rational assumptions, research outcomes and good structural detailing.

If triaxial stress state analysis is undertaken, the following failure condition (developed from the von Mises yield criterion) is generally applied where an equivalent stress, σ_{eq}, is established and must be less than or equal to a factored limit which embraces the commonly used yield stress, f_y, from a uniaxial tensile test:

$$\sigma_{eq}^2 = (\sigma_x - \sigma_y)^2 + (\sigma_y - \sigma_z)^2 + (\sigma_z - \sigma_x)^2 + 3\tau_{xy}^2 + 3\tau_{yz}^2 + 3\tau_{zx}^2 \leq 2f_y^2$$

where σ_x, σ_y, σ_z are the normal stresses and τ_{xy}, τ_{yz}, τ_{zx} are the shear stresses for the element/point under consideration. If the principal axis stresses (σ_1, σ_2, σ_3) are only considered (i.e. shear stresses are zero) the above equation reduces to:

$$\sigma_{eq}^2 = (\sigma_1 - \sigma_2)^2 + (\sigma_2 - \sigma_3)^2 + (\sigma_3 - \sigma_1)^2 \leq 2f_y^2$$

For elements subject to a biaxial stress state (i.e. with normal stresses in only two directions such as steel plates subject to bending and membrane stresses, thin pressure vessels, etc), the above non-principal axes criterion can be reduced with minor modification to the following with a permissible stress limit (Lay [1982b], Trahair & Bradford [1998]):

$$\sigma_{eq} = \sqrt{(\sigma_x^2 + \sigma_y^2 - \sigma_x\sigma_y + 3\tau_{xy}^2)} \leq \Omega f_y$$

Depending on the text being referenced, the value of Ω is between 0.60 to 0.66. The above equation is the basis for fillet weld design in AS 3990.

From the above, for the condition of uniaxial normal and shear stress, the following may be used in permissible stress design:

$$\sigma_{eq} = \sqrt{(\sigma_x^2 + 3\tau^2)} \leq \Omega f_y$$

This can now be used to develop the permissible stress limits for the case of pure uniaxial stress ($\tau = 0$) and pure shear stress present ($\sigma_{eq} = 0$).

The critical aspect of biaxial and triaxial stress states is to relate it to a failure condition. Many texts on strength of materials consider several theories on these types of conditions. Some of these include:

- *Maximum normal stress theory*—states that the failure condition of an element subject to biaxial or triaxial stresses is reached when the maximum normal stress attains its uniaxial stress limiting condition. This theory is in good agreement with testing of brittle materials and failure generally manifests itself as yielding or fracture.
- *Maximum shearing stress theory*—considers the failure condition of an element subject to biaxial or triaxial stresses to occur when the maximum shearing stress attains the value of shear stress failure from a simple axial tension/compression test. Though more complex it gives a good reflection of ductile material behaviour.
- *von Mises Yield Criterion*—also called Huber or Hencky Yield Criterion or Maximum energy of distortion theory. This is the most commonly used failure condition (noted above) and provides a good assessment of ductile materials.

B.5 Stresses in connection elements

Connection elements consist of gussets, web and end plates, eye plates, connection angles and a variety of other details. These elements are not easy to design correctly and often fall into the 'too hard' basket. Judging by published reports, the number of collapses caused by inadequate connections is significant.

The force distribution from the fasteners (bolts, welds) into the gusset plate follows the dispersion rule, so named because the highly stressed areas widen down the load path. The angle of dispersion in working stress design is 30-45 degrees from the member axis.

Interestingly, apart from bolt shear design, fillet/incomplete penetration butt weld design and bearing design, most of the AS 4100 limit state connection design provisions and associated (separately published) connection design models are "soft-converted" from permissible stress methods. This is due to the rational basis and simplicity of these methods in lieu of the detailed research, testing and application required of limit states connection models.

B.6 Unsymmetrical bending

The term 'unsymmetrical bending' has been used in permissible stress design and refers to bending about an axis other than the principal axes. For doubly symmetrical sections, this means the load plane passes through the centroid but is inclined to the principal axes. In monosymmetrical sections (I-beam with unequal flanges) and in channel sections, two conditions should be distinguished:

(a) The load plane passes through the shear centre which is the centre of twisting, with the result of bending and no twisting occuring.

(b) The load plane does not pass through the shear centre, and bending is accompanied by torsion.

Figure B.2 illustrates the two conditions.

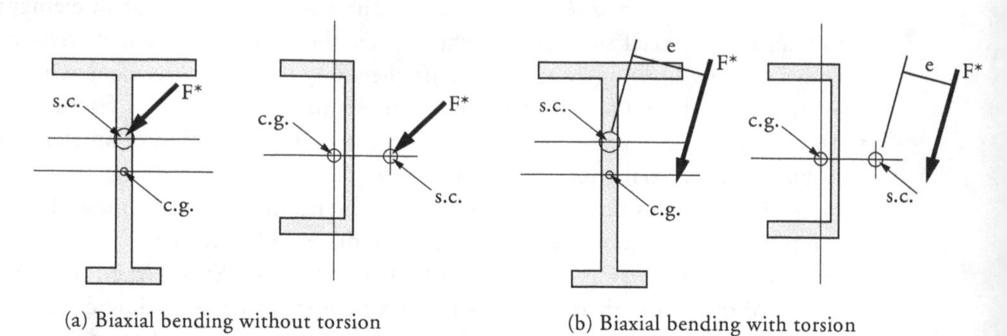

(a) Biaxial bending without torsion (b) Biaxial bending with torsion

Legend: s.c = shear centre; c.g. = centre of gravity

Figure B.2 *Biaxial bending of monosymmetrical sections*

Unsymmetrical bending without torsion is treated differently in the strength limit state design method of AS 4100. The loads are resolved into components parallel with the principal planes and the bending moments determined accordingly. The section capacities are determined for each principal plane, and a check is made for biaxial bending capacity using an interaction equation as described in Chapter 6. See Section B.7 for the interaction of bending and torsion.

B.7 Beams subject to torsion

B.7.1 General

The most optimal sections for torsional loads are closed sections (i.e. hollow sections or solid circular) which don't warp. Of these, circular sections are the most optimal and are initially used to theoretically determine torsional behaviour. Like members subject to bending, the behaviour of circular members when subject to torsional loads are assumed to have plane sections that remain plane but rotate (about the longitudinal axis) with respect to each other. Additionally, within a circular cross-section there is the assumption that shear strain (and hence shear stress) varies in direct proportion to the radius in the cross-section—i.e. no shearing at the centre and highest shear strain at the circumference. For circular and closed section members, torsional loadings are resisted by uniform torsion.

Some modification of torsional behaviour theory is required for non-circular sections. For non-circular hollow sections (e.g. RHS/SHS), allowance must be made for the higher shear stresses at the corners. A method for this is described in ASI [2004] and OneSteel [2012b]. A significant change in torsional behaviour occurs with open sections which resist torsional loads by a combination of uniform torsion and warping torsion as noted in Figure B.3. From this it can be surmised that uniform torsion embrace cross-sectional shear stresses whereas warping torsion embrace shear and normal (out-of-plane) stresses. Due to this complexity, there are no readily available "manual" or closed form solutions in structural codes for routine design of open sections subject to torsional loadings.

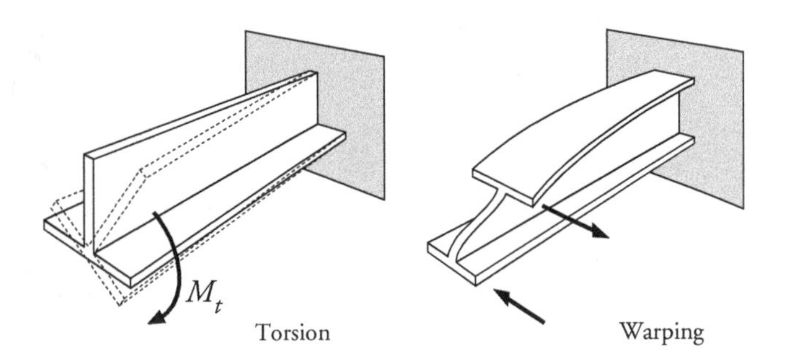

Figure B.3 *Uniform torsion and warping of open sections*

Consequently, there are no provisions in AS 4100 for dealing with members subject either to pure torsion or to torsion with bending. However, the AS 4100 Commentary gives a general outline of the problem and some specific recommendations for determining the capacity of members subject to bending and torsion.

Basically, there are three types of behaviour of members subject to torsion:

- self-limiting or secondary torsion, which occurs when there are lateral restraint members that are able to prevent excessive angle of twist
- free torsion, where there are restraints against lateral displacements but no restraints against twisting
- bending with destabilising torsion, where no restraints are available and the torsion interacts unfavourably with flexural-torsional bending.

Torsion of the self-limiting type occurs quite often. For example, a floor beam connected to a girder using a web cleat generates torsion in the girder, which diminishes as the slack in the bolts is taken up, without causing any significant torsion effects in the girder. In such cases torsion can easily be neglected in capacity verification.

Free and destabilising torsion, in contrast, do not diminish with the increased angle of twist and therefore cannot be neglected. There are no detailed provisions for computing torsional effects in AS 4100 (although the AS 4100 Commentary offers useful advice). This is because the subject of torsion with bending is too broad and complex for (building) structural applications. In many cases the designer may have no choice but to use the elastic solutions described in the literature on the subject (Terrington [1970]) or refer to more recent references (Trahair & Pi [1996]). Table B.3 provides some member behaviour information for free torsion.

Table B.3 Angle of twist, θ, and torsion shear stress, τ

Section	Characteristic dimensions	Angle of twist/ Shear stress
Round bar	r = radius	$\theta = 2\frac{M_t L}{(G\pi r^4)}$
	d = diameter	$= \frac{32\ M_t L}{(G\pi d^4)}$
		$\tau = \frac{2\ M_t}{(\pi r^3)}$
		$= \frac{16\ M_t}{(\pi d^3)}$
Circular Tube	r_o = external radius	$\theta = \frac{2\ M_t L}{[G\pi(r_o^4 - r_i^4)]}$
	r_i = internal radius	$= \frac{32\ M_t L}{[G\pi(d_o^4 - d_i^4)]}$
	d_o = external diameter	$\tau = \frac{2\ M_t r_o}{[\pi(r_o^4 - r_i^4)]}$
	d_i = internal diameter	$= \frac{16\ M_t d_o}{[\pi(d_o^4 - d_i^4)]}$
Square bar	s = side	$\theta = \frac{7.10\ M_t L}{(Gs^4)}$
		$\tau = \frac{4.81\ M_t}{s^3}$
Plate	b = long side	$\theta = \frac{M_t L}{(G\alpha_1 t^3 b)}$
	t = thickness	$\tau = \frac{M_t}{(\alpha_2 t^2 b)}$

$\frac{b}{t}$	1.0	2.0	4.0	6.0	8.0	10.0	>10.0
α_1	0.141	0.229	0.281	0.298	0.307	0.312	0.333
α_2	0.208	0.245	0.282	0.298	0.307	0.312	0.333

Section	Characteristic dimensions	Angle of twist/ Shear stress
I-section	b_i = element width	$\theta = \frac{M_t L}{(JG)}$
	t_i = element thickness	$\tau_{max} = \frac{M_t t_{max}}{J}$
	J = torsion constant = $\Sigma(\alpha_2 t_i^3 b_i)$	
	α_2 – see for Plate	
Box, other sections	See Bridge & Trahair [1981]	

Notes:
1. Above sections are for members loaded only by equal and opposite twisting moments, M_t, and end sections are free to warp.
2. θ = angles of twist.
3. τ = torsional shear stress (for square bar/plate this is at the mid-point of the longer side and for non-circular sections this may not be the peak shear stress).
4. L = length of member free to rotate without warping restraints between points of M_t being applied.
5. G = shear modulus of elasticity (taken as 80×10^3 MPa for steel).
6. For RHS and SHS see Section 3 of ASI [2004] or OneSteel [2012b].

B.7.2 Torsion without bending

The torsional resistance of a beam twisting about its (longitudinal) z-axis is usually divided into: uniform (St Venant) torsion and non-uniform (Warping) torsion—see Figures B.3, B.4 and B.7. Uniform-torsion occurs in bars, flats, angles, box-sections and hollow sections. Warping generally occurs when the (cross-) section is no longer planar (i.e. parts of the section do not remain in its original plane) during the plane's displacement as the member twists. The warping resistance against twisting (or torsion) develops if the displacement is (wholly or partially) prevented, (e.g. by an immovable or resisting contiguous section of the beam, or construction). Increased warping resistance improves member moment capacity (resistance). Non-uniform torsion arises in parallel-flanged open-sections (e.g. I-sections and channels). Warping resistance results from the differential in-plane bending of parallel plate elements such as the beam flanges. On the other hand, open sections consisting of one plate, or having two intersecting plates (e.g. angles and T sections) have negligible warping resistance.

B.7.2.1 Uniform torsion

In a member under uniform torsion with none of the cross-sections restrained against warping, the torsional moment M_z is:

$$M_z = \frac{GJ\phi}{L}$$

where ϕ is the angle of twist over the length L and J is the torsional constant.

For an open section consisting of rectangular elements of width b and thickness t:

$$J = \Sigma\left(\frac{bt^3}{3}\right)$$

where each $\left(\frac{bt^3}{3}\right)$ term is the torsion constant of that element. For example, J for an I-section becomes:

$$J = \frac{(2\, b_f t_f^3 + d_1 t_w{}^3)}{3}$$

where $d_1 = d - 2t_f$ (see also Table B.3 for a more precise evaluation of J).

For a box section of any shape and enclosing only one internal cell, J is given by:

$$J = \frac{4A_o^{\,2}}{\Sigma\left(\frac{s}{t}\right)}$$

where s/t is the length-to-thickness ratio of the component walls along the periphery of the section. In particular, a thin-walled round hollow section has:

$$A_o = \frac{\pi d^2}{4},$$

$$\Sigma\left(\frac{s}{t}\right) = \frac{\pi d}{t},$$

$$J = \frac{\pi d^3 t}{4}$$

The concept of the enclosed area A_o can also be extended to closed box tubes, where:

$$A_o = (D - t)\,(B - t)$$

In this expression, D is the overall depth of the box and B is its width.

The above expressions are valid only for closed sections such as tubes and box sections, because their sections after twisting remain in their plane within practical limits of accuracy, and the torsional resistance contributed by the parts of the cross-section is proportional to their distance from the centre of twist.

For an I-section member under uniform torsion such that flange warping is unrestrained (see Figure B.5), the pattern of shear stress takes the form shown in Figure B.4.

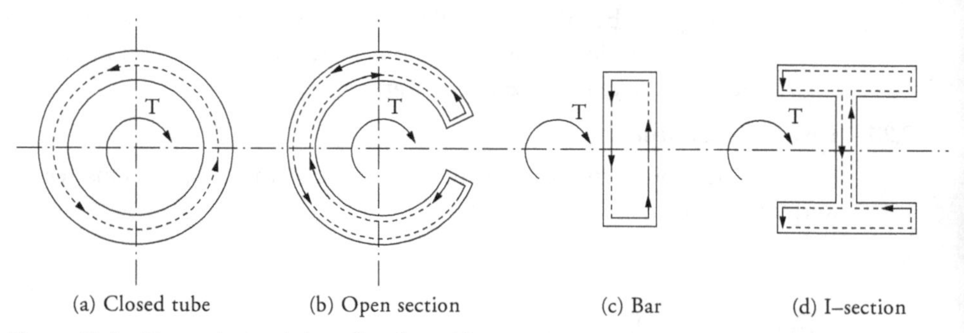

Figure B.4 *Torque-induced shear flow for uniform torsion*

Uniform torsion can also arise in I-beams, channels and other open sections where flanges are not restrained against warping. Open sections are substantially less rigid torsionally than sections of the same overall dimensions and thickness with flanges restrained against warping.

The torsional rigidity of a member is GJ, where G is the shear modulus of elasticity (80 000 MPa for steel) and J is the torsion constant. A circular section, whether solid or hollow, is the only instance in which J takes the same value as the polar second moment of area. For other sections, J is less than the polar second moment of area and may be only a very small fraction of it.

As noted in AS 3990, the verification of torsional capacity can be based on the elastic design method by limiting the shear stresses under design loads (factored loads) to:

$$f_v \leqslant 0.45\, f_y$$

This is conservative because plastic theory predicts higher shear capacity. Many references (e.g. Trahair & Pi [1996]) suggest a 28% increase in capacity by using the von Mises Yield Criterion:

$$f_v \leqslant 0.60\, f_y$$

For an open section or a circular tube under uniform torsion, the maximum shear stress, f_v, is:

$$f_v = \frac{M_z t}{J}$$

in which the maximum f_v occurs in the thickest part of the cross-section.

The tangential shear stress in a thin-walled tube can be assumed to be constant through the wall thickness, and can be computed from:

$$f_v = \frac{M_z}{(2\pi r^2 t)}$$
$$= \frac{M_z}{(2A_o t)}$$

where t is the thickness of tube, r is the mean radius, and A_o is the area enclosed by the mean circumference. The tangential shear stress, f_v, must not exceed the value of the maximum permissible shear stress.

B.7.2.2 Non-uniform or warping torsion

The torsional resistance of a member is made up of a combination of warping torsion and St Venant torsion such that the contribution of each type of torsion is not uniform along the axis of the beam. The warping constant, I_w, is calculated in accordance with theory and for standard sections is listed in sectional property tables (e.g. ASI [2009a], OneSteel [2011], EHA [1986]).

For doubly-symmetric (equal flanged) I-sections, the warping constant is given as:

$$I_w = \frac{I_y d_f^2}{4}$$

See Clause H4 of AS 4100 for the evaluation of I_w and J for other section types. Figure B.6 notes some typical warping restraints for beam ends.

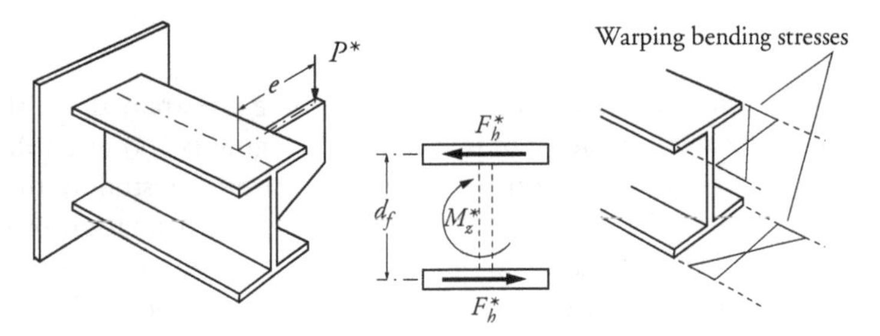

Figure B.5 *I-section beam subject to torsion and warping stresses*

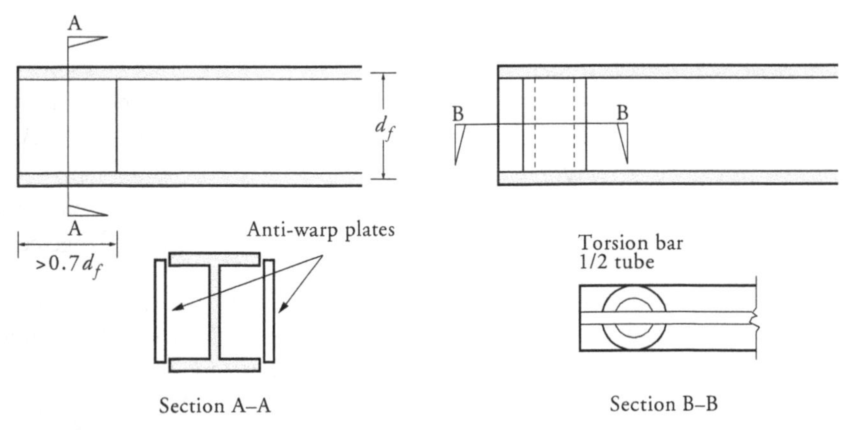

Figure B.6 *Types of warping restraint at beam ends*

B.7.3 Torsion with bending—working stress method

A method of analysis to determine the normal stress due to torsion is given by Terrington [1970]. When warping of the flanges is totally prevented (Figure B.6), the shear stress through the thickness of the flange is practically constant and takes the distribution shown in Figure B.7(b). Under these conditions each flange is subjected to a shear flow like that in an ordinary rectangular section beam carrying a horizontal transverse load. The most effective way of achieving complete restraint against warping of the flanges of an I-section is to box in the section by the addition of plates welded to the tips of the flanges (Figure B.6 again).

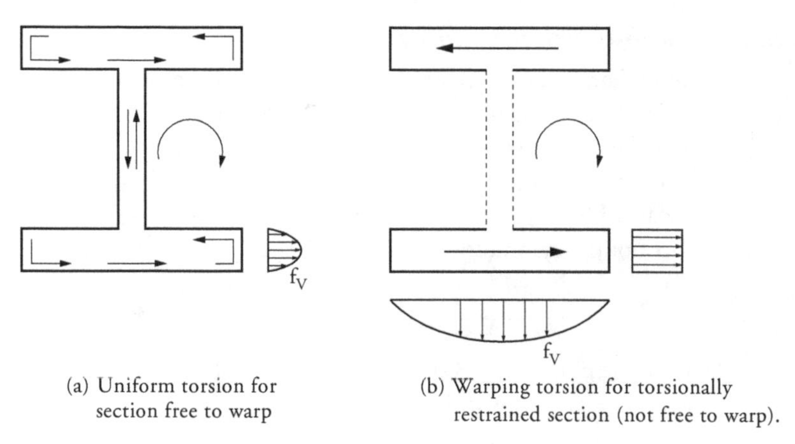

Figure B.7 *Shear stress distributions*

Away from a location in which flange warping is restrained, the shear stress distribution is generally a combination of that shown in Figure B.7(a) and (b). The farther away the section of a beam is from a location of warping restraint, the more similar is the distribution to that in Figure B.7(a). A rough approximation of the distance necessary from a position of flange warping restraint to a point along the beam in which the influence of restraint is negligible (and thus approaching the distribution in Figure B.7(a)) is given by:

$$a = \sqrt{\left(\frac{EI_w}{GJ}\right)}$$

where a is the torsion bending constant, and I_w is the warping constant (see Sections 5.5.3.3, 5.6, B.7.2.2 of this Handbook and Appendix H of AS 4100), that is:

$$I_w = \frac{I_y h^2}{4}$$

, approximately, for an equal-flanged I-section or a channel where $h = d_f$ = distance between flange centroids.

The torsion bending constant, a, is combined with the span, L, into a dimensionless parameter L/a for use in Tables B.4(a) and B.4(b), which have been adapted from Johnston et al. [1986]. The influence of warping and uniform torsion based on beam distance from a warping restraint is shown in Figure B.8.

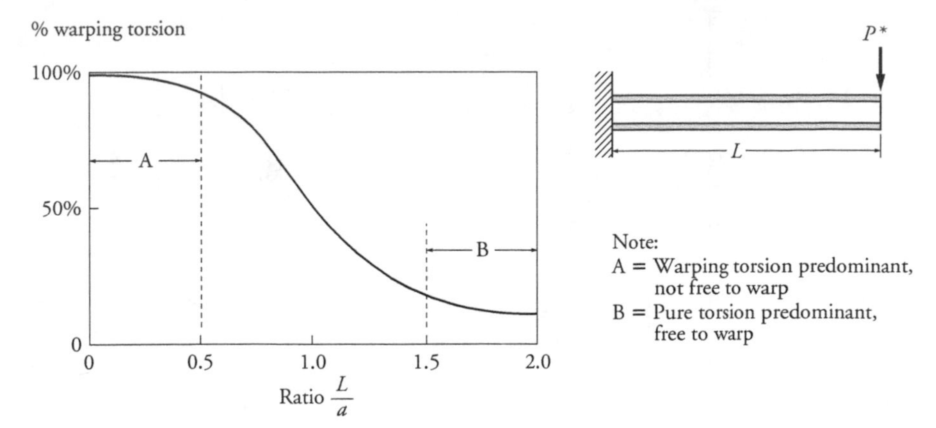

Figure B.8 *Plot of percentage of warping torsion as a function of* $\frac{L}{a}$ *(from Johnston et al. [1986])*

Although box sections are most suitable, sometimes short stocky lengths of I-sections may be adequate in strength and stiffness for use in combined bending and torsion. If the span, L, of a cantilever beam is less than $0.5a$ (Figure B.8), the pair of flanges act like twin beams loaded in opposite directions to take all the applied torsional moment, M_z^* (Figure B.5). For an I-section cantilever beam, carrying a load at the free tip with an eccentricity, e, to the centre of the beam, then from Figure B.5 the horizontal force, F_h^*, acting on each flange is given by:

$$F_h^* = \frac{M_z^*}{h} = \frac{P^*e}{h}$$

Tables B.4(a) and B.4(b) include simple approximate expressions for the maximum flange moment, M_{fy}, due to the flange warping restraint and the maximum angle of twist, ϕ_t, for some common cases of torsional loading. Short and long beams are covered directly by the formulae quoted, while beams of intermediate length require the use of coefficients that may be interpolated.

Based on the selected section, geometric configuration (load type, beam length, support type), load eccentricity and eccentric load magnitude, the maximum (or total) angle of twist, ϕ_t , can be obtained directly from Tables B.4(a) or B.4(b). These tables also provide the maximum flange moment due to warping in the normal direction (i.e. about the x-axis), M_{fy}, which must then be added to the x-axis moment, M_x, based on no eccentricity (i.e. $e = 0$). These combined moments when calculated to a stress format must then be below the permissible stress for that section based on $e = 0$.

Intermediate and long beams using open I-sections in combined bending and torsion are feasible only if the applied twisting moment, M_z^*, is small. Large moments produce excessive twisting, which becomes unacceptable.

A more accurate solution may be obtained analytically by using the finite element method or by solving the differential equation:

$$M_z = GJ\phi' - EI_w\phi'''$$

where $\phi' = d\theta/dz$ which is the change of angle of twist along the member and $\phi''' = d^3\theta/dz^3$.

Table B.4(a) Approximate flange bending moment, M_{fy}, about section y-axis of one flange and total angle of twist, ϕ_t, of beam for concentrated loads at an eccentricity, e, to the shear centre vertical axis. Beams are torsionally restrained at supports: $a = \sqrt{\frac{EI_w}{GJ}}$, $h = d - t_f$, $P = P^*$ (as noted in Section 8.7.3)

Case			
1 Cantilever (Section also applies to cases 2 and 3)	$\frac{L}{a} < 0.5$	$0.5 \leqslant \frac{L}{a} \leqslant 2.0$	$\frac{L}{a} > 2.0$
	$M_{fy} = \frac{PeL}{h}$	$M_{fy} = \frac{Pea}{h} k_1$; $\phi_t = \frac{Pea}{GJ} k_2$; $a = \sqrt{\frac{EI_w}{GJ}}$	$M_{fy} = \frac{Pea}{h}$
	$\phi_t = 0.32 \frac{Pea}{GJ} \left[\frac{L}{a}\right]^3$		$\phi_t = \frac{Pe}{GJ} [L - a]$
2 Simple beam	$\frac{L}{a} < 1.0$	$1.0 \leqslant \frac{L}{a} \leqslant 4.0$	$\frac{L}{a} > 4.0$
	$M_{fy} = \frac{PeL}{4h}$	$M_{fy} = \frac{Pea}{2h} k_1$; $\phi_t = \frac{Pea}{2GJ} k_2$	$M_{fy} = \frac{Pea}{2h}$
	$\phi_t = 0.16 \frac{Pea}{GJ} \left[\frac{L}{2a}\right]^3$		$\phi_t = \frac{Pe}{2GJ} \left[\frac{L}{2} - a\right]$
3 Continuous spans, equal length, equally loaded	$\frac{L}{a} < 2.0$	$2.0 \leqslant \frac{L}{a} \leqslant 8.0$	$\frac{L}{a} > 8.0$
	$M_{fy} = \frac{PeL}{8h}$	$M_{fy} = \frac{Pea}{2h} k_1$; $\phi_t = \frac{Pea}{GJ} k_2$	$M_{fy} = \frac{Pea}{2h}$
	$\phi_t = 0.32 \frac{Pea}{GJ} \left[\frac{L}{4a}\right]^3$		$\phi_t = \frac{Pe}{GJ} \left[\frac{L}{4} - a\right]$

Case 1 ($0.5 \leqslant \frac{L}{a} \leqslant 2.0$):

$\frac{L}{a}$	0.5	1.0	1.5	2.0
k_1	0.460	0.750	0.920	0.970
k_2	0.038	0.237	0.570	1.000

Case 2 ($1.0 \leqslant \frac{L}{a} \leqslant 4.0$):

$\frac{L}{a}$	1.0	1.5	2.0	2.5	3.0	3.5	4.0
k_1	0.460	0.620	0.750	0.850	0.920	0.960	0.970
k_2	0.038	0.121	0.237	0.387	0.570	0.786	1.000

Case 3 ($2.0 \leqslant \frac{L}{a} \leqslant 8.0$):

$\frac{L}{a}$	2.0	3.0	4.0	5.0	6.0	7.0	8.0
k_1	0.460	0.620	0.750	0.850	0.920	0.960	0.970
k_2	0.038	0.121	0.237	0.387	0.570	0.786	1.000

Table B.4(b) Approximate flange bending moment, M_{fy}, about section y-axis of one flange and total angle of twist, ϕ_t, of beam for a UDL at an eccentricity, e, to the shear centre vertical axis. Beams are torsionally restrained at supports: $a = \sqrt{\frac{EI_w}{GJ}}$, $h = d - t_f$, $w = w^* N/mm$, $W = wL$.

Case	Range 1	Range 2	Range 3
1 Cantilever (w kN/m; W; t_f; d; e; L; M_{fy}; ϕ_t). Section also applies to cases 2 and 3	$\frac{L}{a} < 0.5$	$0.5 \leqslant \frac{L}{a} \leqslant 3.0$	$\frac{L}{a} > 3.0$
	$M_{fy} = \frac{wL^2e}{2h}$	$M_{fy} = \frac{wLea}{h} k_1$; $\phi_t = \frac{wLea}{GJ} k_2$	$M_{fy} = \frac{wLea}{h}\left[1 - \frac{a}{L}\right]$
	$\phi_t = 0.114 \frac{wLea}{GJ}\left[\frac{L}{a}\right]^3$	see k_1, k_2 values below	$\phi_t = \frac{wLea}{GJ}\left[\frac{L}{2a} - 1 + \frac{a}{L}\right]$
2 Simple beam (w kN/m; L; M_{fy}; ϕ_t)	$\frac{L}{a} < 1.0$	$1.0 \leqslant \frac{L}{a} \leqslant 6.0$	$\frac{L}{a} > 6.0$
	$M_{fy} = \frac{wL^2e}{8h}$	$M_{fy} = \frac{wLea}{h} k_1$; $\phi_t = \frac{wLea}{GJ} k_2$	$M_{fy} = \frac{wea^2}{h}$
	$\phi_t = 0.094 \frac{wLea}{GJ}\left[\frac{L}{2a}\right]^3$	see k_1, k_2 values below	$\phi_t = \frac{wLea}{GJ}\left[\frac{L}{8a} - \frac{a}{L}\right]$
3 Continuous spans, equal length, equally loaded (w kN/m; L; M_{fy}; ϕ_t)	$\frac{L}{a} < 2.0$	$2.0 \leqslant \frac{L}{a} \leqslant 8.0$	$\frac{L}{a} > 8.0$
	$M_{fy} = \frac{wL^2e}{12h}$	$M_{fy} = \frac{wLea}{h} k_1$; $\phi_t = \frac{wLea}{2GJ} k_2$	$M_{fy} = \frac{wLea}{h}\left[\frac{1}{2} - \frac{a}{L}\right]$
	$\phi_t = 0.151 \frac{wLea}{GJ}\left[\frac{L}{4a}\right]^3$	see k_1, k_2 values below	$\phi_t = \frac{wLea}{GJ}\left[\frac{L}{8a} - \frac{1}{2}\right]$

Case 1, $0.5 \leqslant \frac{L}{a} \leqslant 3.0$:

$\frac{L}{a}$	0.5	1.0	1.5	2.0	2.5	3.0
k_1	0.236	0.396	0.523	0.615	0.674	0.698
k_2	0.013	0.092	0.214	0.379	0.587	0.838

Case 2, $1.0 \leqslant \frac{L}{a} \leqslant 6.0$:

$\frac{L}{a}$	1.0	2.0	3.0	4.0	5.0	6.0
k_1	0.138	0.166	0.181	0.183	0.173	0.150
k_2	0.012	0.082	0.178	0.300	0.448	0.622

Case 3, $2.0 \leqslant \frac{L}{a} \leqslant 8.0$:

$\frac{L}{a}$	2.0	3.0	4.0	5.0	6.0	7.0	8.0
k_1	0.157	0.218	0.269	0.311	0.343	0.365	0.377
k_2	0.038	0.121	0.237	0.387	0.570	0.786	1.000

The appropriate boundary conditions should be applied to find the function ϕ involving hyperbolic and exponential functions. Standard solutions for the function ϕ for a variety of cases are published. Many aids are available to shorten the labour of computation. Three idealised boundary conditions are possible:

(a) Free end, in which the end of the beam is free to twist and also free to warp. An example is the free tip of a cantilever beam:

$$\phi \neq 0, \phi' \neq 0, \phi'' = 0$$

(b) Pinned end, in which the end of the beam is free to twist and not free to warp:

$$\phi \neq 0, \phi' \neq 0, \phi'' \neq 0$$

Some examples are beam-to-column connections with reasonably shallow web side plates, and flexible end plate connections.

(c) Fixed end, in which the end of the beam is not free to twist and not free to warp:

$$\phi = 0, \phi' = 0, \phi'' \neq 0$$

An example of such a connection is a moment connection.

Charts are available for the direct reading of the values of the torsion functions, ϕ, ϕ', ϕ'' and ϕ''' for given values of L/a, and the position along the beam as a fraction of the span z/L. See Terrington [1970].

B.7.4 Torsion—plastic & limit states design

Procedures have been developed for plastic analysis and design of beams subject to bending and torsion. See Trahair & Pi [1996] and Pi & Trahair [1994] for further information or Centre for Advanced Structural Engineering, The University of Sydney, research reports R683, R685 and R686 (CASE [1993, 1994 a,b]).

Though somewhat mentioned earlier in this Appendix, there are no specific provisions on torsion design in AS 4100, AS 3990 and its predecessor AS 1250. This is reflective of many other similar national codes. Further information on the reasons for the current situation on torsion design can be found in Clause C8.5 of the AS 4100 Commentary. This publication also gives some very good guidance on dealing with torsional loadings (either singly or in combination with bending) in a limit states design format. It also highlights some other references (as noted above) for practical design problems calculated in limit states that are compatible with AS 4100.

B.8 Further reading

- AS 3990 (previously AS 1250) should be consulted for elastic design using permissible stress principles. Lay [1982b] is good reference to reflect on the background of this Standard.
- Gorenc & Tinyou (1984) also provide much useful information on this topic.
- EHA [1986] is a handbook for engineers on *aluminium* in Australia. It includes useful tables, formulae and brief notes on shear centre, torsion constants, maximum stress,

ultimate torque, and warping constants, over a largish range of (unusual) structural shapes (sections), roots and bulbs (as welds).

- Additional references for torsional analysis and design are Schmith et al. [1970] and Hens & Seaberg [1983].
- Various formulae for stress and strain analysis in uniaxial, biaxial and triaxial stress systems can be found in Young & Budynas [2002].
- There are numerous texts on behaviour/strength of materials and associated failure criterion for structural elements subject to biaxial and triaxial stresses. Some suggested good base texts on the topic include Popov [1978], Crandall et al. [1978] and Hall [1984].

Appendix C

Design Aids

C.1 Contents

Appendix C contains the following sub-sections, information and design aids:

C.2 Beam formulae: Moments, shear forces & deflections

Table C.2.1 (a) Moments, shear forces and deflections—cantilevers

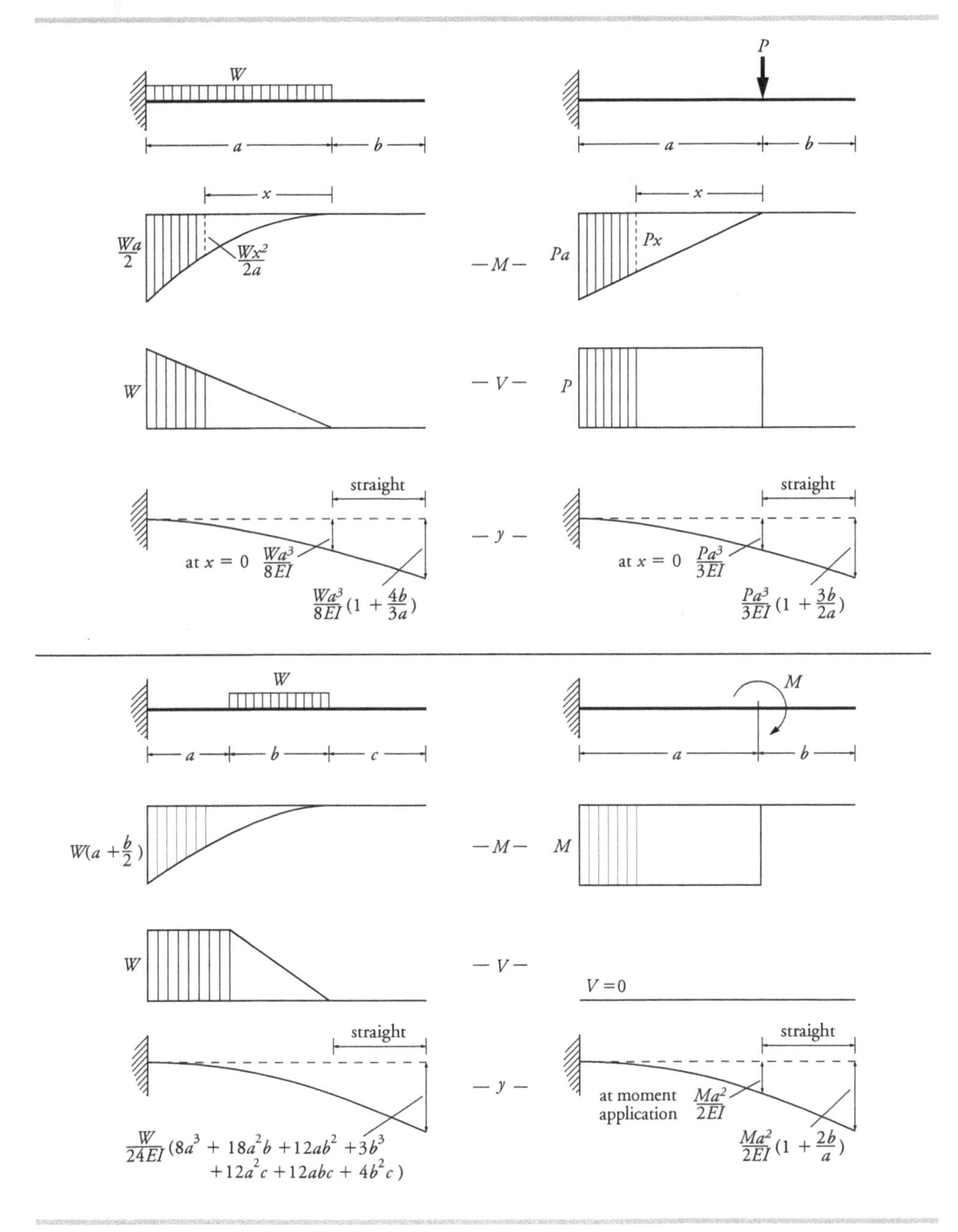

Note: M = bending moment; V = shear force; y = deflection. See also Syam [1992].

Table C.2.1 (b) Moments, shear forces and deflections—simply supported beams

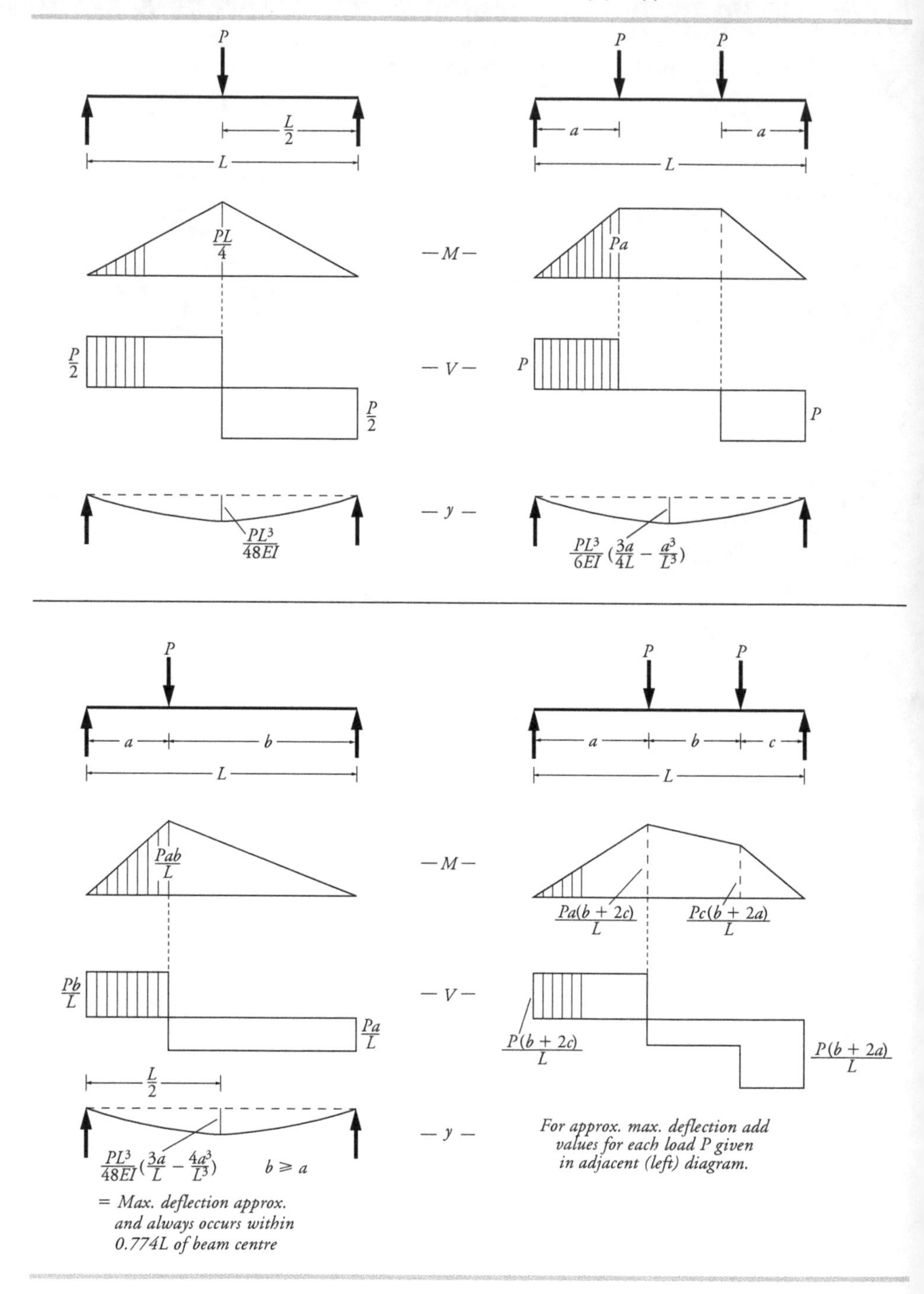

Note: M = bending moment; V = shear force; y = deflection. See also Syam [1992].

Table C.2.1 (b) Moments, shear forces and deflections—simply supported beams (continued)

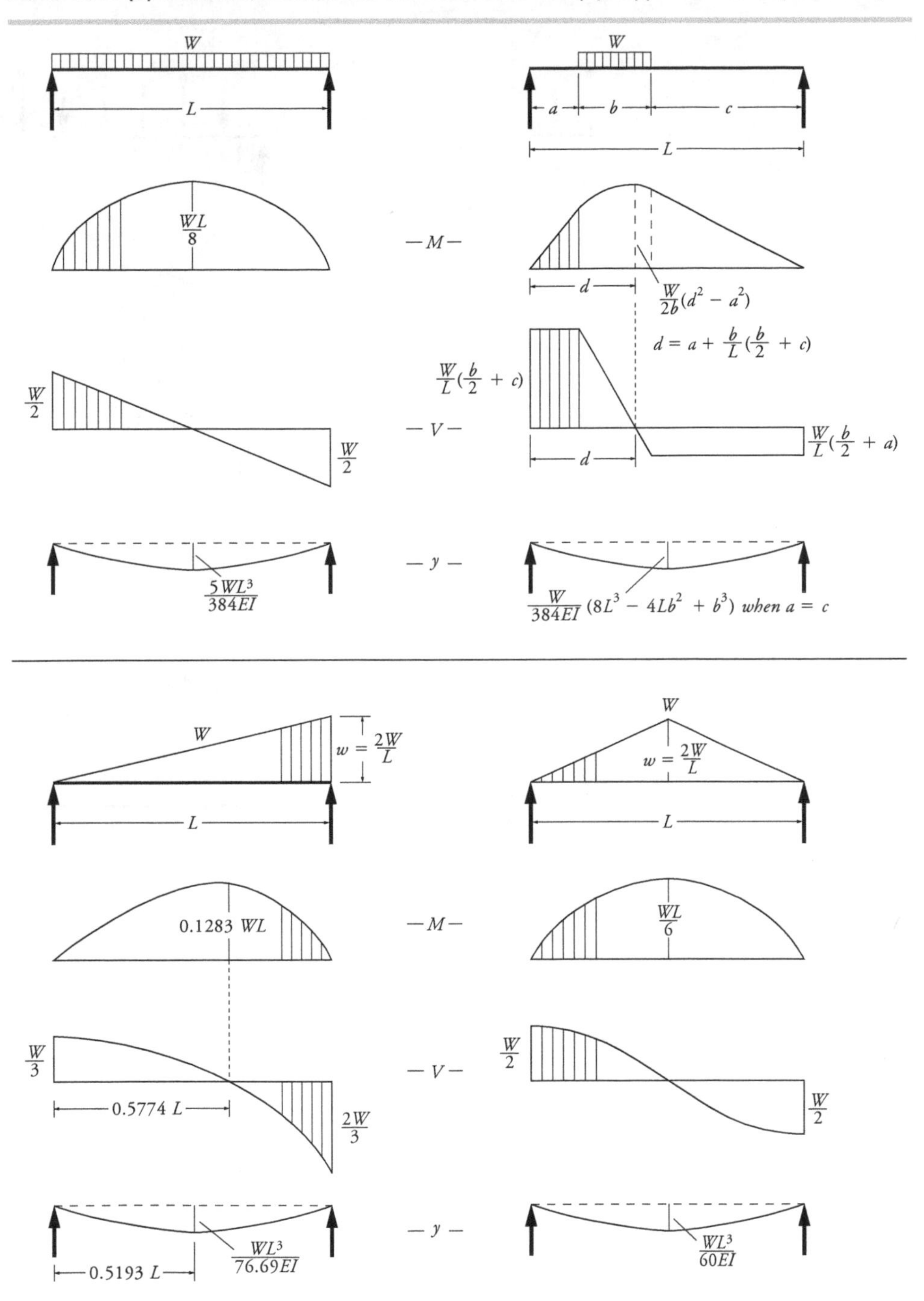

Note: M = bending moment; V = shear force; y = deflection. See also Syam [1992].

Table C.2.1 (b) Moments, shear forces and deflections—simply supported beams (continued)

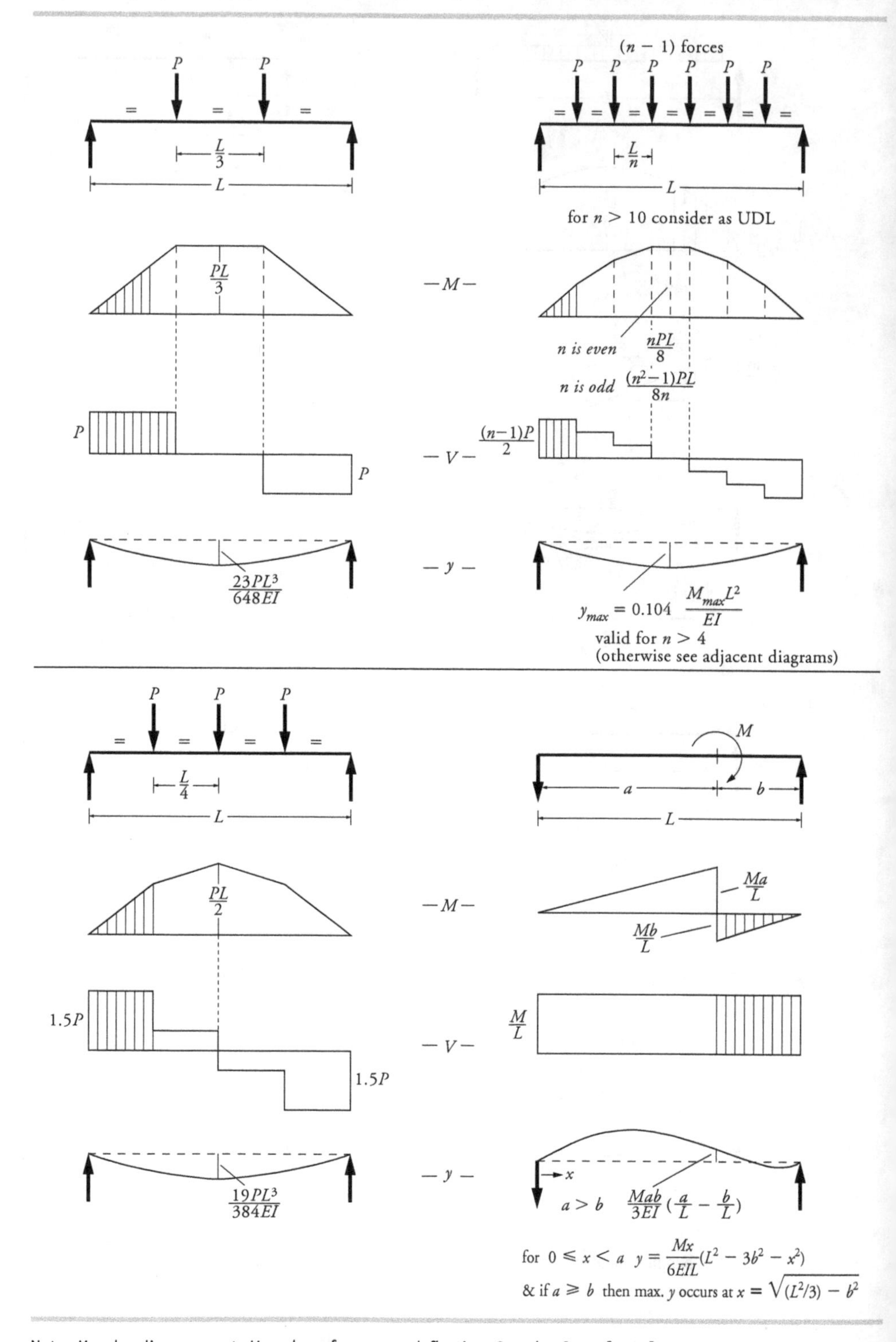

Note: M = bending moment; V = shear force; y = deflection. See also Syam [1992].

Table C.2.1 (c) Moments, shear forces and deflections—simply supported beams with overhang

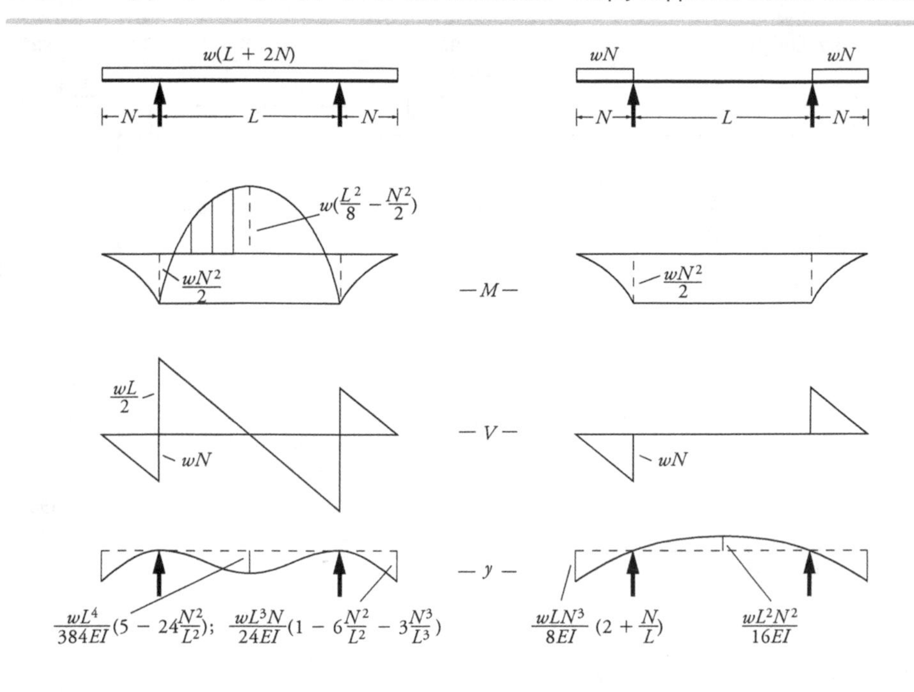

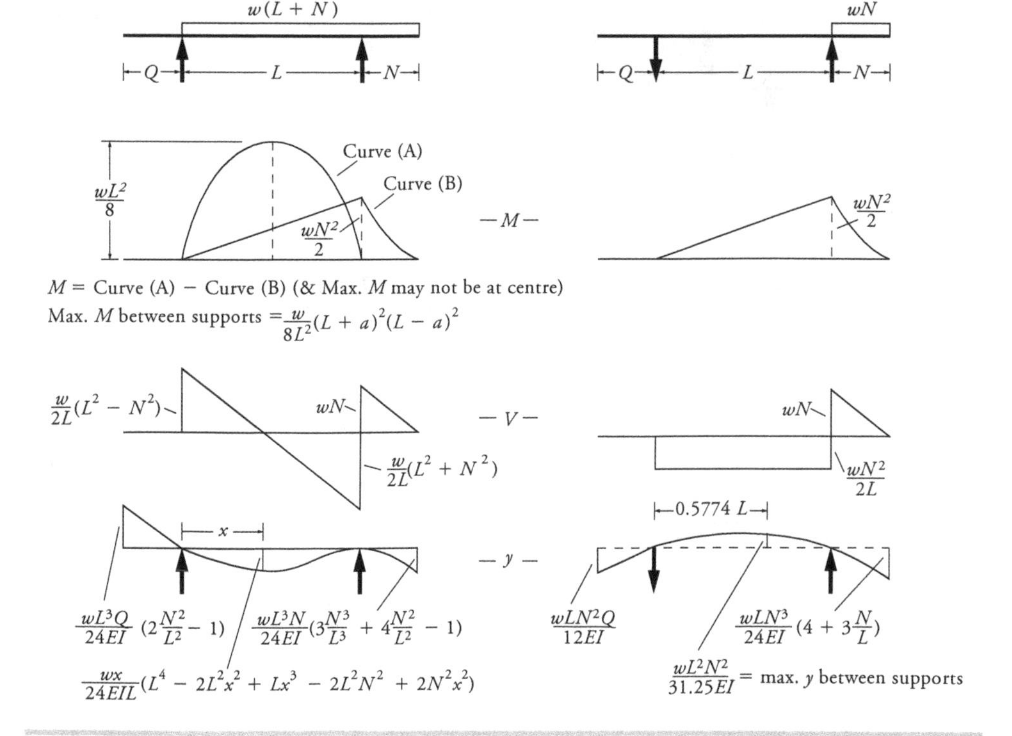

Note: M = bending moment; V = shear force; y = deflection. See also Syam [1992].

Table C.2.1 (d) Moments, shear forces and deflections—continuous beams

Loaded spans	Reactions, moments, deflections	Loading pattern on loaded spans: W (uniform)	W (central point load)	W W (two point loads)
1 2, L L, A B C	A	0.375	0.313	0.667
	B	1.250	1.375	2.667
	M_1	0.070 at 0.375 from A	0.156	0.222
	M_B	−0.125	−0.188	−0.333
	Δ_1	0.0054 at 0.421 from A		
A 1 B 2 C	A	0.438	0.406	0.833
	B	0.625	0.688	1.337
	C	−0.063	−0.094	0.167
	M_1	0.096 at 0.438 from A	0.203	0.278
	M_B	−0.063	−0.094	−0.167
	Δ_1	0.0092 at 0.472 from A		
1, L L L, A B C D	A	0.400	0.350	0.733
	B	1.100	1.150	2.267
	M_1	0.080 at 0.400 from A	0.175	0.244
	M_B	−0.100	−0.150	−0.267
	Δ_1	0.0069 at 0.446 from A		
A 1 B C D	A	0.450	0.425	0.867
	B	0.550	0.575	1.133
	M_1	0.101 at 0.450 from A	0.213	0.289
	M_B	−0.050	−0.075	−0.133
	Δ_1	0.0099 at 0.479 from A		
A B 2 C D	A	−0.050	−0.075	−0.133
	B	0.550	0.575	1.133
	M_2	0.075 at 0.5 from B	0.175	0.200
	M_B	−0.050	−0.075	−0.133
	Δ_2	0.0068 at 0.5 from B		
A 1 B 2 C D	A	0.383	0.325	0.690
	B	1.200	1.300	2.533
	C	0.450	0.425	0.866
	D	−0.033	−0.050	−0.090
	M_1	0.073 at 0.383 from A	0.163	0.230
	M_2	0.054 at 0.583 from B	0.138	0.170
	M_B	−0.117	−0.175	−0.311
	M_C	−0.033	−0.050	−0.089
	Δ_1	0.0059 at 0.43 from A		
A 1 B C D	A	0.433	0.400	0.822
	B	0.650	0.725	1.400
	C	−0.100	−0.150	−0.266
	D	0.017	0.025	0.044
	M_1	0.094 at 0.433 from A	0.200	0.274
	M_B	−0.067	−0.100	−0.178
	M_C	0.017	0.025	0.044
	Δ_1	0.0089 at 0.471 from A		

Note: Reactions A, B, C, etc are in terms of W. Moments M are in terms of WL. Deflections Δ are in terms of WL^3/EI. Locations of M_1, M_2, etc and Δ are in terms of L. L = span. See also Syam [1992].

Table C.2.1 (d) Moments, shear forces and deflections—continuous beams (continued)

Loaded spans	Reactions, moments, deflections	Loading pattern on loaded spans		
		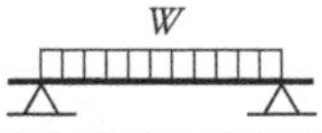 W	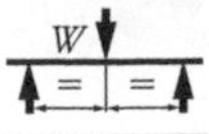 W	 W W
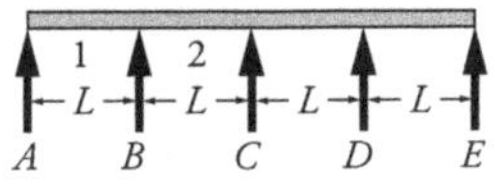 1 2 — L L L L — A B C D E	A	0.393	0.339	0.714
	B	1.143	1.214	2.381
	C	0.929	0.892	1.810
	M_1	0.077 at 0.393 from A	0.170	0.238
	M_2	0.036 at 0.536 from B	0.116	0.111
	M_B	−0.107	−0.161	−0.286
	M_C	−0.071	−0.107	−0.190
	Δ_1	0.0065 at 0.44 from A		
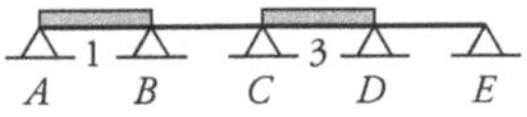 1 3 — A B C D E	A	0.446	0.420	0.857
	B	0.572	0.607	1.192
	C	0.464	0.446	0.904
	D	0.572	0.607	1.192
	E	−0.054	−0.080	−0.144
	M_1	0.098 at 0.446 from A	0.210	0.286
	M_3	0.081 at 0.482 from C	0.183	0.222
	M_B	−0.054	−0.080	−0.143
	M_C	−0.036	−0.054	−0.096
	M_D	−0.054	−0.080	−0.143
	Δ_1	0.0097 at 0.477 from A		
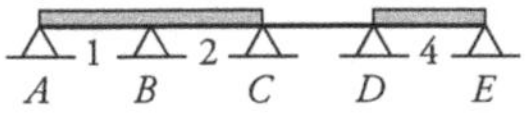 1 2 4 — A B C D E	A	0.380	0.319	0.680
	B	1.223	1.335	2.595
	C	0.357	0.286	0.618
	D	0.598	0.647	1.262
	E	0.442	0.413	0.846
	M_1	0.072 at 0.38 from A	0.160	0.226
	M_2	0.061 at 0.603 from B	0.146	0.194
	M_4	0.098 at 0.558 from D	0.207	0.282
	M_B	−0.121	−0.181	−0.321
	M_C	−0.018	−0.027	−0.048
	M_D	−0.058	−0.087	−0.155
	Δ_4	0.0094 at 0.525 from D		
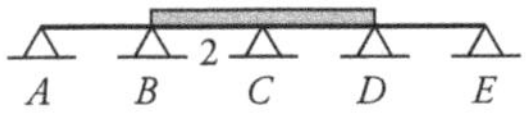 2 — A B C D E	A	−0.036	−0.054	−0.096
	B	0.464	0.446	0.906
	C	1.143	1.214	2.381
	M_2	0.056	0.143	0.174
	M_B	−0.036	−0.054	−0.095
	M_C	−0.107	−0.161	−0.286
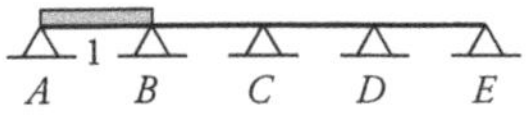 1 — A B C D E	A	0.433	0.400	0.822
	B	0.652	0.728	1.404
	C	−0.107	−0.161	−0.286
	D	0.027	0.040	0.072
	E	−0.005	−0.007	−0.012
	M_1	0.094	0.200	0.274
	M_B	−0.067	−0.100	−0.178
	M_C	0.018	0.027	0.048
	M_D	−0.005	−0.007	−0.012
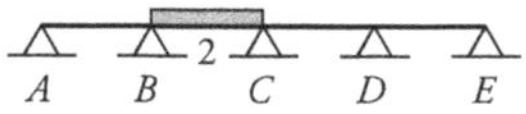 2 — A B C D E	A	−0.049	−0.074	−0.132
	B	0.545	0.567	1.120
	C	0.571	0.607	1.190
	D	−0.080	−0.121	−0.214
	E	0.013	0.020	0.036
	M_2	0.074	0.173	0.198
	M_B	−0.049	−0.074	−0.131
	M_C	−0.054	−0.080	−0.143
	M_D	0.013	0.020	0.036

Note: Reactions A, B, C, etc are in terms of W. Moments M are in terms of WL. Deflections Δ are in terms of WL^3/EI. Locations of M_1, M_2, etc and Δ are in terms of L. L = span. See also Syam [1992].

Table C.2.2 Bending moment values for various load cases

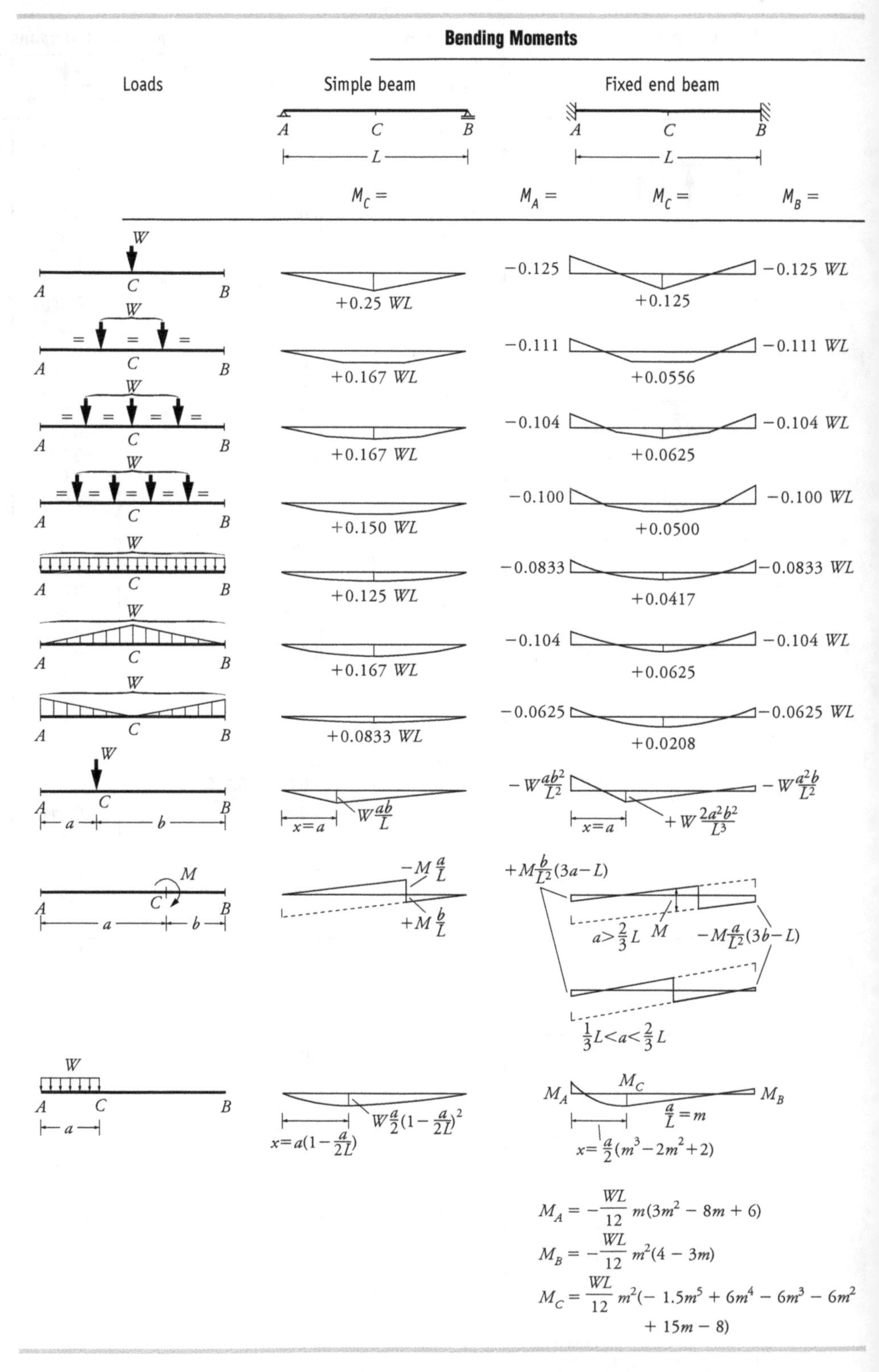

Table C.2.3 Rapid deflection calculation of symmetrical beam sections

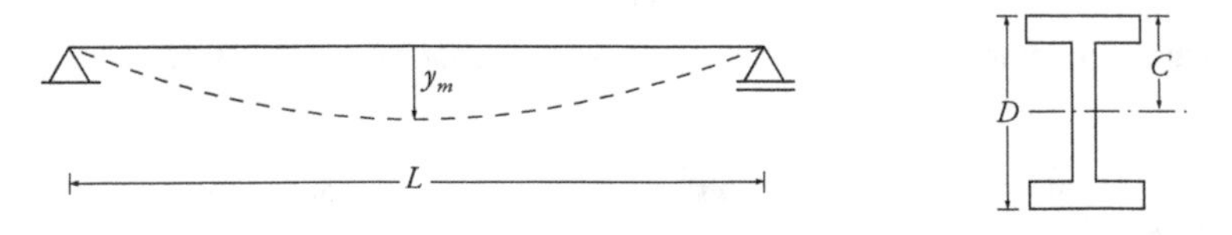

$$y_m = \frac{521 \times 10^3 ML^2}{I} K_1 \text{ ,mm}$$

$$= \frac{521 \times 10^3 ML^2}{ZC} K_1 \text{ ,mm}$$

M = maximum bending moment in table below, kNm
L = span of beam, m
K_1 = deflection coefficient in table below
I = second moment of area, mm^4 (note the units) = ZC
Z = elastic section modulus, mm^3 (note the units)
C = length from neutral axis to outer fibre, mm (see figure)
E = Young's modulus of elasticity
= 200×10^6 kN/m^2 (assumed in the equation)

Case	Load diagram	B.M. diagram	Case	M	K_1
1	W	M	1	$\frac{WL}{8}$	1.00
2	P P P, = = = =	M	2	$\frac{PL}{2}$	0.950
3	P P, = = =	M	3	$\frac{PL}{3}$	1.02
4	P P, = = = =	M	4	$\frac{PL}{4}$	1.10
5	P, = =	M	5	$\frac{PL}{4}$	0.800
6	P, = = =	M	6	$\frac{2PL}{9}$	0.767
7	P, = = = =	M	7	$\frac{3PL}{16}$	0.733
8	M	M	8	M	0.595
9	M M	M	9	M	1.20

Note: To use the Table –

1. Determine the Case type.
2. Evaluate M (in kNm) and K_1 for the respective Case.
3. Obtain I (or ZC) (in mm multiples) from external references/evaluation.
4. Calculate the maximum beam deflection y_m from the above equation.

C.3 Section properties & AS 4100 design section capacities

Table C.3.1 Section properties and AS 4100 design section capacities: WB—Grade 300

	Section Properties										
Designation & Mass per m	Gross Area of Cross-Section	About *x*-axis				About *y*-axis				Torsion Constant	Warping Constant
	A_g	I_x	Z_x	S_x	r_x	I_y	Z_y	S_y	r_y	J	I_w
kg/m	mm^2	10^6mm^4	10^3mm^3	10^3mm^3	mm	10^6mm^4	10^3mm^3	10^3mm^3	mm	10^3mm^4	10^9mm^6
1200 WB 455	57900	15300	25600	28200	515	834	3330	5070	120	22000	280000
423	53900	13900	23300	25800	508	750	3000	4570	118	16500	251000
392	49900	12500	21100	23400	500	667	2670	4070	116	12100	221000
342	43500	10400	17500	19800	488	342	1710	2630	88.6	9960	113000
317	40300	9250	15700	17900	479	299	1500	2310	86.1	7230	98500
278	35400	7610	13000	15000	464	179	1020	1600	71.1	5090	58700
249	31700	6380	10900	12900	449	87.0	633	1020	52.4	4310	28500
1000 WB 322	41000	7480	14600	16400	427	342	1710	2620	91.3	9740	84100
296	37800	6650	13100	14800	420	299	1490	2300	89.0	7010	73000
258	32900	5430	10700	12300	406	179	1020	1590	73.8	4870	43400
215	27400	4060	8120	9570	385	90.3	602	961	57.5	2890	21700
900 WB 282	35900	5730	12400	13600	399	341	1710	2590	97.5	8870	67900
257	32700	5050	11000	12200	393	299	1490	2270	95.6	6150	58900
218	27800	4060	8930	9960	382	179	1020	1560	80.2	4020	35000
175	22300	2960	6580	7500	364	90.1	601	931	63.5	2060	17400
800 WB 192	24400	2970	7290	8060	349	126	840	1280	71.9	4420	19600
168	21400	2480	6140	6840	341	86.7	631	964	63.7	2990	13400
146	18600	2040	5100	5730	331	69.4	505	775	61.1	1670	10600
122	15600	1570	3970	4550	317	41.7	334	519	51.7	921	6280
700 WB 173	22000	2060	5760	6390	306	97.1	706	1080	66.4	4020	11500
150	19100	1710	4810	5370	299	65.2	521	798	58.4	2690	7640
130	16600	1400	3990	4490	290	52.1	417	642	56.0	1510	6030
115	14600	1150	3330	3790	281	41.7	334	516	53.5	888	4770

Note: For dimensions, other design capacities and related information see ASI [2009a] or OneSteel [2011].

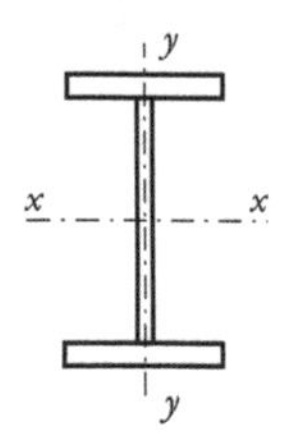

Properties for Design to AS 4100							Design Section Capacities to AS 4100					
Yield Stress		**Form Factor**	**About *x*-axis**		**About *y*-axis**		**Design Section Axial Capacities**		**Design Section Moment Capacity about**		**Design Shear Capacity along**	**Designation & Mass per m**
Flange	Web		Compact-ness		Compact-ness		Tens	Comp	*x*-axis	*y*-axis	*y*-axis	
f_{yf}	f_{yw}	k_f		Z_{ex}		Z_{ey}	ϕN_t	ϕN_s	ϕM_{sx}	ϕM_{sy}	ϕV_{vx}	
MPa	MPa	–	(C,N,S)	10^3mm^3	(C,N,S)	10^3mm^3	kN	kN	kNm	kNm	kN	kg/m
280	300	0.837	C	28200	C	5000	14600	12200	7110	1260	2900	1200 WB 455
280	300	0.825	C	25800	C	4500	13600	11200	6510	1130	2900	423
280	300	0.811	C	23400	N	4000	12600	10200	5910	1010	2900	392
280	300	0.783	C	19800	C	2560	11000	8580	4980	646	2900	342
280	300	0.766	C	17900	C	2240	10200	7780	4500	565	2900	317
280	300	0.733	C	15000	C	1530	8930	6540	3790	387	2900	278
280	300	0.701	C	12900	C	949	7980	5600	3250	239	2900	249
280	300	0.832	C	16400	C	2560	10300	8580	4130	646	2490	1000 WB 322
280	300	0.817	C	14800	C	2240	9520	7780	3720	565	2490	296
280	300	0.790	C	12300	C	1530	8280	6540	3100	387	2490	258
300	300	0.738	C	9570	C	903	7390	5450	2580	244	2490	215
280	310	0.845	C	13600	C	2560	9050	7650	3440	645	1730	900 WB 282
280	310	0.830	C	12200	C	2240	8250	6840	3070	565	1730	257
280	310	0.800	C	9960	C	1530	7010	5610	2510	386	1730	218
300	310	0.744	C	7500	C	901	6030	4480	2020	243	1730	175
280	310	0.824	C	8060	C	1260	6150	5070	2030	318	1190	800 WB 192
280	310	0.799	C	6840	C	946	5380	4300	1720	238	1190	168
300	310	0.763	N	5710	C	757	5020	3830	1540	204	1190	146
300	310	0.718	N	4530	N	498	4210	3020	1220	135	1190	122
280	310	0.850	C	6390	C	1060	5540	4710	1610	267	1100	700 WB 173
280	310	0.828	C	5370	C	782	4810	3980	1350	197	1100	150
300	310	0.795	C	4490	C	626	4480	3560	1210	169	1100	130
300	310	0.767	C	3790	N	498	3940	3020	1020	134	1100	115

Table C.3.2 Section properties and AS 4100 design section capacities: WC - Grade 300

	Section Properties										
Designation & Mass per m	**Gross Area of Cross-Section**	**About *x*-axis**				**About *y*-axis**				**Torsion Constant**	**Warping Constant**
	A_g	I_x	Z_x	S_x	r_x	I_y	Z_y	S_y	r_y	J	I_w
kg/m	mm²	10⁶mm⁴	10³mm³	10³mm³	mm	10⁶mm⁴	10³mm³	10³mm³	mm	10³mm⁴	10⁹mm⁶
500 WC 440	56000	2150	8980	10400	196	835	3340	5160	122	30100	40400
414	52800	2110	8800	10100	200	834	3340	5100	126	25400	40400
383	48800	1890	7990	9130	197	751	3000	4600	124	19900	35700
340	43200	2050	7980	8980	218	667	2670	4070	124	13100	38800
290	37000	1750	6930	7700	218	584	2330	3540	126	8420	33300
267	34000	1560	6250	6950	214	521	2080	3170	124	6370	29400
228	29000	1260	5130	5710	208	417	1670	2540	120	3880	23000
400 WC 361	46000	1360	6340	7460	172	429	2140	3340	96.5	24800	16300
328	41800	1320	6140	7100	178	427	2140	3270	101	19200	16200
303	38600	1180	5570	6420	175	385	1920	2950	99.8	14800	14300
270	34400	1030	4950	5660	173	342	1710	2610	99.8	10400	12500
212	27000	776	3880	4360	169	267	1330	2040	99.4	5060	9380
181	23000	620	3180	3570	164	214	1070	1640	96.4	3080	7310
144	18400	486	2550	2830	163	171	854	1300	96.3	1580	5720
350 WC 280	35700	747	4210	4940	145	286	1640	2500	89.6	16500	7100
258	32900	661	3810	4450	142	258	1470	2260	88.5	12700	6230
230	29300	573	3380	3910	140	229	1310	2000	88.4	8960	5400
197	25100	486	2940	3350	139	200	1140	1740	89.3	5750	4600

Note: For dimensions, other design capacities and related information see ASI [2009a] or OneSteel [2011].

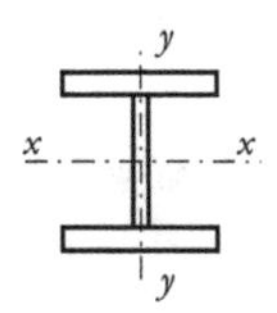

Properties for Design to AS 4100							Design Section Capacities to AS 4100						
Yield Stress		Form Factor	About *x*-axis		About *y*-axis		Design Section Axial Capacities		Design Section Moment Capacity about		Design Shear Capacity along	Designation & Mass per m
Flange	Web		Compactness		Compactness		Tens	Comp	*x*-axis	*y*-axis	*y*-axis	
f_{yf}	f_{yw}	k_f		Z_{ex}		Z_{ey}	ϕN_t	ϕN_s	ϕM_{sx}	ϕM_{sy}	ϕV_{vx}	
MPa	MPa	–	(C,N,S)	10^3mm^3	(C,N,S)	10^3mm^3	kN	kN	kNm	kNm	kN	kg/m
280	280	1.00	C	10400	C	5010	14100	14100	2620	1260	2420	500 WC 440
280	280	1.00	C	10100	C	5010	13300	13300	2540	1260	1940	414
280	280	1.00	C	9130	C	4510	12300	12300	2300	1140	1940	383
280	280	1.00	C	8980	C	4000	10900	10900	2260	1010	1700	340
280	300	1.00	N	7570	N	3410	9320	9320	1910	860	1460	290
280	300	1.00	N	6700	N	2970	8570	8570	1690	747	1460	267
300	300	1.00	N	5210	N	2200	7830	7830	1410	593	1460	228
280	280	1.00	C	7470	C	3210	11600	11600	1880	810	2120	400 WC 361
280	280	1.00	C	7100	C	3200	10500	10500	1790	808	1480	328
280	280	1.00	C	6420	C	2880	9730	9730	1620	727	1480	303
280	280	1.00	C	5660	C	2560	8660	8660	1430	646	1320	270
280	300	1.00	N	4360	N	2000	6800	6800	1100	504	1130	212
300	300	1.00	N	3410	N	1510	6210	6210	922	408	1130	181
300	300	1.00	N	2590	N	1120	4970	4970	698	303	907	144
280	280	1.00	C	4940	C	2450	9000	9000	1240	618	1160	350 WC 280
280	280	1.00	C	4450	C	2210	8290	8290	1120	557	1160	258
280	280	1.00	C	3910	C	1960	7380	7380	986	495	1040	230
280	300	1.00	C	3350	C	1720	6330	6330	844	433	891	197

Table C.3.3 Section properties and AS 4100 design section capacities: UB - Grade 300

	Section Properties										
Designation & Mass per m	**Gross Area of Cross-Section**	**About x-axis**				**About y-axis**				**Torsion Constant**	**Warping Constant**
	A_g	I_x	Z_x	S_x	r_x	I_y	Z_y	S_y	r_y	J	I_w
kg/m	mm^2	$10^6 mm^4$	$10^3 mm^3$	$10^3 mm^3$	mm	$10^6 mm^4$	$10^3 mm^3$	$10^3 mm^3$	mm	$10^3 mm^4$	$10^9 mm^6$
610 UB 125	16000	986	3230	3680	249	39.3	343	536	49.6	1560	3450
113	14500	875	2880	3290	246	34.3	300	469	48.7	1140	2980
101	13000	761	2530	2900	242	29.3	257	402	47.5	790	2530
530 UB 92.4	11800	554	2080	2370	217	23.8	228	355	44.9	775	1590
82.0	10500	477	1810	2070	213	20.1	193	301	43.8	526	1330
460 UB 82.1	10500	372	1610	1840	188	18.6	195	303	42.2	701	919
74.6	9520	335	1460	1660	188	16.6	175	271	41.8	530	815
67.1	8580	296	1300	1480	186	14.5	153	238	41.2	378	708
410 UB 59.7	7640	216	1060	1200	168	12.1	135	209	39.7	337	467
53.7	6890	188	933	1060	165	10.3	115	179	38.6	234	394
360 UB 56.7	7240	161	899	1010	149	11.0	128	198	39.0	338	330
50.7	6470	142	798	897	148	9.60	112	173	38.5	241	284
44.7	5720	121	689	777	146	8.10	94.7	146	37.6	161	237
310 UB 46.2	5930	100	654	729	130	9.01	109	166	39.0	233	197
40.4	5210	86.4	569	633	129	7.65	92.7	142	38.3	157	165
32.0	4080	63.2	424	475	124	4.42	59.3	91.8	32.9	86.5	92.9
250 UB 37.3	4750	55.7	435	486	108	5.66	77.5	119	34.5	158	85.2
31.4	4010	44.5	354	397	105	4.47	61.2	94.2	33.4	89.3	65.9
25.7	3270	35.4	285	319	104	2.55	41.1	63.6	27.9	67.4	36.7
200 UB 29.8	3820	29.1	281	316	87.3	3.86	57.5	88.4	31.8	105	37.6
25.4	3230	23.6	232	260	85.4	3.06	46.1	70.9	30.8	62.7	29.2
22.3	2870	21.0	208	231	85.5	2.75	41.3	63.4	31.0	45.0	26.0
18.2	2320	15.8	160	180	82.6	1.14	23.0	35.7	22.1	38.6	10.4
180 UB 22.2	2820	15.3	171	195	73.6	1.22	27.1	42.3	20.8	81.6	8.71
18.1	2300	12.1	139	157	72.6	0.975	21.7	33.7	20.6	44.8	6.80
16.1	2040	10.6	123	138	72.0	0.853	19.0	29.4	20.4	31.5	5.88
150 UB 18.0	2300	9.05	117	135	62.8	0.672	17.9	28.2	17.1	60.5	3.56
14.0	1780	6.66	88.8	102	61.1	0.495	13.2	20.8	16.6	28.1	2.53

Note: For dimensions, other design capacities and related information see ASI [2009a] or OneSteel [2011].

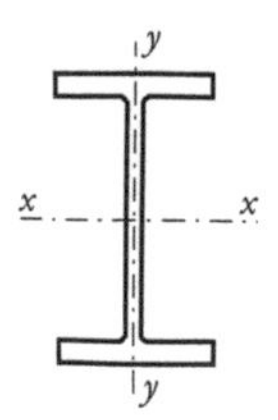

Properties for Design to AS 4100							Design Section Capacities to AS 4100					
Yield Stress		Form Factor	About *x*-axis		About *y*-axis		Design Section Axial Capacities		Design Section Moment Capacity about		Design Shear Capacity along	Designation & Mass per m
Flange	Web		Compact-ness		Compact-ness		Tens	Comp	*x*-axis	*y*-axis	*y*-axis	
f_{yf}	f_{yw}	k_f		Z_{ex}		Z_{ey}	ϕN_t	ϕN_s	ϕM_{sx}	ϕM_{sy}	ϕV_{vx}	
MPa	MPa	—	(C,N,S)	10^3mm^3	(C,N,S)	10^3mm^3	kN	kN	kNm	kNm	kN	kg/m
280	300	0.950	C	3680	C	515	4020	3820	927	130	1180	610 UB 125
280	300	0.926	C	3290	C	451	3650	3370	829	114	1100	113
300	320	0.888	C	2900	C	386	3510	3110	782	104	1100	101
300	320	0.928	C	2370	C	342	3190	2960	640	92.2	939	530 UB 92.4
300	320	0.902	C	2070	C	289	2840	2560	558	78.0	876	82.0
300	320	0.979	C	1840	C	292	2830	2770	496	79.0	788	460 UB 82.1
300	320	0.948	C	1660	C	262	2570	2440	449	70.8	719	74.6
300	320	0.922	C	1480	C	230	2320	2130	399	62.0	667	67.1
300	320	0.938	C	1200	C	203	2060	1940	324	54.8	548	410 UB 59.7
320	320	0.913	C	1060	C	173	1980	1810	304	49.8	529	53.7
300	320	0.996	C	1010	C	193	1960	1950	273	52.0	496	360 UB 56.7
300	320	0.963	C	897	C	168	1750	1680	242	45.5	449	50.7
320	320	0.930	N	770	N	140	1650	1530	222	40.4	420	44.7
300	320	0.991	C	729	C	163	1600	1590	197	44.0	356	310 UB 46.2
320	320	0.952	C	633	C	139	1500	1430	182	40.0	320	40.4
320	320	0.915	N	467	N	86.9	1180	1070	134	25.0	283	32.0
320	320	1.00	C	486	C	116	1370	1370	140	33.5	283	250 UB 37.3
320	320	1.00	N	395	N	91.4	1150	1150	114	26.3	265	31.4
320	320	0.949	C	319	C	61.7	941	893	92.0	17.8	214	25.7
320	320	1.00	C	316	C	86.3	1100	1100	90.9	24.9	225	200 UB 29.8
320	320	1.00	N	259	N	68.8	930	930	74.6	19.8	204	25.4
320	320	1.00	N	227	N	60.3	826	826	65.3	17.4	174	22.3
320	320	0.990	C	180	C	34.4	668	661	51.8	9.92	154	18.2
320	320	1.00	C	195	C	40.7	813	813	56.2	11.7	186	180 UB 22.2
320	320	1.00	C	157	C	32.5	663	663	45.2	9.36	151	18.1
320	320	1.00	C	138	C	28.4	589	589	39.8	8.19	135	16.1
320	320	1.00	C	135	C	26.9	661	661	38.9	7.74	161	150 UB 18.0
320	320	1.00	C	102	C	19.8	514	514	29.3	5.70	130	14.0

Table C.3.4 Section properties and AS 4100 design section capacities: UC - Grade 300

	Section Properties										
Designation & Mass per m	**Gross Area of Cross-Section**	**About *x*-axis**				**About *y*-axis**				**Torsion Constant**	**Warping Constant**
	A_g	I_x	Z_x	S_x	r_x	I_y	Z_y	S_y	r_y	J	I_w
kg/m	mm²	10⁶mm⁴	10³mm³	10³mm³	mm	10⁶mm⁴	10³mm³	10³mm³	mm	10³mm⁴	10⁹mm⁶
310 UC 158	20100	388	2370	2680	139	125	807	1230	78.9	3810	2860
137	17500	329	2050	2300	137	107	691	1050	78.2	2520	2390
118	15000	277	1760	1960	136	90.2	588	893	77.5	1630	1980
96.8	12400	223	1450	1600	134	72.9	478	725	76.7	928	1560
250 UC 89.5	11400	143	1100	1230	112	48.4	378	575	65.2	1040	713
72.9	9320	114	897	992	111	38.8	306	463	64.5	586	557
200 UC 59.5	7620	61.3	584	656	89.7	20.4	199	303	51.7	477	195
52.2	6660	52.8	512	570	89.1	17.7	174	264	51.5	325	166
46.2	5900	45.9	451	500	88.2	15.3	151	230	51.0	228	142
150 UC 37.2	4730	22.2	274	310	68.4	7.01	91.0	139	38.5	197	39.6
30.0	3860	17.6	223	250	67.5	5.62	73.4	112	38.1	109	30.8
23.4	2980	12.6	166	184	65.1	3.98	52.4	80.2	36.6	50.2	21.1
100 UC 14.8	1890	3.18	65.6	74.4	41.1	1.14	22.9	35.2	24.5	34.9	2.30

Note: For dimensions, other design capacities and related information see ASI [2009a] or OneSteel [2011].

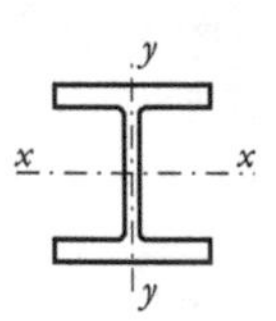

Properties for Design to AS 4100							**Design Section Capacities to AS 4100**					
Yield Stress		**Form Factor**	**About x-axis**		**About y-axis**		**Design Section Axial Capacities**		**Design Section Moment Capacity about**		**Design Shear Capacity along**	**Designation & Mass per m**
Flange	Web		Compact-ness		Compact-ness		Tens	Comp	x-axis	y-axis	y-axis	
f_{yf}	f_{yw}	k_f		Z_{ex}		Z_{ey}	ϕN_t	ϕN_s	ϕM_{sx}	ϕM_{sy}	ϕV_{vx}	
MPa	MPa	–	(C,N,S)	10^3mm^3	(C,N,S)	10^3mm^3	kN	kN	kNm	kNm	kN	kg/m
280	300	1.00	C	2680	C	1210	5070	5070	676	305	832	310 UC 158
280	300	1.00	C	2300	C	1040	4400	4400	580	261	717	137
280	300	1.00	C	1960	C	882	3780	3780	494	222	606	118
300	320	1.00	N	1560	N	694	3340	3340	422	187	527	96.8
280	320	1.00	C	1230	C	567	2870	2870	309	143	472	250 UC 89.5
300	320	1.00	N	986	N	454	2520	2520	266	123	377	72.9
300	320	1.00	C	656	C	299	2060	2060	177	80.6	337	200 UC 59.5
300	320	1.00	C	570	C	260	1800	1800	154	70.3	285	52.2
300	320	1.00	N	494	N	223	1590	1590	133	60.3	257	46.2
300	320	1.00	C	310	C	137	1280	1280	83.6	36.9	226	150 UC 37.2
320	320	1.00	C	250	C	110	1110	1110	71.9	31.7	180	30.0
320	320	1.00	N	176	N	73.5	859	859	50.7	21.2	161	23.4
320	320	1.00	C	74.4	C	34.4	543	543	21.4	9.91	83.8	100 UC 14.8

Table C.3.5(a) Section properties and AS 4100 design section capacities: PFC - Grade 300

		Section Properties														
Designation	Mass per metre	Gross Area of Cross-Section	Coordinate of Centroid	Coordinate of Shear Centre	About x-axis				About y-axis					Torsion Constant	Warping Constant
d		A_g	x_L	x_o	I_x	Z_x	S_x	r_x	I_y	Z_{yL}	Z_{yR}	S_y	r_y	J	I_w
mm	kg/m	mm²	mm	mm	10⁶mm⁴	10³mm³	10³mm³	mm	10⁶mm⁴	10³mm³	10³mm³	10³mm³	mm	10³mm⁴	10⁹mm⁶
380 PFC	55.2	7030	27.5	56.7	152	798	946	147	6.48	236	89.4	161	30.4	472	151
300 PFC	40.1	5110	27.2	56.1	72.4	483	564	119	4.04	148	64.4	117	28.1	290	58.2
250 PFC	35.5	4520	28.6	58.5	45.1	361	421	99.9	3.64	127	59.3	107	28.4	238	35.9
230 PFC	25.1	3200	22.6	46.7	26.8	233	271	91.4	1.76	77.8	33.6	61.0	23.5	108	15.0
200 PFC	22.9	2920	24.4	50.5	19.1	191	221	80.9	1.65	67.8	32.7	58.9	23.8	101	10.6
180 PFC	20.9	2660	24.5	50.3	14.1	157	182	72.9	1.51	61.5	29.9	53.8	23.8	81.4	7.82
150 PFC	17.7	2250	24.9	51.0	8.34	111	129	60.8	1.29	51.6	25.7	46.0	23.9	54.9	4.59
125 PFC	11.9	1520	21.8	45.0	3.97	63.5	73.0	51.1	0.658	30.2	15.2	27.2	20.8	23.1	1.64
100 PFC	8.33	1060	16.7	33.9	1.74	34.7	40.3	40.4	0.267	16.0	8.01	14.4	15.9	13.2	0.424
75 PFC	5.92	754	13.7	27.2	0.683	18.2	21.4	30.1	0.120	8.71	4.56	8.20	12.6	8.13	0.106

Note:

(1) For dimensions, other design capacities and related information see ASI [2009a] or OneSteel [2011].

(2) x_L is the distance from the back of web to centroid and x_o is from the centroid to the shear centre along the x-axis.

(3) Z_{yL} is the elastic section modulus about the y-axis to the PFC web and Z_{yR} is to the PFC toes.

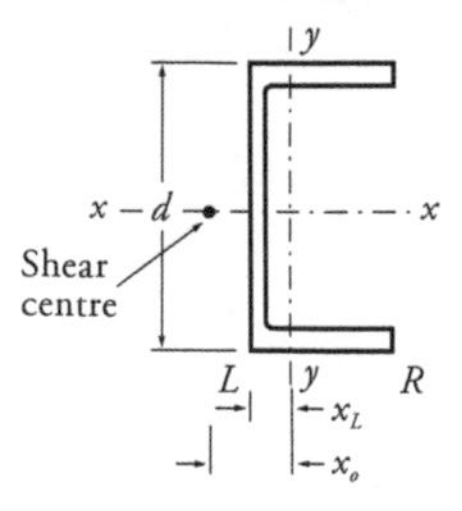

Table C.3.5(b) Section properties and AS 4100 design section capacities: PFC - Grade 300

		Properties for Design to AS 4100						Design Section Capacities to AS 4100					
Designation	Mass per metre	Yield Stress		Form Factor	About x-axis	About y-axis		Design Section Axial Capacities		Design Section Moment Capacity about			Design Shear Capacity along
		Flange	Web					Tens	Comp	x-axis	y-axis		y-axis
d		f_{yf}	f_{yw}	k_f	Z_{ex}	Z_{eyL}	Z_{eyR}	ϕN_t	ϕN_s	ϕM_{sx}	ϕM_{syL}	ϕM_{syR}	ϕV_{vx}
mm	kg/m	MPa	MPa	-	10³mm³	10³mm³	10³mm³	kN	kN	kNm	kNm	kNm	kN
380 PFC	55.2	280	320	1.00	946	115	134	1770	1770	238	28.9	33.8	657
300 PFC	40.1	300	320	1.00	564	82.3	96.6	1380	1380	152	22.2	26.1	415
250 PFC	35.5	300	320	1.00	421	88.7	89.0	1220	1220	114	24.0	24.0	346
230 PFC	25.1	300	320	1.00	271	45.1	50.4	864	864	73.3	12.2	13.6	258
200 PFC	22.9	300	320	1.00	221	46.7	49.1	788	788	59.7	12.6	13.2	207
180 PFC	20.9	300	320	1.00	182	44.9	44.8	718	718	49.0	12.1	12.1	187
150 PFC	17.7	320	320	1.00	129	38.5	38.5	645	649	37.0	11.1	11.1	156
125 PFC	11.9	320	320	1.00	72.8	22.8	22.8	435	438	21.0	6.58	6.56	102
100 PFC	8.33	320	320	1.00	40.3	12.0	12.0	304	306	11.6	3.46	3.46	72.6
75 PFC	5.92	320	320	1.00	21.4	6.84	6.84	216	217	6.16	1.97	1.97	49.2

Table C.3.6(a) Section properties and AS 4100 design section capacities: CHS - Grade C350L0

Dimensions and Ratios		Section Properties						Properties for Design to AS 4100			Design Section Capacities to AS 4100				
Designation	Mass per m	Gross Section Area	About any axis				Torsion Constant	Form Factor	About any axis Compactness		Design Section Axial Capacities		Des. Section Moment Capacity	Design Shear Capacity	Torsion
d_o t		A_g	I	Z	S	r	J	k_f		Z_e	ϕN_t	ϕN_s	ϕM_s	ϕV_v	ϕM_z
mm mm	kg/m	mm²	10⁶mm⁴	10³mm³	10³mm³	mm	10⁶mm⁴		(C,N,S)	10³mm³	kN	kN	kNm	kN	kNm
508.0 × 12.7 CHS	155	19800	606	2390	3120	175	1210	1.00	N	3050	6220	6220	962	2240	902
9.5 CHS	117	14900	462	1820	2360	176	925	1.00	N	2170	4690	4690	683	1690	688
6.4 CHS	79.2	10100	317	1250	1610	177	634	0.857	N	1290	3180	2720	408	1140	472
457.0 × 12.7 CHS	139	17700	438	1920	2510	157	876	1.00	N	2500	5580	5580	789	2010	724
9.5 CHS	105	13400	334	1460	1900	158	669	1.00	N	1790	4210	4210	565	1510	553
6.4 CHS	71.1	9060	230	1010	1300	159	460	0.904	N	1090	2850	2580	343	1030	380
406.4 × 12.7 CHS	123	15700	305	1500	1970	139	609	1.00	C	1970	4950	4950	620	1780	567
9.5 CHS	93.0	11800	233	1150	1500	140	467	1.00	N	1450	3730	3730	456	1340	434
6.4 CHS	63.1	8040	161	792	1020	141	322	0.960	N	895	2530	2430	282	912	299
355.6 × 12.7 CHS	107	13700	201	1130	1490	121	403	1.00	C	1490	4310	4310	471	1550	428
9.5 CHS	81.1	10300	155	871	1140	122	310	1.00	N	1130	3250	3250	356	1170	329
6.4 CHS	55.1	7020	107	602	781	123	214	1.00	N	710	2210	2210	224	796	228
323.9 × 12.7 CHS	97.5	12400	151	930	1230	110	301	1.00	C	1230	3910	3910	388	1410	351
9.5 CHS	73.7	9380	116	717	939	111	232	1.00	C	939	2960	2960	296	1060	271
6.4 CHS	50.1	6380	80.5	497	645	112	161	1.00	N	601	2010	2010	189	724	188
273.1 × 12.7 CHS	81.6	10400	88.3	646	862	92.2	177	1.00	C	862	3270	3270	271	1180	244
9.3 CHS	60.5	7710	67.1	492	647	93.3	134	1.00	C	647	2430	2430	204	874	186
6.4 CHS	42.1	5360	47.7	349	455	94.3	95.4	1.00	N	441	1690	1690	139	608	132
4.8 CHS	31.8	4050	36.4	267	346	94.9	72.8	1.00	N	312	1270	1270	98.3	459	101
219.1 × 8.2 CHS	42.6	5430	30.3	276	365	74.6	60.5	1.00	C	365	1710	1710	115	616	104
6.4 CHS	33.6	4280	24.2	221	290	75.2	48.4	1.00	C	290	1350	1350	91.2	485	83.5
4.8 CHS	25.4	3230	18.6	169	220	75.8	37.1	1.00	N	210	1020	1020	66.3	366	64.0
168.3 × 7.1 CHS	28.2	3600	11.7	139	185	57.0	23.4	1.00	C	185	1130	1130	58.2	408	52.6
6.4 CHS	25.6	3260	10.7	127	168	57.3	21.4	1.00	C	168	1030	1030	52.9	369	48.0
4.8 CHS	19.4	2470	8.25	98.0	128	57.8	16.5	1.00	C	128	777	777	40.4	280	37.0

Note: For dimensions, other design capacities, availability and related information see ASI [2004], OneSteel [2012b].

Table C.3.6(b) Section properties and AS 4100 design section capacities: CHS - Grade C350L0

Dimensions and Ratios			Section Properties						Properties for Design to AS 4100			Design Section Capacities to AS 4100				
Designation		Mass per m	Gross Section Area	About any axis				Torsion Constant	Form Factor	About any axis Compactness		Design Section Axial Capacities		Des. Section Moment Capacity	Design Shear Capacity	Torsion
d_o	t		A_g	I	Z	S	r	J	k_f		Z_e	ϕN_t	ϕN_s	ϕM_s	ϕV_v	ϕM_z
mm	mm	kg/m	mm²	10⁶mm⁴	10³mm³	10³mm³	mm	10⁶mm⁴		(C,N,S)	10³mm³	kN	kN	kNm	kN	kNm
165.1 ×	3.5 CHS	13.9	1780	5.80	70.3	91.4	57.1	11.6	1.00	N	86.6	560	560	27.3	201	26.6
	3.0 CHS	12.0	1530	5.02	60.8	78.8	57.3	10.0	1.00	N	71.9	481	481	22.6	173	23.0
139.7 ×	3.5 CHS	11.8	1500	3.47	49.7	64.9	48.2	6.95	1.00	N	63.7	472	472	20.1	170	18.8
	3.0 CHS	10.1	1290	3.01	43.1	56.1	48.3	6.02	1.00	N	53.3	406	406	16.8	146	16.3
114.3 ×	3.6 CHS	9.83	1250	1.92	33.6	44.1	39.2	3.84	1.00	C	44.1	394	394	13.9	142	12.7
	3.2 CHS	8.77	1120	1.72	30.2	39.5	39.3	3.45	1.00	N	39.5	352	352	12.4	127	11.4
101.6 ×	3.2 CHS	7.77	989	1.20	23.6	31.0	34.8	2.40	1.00	C	31.0	312	312	9.76	112	8.92
	2.6 CHS	6.35	809	0.991	19.5	25.5	35.0	1.98	1.00	N	25.1	255	255	7.90	91.7	7.38
88.9 ×	3.2 CHS	6.76	862	0.792	17.8	23.5	30.3	1.58	1.00	C	23.5	271	271	7.41	97.7	6.74
	2.6 CHS	5.53	705	0.657	14.8	19.4	30.5	1.31	1.00	C	19.4	222	222	6.10	79.9	5.59
76.1 ×	3.2 CHS	5.75	733	0.488	12.8	17.0	25.8	0.976	1.00	C	17.0	231	231	5.36	83.1	4.85
	2.3 CHS	4.19	533	0.363	9.55	12.5	26.1	0.727	1.00	C	12.5	168	168	3.95	60.5	3.61
60.3 ×	2.9 CHS	4.11	523	0.216	7.16	9.56	20.3	0.432	1.00	C	9.56	165	165	3.01	59.3	2.71
	2.3 CHS	3.29	419	0.177	5.85	7.74	20.5	0.353	1.00	C	7.74	132	132	2.44	47.5	2.21
48.3 ×	2.9 CHS	3.25	414	0.107	4.43	5.99	16.1	0.214	1.00	C	5.99	130	130	1.89	46.9	1.67
	2.3 CHS	2.61	332	0.0881	3.65	4.87	16.3	0.176	1.00	C	4.87	105	105	1.53	37.7	1.38
42.4 ×	2.6 CHS	2.55	325	0.0646	3.05	4.12	14.1	0.129	1.00	C	4.12	102	102	1.30	36.9	1.15
	2.0 CHS	1.99	254	0.0519	2.45	3.27	14.3	0.104	1.00	C	3.27	80.0	80.0	1.03	28.8	0.926
33.7 ×	2.6 CHS	1.99	254	0.0309	1.84	2.52	11.0	0.0619	1.00	C	2.52	80.0	80.0	0.794	28.8	0.694
	2.0 CHS	1.56	199	0.0251	1.49	2.01	11.2	0.0502	1.00	C	2.01	62.7	62.7	0.634	22.6	0.563
26.9 ×	2.3 CHS	1.40	178	0.0136	1.01	1.40	8.74	0.0271	1.00	C	1.40	56.0	56.0	0.440	20.2	0.381
	2.0 CHS	1.23	156	0.0122	0.907	1.24	8.83	0.0244	1.00	C	1.24	49.3	49.3	0.391	17.7	0.343

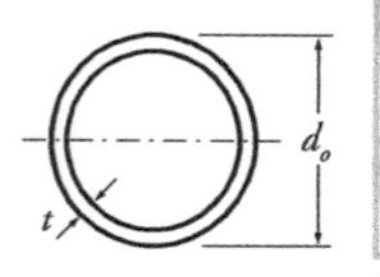

Note: For dimensions, other design capacities, availability and related information see ASI [2004], OneSteel [2012b].

Table C.3.6(c) Section properties and AS 4100 design section capacities: CHS - Grade C250L0

Dimensions and Ratios			Section Properties						Properties for Design to AS 4100			Design Section Capacities to AS 4100				
Designation		Mass per m	Gross Section Area	About any axis				Torsion Constant	Form Factor	About any axis Compactness		Design Section Axial Capacities		Des. Section Moment Capacity	Design Shear Capacity	Torsion
d_o	t		A_g	I	Z	S	r	J	k_f		Z_e	ϕN_t	ϕN_s	ϕM_s	ϕV_v	ϕM_z
mm	mm	kg/m	mm²	10^6mm⁴	10^3mm³	10^3mm³	mm	10^6mm⁴		(C,N,S)	10^3mm³	kN	kN	kNm	kN	kNm
165.1 ×	5.4 CHS	21.3	2710	8.65	105	138	56.5	17.3	1.00	C	138	610	610	31.0	219	28.3
	5.0 CHS	19.7	2510	8.07	97.7	128	56.6	16.1	1.00	C	128	566	566	28.8	204	26.4
139.7 ×	5.4 CHS	17.9	2280	5.14	73.7	97.4	47.5	10.3	1.00	C	97.4	513	513	21.9	185	19.9
	5.0 CHS	16.6	2120	4.81	68.8	90.8	47.7	9.61	1.00	C	90.8	476	476	20.4	171	18.6
114.3 ×	5.4 CHS	14.5	1850	2.75	48.0	64.1	38.5	5.49	1.00	C	64.1	416	416	14.4	150	13.0
	4.5 CHS	12.2	1550	2.34	41.0	54.3	38.9	4.69	1.00	C	54.3	349	349	12.2	126	11.1
101.6 ×	5.0 CHS	11.9	1520	1.77	34.9	46.7	34.2	3.55	1.00	C	46.7	341	341	10.5	123	9.43
	4.0 CHS	9.63	1230	1.46	28.8	38.1	34.5	2.93	1.00	C	38.1	276	276	8.58	99.3	7.77
88.9 ×	5.9 CHS	12.1	1540	1.33	30.0	40.7	29.4	2.66	1.00	C	40.7	346	346	9.16	125	8.09
	5.0 CHS	10.3	1320	1.16	26.2	35.2	29.7	2.33	1.00	C	35.2	297	297	7.93	107	7.07
	4.0 CHS	8.38	1070	0.963	21.7	28.9	30.0	1.93	1.00	C	28.9	240	240	6.49	86.4	5.85
76.1 ×	5.9 CHS	10.2	1300	0.807	21.2	29.1	24.9	1.61	1.00	C	29.1	293	293	6.56	105	5.73
	4.5 CHS	7.95	1010	0.651	17.1	23.1	25.4	1.30	1.00	C	23.1	228	228	5.20	82.0	4.62
	3.6 CHS	6.44	820	0.540	14.2	18.9	25.7	1.08	1.00	C	18.9	184	184	4.26	66.4	3.83
60.3 ×	5.4 CHS	7.31	931	0.354	11.8	16.3	19.5	0.709	1.00	C	16.3	210	210	3.67	75.4	3.17
	4.5 CHS	6.19	789	0.309	10.2	14.0	19.8	0.618	1.00	C	14.0	177	177	3.16	63.9	2.77
	3.6 CHS	5.03	641	0.259	8.58	11.6	20.1	0.517	1.00	C	11.6	144	144	2.61	51.9	2.32
48.3 ×	4.0 CHS	4.37	557	0.138	5.70	7.87	15.7	0.275	1.00	C	7.87	125	125	1.77	45.1	1.54
	3.2 CHS	3.56	453	0.116	4.80	6.52	16.0	0.232	1.00	C	6.52	102	102	1.47	36.7	1.30
42.4 ×	4.0 CHS	3.79	483	0.0899	4.24	5.92	13.6	0.180	1.00	C	5.92	109	109	1.33	39.1	1.15
	3.2 CHS	3.09	394	0.0762	3.59	4.93	13.9	0.152	1.00	C	4.93	88.7	88.7	1.11	31.9	0.970
33.7 ×	4.0 CHS	2.93	373	0.0419	2.49	3.55	10.6	0.0838	1.00	C	3.55	84.0	84.0	0.799	30.2	0.671
	3.2 CHS	2.41	307	0.0360	2.14	2.99	10.8	0.0721	1.00	C	2.99	69.0	69.0	0.672	24.8	0.578
26.9 ×	4.0 CHS	2.26	288	0.0194	1.45	2.12	8.22	0.0389	1.00	C	2.12	64.7	64.7	0.477	23.3	0.390
	3.2 CHS	1.87	238	0.0170	1.27	1.81	8.46	0.0341	1.00	C	1.81	53.6	53.6	0.407	19.3	0.342
	2.6 CHS	1.56	198	0.0148	1.10	1.54	8.64	0.0296	1.00	C	1.54	44.7	44.7	0.347	16.1	0.297

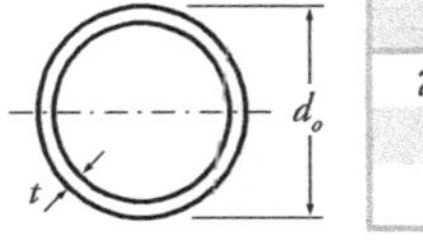

Note: 1. The CHS considered in this table **ARE GRADE C250L0 to AS/NZS 1163.**

2. For dimensions, other design capacities, availability and related information see ASI [2004], OneSteel [2012b].

Table C.3.7(a) Section properties and AS 4100 design section capacities: RHS - Grade C450L0 (C450Plus)

Dimensions			Section Properties										
Designation			Mass per m	Gross Section Area	About x-axis				About y-axis				Torsion Constant
d	b	t		A_g	I_x	Z_x	S_x	r_x	I_y	Z_y	S_y	r_y	J
mm	mm	mm	kg/m	mm²	10^6mm⁴	10^3mm³	10^3mm³	mm	10^6mm⁴	10^3mm³	10^3mm³	mm	10^6mm⁴
400 × 300 ×		16.0 RHS	161	20500	453	2260	2750	149	290	1940	2260	119	586
		12.5 RHS	128	16300	370	1850	2230	151	238	1590	1830	121	471
		10.0 RHS	104	13300	306	1530	1820	152	197	1320	1500	122	384
		8.0 RHS	84.2	10700	251	1260	1490	153	162	1080	1220	123	312
400 × 200 ×		16.0 RHS	136	17300	335	1670	2140	139	113	1130	1320	80.8	290
		12.5 RHS	109	13800	277	1380	1740	141	94.0	940	1080	82.4	236
		10.0 RHS	88.4	11300	230	1150	1430	143	78.6	786	888	83.6	194
		8.0 RHS	71.6	9120	190	949	1170	144	65.2	652	728	84.5	158
350 × 250 ×		16.0 RHS	136	17300	283	1620	1990	128	168	1340	1580	98.5	355
		12.5 RHS	109	13800	233	1330	1620	130	139	1110	1290	100	287
		10.0 RHS	88.4	11300	194	1110	1330	131	116	927	1060	101	235
		8.0 RHS	71.6	9120	160	914	1090	132	95.7	766	869	102	191
		6.0 RHS	54.4	6930	124	706	837	134	74.1	593	667	103	146
300 × 200 ×		16.0 RHS	111	14100	161	1080	1350	107	85.7	857	1020	78.0	193
		12.5 RHS	89.0	11300	135	899	1110	109	72.0	720	842	79.7	158
		10.0 RHS	72.7	9260	113	754	921	111	60.6	606	698	80.9	130
		8.0 RHS	59.1	7520	93.9	626	757	112	50.4	504	574	81.9	106
		6.0 RHS	45.0	5730	73.0	487	583	113	39.3	393	443	82.8	81.4
250 × 150 ×		16.0 RHS	85.5	10900	80.2	641	834	85.8	35.8	478	583	57.3	88.2
		12.5 RHS	69.4	8840	68.5	548	695	88.0	30.8	411	488	59.0	73.4
		10.0 RHS	57.0	7260	58.3	466	582	89.6	26.3	351	409	60.2	61.2
		9.0 RHS	51.8	6600	53.7	430	533	90.2	24.3	324	375	60.7	56.0
		8.0 RHS	46.5	5920	48.9	391	482	90.8	22.2	296	340	61.2	50.5
		6.0 RHS	35.6	4530	38.4	307	374	92.0	17.5	233	264	62.2	39.0
		5.0 RHS	29.9	3810	32.7	262	317	92.6	15.0	199	224	62.6	33.0
200 × 100 ×		10.0 RHS	41.3	5260	24.4	244	318	68.2	8.18	164	195	39.4	21.5
		9.0 RHS	37.7	4800	22.8	228	293	68.9	7.64	153	180	39.9	19.9
		8.0 RHS	33.9	4320	20.9	209	267	69.5	7.05	141	165	40.4	18.1
		6.0 RHS	26.2	3330	16.7	167	210	70.8	5.69	114	130	41.3	14.2
		5.0 RHS	22.1	2810	14.4	144	179	71.5	4.92	98.3	111	41.8	12.1
		4.0 RHS	17.9	2280	11.9	119	147	72.1	4.07	81.5	91.0	42.3	9.89
152 × 76 ×		6.0 RHS	19.4	2470	6.91	90.9	116	52.9	2.33	61.4	71.5	30.7	5.98
		5.0 RHS	16.4	2090	6.01	79.0	99.8	53.6	2.04	53.7	61.6	31.2	5.13
150 × 100 ×		10.0 RHS	33.4	4260	11.6	155	199	52.2	6.14	123	150	38.0	14.3
		9.0 RHS	30.6	3900	10.9	145	185	52.9	5.77	115	140	38.5	13.2
		8.0 RHS	27.7	3520	10.1	134	169	53.5	5.36	107	128	39.0	12.1
		6.0 RHS	21.4	2730	8.17	109	134	54.7	4.36	87.3	102	40.0	9.51
		5.0 RHS	18.2	2310	7.07	94.3	115	55.3	3.79	75.7	87.3	40.4	8.12
		4.0 RHS	14.8	1880	5.87	78.2	94.6	55.9	3.15	63.0	71.8	40.9	6.64

Note: For dimensions, other design capacities, availability and related information see ASI [2004] and OneSteel [2012b].

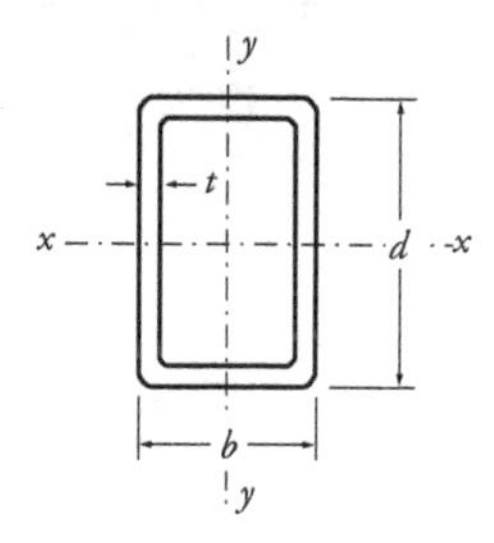

Properties for Design to AS 4100 (f_y = 450 MPa)					Design Section Capacities to AS 4100 (f_y = 450 MPa)							Dimensions
Form Factor	About x-axis		About y-axis		Design Section Axial Capacities		Design Section Moment Capacity about		Design Shear Capacities along		Design Torsion Capacity	Designation
	Compact-ness		Compact-ness		Tens	Comp	x-axis	y-axis	y-axis	x-axis		
k_f		Z_{ex}		Z_{ey}	ϕN_t	ϕN_s	ϕM_{sx}	ϕM_{sy}	ϕV_{vx}	ϕV_{vy}	ϕM_z	d b t
–	(C,N,S)	10^3mm^3	(C,N,S)	10^3mm^3	kN	kN	kNm	kNm	kN	kN	kNm	mm mm mm
1.00	C	2750	N	2230	7840	8300	1110	905	2790	2080	771	400 × 300 × 16.0 RHS
0.996	C	2230	S	1580	6250	6590	901	641	2220	1670	628	12.5 RHS
0.877	N	1600	S	1120	5070	4710	649	454	1800	1360	518	10.0 RHS
0.715	S	1140	S	800	4100	3110	463	324	1450	1100	425	8.0 RHS
1.00	C	2140	N	1300	6620	7010	866	527	2730	1310	485	400 × 200 × 16.0 RHS
0.996	C	1740	S	936	5290	5580	705	379	2170	1060	401	12.5 RHS
0.855	C	1430	S	658	4310	3900	581	266	1760	875	334	10.0 RHS
0.745	N	1150	S	464	3490	2750	467	188	1420	715	275	8.0 RHS
1.00	C	1990	C	1580	6620	7010	807	641	2400	1700	543	350 × 250 × 16.0 RHS
1.00	C	1620	N	1200	5290	5600	657	487	1920	1370	446	12.5 RHS
0.943	N	1320	S	865	4310	4300	533	350	1560	1120	370	10.0 RHS
0.833	N	928	S	614	3490	3080	376	249	1260	910	304	8.0 RHS
0.622	S	611	S	399	2650	1750	247	162	957	694	235	6.0 RHS
1.00	C	1350	C	1020	5390	5710	548	414	2020	1310	354	300 × 200 × 16.0 RHS
1.00	C	1110	C	842	4340	4590	450	341	1620	1060	294	12.5 RHS
1.00	C	921	N	628	3540	3750	373	254	1320	875	246	10.0 RHS
0.903	N	746	S	447	2880	2750	302	181	1070	715	204	8.0 RHS
0.753	S	474	S	288	2190	1750	192	116	813	548	158	6.0 RHS
1.00	C	834	C	583	4170	4410	338	236	1630	918	203	250 × 150 × 16.0 RHS
1.00	C	695	C	488	3380	3580	282	198	1320	759	173	12.5 RHS
1.00	C	582	N	404	2780	2940	236	164	1080	632	146	10.0 RHS
1.00	C	533	N	352	2520	2670	216	143	976	577	135	9.0 RHS
1.00	C	482	N	299	2270	2400	195	121	875	521	122	8.0 RHS
0.843	N	368	S	191	1730	1550	149	77.5	668	402	96.0	6.0 RHS
0.762	N	275	S	144	1460	1180	111	58.5	561	340	81.8	5.0 RHS
1.00	C	318	C	195	2010	2130	129	79.1	833	389	71.0	200 × 100 × 10.0 RHS
1.00	C	293	C	180	1840	1940	119	73.1	758	359	66.0	9.0 RHS
1.00	C	267	N	163	1650	1750	108	65.9	681	327	60.7	8.0 RHS
0.967	C	210	S	110	1270	1310	85.1	44.4	522	257	48.5	6.0 RHS
0.855	C	179	S	82.2	1080	974	72.6	33.3	440	219	41.7	5.0 RHS
0.745	N	144	S	58.0	873	688	58.4	23.5	355	179	34.4	4.0 RHS
1.00	C	116	N	70.2	944	1000	47.0	28.4	389	187	26.4	152 × 76 × 6.0 RHS
1.00	C	99.8	N	55.2	801	848	40.4	22.3	329	160	22.9	5.0 RHS
1.00	C	199	C	150	1630	1720	80.7	60.9	611	389	51.3	150 × 100 × 10.0 RHS
1.00	C	185	C	140	1490	1580	74.8	56.5	559	359	47.9	9.0 RHS
1.00	C	169	C	128	1350	1430	68.5	51.8	504	327	44.2	8.0 RHS
1.00	C	134	N	101	1050	1110	54.4	40.7	389	257	35.6	6.0 RHS
1.00	C	115	N	78.5	885	937	46.6	31.8	329	219	30.7	5.0 RHS
0.903	N	93.2	S	55.9	720	688	37.8	22.6	267	179	25.5	4.0 RHS

Table C.3.7(b) Section properties and AS 4100 design section capacities: RHS - Grade C450L0 (C450Plus)

Dimensions			Section Properties								
Designation	Mass per m	Gross Section Area	About *x*-axis				About *y*-axis				Torsion Constant
d *b* *t*		A_g	I_x	Z_x	S_x	r_x	I_y	Z_y	S_y	r_y	J
mm mm mm	kg/m	mm²	10⁶mm⁴	10³mm³	10³mm³	mm	10⁶mm⁴	10³mm³	10³mm³	mm	10⁶mm⁴
150 × 50 × 5.0 RHS	14.2	1810	4.44	59.2	78.9	49.5	0.765	30.6	35.7	20.5	2.30
4.0 RHS	11.6	1480	3.74	49.8	65.4	50.2	0.653	26.1	29.8	21.0	1.93
3.0 RHS	8.96	1140	2.99	39.8	51.4	51.2	0.526	21.1	23.5	21.5	1.50
127 × 51 × 6.0 RHS	14.7	1870	3.28	51.6	68.9	41.9	0.761	29.8	35.8	20.2	2.20
5.0 RHS	12.5	1590	2.89	45.6	59.9	42.6	0.679	26.6	31.3	20.6	1.93
3.5 RHS	9.07	1150	2.20	34.7	44.6	43.7	0.526	20.6	23.4	21.3	1.44
125 × 75 × 6.0 RHS	16.7	2130	4.16	66.6	84.2	44.2	1.87	50.0	59.1	29.6	4.44
5.0 RHS	14.2	1810	3.64	58.3	72.7	44.8	1.65	43.9	51.1	30.1	3.83
4.0 RHS	11.6	1480	3.05	48.9	60.3	45.4	1.39	37.0	42.4	30.6	3.16
3.0 RHS	8.96	1140	2.43	38.9	47.3	46.1	1.11	29.5	33.3	31.1	2.43
2.5 RHS	7.53	959	2.07	33.0	40.0	46.4	0.942	25.1	28.2	31.4	2.05
102 × 76 × 6.0 RHS	14.7	1870	2.52	49.4	61.9	36.7	1.59	42.0	50.5	29.2	3.38
5.0 RHS	12.5	1590	2.22	43.5	53.7	37.3	1.41	37.0	43.9	29.7	2.91
3.5 RHS	9.07	1150	1.68	33.0	39.9	38.2	1.07	28.2	32.6	30.5	2.14
100 × 50 × 6.0 RHS	12.0	1530	1.71	34.2	45.3	33.4	0.567	22.7	27.7	19.2	1.53
5.0 RHS	10.3	1310	1.53	30.6	39.8	34.1	0.511	20.4	24.4	19.7	1.35
4.0 RHS	8.49	1080	1.31	26.1	33.4	34.8	0.441	17.6	20.6	20.2	1.13
3.5 RHS	7.53	959	1.18	23.6	29.9	35.1	0.400	16.0	18.5	20.4	1.01
3.0 RHS	6.60	841	1.06	21.3	26.7	35.6	0.361	14.4	16.4	20.7	0.886
2.5 RHS	5.56	709	0.912	18.2	22.7	35.9	0.311	12.4	14.0	20.9	0.754
2.0 RHS	4.50	574	0.750	15.0	18.5	36.2	0.257	10.3	11.5	21.2	0.616
76 × 38 × 4.0 RHS	6.23	793	0.527	13.9	18.1	25.8	0.176	9.26	11.1	14.9	0.466
3.0 RHS	4.90	625	0.443	11.7	14.8	26.6	0.149	7.82	9.09	15.4	0.373
2.5 RHS	4.15	529	0.383	10.1	12.7	26.9	0.129	6.81	7.81	15.6	0.320
75 × 50 × 6.0 RHS	9.67	1230	0.800	21.3	28.1	25.5	0.421	16.9	21.1	18.5	1.01
5.0 RHS	8.35	1060	0.726	19.4	24.9	26.1	0.384	15.4	18.8	19.0	0.891
4.0 RHS	6.92	881	0.630	16.8	21.1	26.7	0.335	13.4	16.0	19.5	0.754
3.0 RHS	5.42	691	0.522	13.9	17.1	27.5	0.278	11.1	12.9	20.0	0.593
2.5 RHS	4.58	584	0.450	12.0	14.6	27.7	0.240	9.60	11.0	20.3	0.505
2.0 RHS	3.72	474	0.372	9.91	12.0	28.0	0.199	7.96	9.06	20.5	0.414
1.6 RHS	3.01	383	0.305	8.14	9.75	28.2	0.164	6.56	7.40	20.7	0.337
75 × 25 × 2.5 RHS	3.60	459	0.285	7.60	10.1	24.9	0.0487	3.89	4.53	10.3	0.144
2.0 RHS	2.93	374	0.238	6.36	8.31	25.3	0.0414	3.31	3.77	10.5	0.120
1.6 RHS	2.38	303	0.197	5.26	6.81	25.5	0.0347	2.78	3.11	10.7	0.0993
65 × 35 × 4.0 RHS	5.35	681	0.328	10.1	13.3	22.0	0.123	7.03	8.58	13.4	0.320
3.0 RHS	4.25	541	0.281	8.65	11.0	22.8	0.106	6.04	7.11	14.0	0.259
2.5 RHS	3.60	459	0.244	7.52	9.45	23.1	0.0926	5.29	6.13	14.2	0.223
2.0 RHS	2.93	374	0.204	6.28	7.80	23.4	0.0778	4.44	5.07	14.4	0.184
50 × 25 × 3.0 RHS	3.07	391	0.112	4.47	5.86	16.9	0.0367	2.93	3.56	9.69	0.0964
2.5 RHS	2.62	334	0.0989	3.95	5.11	17.2	0.0328	2.62	3.12	9.91	0.0843
2.0 RHS	2.15	274	0.0838	3.35	4.26	17.5	0.0281	2.25	2.62	10.1	0.0706
1.6 RHS	1.75	223	0.0702	2.81	3.53	17.7	0.0237	1.90	2.17	10.3	0.0585

Notes:

1. 75 x 25 RHS and smaller sizes are typically available in Grade C350L0—see Note (2).

2. For dimensions, other design capacities, availability and related information see ASI [2004] and OneSteel [2012b].

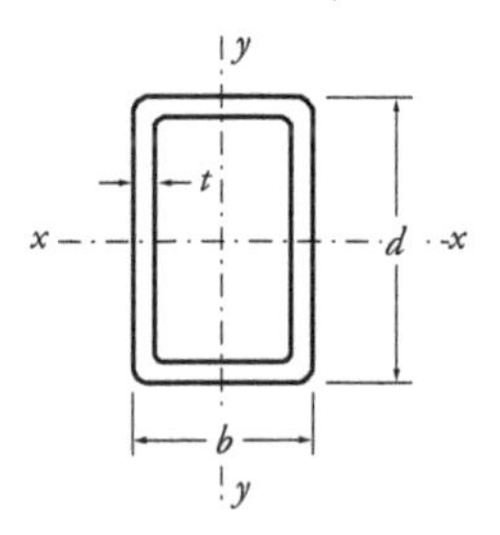

Properties for Design to AS 4100 (f_y = 450 MPa)					Design Section Capacities to AS 4100 (f_y = 450 MPa)							Dimensions
Form Factor	About *x*-axis		About *y*-axis		Design Section Axial Capacities		Design Section Moment Capacity about		Design Shear Capacities along		Design Torsion Capacity	Designation
	Compact-ness		Compact-ness		Tens	Comp	*x*-axis	*y*-axis	*y*-axis	*x*-axis		
k_f		Z_{ex}		Z_{ey}	ϕN_t	ϕN_s	ϕM_{sx}	ϕM_{sy}	ϕV_{vx}	ϕV_{vy}	ϕM_z	*d* *b* *t*
–	(C,N,S)	10^3mm^3	(C,N,S)	10^3mm^3	kN	kN	kNm	kNm	kN	kN	kNm	mm mm mm
1.00	C	78.9	N	31.8	694	735	31.9	12.9	316	97.2	13.8	150 × 50 × 5.0 RHS
0.877	C	65.4	S	22.7	567	526	26.5	9.19	257	81.6	11.7	4.0 RHS
0.713	C	51.4	S	14.5	436	329	20.8	5.89	195	64.2	9.30	3.0 RHS
1.00	C	68.9	C	35.8	715	757	27.9	14.5	315	114	13.3	127 × 51 × 6.0 RHS
1.00	C	59.9	N	30.6	610	646	24.3	12.4	267	99.6	11.8	5.0 RHS
0.905	C	44.6	S	18.5	442	423	18.1	7.49	192	74.8	9.04	3.5 RHS
1.00	C	84.2	C	59.1	816	864	34.1	23.9	317	184	21.0	125 × 75 × 6.0 RHS
1.00	C	72.7	N	50.5	694	735	29.5	20.5	269	158	18.3	5.0 RHS
1.00	C	60.3	N	37.4	567	600	24.4	15.1	219	130	15.3	4.0 RHS
0.845	N	46.5	S	24.2	436	390	18.8	9.80	167	101	12.0	3.0 RHS
0.763	N	34.7	S	18.2	367	296	14.1	7.39	140	85.1	10.2	2.5 RHS
1.00	C	61.9	C	50.5	715	757	25.1	20.5	255	187	17.0	102 × 76 × 6.0 RHS
1.00	C	53.7	C	43.9	610	646	21.7	17.8	218	160	14.9	5.0 RHS
1.00	C	39.9	N	29.8	442	468	16.1	12.1	157	117	11.2	3.5 RHS
1.00	C	45.3	C	27.7	586	621	18.4	11.2	244	111	9.94	100 × 50 × 6.0 RHS
1.00	C	39.8	C	24.4	503	532	16.1	9.88	208	97.2	8.87	5.0 RHS
1.00	C	33.4	N	20.3	414	438	13.5	8.23	170	81.6	7.58	4.0 RHS
1.00	C	29.9	N	17.1	367	388	12.1	6.92	151	73.1	6.85	3.5 RHS
0.967	C	26.7	S	13.9	322	329	10.8	5.63	131	64.2	6.08	3.0 RHS
0.856	C	22.7	S	10.4	271	246	9.18	4.22	110	54.7	5.22	2.5 RHS
0.746	N	18.2	S	7.33	219	173	7.37	2.97	88.9	44.7	4.31	2.0 RHS
1.00	C	18.1	C	11.1	303	321	7.34	4.50	126	58.3	4.03	76 × 38 × 4.0 RHS
1.00	C	14.8	N	8.92	239	253	6.00	3.61	97.2	46.7	3.31	3.0 RHS
1.00	C	12.7	N	7.00	202	214	5.14	2.83	82.2	40.1	2.87	2.5 RHS
1.00	C	28.1	C	21.1	471	499	11.4	8.56	178	111	7.11	75 × 50 × 6.0 RHS
1.00	C	24.9	C	18.8	407	431	10.1	7.61	153	97.2	6.41	5.0 RHS
1.00	C	21.1	C	16.0	337	357	8.56	6.47	126	81.6	5.52	4.0 RHS
1.00	C	17.1	N	12.8	264	280	6.92	5.17	97.4	64.2	4.47	3.0 RHS
1.00	C	14.6	N	9.95	223	236	5.91	4.03	82.3	54.7	3.85	2.5 RHS
0.904	N	11.8	S	7.07	181	173	4.77	2.86	66.8	44.7	3.19	2.0 RHS
0.799	N	8.26	S	5.01	147	124	3.34	2.03	54.0	36.4	2.62	1.6 RHS
1.00	C	10.1	N	4.05	176	186	4.07	1.64	79.1	24.3	1.73	75 × 25 × 2.5 RHS
0.878	C	8.31	S	2.88	143	133	3.36	1.17	64.2	20.4	1.47	2.0 RHS
0.746	C	6.81	S	2.02	116	91.6	2.76	0.816	51.9	17.0	1.23	1.6 RHS
1.00	C	13.3	C	8.58	261	276	5.38	3.48	106	52.5	3.05	65 × 35 × 4.0 RHS
1.00	C	11.0	C	7.11	207	219	4.45	2.88	82.3	42.3	2.54	3.0 RHS
1.00	C	9.45	N	5.95	176	186	3.83	2.41	69.7	36.5	2.21	2.5 RHS
0.985	C	7.80	S	4.37	143	149	3.16	1.77	56.7	30.1	1.85	2.0 RHS
1.00	C	5.86	C	3.56	149	158	2.37	1.44	61.1	27.7	1.26	50 × 25 × 3.0 RHS
1.00	C	5.11	C	3.12	128	135	2.07	1.26	52.1	24.3	1.12	2.5 RHS
1.00	C	4.26	N	2.58	105	111	1.73	1.05	42.6	20.4	0.952	2.0 RHS
1.00	C	3.53	N	1.92	85.4	90.4	1.43	0.777	34.7	17.0	0.800	1.6 RHS

Table C.3.8(a) Section properties and AS 4100 design section capacities: SHS - Grade C450L0 (C450Plus)

Dimensions		Section Properties						
Designation	Mass per m	Gross Section Area	About *x*-,*y*- and *n*-axis					Torsion Constant
d *b* *t*		A_g	I_x	Z_x	Z_n	S_x	r_x	J
mm mm mm	kg/m	mm²	10⁶mm⁴	10³mm³	10³mm³	10³mm³	mm	10⁶mm⁴
400 × 400 × 16.0 SHS	186	23700	571	2850	2140	3370	155	930
12.5 SHS	148	18800	464	2320	1720	2710	157	744
10.0 SHS	120	15300	382	1910	1400	2210	158	604
350 × 350 × 16.0 SHS	161	20500	372	2130	1610	2530	135	614
12.5 SHS	128	16300	305	1740	1300	2040	137	493
10.0 SHS	104	13300	252	1440	1060	1670	138	401
8.0 SHS	84.2	10700	207	1180	865	1370	139	326
300 × 300 × 16.0 SHS	136	17300	226	1510	1160	1810	114	378
12.5 SHS	109	13800	187	1240	937	1470	116	305
10.0 SHS	88.4	11300	155	1030	769	1210	117	250
8.0 SHS	71.6	9120	128	853	628	991	118	203
250 × 250 × 16.0 SHS	111	14100	124	992	774	1210	93.8	212
12.5 SHS	89.0	11300	104	830	634	992	95.7	173
10.0 SHS	72.7	9260	87.1	697	523	822	97.0	142
9.0 SHS	65.9	8400	79.8	639	477	750	97.5	129
8.0 SHS	59.1	7520	72.3	578	429	676	98.0	116
6.0 SHS	45.0	5730	56.2	450	330	521	99.0	88.7
200 × 200 × 16.0 SHS	85.5	10900	58.6	586	469	728	73.3	103
12.5 SHS	69.4	8840	50.0	500	389	607	75.2	85.2
10.0 SHS	57.0	7260	42.5	425	324	508	76.5	70.7
9.0 SHS	51.8	6600	39.2	392	297	465	77.1	64.5
8.0 SHS	46.5	5920	35.7	357	268	421	77.6	58.2
6.0 SHS	35.6	4530	28.0	280	207	327	78.6	44.8
5.0 SHS	29.9	3810	23.9	239	175	277	79.1	37.8
150 × 150 × 10.0 SHS	41.3	5260	16.5	220	173	269	56.1	28.4
9.0 SHS	37.7	4800	15.4	205	159	248	56.6	26.1
8.0 SHS	33.9	4320	14.1	188	144	226	57.1	23.6
6.0 SHS	26.2	3330	11.3	150	113	178	58.2	18.4
5.0 SHS	22.1	2810	9.70	129	96.2	151	58.7	15.6
125 × 125 × 10.0 SHS	33.4	4260	8.93	143	114	178	45.8	15.7
9.0 SHS	30.6	3900	8.38	134	106	165	46.4	14.5
8.0 SHS	27.7	3520	7.75	124	96.8	151	46.9	13.3
6.0 SHS	21.4	2730	6.29	101	76.5	120	48.0	10.4
5.0 SHS	18.2	2310	5.44	87.1	65.4	103	48.5	8.87
4.0 SHS	14.8	1880	4.52	72.3	53.6	84.5	49.0	7.25
100 × 100 × 10.0 SHS	25.6	3260	4.11	82.2	68.1	105	35.5	7.50
9.0 SHS	23.5	3000	3.91	78.1	63.6	98.6	36.1	7.00
8.0 SHS	21.4	2720	3.66	73.2	58.6	91.1	36.7	6.45
6.0 SHS	16.7	2130	3.04	60.7	47.1	73.5	37.7	5.15
5.0 SHS	14.2	1810	2.66	53.1	40.5	63.5	38.3	4.42
4.0 SHS	11.6	1480	2.23	44.6	33.5	52.6	38.8	3.63
3.0 SHS	8.96	1140	1.77	35.4	26.0	41.2	39.4	2.79
2.5 SHS	7.53	959	1.51	30.1	21.9	34.9	39.6	2.35

Note: For dimensions, other design capacities, availability and related information see ASI [2004] and OneSteel [2012b].

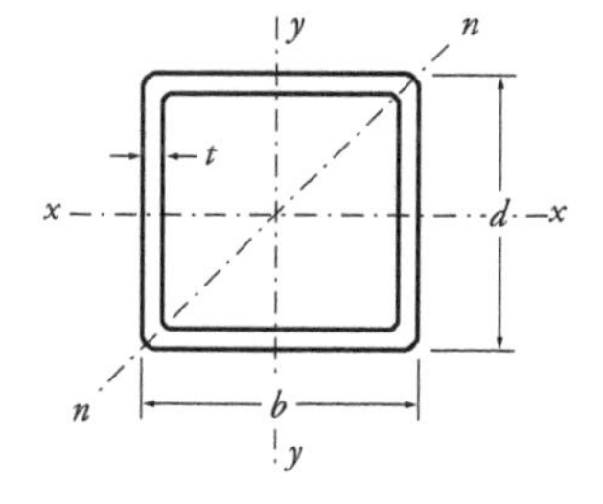

Properties for Design to AS 4100			Design Section Capacities to AS 4100 (f_y = 450 MPa)					Dimensions
Form Factor	About x- and y-axis		Design Section Axial Capacities		Des. Section Moment Cap'y about x-axis	Design Shear Capacity along y-axis	Design Torsion Capacity	Designation
k_f	Compact-ness	Z_{ex}	Tens ϕN_t	Comp ϕN_s	ϕM_{sx}	ϕV_{vx}	ϕM_z	d b t
–	(C,N,S)	$10^3 mm^3$	kN	kN	kNm	kN	kNm	mm mm mm
1.00	N	3320	9060	9600	1350	2830	1060	400 × 400 × 16.0 SHS
0.994	S	2310	7210	7580	937	2250	856	12.5 SHS
0.785	S	1650	5840	4850	670	1820	703	10.0 SHS
1.00	C	2530	7840	8300	1020	2440	790	350 × 350 × 16.0 SHS
1.00	N	1900	6250	6620	768	1950	644	12.5 SHS
0.904	S	1350	5070	4850	548	1580	530	10.0 SHS
0.715	S	971	4100	3110	393	1280	434	8.0 SHS
1.00	C	1810	6620	7010	732	2060	562	300 × 300 × 16.0 SHS
1.00	C	1470	5290	5600	596	1650	461	12.5 SHS
1.00	N	1080	4310	4560	436	1340	382	10.0 SHS
0.840	S	768	3490	3110	311	1090	314	8.0 SHS
1.00	C	1210	5390	5710	489	1670	373	250 × 250 × 16.0 SHS
1.00	C	992	4340	4590	402	1350	309	12.5 SHS
1.00	N	811	3540	3750	329	1100	258	10.0 SHS
1.00	N	699	3210	3400	283	1000	236	9.0 SHS
1.00	N	586	2880	3050	237	899	213	8.0 SHS
0.753	S	380	2190	1750	154	685	165	6.0 SHS
1.00	C	728	4170	4410	295	1290	222	200 × 200 × 16.0 SHS
1.00	C	607	3380	3580	246	1050	188	12.5 SHS
1.00	C	508	2780	2940	206	864	158	10.0 SHS
1.00	C	465	2520	2670	188	786	146	9.0 SHS
1.00	N	415	2270	2400	168	707	132	8.0 SHS
0.952	S	272	1730	1750	110	541	103	6.0 SHS
0.785	S	207	1460	1210	83.8	456	87.9	5.0 SHS
1.00	C	269	2010	2130	109	624	82.9	150 × 150 × 10.0 SHS
1.00	C	248	1840	1940	101	570	76.8	9.0 SHS
1.00	C	226	1650	1750	91.5	515	70.2	8.0 SHS
1.00	N	175	1270	1350	71.0	397	55.7	6.0 SHS
1.00	N	135	1080	1140	54.6	336	47.8	5.0 SHS
1.00	C	178	1630	1720	72.0	504	54.2	125 × 125 × 10.0 SHS
1.00	C	165	1490	1580	66.8	462	50.6	9.0 SHS
1.00	C	151	1350	1430	61.2	419	46.6	8.0 SHS
1.00	C	120	1050	1110	48.6	325	37.4	6.0 SHS
1.00	N	101	885	937	41.1	276	32.3	5.0 SHS
1.00	N	73.2	720	762	29.7	225	26.7	4.0 SHS
1.00	C	105	1250	1320	42.6	384	31.6	100 × 100 × 10.0 SHS
1.00	C	98.6	1150	1210	39.9	354	29.8	9.0 SHS
1.00	C	91.1	1040	1100	36.9	323	27.8	8.0 SHS
1.00	C	73.5	816	864	29.8	253	22.7	6.0 SHS
1.00	C	63.5	694	735	25.7	216	19.8	5.0 SHS
1.00	N	51.9	567	600	21.0	177	16.5	4.0 SHS
0.952	S	34.4	436	440	13.9	135	12.9	3.0 SHS
0.787	S	26.1	367	305	10.6	114	11.0	2.5 SHS

Table C.3.8(b) Section properties and AS 4100 design section capacities: SHS - Grade C450L0 (C450Plus)

Dimensions		Section Properties							
Designation	Mass per m	Gross Section Area	About *x*-,*y*- and *n*-axis						Torsion Constant
d *b* *t*		A_g	I_x	Z_x	Z_n	S_x	r_x		J
mm mm mm	kg/m	mm²	10⁶mm⁴	10³mm³	10³mm³	10³mm³	mm		10⁶mm⁴
89 × 89 × 6.0 SHS	14.7	1870	2.06	46.4	36.4	56.7	33.2		3.55
5.0 SHS	12.5	1590	1.82	40.8	31.5	49.2	33.8		3.06
3.5 SHS	9.07	1150	1.38	31.0	23.3	36.5	34.6		2.25
2.0 SHS	5.38	686	0.858	19.3	14.0	22.3	35.4		1.33
75 × 75 × 6.0 SHS	12.0	1530	1.16	30.9	24.7	38.4	27.5		2.04
5.0 SHS	10.3	1310	1.03	27.5	21.6	33.6	28.0		1.77
4.0 SHS	8.49	1080	0.882	23.5	18.1	28.2	28.6		1.48
3.5 SHS	7.53	959	0.797	21.3	16.1	25.3	28.8		1.32
3.0 SHS	6.60	841	0.716	19.1	14.2	22.5	29.2		1.15
2.5 SHS	5.56	709	0.614	16.4	12.0	19.1	29.4		0.971
2.0 SHS	4.50	574	0.505	13.5	9.83	15.6	29.7		0.790
65 × 65 × 6.0 SHS	10.1	1290	0.706	21.7	17.8	27.5	23.4		1.27
5.0 SHS	8.75	1110	0.638	19.6	15.6	24.3	23.9		1.12
4.0 SHS	7.23	921	0.552	17.0	13.2	20.6	24.5		0.939
3.0 SHS	5.66	721	0.454	14.0	10.4	16.6	25.1		0.733
2.5 SHS	4.78	609	0.391	12.0	8.91	14.1	25.3		0.624
2.0 SHS	3.88	494	0.323	9.94	7.29	11.6	25.6		0.509
1.6 SHS	3.13	399	0.265	8.16	5.94	9.44	25.8		0.414
50 × 50 × 6.0 SHS	7.32	932	0.275	11.0	9.45	14.5	17.2		0.518
5.0 SHS	6.39	814	0.257	10.3	8.51	13.2	17.8		0.469
4.0 SHS	5.35	681	0.229	9.15	7.33	11.4	18.3		0.403
3.0 SHS	4.25	541	0.195	7.79	5.92	9.39	19.0		0.321
2.5 SHS	3.60	459	0.169	6.78	5.09	8.07	19.2		0.275
2.0 SHS	2.93	374	0.141	5.66	4.20	6.66	19.5		0.226
1.6 SHS	2.38	303	0.117	4.68	3.44	5.46	19.6		0.185
40 × 40 × 4.0 SHS	4.09	521	0.105	5.26	4.36	6.74	14.2		0.192
3.0 SHS	3.30	421	0.0932	4.66	3.61	5.72	14.9		0.158
2.5 SHS	2.82	359	0.0822	4.11	3.13	4.97	15.1		0.136
2.0 SHS	2.31	294	0.0694	3.47	2.61	4.13	15.4		0.113
1.6 SHS	1.88	239	0.0579	2.90	2.15	3.41	15.6		0.0927
35 × 35 × 3.0 SHS	2.83	361	0.0595	3.40	2.67	4.23	12.8		0.102
2.5 SHS	2.42	309	0.0529	3.02	2.33	3.69	13.1		0.0889
2.0 SHS	1.99	254	0.0451	2.58	1.95	3.09	13.3		0.0741
1.6 SHS	1.63	207	0.0379	2.16	1.62	2.57	13.5		0.0611
30 × 30 × 3.0 SHS	2.36	301	0.0350	2.34	1.87	2.96	10.8		0.0615
2.5 SHS	2.03	259	0.0316	2.10	1.65	2.61	11.0		0.0540
2.0 SHS	1.68	214	0.0272	1.81	1.39	2.21	11.3		0.0454
1.6 SHS	1.38	175	0.0231	1.54	1.16	1.84	11.5		0.0377
25 × 25 × 3.0 SHS	1.89	241	0.0184	1.47	1.21	1.91	8.74		0.0333
2.5 SHS	1.64	209	0.0169	1.35	1.08	1.71	8.99		0.0297
2.0 SHS	1.36	174	0.0148	1.19	0.926	1.47	9.24		0.0253
1.6 SHS	1.12	143	0.0128	1.02	0.780	1.24	9.44		0.0212
20 × 20 × 2.0 SHS	1.05	134	0.00692	0.692	0.554	0.877	7.20		0.0121
1.6 SHS	0.873	111	0.00608	0.608	0.474	0.751	7.39		0.0103

Notes:

1. 50 x 50 x 5.0 SHS and smaller sizes are typically available in Grade C350L0—see Note (2).

2. For dimensions, other design capacities, availability and related information see ASI [2004] and OneSteel [2012b].

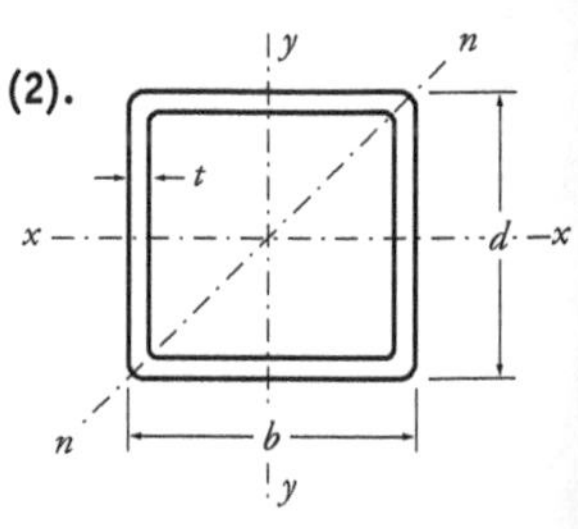

Properties for Design to AS 4100			Design Section Capacities to AS 4100 (f_y = 450 MPa)					Dimensions
Form Factor	About *x*- and *y*-axis		Design Section Axial Capacities		Des. Section Moment Cap'y about *x*-axis	Design Shear Capacity along *y*-axis	Design Torsion Capacity	Designation
	Compact-ness		Tens	Comp				
k_f		Z_{ex}	ϕN_t	ϕN_s	ϕM_{sx}	ϕV_{vx}	ϕM_z	*d* *b* *t*
–	(C,N,S)	$10^3 mm^3$	kN	kN	kNm	kN	kNm	mm mm mm
1.00	C	56.7	715	757	23.0	222	17.4	89 × 89 × 6.0 SHS
1.00	C	49.2	610	646	19.9	190	15.3	5.0 SHS
1.00	N	35.8	442	468	14.5	138	11.5	3.5 SHS
0.704	S	15.7	262	196	6.37	81.6	7.04	2.0 SHS
1.00	C	38.4	586	621	15.6	181	11.7	75 × 75 × 6.0 SHS
1.00	C	33.6	503	532	13.6	156	10.4	5.0 SHS
1.00	C	28.2	414	438	11.4	129	8.78	4.0 SHS
1.00	C	25.3	367	388	10.2	114	7.90	3.5 SHS
1.00	N	22.2	322	341	8.99	99.4	6.98	3.0 SHS
1.00	N	17.0	271	287	6.90	84.0	5.98	2.5 SHS
0.841	S	12.1	219	196	4.91	68.2	4.91	2.0 SHS
1.00	C	27.5	494	523	11.1	153	8.31	65 × 65 × 6.0 SHS
1.00	C	24.3	426	451	9.85	132	7.43	5.0 SHS
1.00	C	20.6	352	373	8.34	109	6.36	4.0 SHS
1.00	C	16.6	276	292	6.71	85.0	5.11	3.0 SHS
1.00	N	13.7	233	247	5.54	72.0	4.40	2.5 SHS
0.978	S	9.80	189	196	3.97	58.6	3.63	2.0 SHS
0.774	S	7.01	153	125	2.84	47.5	2.98	1.6 SHS
1.00	C	14.5	357	378	5.89	109	4.30	50 × 50 × 6.0 SHS
1.00	C	13.2	311	330	5.33	96.0	3.95	5.0 SHS
1.00	C	11.4	261	276	4.61	80.6	3.47	4.0 SHS
1.00	C	9.39	207	219	3.80	63.4	2.86	3.0 SHS
1.00	C	8.07	176	186	3.27	54.0	2.48	2.5 SHS
1.00	N	6.58	143	151	2.66	44.2	2.07	2.0 SHS
1.00	N	4.74	116	123	1.92	35.9	1.71	1.6 SHS
1.00	C	6.74	199	211	2.73	61.4	2.02	40 × 40 × 4.0 SHS
1.00	C	5.72	161	170	2.32	49.0	1.72	3.0 SHS
1.00	C	4.97	137	145	2.01	42.0	1.51	2.5 SHS
1.00	C	4.13	112	119	1.67	34.6	1.27	2.0 SHS
1.00	N	3.37	91.5	96.9	1.36	28.3	1.06	1.6 SHS
1.00	C	4.23	138	146	1.71	41.8	1.26	35 × 35 × 3.0 SHS
1.00	C	3.69	118	125	1.50	36.0	1.11	2.5 SHS
1.00	C	3.09	97.0	103	1.25	29.8	0.945	2.0 SHS
1.00	C	2.57	79.2	83.9	1.04	24.4	0.792	1.6 SHS
1.00	C	2.96	115	122	1.20	34.6	0.869	30 × 30 × 3.0 SHS
1.00	C	2.61	99.0	105	1.06	30.0	0.778	2.5 SHS
1.00	C	2.21	81.7	86.5	0.893	25.0	0.667	2.0 SHS
1.00	C	1.84	67.0	70.9	0.746	20.6	0.564	1.6 SHS
1.00	C	1.91	92.1	97.5	0.776	27.4	0.553	25 × 25 × 3.0 SHS
1.00	C	1.71	79.9	84.6	0.694	24.0	0.503	2.5 SHS
1.00	C	1.47	66.4	70.3	0.594	20.2	0.438	2.0 SHS
1.00	C	1.24	54.8	58.0	0.500	16.7	0.375	1.6 SHS
1.00	C	0.877	51.1	54.1	0.355	15.4	0.258	20 × 20 × 2.0 SHS
1.00	C	0.751	42.5	45.0	0.304	12.9	0.224	1.6 SHS

C.4 Miscellaneous cross-section parameters

Table C.4.1 Geometrical properties of plane sections

Section	Area	Centroidal distance	Second moment of inertia		Elastic section modulus		
		c	I_x, I_1	I_y	Z_x	Z_y	r_x
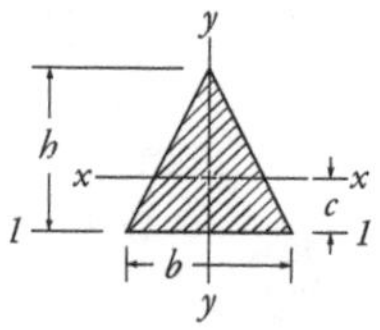 Triangle	$\frac{bh}{2}$	$\frac{h}{3}$	$I_x = \frac{bh^3}{36}$ $I_1 = \frac{bh^3}{12}$	$\frac{hb^3}{48}$	Apex $\frac{bh^2}{24}$ Base $\frac{bh^2}{12}$	$\frac{bh^2}{24}$	$4.24h$
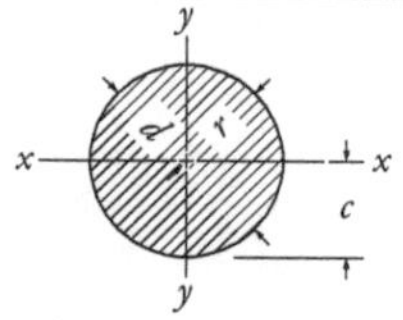 Circle	$3.14r^2$ $(= 0.785d^2)$	$r\left(=\frac{d}{2}\right)$	$0.785r^4$ $(= 0.0491d^4)$	$0.785r^4$	$0.785r^3$ $(= 0.0982d^3)$	$0.785r^3$	$0.5r$ $(= 0.25d)$
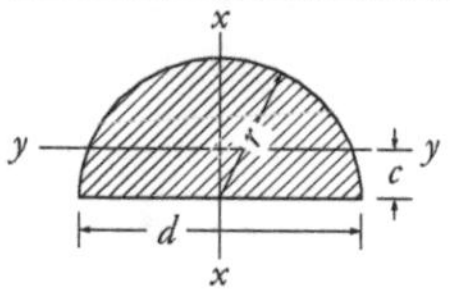 Semicircle	$1.57r^2$	$0.424r$	$0.393r^4$	$0.110r^4$	$0.393r^3$	Crown $0.191r^3$ Base $0.259r^3$	$0.264r$
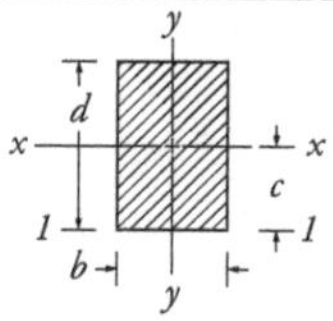 Rectangle	bd	$\frac{d}{2}$	$I_x = \frac{bd^3}{12}$ $I_1 = \frac{bd^3}{3}$	$\frac{db^3}{12}$	$\frac{bd^2}{6}$	$\frac{db^2}{6}$	$0.289d$

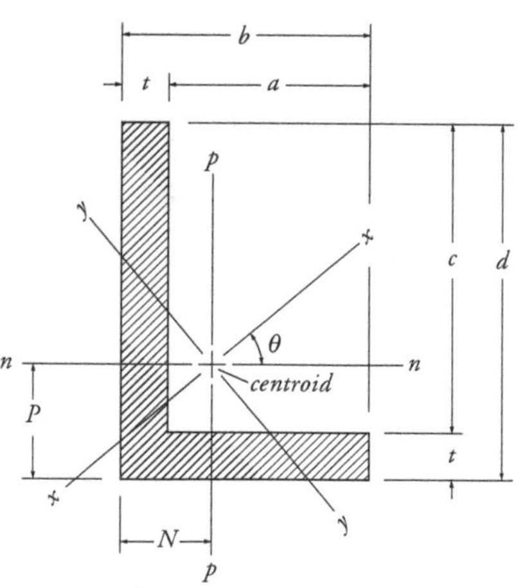

x- and y-axis are the major and minor principal axis respectively (with $I_{xy} = 0$). Minimum $I = I_y$. The product second moment of inertia about the n-, p-axis (I_{np}) is –ve when the heel of the angle (with respect to the centroid) is in the 1st (top right) or 3rd (bottom left) quadrants and positive otherwise.

$$\tan 2\theta = -\frac{2I_{np}}{I_n - I_p} \qquad A = t(b + c)$$

$$N = \frac{b^2 + ct}{2(b + c)} \qquad P = \frac{d^2 + at}{2(b + c)}$$

$$I_{np} = \pm \frac{abcdt}{4(b + c)}$$

$$I_n = \frac{1}{3}\left[t(d - P)^3 + bP^3 - a(P - t)^3\right]$$

$$I_p = \frac{1}{3}\left[t(b - N)^3 + dN^3 - c(N - t)^3\right]$$

$$I_x = \frac{I_n + I_p}{2} + \frac{I_n - I_p}{2\cos 2\theta}$$

$$I_y = \frac{I_n + I_p}{2} - \frac{I_n - I_p}{2\cos 2\theta}$$

Table C.4.1 Geometrical properties of plane sections (continued)

Section	Area A and Centroidal dist.	Second moment of area	Radius of gyration	Elastic section modulus
	A, c	I_x	r_x	Z_x
c, x, x, d_o, d	$A = \frac{\pi}{4}(d^2 - d_0^2)$ $c = \frac{d}{2}$	$I_x = \frac{\pi}{64}(d^4 - d_0^4)$	$r_x = \frac{1}{4}\sqrt{(d^2 + d_0^2)}$	$Z = \frac{\pi}{32d}(d^4 - d_0^4)$
t, c, x, x, d	$A = \pi(d - t)t$ $c = \frac{d}{2}$	$I_x \approx \frac{\pi}{8}(d - t)^3 t$ $\approx 0.393(d - t)^3 t$	$r_x \approx 0.354d$	$Z = \frac{\pi}{4}\frac{(d - t)^3 t}{d}$
c, x, d_o, d, x, b_o, b	$A = bd - b_0 d_0$ $c = \frac{d}{2}$	$I_x = \frac{1}{12}(bd^3 - b_0 d_0^3)$	$r_x = \sqrt{\frac{I_x}{A}}$	$Z_x = \frac{1}{6d}(bd^3 - b_0 d_0^3)$
c, t, x, d_o, d, x, t, b	$A = 2bt$ $c = \frac{d}{2}$	$I_x = \frac{b}{12}(d^3 - d_0^3)$	$r_x = \sqrt{\frac{I_x}{A}}$	$Z_x = \frac{b}{6d}(d^3 - d_0^3)$
b_1, c_1, t_1, y_1, d, x, c_2, t_2, y_2, b_2 $y_1 = \left(\frac{c_1 - t_1}{2}\right)$ $y_2 = \left(\frac{c_2 - t_2}{2}\right)$	$A = b_1 t_1 + b_2 t_2$ $c_1 = \frac{\frac{1}{2}b_1 t_1^2 + b_2 t_2\left(d - \frac{1}{2}t_2\right)}{A}$ $c_2 = d - c_1$	$I_x = \frac{b_1 t_1^3}{12} + \frac{b_2 t_2^3}{12} + b_1 t_1 y_1^2 + b_2 t_2 y_2^2$	$r_x = \sqrt{\frac{I_x}{A}}$	$Z_{top} = \frac{I_x}{c_1}$ $Z_{btm} = \frac{I_x}{c_2}$
c, T, x, t, d_o, d, x, T, b	$A = 2bT + (d - 2T)t$ $c = \frac{d}{2}$	$d_0 = d - 2T$ $I_x = \frac{b}{12}(d^3 - d_0^3) + \frac{1}{12}d_0^3 t$	$r_x = \sqrt{\frac{I_x}{A}}$	$Z_x = \frac{I_x}{c}$

C.5 Information on other construction materials

Table C.5.1 Cross-section area (mm²) of D500N reinforcing bars to AS/NZS 4671

No.	Bar size (mm) 10	12	16	20	24	28	32	36
1	80	110	200	310	450	620	800	1020
2	160	220	400	620	900	1240	1600	2040
3	240	330	600	930	1350	1860	2400	3060
4	320	440	800	1240	1800	2480	3200	4080
5	400	550	1000	1550	2250	3100	4000	5100
6	480	660	1200	1860	2700	3720	4800	6120
7	560	770	1400	2170	3150	4340	5600	7140
8	640	880	1600	2480	3600	4960	6400	8160
9	720	990	1800	2790	4050	5580	7200	9180
10	800	1100	2000	3100	4500	6200	8000	10200
11	880	1210	2200	3410	4950	6820	8800	11220
12	960	1320	2400	3720	5400	7440	9600	12240
13	1040	1430	2600	4030	5850	8060	10400	13260
14	1120	1540	2800	4340	6300	8680	11200	14280
15	1200	1650	3000	4650	6750	9300	12000	15300
16	1280	1760	3200	4960	7200	9920	12800	16320
17	1360	1870	3400	5270	7650	10540	13600	17340
18	1440	1980	3600	5580	8100	11160	14400	18360
19	1520	2090	3800	5890	8550	11780	15200	19380
20	1600	2200	4000	6200	9000	12400	16000	20400

Table C.5.2 Cross-sectional area of D500N bars per metre width (mm²/m) to AS/NZS 4671

Bar spacing (mm)	Bar size (mm) 10	12	16	20	24	28	32	36
50	1600	2200						
75	1067	1467	2667					
100	800	1100	2000	3100	4500			
125	640	880	1600	2480	3600	4960		
150	533	733	1333	2067	3000	4133	5338	6800
175	457	629	1143	1771	2571	3543	4571	5828
200	400	550	1000	1550	2250	3100	4000	5100
250	320	440	800	1240	1800	2480	3200	4080
300	267	367	667	1033	1500	2067	2667	3400
350	229	314	571	886	1286	1771	2286	2914
400	200	275	500	775	1125	1550	2000	2550
500	160	220	400	620	900	1240	1600	2040
600	133	183	333	517	750	1033	1333	1700
1000	80	110	200	310	450	620	800	1020

Table C.5.3 Reinforcing fabric to AS/NZS 4671

Ref. No.	Longitudinal Wires		Cross Wires		Area of Cross-Section		Mass per Unit Area
	Size	Pitch	Size	Pitch	Long l	Cross	
	mm	mm	mm	mm	mm²/m	mm²/m	kg/m²
Rectangular Meshes							
RL1218	11.90	100	7.6	200	1112	227	10.9
RL1118	10.65	100	7.6	200	891	227	9.1
RL1018	9.50	100	7.6	200	709	227	7.6
RL918	8.55	100	7.6	200	574	227	6.5
RL818	7.60	100	7.6	200	454	227	5.6
RL718	6.75	100	7.6	200	358	227	4.7
Square Meshes							
SL81	7.60	100	7.60	100	454	454	7.3
SL102	9.50	200	9.50	200	354	354	5.6
SL92	8.55	200	8.55	200	287	287	4.5
SL82	7.60	200	7.60	200	227	227	3.6
SL72	6.75	200	6.75	200	179	179	2.8
SL62	6.00	200	6.00	200	141	141	2.3
SL52	4.75	200	4.75	200	89	89	1.5
SL42	4.0	200	4.0	200	63	63	1.0

Table C.5.4 Dimensions of ribbed hard-drawn reinforcing wire (D500L) to AS/NZS 4671

Size mm	Area mm²	Mass per Unit Length, kg/m
4.0	12.6	0.099
4.75	17.7	0.139
6.00	28.3	0.222
6.75	35.8	0.251
7.60	45.4	0.356
8.55	57.4	0.451
9.50	70.9	0.556
10.65	89.1	0.699
11.90	111.2	0.873

Table C.5.5 Metric brickwork measurements

Metric standard brick
Size: 230 × 110 × 76 mm

No. of Bricks	Length of Wall	Width of Opening	No. of Courses	Height of Brickwork
1	230	250	1	86
$1\frac{1}{2}$	350	370	2	172
2	470	490	3	258
$2\frac{1}{2}$	590	610	4	344
			5	430
3	710	730	6	515
$3\frac{1}{2}$	830	850	7	600
4	950	970		
$4\frac{1}{2}$	1070	1090	8	686
5	1190	1210	9	772
			10	858
$5\frac{1}{2}$	1310	1330	11	944
6	1430	1450	12	1030
$6\frac{1}{2}$	1550	1570	13	1115
7	1670	1690	14	1200
$7\frac{1}{2}$	1790	1810		
			15	1286
8	1910	1930	16	1372
$8\frac{1}{2}$	2030	2050	17	1458
9	2150	2170	18	1544
$9\frac{1}{2}$	2270	2290	19	1630
10	2390	2410	20	1715
			21	1800
$10\frac{1}{2}$	2510	2530		
11	2630	2650	22	1886
$11\frac{1}{2}$	2750	2770	23	1972
12	2870	2890	24	2058
$12\frac{1}{2}$	2990	3010	25	2144
13	3110	3130	26	2230
$13\frac{1}{2}$	3230	3250	27	2315
14	3350	3370	28	2400
$14\frac{1}{2}$	3470	3490	29	2486
15	3590	3610	30	2572
$15\frac{1}{2}$	3710	3730	31	2658
16	3830	3850	32	2744
$16\frac{1}{2}$	3950	3970	33	2830
17	4070	4090	34	2915
$17\frac{1}{2}$	4190	4210	35	3000

Metric 'modular' brick
Size: 290 × 90 × 90 mm

No. of Bricks	Length of Wall	Width of Opening	No. of Courses	Height of Brickwork
1	290	310	1	100
$1\frac{1}{3}$	390	410	2	200
$1\frac{2}{3}$	490	510	3	300
2	590	610	4	400
			5	500
$2\frac{1}{3}$	690	710	6	600
$2\frac{2}{3}$	790	810		
3	890	910	7	700
$3\frac{1}{3}$	990	1010	8	800
$3\frac{2}{3}$	1090	1110	9	900
4	1190	1210	10	1000
			11	1100
$4\frac{1}{3}$	1290	1310	12	1200
$4\frac{2}{3}$	1390	1410		
5	1490	1510	13	1300
$5\frac{1}{3}$	1590	1610	14	1400
$5\frac{2}{3}$	1690	1710	15	1500
6	1790	1810	16	1600
			17	1700
$6\frac{1}{3}$	1890	1910	18	1800
$6\frac{2}{3}$	1990	2010		
7	2090	2110	19	1900
$7\frac{1}{3}$	2190	2210	20	2000
$7\frac{2}{3}$	2290	2310	21	2100
8	2390	2410	22	2200
			23	2300
$8\frac{1}{3}$	2490	2510	24	2400
$8\frac{2}{3}$	2590	2610		
9	2690	2710	25	2500
$9\frac{1}{3}$	2790	2810	26	2600
$9\frac{2}{3}$	2890	2910	27	2700
10	2990	3010	28	2800
			29	2900
$10\frac{1}{3}$	3090	3110	30	3000
$10\frac{2}{3}$	3190	3210		
11	3290	3310	31	3100
$11\frac{1}{3}$	3390	3410	32	3200
$11\frac{2}{3}$	3490	3510	33	3300
12	3590	3610	34	3400

Notes:

1. Length of wall or pier: $n \times$ (Brick + Joint) − 10 mm.
2. Width of openings: $n \times$ (Brick + Joint) + 10 mm.
3. Height of brickwork: $n \times$ (Brick + Joint).
4. Brick joints are to be 10 mm normal.

C.6 General formulae—miscellaneous

Table C.6.1 Bracing formulae

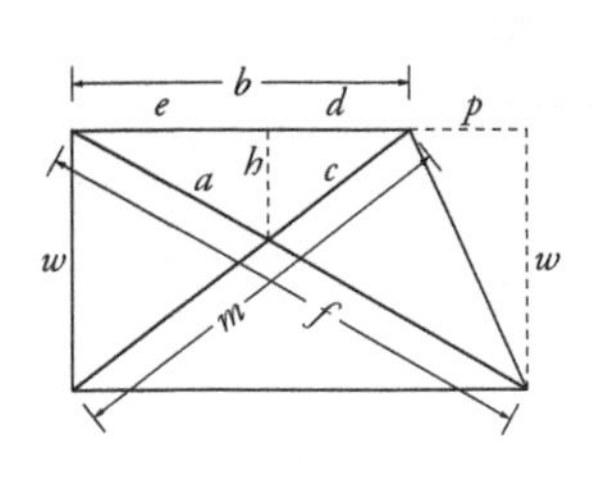

$f = \sqrt{(b + p)^2 + w^2}$

$m = \sqrt{b^2 + w^2}$

$d = b^2 \div (2b + p)$

$e = b(b + p) \div (2b + p)$

$a = bf \div (2b + p)$

$c = bm \div (2b + p)$

$h = bw \div (2b + p)$

$= aw \div f$

$= cw \div m$

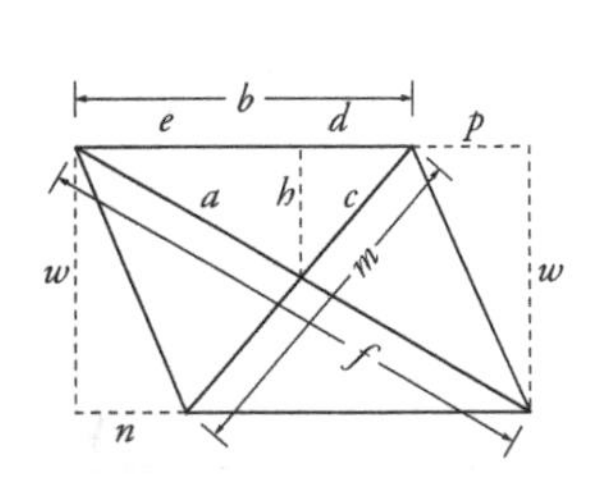

$f = \sqrt{(b + p)^2 + w^2}$

$m = \sqrt{(b - n)^2 + w^2}$

$d = b(b - n) \div (2b + p - n)$

$e = b(b + p) \div (2b + p - n)$

$d = bf \div (2b + p - n)$

$c = bm \div (2b + p - n)$

$h = bw \div (2b + p - n)$

$= aw \div f$

$= cw \div m$

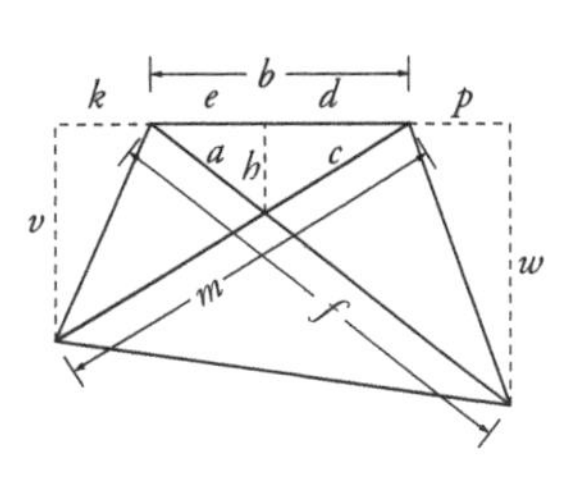

$f = \sqrt{(b + p)^2 + w^2}$

$m = \sqrt{(b + k)^2 + v^2}$

$d = bw(b + k) \div [v(b + p) + w(b + k)]$

$e = bv(b + p) \div [v(b + p) + w(b + k)]$

$a = fbv \div [v(b + p) + w(b + k)]$

$c = bmw \div [v(b + p) + w(b + k)]$

$h = bvw \div [v(b + p) + w(b + k)]$

$= aw \div f$

$= cv \div m$

Table C.6.2 Trigonometric formulae

$$a^2 = c^2 - b^2$$
$$b^2 = c^2 - a^2$$
$$c^2 = a^2 + b^2$$

Known	Required					
	α	β	a	b	c	Area
a, b	$\tan\alpha = \frac{a}{b}$	$\tan\beta = \frac{b}{a}$			$\sqrt{a^2 + b^2}$	$\frac{ab}{2}$
a, c	$\sin\alpha = \frac{a}{c}$	$\cos\beta = \frac{a}{c}$		$\sqrt{c^2 - a^2}$		$\frac{a\sqrt{c^2 - a^2}}{2}$
α, a		$90° - \alpha$		$a \cot\alpha$	$\frac{a}{\sin\alpha}$	$\frac{a^2 \cot\alpha}{2}$
α, b		$90° - \alpha$	$b \tan\alpha$		$\frac{b}{\cos\alpha}$	$\frac{b^2 \tan\alpha}{2}$
α, c		$90° - \alpha$	$c \sin\alpha$	$c \cos\alpha$		$\frac{c^2 \sin 2\alpha}{4}$

Let:

$$s = \frac{a + b + c}{2}$$

then:

$$a^2 = b^2 + c^2 - 2bc\cos\alpha$$
$$b^2 = a^2 + c^2 - 2ac\cos\beta$$
$$c^2 = a^2 + b^2 - 2ab\cos\gamma$$
$$K = \sqrt{\frac{(s - a)(s - b)(s - c)}{s}}$$

Known	Required				
	α	β	γ	b	c
a, b, c	$\tan\frac{\alpha}{2} = \frac{K}{s - a}$	$\tan\frac{\beta}{2} = \frac{K}{s - b}$	$\tan\frac{\gamma}{2} = \frac{K}{s - c}$	—	—
a, α, β	—	—	$180° - (\alpha + \beta)$	$\frac{a\sin\beta}{\sin\alpha}$	$\frac{a\sin\gamma^*}{\sin\alpha}$
a, b, α	—	$\sin\beta = \frac{b\sin\alpha}{a}$	$180° - (\alpha + \beta)^*$	—	$\frac{b\sin\gamma^*}{\sin\beta}$
a, b, γ	$\tan\alpha = \frac{a\sin\gamma}{b - a\cos\gamma}$	$180° - (\alpha + \gamma)^*$	—	—	$\sqrt{a^2 + b^2 - 2ab\cos\gamma}$

$$\text{Area} = sK = \sqrt{s(s - a)(s - b)(s - c)} = \frac{ab\sin\gamma}{2} = \frac{bc\sin\alpha}{2} = \frac{ac\sin\beta}{2}$$

Note: * indicates a non-dependant variable calculated elsewhere in this row is used.

C.7 Conversion factors

Table C.7.1 Conversion factors

Imperial to SI metric			SI metric to Imperial		
Plane angle					
1 degree	=	0.017 453 3 rad	1 rad	=	57.2958 degree
1 minute	=	0.290 888 × 10^{-3} rad	1 rad	=	3437.75 minute
1 second	=	4.848 14 × 10^{-6} rad	1 rad	=	206 265 second
Length					
1 mile	=	1.609 344 km	1 km	=	0.621 371 mile
1 chain	=	20.1168 m	1 km	=	49.7097 chain
1 yd	=	0.9144 m	1 m	=	1.093 61 yd
1 ft	=	0.3048 m	1 m	=	3.280 84 ft
1 in	=	25.4 mm	1 m	=	39.3701 in
Area					
1 $mile^2$	=	2.589 99 km^2	1 km^2	=	0.386 102 $mile^2$
1 acre	=	0.404 686 ha	1 ha	=	2.471 05 acre
1 acre	=	4046.86 m^2	1 m^2	=	0.247 105 × 10^{-3} acres
1 yd^2	=	0.836 127 m^2	1 m^2	=	1.195 99 yd^2
1 ft^2	=	0.092 903 0 m^2	1 m^2	=	10.7639 ft^2
1 in^2	=	645.16 mm^2	1 mm^2	=	0.001 55 in^2
Volume, etc.					
1 acre.ft	=	1233.48 m^3	1 m^3	=	0.810 712 × 10^{-3} acre.ft
1 yd^3	=	0.764 555 m^3	1 m^3	=	1.307 95 yd^3
1 ft^3	=	0.028 316 8 m^3	1 m^3	=	35.3147 ft^3
100 super ft	=	0.235 973 m^3	1 m^3	=	423.776 super ft
1 ft^3	=	28.3168 litre	1 litre	=	0.035 314 7 ft^3
1 gal (imp.)	=	4.546 09 litre	1 litre	=	0.219 969 gal (imp.)
1 gal (US)	=	3.785 litre			
1 in^3	=	16.3871 × 10^3 mm^3	1 mm^3	=	0.061 023 6 × 10^{-3} in^3
1 in^3	=	16.3871 mL	1 mL	=	0.061 023 6 in^3
Second moment of area					
1 in^4	=	0.416 231 × 10^6 mm^4	1 mm^4	=	2.402 51 × 10^{-6} in^4
Mass					
1 ton (imp)	=	1.016 05 t	1 t	=	0.984 206 ton (imp)
1 lb	=	0.453 592 37 kg	1 kg	=	2.204 62 lb
1 oz	=	28.3495 g	1 g	=	0.035 274 oz
Mass/unit length					
1 lb/ft	=	1.488 16 kg/m	1 kg/m	=	0.671 971 lb/ft
1 lb/100 yd	=	4.960 55 g/m	1 g/m	=	0.201 591 lb/100 yd
1 lb/mile	=	0.281 849 g/m	1 g/m	=	3.548 lb/mile
Mass/unit area					
1 lb/ft^2	=	4.882 43 kg/m^2	1 kg/m^2	=	0.204 816 lb/ft^2
1 oz/yd^2	=	33.9057 g/m^2	1 g/m^2	=	0.029 494 oz/yd^2
1 oz/ft^2	=	305.152 g/m^2	1 g/m^2	=	0.003 277 06 oz/ft^2

Table C.7.1 Conversion factors (continued)

Imperial to SI metric			SI metric to Imperial		
Mass/unit time					
1 lb/s	=	0.453 592 kg/s	1 kg/s	=	2.204 62 lb/s
1 ton/h	=	1.016 05 t/h	1 t/h	=	0.984 207 ton/h
Density (mass/unit volume)					
1 lb/ft^3	=	16.0185 kg/m^3	1 kg/m^3	=	0.062 428 lb/ft^3
1 lb/yd^3	=	0.593 278 kg/m^3	1 kg/m^3	=	1.685 56 lb/yd^3
1 ton (imp)/yd^3	=	1.328 94 t/m^3	1 t/m^3	=	0.752 48 ton (imp)/yd^3
Force					
1 lbf	=	4.448 22 N	1 N	=	0.224 809 lbf
1 tonf (imp)	=	9.964 02 kN	1 kN	=	0.100 361 tonf (imp)
Moment of force (torque)					
1 kip.in	=	0.112 985 kN.m	1 kN.m	=	8.850 75 kip.in
1 kip.ft	=	1.355 82 kN.m	1 kN.m	=	0.737 562 kip.ft
1 tonf (imp).ft	=	3.037 04 kN.m	1 kN.m	=	0.329 269 tonf (imp).ft
Force / unit length					
1 lbf/ft	=	14.5939 N/m	1 N/m	=	0.068 522 lbf/ft
1 tonf (imp)/ft	=	32.6903 kN/m	1 kN/m	=	0.030 590 tonf (imp)/ft
Modulus of elasticity, pressure, stress					
1 tonf (imp)/in^2	=	15.4443 MPa	1 MPa	=	0.064 749 tonf (imp)/in^2
1 kip/in^2	=	6.894 76 MPa	1 MPa	=	0.145 038 kip/in^2
1 tonf (imp)/ft^2	=	107.252 kPa	1 MPa	=	9.323 85 tonf (imp)/ft^2
1 kip/ft^2	=	47.8803 kPa	1 kPa	=	0.020 885 4 kip/ft^2
Work, energy, heat					
1 lb.ft	=	1.355 82 J	1 J	=	0.737 562 lbf.ft
1 Btu	=	1055.06 J = 0.293 W.h	1 J	=	0.947 813 $\times 10^{-3}$ Btu
Power, rate of heat flow					
1 hp	=	0.7457 kW	1 kW	=	1.341 02 hp
1 Btu/h	=	0.293 071 W	1 W	=	3.412 14 Btu/h
Thermal conductivity					
1 Btu/(ft.h.°F)	=	1.730 73 W/(m.K)	1 W/(m.K)	=	0.577 789 Btu/(ft.h.°F)
Coefficient of heat transfer					
1 Btu/(ft^2.h.°F)	=	5.678 26 W/(m^2.K)	1 W/(m^2.K)	=	0.176 10 Btu/(ft^2.h.°F)

Table C.7.1 Conversion factors (continued)

Imperial to SI metric			SI metric to Imperial		
Temperature value					
°C	=	$\frac{5}{9}$ (°F − 32)	°F	=	$\frac{9}{5}$°C + 32
Temperature interval					
1°C	=	$\frac{5}{9}$°F	1°F	=	$\frac{9}{5}$°C
Velocity, speed					
1 ft/s	=	0.3048 m/s	1 m/s	=	3.280 84 ft/s
				=	3.600 km/h
1 mile/h	=	1.609 344 km/h	1 km/h	=	0.621 371 mile/h
				=	0.2778 m/s
1 mile/h	=	0.447 04 m/s	1 m/s	=	2.236 94 mile/h
1 knot	=	1.852 km/h	1 km/h	=	0.5340 knot
1 knot	=	0.514 m/s	1 m/s	=	1.943 knot
1 knot	=	1.151 mile/h			
Acceleration					
1 ft/s^2	=	0.3048 m/s^2	1 m/s^2	=	3.280 84 ft/s^2
Volumetric flow					
1 ft^3/s	=	0.028 316 8 m^3/s	1 m^3/s	=	35.3147 ft^3/s
1 ft^3/min	=	0.471 947 litre/s	1 litre/s	=	2.118 88 ft^3/min
1 gal (imp)/min	=	0.075 682 litre/s	1 litre/s	=	13.1981 gal (imp)/min
1 gal (imp)/h	=	1.262 80 × 10^{-3} litre/s	1 litre/s	=	791.888 gal (imp)/h
1 million gal/day	=	0.052 6168 m^3/s	1 m^3/s	=	19.0053 million gal (imp)/day
1 acre ft/s	=	1233.481 m^3/s	1 m^3/s	=	0.8107 × 10^{-3} acre.ft/s

Table C.7.1 Conversion factors (continued)

Conversion from Imperial, or MKS, to SI units	Conversion from SI to Imperial or MKS units
(i) Force	
pound force	*newton*
1 lbf = 4.448 20 N	1 N = 0.224 81 lbf
= 4.448 20 × 10^5 dyne	= 7.233 pdl
= 32.174 pdl	= 0.101 97 kgf
kilopound force	*kilonewton*
1 kip = 4.448 kN	1 kN = 224.81 lbf
1 kip = 1000 lbf	= 0.2248 kip
ton force (long)	= 0.1003 tonf
1 tonf = 9.964 kN	= 101.972 kgf
1 tonf = 2240 lbf	= 0.101 97 megapond
poundal	
1 pdl = 0.138 26 N	
1 pdl = 0.0311 lbf	
kilogram force	
1 kgf = 9.806 65 N	
1 kgf = 2.2046 lbf	
= 70.931 pdl	
megapond (technical unit)	
1 Mp = 9.806 65 kN	
1 Mp = 1000 kgf	
= 2204.6 lbf	
megadyne (c.g.s. unit)	
1 Mdyn = 10.00 N	
1 Mdyn = 2.2481 lbf	
= 72.33 pdl	
(ii) Line load	
pound force per ft run	*newton per metre*
1 lbf/ft = 14.594 N/m	1 N/m = 0.0685 lbf/ft
kilopound per ft run	*kilonewton per metre*
1 kip/ft = 14.594 kN/m	1 kN/m = 68.52 lbf/ft
	= 0.0306 tonf/ft
ton force per ft run	= 0.0685 kip/ft
1 ton/ft = 32.690 kN/m	= 101.97 kgf/m
kilogramforce per metre	
1 kgf/m = 9.807 N/m	
= 0.6720 lbf/ft	

Table C.7.1 Conversion factors (continued)

Conversion from Imperial, or MKS, to SI units	Conversion from SI to Imperial or MKS units
(iii) Stress, pressure, distributed load	
pound force per sq in 1 lbf/sq in = 6.895 kPa 1 lbf/sq in = 0.006895 MPa *ton force per sq in* (long ton) 1 tonf (imp)/sq in = 15.44 MPa = 2240 lbf/sq in *ton force per sq ft* 1 tonf (imp)/sq ft = 107.25 kPa *kilopound force per sq in* 1 ksi = 6.895 MPa *kilopound force per sq ft* 1 ksf = 47.88 kPa *kilogram force per sq cm* 1 kgf/cm^2 = 98.07 kPa = 14.223 lbf/sq in *atmosphere* 1 atm = 101.3 kPa = 14.696 lbf/sq in *bar* 1 bar = 14.504 lbf/sq in = 100 kPa = 1.02 kgf/cm^2	*megapascal* *(meganewton per sq metre)* 1 MPa = 145.04 lbf/sq in = 0.145 ksi = 20.885 ksf = 9.32 tonf (imp)/sq ft *kilopascal* 1 kPa = 0.020 89 ksf = 0.009 32 tonf (imp)/sq ft *pascal* 1 Pa = 1.45×10^{-4} lbf/sq in = 1 N/m^2
(iv) Bending, moment, torque	
pound force inch 1 lbf.in = 0.1130 N.m *pound force foot* 1 lbf.ft = 1.356 N.m *ton force (imp) inch* 1 tonf.in = 0.2531 kN.m *ton force (imp) foot* 1 tonf.ft = 3.0372 kN.m *kilopound force inch* 1 kip.in = 0.1130 kN.m *kilopound force foot* 1 kip.ft = 1.356 kN.m *kilogram force centimetre* 1 kgf.cm = 0.0981 N.m = 0.0723 lbf.ft *megapond metre* 1 Mp.m = 9.81 kN.m	*newton metre* 1 N.m = 8.851 lbf.in = 0.7376 lbf.ft = 10.20 kgf.cm *kilonewton metre* 1 kN.m = 737.6 lbf.ft = 3.951 tonf (imp).in = 0.3293 tonf (imp).ft = 8.851 kip.in = 0.7375 kip.ft = 0.1020 Mp.m

Appendix D

Amendment No. 1 to AS 4100 – 1998 (AS 4100 AMD 1)

As noted in Section 1.14, this Appendix contains Amendment No.1 to AS 4100-1998 (AS 4100 AMD 1) which was published on 29 February 2012. In conjunction with the rest of the Handbook, the inclusion of AS 4100 AMD 1 allows the reader to fully consider the amendments to the last major revision of AS 4100.

The Authors and Publishers gratefully acknowledge SAI Global Ltd under Licence 1206-c009 for providing permission to reproduce AS 4100 AMD 1 in this Handbook. For the information of readers, copies of AS 4100 can be purchased online at http://www.saiglobal.com .

AS 4100/Amdt 1/2012-02-29

STANDARDS AUSTRALIA

Amendment No. 1
to
AS 4100—1998
Steel structures

REVISED TEXT

The 1998 edition of AS 4100 is amended as follows; the amendments should be inserted in the appropriate places.

SUMMARY: This Amendment applies to the Preface, Clauses 1.1.1, 1.1.2, 1.3, 1.4, 2.2.1, 2.2.2, 2.3.1, 2.3.3, 2.3.4, 3.2.1, 3.2.3, 3.2.4, 3.2.5(new), 3.3, 3.11, 4.1.1, 5.13.1, 5.13.3, 5.13.4, 5.13.5, 6.3.3, 8.3.2, 8.3.3, 8.3.4, 8.4.2.2, 9.1.4, 9.1.9, 9.3.1, 9.3.2.5, 9.3.3.3, 9.7.1.1, 9.7.1.3, 9.7.2.1, 9.7.2.3, 9.7.2.7, 10.2, 10.3.1, 10.3.2, 10.3.3(new), 10.4.2, 10.4.3.1, 10.4.3.2, 10.4.3.3, 10.4.3.4(new), 10.5, 11.1.5, 11.2, 11.8.2, 12.5, 12.6.1, 12.6.2.2, Section 13, Clauses 14.3.3, 14.3.4, 14.3.5.1, 14.3.5.2, 14.4.3, 15.2.1 and 15.2.5.3, Appendices A, B and C, Tables 2.1, 3.4, 5.6.1, 6.3.3(1), 6.3.3(3), 9.3.1, 9.6.2, 9.7.3.10(1), 10.4.1, 10.4.4, 11.5.1(4) and 15.2.5.1, Figures 4.6.3.2 and 9.7.3.1, and Index.

Published on 29 February 2012.

AMDT No. 1 FEB 2012

Preface

Insert the following text before the last paragraph:

Amendment No. 1—2012 to the 1998 edition includes the following major changes:

(a) Revisions to AS/NZS 1163, AS/NZS 3678, AS/NZS 3679.1 and AS/NZS 3679.2 reflected by amendments to Sections 2 and 10.

(b) Revisions to AS/NZS 1554.1, AS/NZS 1554.4 and AS/NZS 1554.5 reflected by amendments to Sections 9 and 10.

(c) Section 13 brought into line with revisions to AS 1170.4.

(d) Quenched and tempered steels included by adding 'AS 3597' to listed material Standards in Section 2.

(e) Typographical errors corrected.

AMDT No. 1 FEB 2012

Clause 1.1.1

1 *Delete* third paragraph relating to bridges.

2 Item (a), *delete* 'AS 1163' and *replace* with 'AS/NZS 1163'.

3 *Delete* Item (b) and *replace* with the following:

(b) Steel members for which the value of the yield stress used in design (f_y) exceeds 690 MPa.

4 Item (c), *delete* 'AS 1163' and *replace* with 'AS/NZS 1163'.

5 *Add* new Item (e) as follows:

(e) Road, railway and pedestrian bridges, which shall be designed in accordance with AS 5100.1, AS 5100.2 and AS 5100.6.

AMDT No. 1 FEB 2012

Clause 1.1.2

Delete title and text of clause.

AMDT No. 1 FEB 2012

Clause 1.3

1 *Delete* the following:

Building frame system—see AS 1170.4.

Concentric braced frame—see AS 1170.4.

Design action or design load—the combination of the nominal actions or loads and the load factors, as specified in AS 1170.1, AS 1170.2, AS 1170.3 or AS 1170.4.

Drift—see AS 1170.4.

Dual system—see AS 1170.4.

Earthquake design category—see AS 1170.4.

Earthquake resisting system—see AS 1170.4.

Eccentric braced frame—see AS 1170.4.

Intermediate moment resisting frame—see AS 1170.4.

Moment resisting frame system—see AS 1170.4.

Ordinary moment resisting frame—see AS 1170.4.

Special moment resisting frame—see AS 1170.4.

2 For '*Ductility*' *add* '(*of structure*)'.

3 *Add* the following terms and definitions:

Braced frame, concentric—see AS 1170.4.

Braced frame, eccentric—see AS 1170.4.

Design action or design load—the combination of the nominal actions or loads and the load factors specified in AS/NZS 1170.0, AS/NZS 1170.1, AS/NZS 1170.2, AS/NZS 1170.3, AS 1170.4 or other standards referenced in Clause 3.2.1.

Moment-resisting frame—see AS 1170.4.

Moment-resisting frame, intermediate—see AS 1170.4.

Moment-resisting frame, ordinary—see AS 1170.4.

Moment-resisting frame, special—see AS 1170.4.

Quenched and tempered steel—high strength steel manufactured by heating, quenching, tempering and levelling steel plate.

Structural ductility factor—see AS 1170.4.

Structural performance factor—see AS 1170.4.

AMDT No. 1 FEB 2012

Clause 1.4

1 *Replace* the definitions of the following symbols:

A_e = effective sectional area of a hollow section in shear; *or*

= effective area of a compression member

A_{ep} = area of an end post

A_n	=	net area of a cross-section
A_o	=	nominal plain shank area of a bolt
b_{fo}	=	half the clear distance between the webs; *or*
	=	least of 3 dimensions defined in Clause 5.11.5.2
c_m	=	factor for unequal end moments
d	=	depth of a section; *or*
	=	depth of preparation for incomplete penetration butt weld; *or*
	=	maximum cross-sectional dimension of a built-up compression member
d_b	=	lateral distance between centroids of the welds or fasteners connecting battens to main components
d_5	=	flat width of web of hollow sections
i	=	number of loading event (Section 11)
k_l	=	effective length factor for load height
k_t	=	effective length factor for twist restraints; *or*
	=	correction factor for distribution of forces in a tension member
k_v	=	ratio of flat width of web (d_5) to thickness (t) of hollow section
l	=	span; or
	=	member length; *or*
	=	member length from centre to centre of its intersections with supporting members; *or*
	=	segment or sub-segment length
N_c	=	nominal member capacity in axial compression
N_{om}	=	elastic buckling load
R_{bb}	=	nominal bearing buckling capacity of a web
R_{by}	=	nominal bearing yield capacity of a web
r	=	radius of gyration
r_{ext}	=	outside radius of hollow section
t	=	thickness; *or*
	=	element thickness; *or*
	=	thickness of thinner part joined; *or*
	=	wall thickness of a circular hollow section; *or*
	=	thickness of an angle section; *or*
	=	time
t_p	=	thickness of a ply; *or*
	=	thickness of thinner ply connected; *or*
	=	thickness of a plate
	=	connecting plate thickncss(es) at a pin

t_w = thickness of a web or web panel

t_w, t_{w1}, t_{w2} = leg lengths of a fillet weld used to define the size of a fillet weld

V^* = design shear force; *or*

= design horizontal storey shear force at column ends; *or*

= design transverse shear force

γ = index used in Clause 8.3.4; *or*

= factor for transverse stiffener arrangement in stiffened web (Clause 5.15.3)

μ = slip factor

= structural ductility factor

2 *Delete* the following:

R_f = structural response factor

3 *Add* the following symbols, including definitions:

A_{gv} = gross area subject to shear at rupture

A_{nt} = net area subject to tension at rupture

A_{nv} = net area subject to shear at rupture

f_{uc} = minimum tensile strength of connection element

f_{yc} = yield stress of connection element

k_{bs} = a factor to account for the effect of eccentricity on the block shear capacity

R_{bs} = nominal design capacity in block shear

R^*_{bs} = design reaction

ϕR_u = design capacity

λ_w, λ_{ew} = values of λ_e and λ_{ey} for the web

μ_i = individual test result from test for slip factor

β_{tf} = thickness correction factor for fatigue

S_p = structural performance factor

4 Symbol α_p, definition, *delete* 'AS 1163' and *replace* with AS/NZS 1163.

AMDT No. 1 FEB 2012

Clause 2.2.1

1 *Add* the following Standard to the list:

AS 3597 Structural and pressure vessel steel—Quenched and tempered plate

2 *Delete* 'AS 1163 Structural steel hollow sections' and *replace* with the following:

AS/NZS 1163 Cold-formed structural steel hollow sections

AMDT No. 1 FEB 2012

Clause 2.2.2

Delete the text of the clause and *replace* with the following:

Test reports or test certificates that comply with the minimum requirements of the appropriate Standard listed in Clause 2.2.1 shall constitute sufficient evidence of compliance of the steel with the Standards listed in Clause 2.2.1. The test reports or test certificates shall be provided by the manufacturer or an independent laboratory accredited by signatories to the International Laboratory Accreditation Corporation (Mutual Recognition Arrangement) ILAC MRA or the Asia Pacific Laboratory Accreditation Cooperation (APLAC) on behalf of the manufacturer. In the event of a dispute as to the compliance of the steel with any of the Standards listed in Clause 2.2.1, the reference testing shall be carried out by independent laboratories accredited by signatories to ILAC MRA or APLAC.

AMDT No. 1 FEB 2012

Clause 2.3.1

1 *Delete* referenced documents AS/NZS 1110, AS/NZS 1111, AS/NZS 1112, AS/NZS 1252 and AS/NZS 1559 and *replace* as follows:

AS 1110 ISO metric hexagon bolts and screws—Product grades A and B (series)

AS 1111 ISO metric hexagon bolts and screws—Product grade C (series)

AS 1112 ISO metric hexagon nuts (series)

AS/NZS 1252 High strength steel bolts with associated nuts and washers for structural engineering

AS/NZS 1559 Hot-dip galvanized steel bolts with associated nuts and washers for tower construction

2 *Add* the following paragraph and Note to follow the listed Standards:

'Test certificates that state that the bolts, nuts and washers comply with all the provisions of the appropriate Standard listed in Clause 2.3.1 shall constitute sufficient evidence of compliance with the appropriate Standard. Such test reports shall be provided by the bolt manufacturer or bolt importer and shall be carried out by an independent laboratory accredited by signatories to the International Laboratory Accreditation Corporation (Mutual Recognition Arrangement) ILAC MRA or the Asia Pacific Laboratory Accreditation Cooperation (APLAC) on behalf of the manufacturer, importer or customer. In the event of a dispute as to the compliance of the bolt, nut or washer with any of the Standards listed in Clause 2.3.1, the reference testing shall be carried out by independent laboratories accredited by signatories to ILAC MRA or APLAC.

NOTE: Acceptable bolts and associated bolting categories are specified in Table 9.3.1.

AMDT No. 1 FEB 2012

Clause 2.3.3

Delete the text of the clause and *replace* with the following:

All welding consumables and deposited weld metal for steel parent material with a specified yield strength ≤ 500 MPa shall comply with AS/NZS 1554.1 except when welding to quenched and tempered steel according to AS 3597, where the welding consumables and deposited weld metal for steel parent material with a specified yield strength ≤ 690 MPa shall comply with AS/NZS 1554.4. Where required by Clause 11.1.5, the welds shall comply with AS/NZS 1554.5.

AMDT No. 1 FEB 2012

Clause 2.3.4

Delete 'AS 1554.2' and *replace* with 'AS/NZS 1554.2'.

AMDT No. 1 FEB 2012

Table 2.1

Delete Table 2.1 including the footnote and Note 2 (but excluding Note 1) and *replace* with the following:

TABLE 2.1

STRENGTHS OF STEELS COMPLYING WITH AS/NZS 1163, AS/NZS 1594, AS/NZS 3678, AS/NZS 3679.1, AS/NZS 3679.2 (Note 2) AND AS 3597

Steel Standard	Form	Steel grade	Thickness of material, t mm	Yield stress (f_y) MPa	Tensile strength (f_u) MPa
AS/NZS 1163 (Note 3)	Hollow sections	C450	All	450	500
		C350	All	350	430
		C250	All	250	320
AS/NZS 1594	Plate, strip, sheet floorplate	HA400	All	380	460
		HW350	All	340	450
		HA350	All	350	430
		HA300/1 HU300/1	All	300	430
		HA300 HU300	All	300	400
		HA250 HA250/1 HU250	All	250	350
		HA200	All	200	300
	Plate and strip	HA4N	All	170	280
		HA3	All	200	300
		HA1	All	(See Note 1)	(See Note 1)
		XF500	$t \leq 8$	480	570
		XF400	$t \leq 8$	380	460
		XF300	All	300	440
AS/NZS 3678 (Note 2 and 3)	Plate and floorplate	450	$t \leq 20$	450	520
		450	$20 < t \leq 32$	420	500
		450	$32 < t \leq 50$	400	500
		400	$t \leq 12$	400	480
		400	$12 < t \leq 20$	380	480
		400	$20 < t \leq 80$	360	480
		350	$t \leq 12$	360	450
		350	$12 < t \leq 20$	350	450
		350	$20 < t \leq 80$	340	450

(continued)

TABLE 2.1 (*continued*)

Steel Standard	Form	Steel grade	Thickness of material, t mm	Yield stress (f_y) MPa	Tensile strength (f_u) MPa
AS/NZS 3678 (Note 3)		350	$80 < t \leq 150$	330	450
		WR350	$t \leq 50$	340	450
		300	$t \leq 8$	320	430
		300	$8 < t \leq 12$	310	430
		300	$12 < t \leq 20$	300	430
		300	$20 < t \leq 50$	280	430
		300	$50 < t \leq 80$	270	430
		300	$80 < t \leq 150$	260	430
		250	$t \leq 8$	280	410
		250	$8 < t \leq 12$	260	410
		250	$12 < t \leq 50$	250	410
		250	$50 < t \leq 80$	240	410
		250	$80 < t \leq 150$	230	410
		200	$t \leq 12$	200	300
AS/NZS 3679.1 (Note 3)	Flats and sections	350	$t \leq 11$	360	480
		350	$11 < t < 40$	340	480
		350	$t \geq 40$	330	480
		300	$t < 11$	320	440
		300	$11 \leq t \leq 17$	300	440
		300	$t > 17$	280	440
	Hexagons, rounds and squares	350	$t \leq 50$	340	480
		350	$50 < t < 100$	330	480
		350	$t \geq 100$	320	480
		300	$t \leq 50$	300	440
		300	$50 < t < 100$	290	440
		300	$t \geq 100$	280	440
AS 3597	Plate	500	$5 \leq t \leq 110$	500	590
		600	$5 \leq t \leq 110$	600	690
		700	$t \leq 5$	650	750
		700	$5 < t \leq 65$	690	790
		700	$65 < t \leq 110$	620	720

AMDT No. 1 FEB 2012

Notes to Table 2.1

Following Note 1, *add* the following additional Notes 2 and 3:

2 Welded I-sections complying with AS/NZS 3679.2 are manufactured from hot-rolled structural steel plates complying with AS/NZS 3678, so the values listed for steel grades to AS/NZS 3678 shall be used for welded I-sections to AS/NZS 3679.2.

3 AS/NZS 3678, AS/NZS 3679.1 and AS/NZS 1163 all contain, within each grade, a variety of impact grades not individually listed in the Table. All impact tested grades within the one grade have the same yield stress and tensile strength as the grade listed.

AMDT No. 1 FEB 2012

Clause 3.2.1

1 *Delete* Item (a) and *replace* with the following:

(a) Dead, live, wind, snow, ice and earthquake loads specified in AS/NZS 1170.1, AS/NZS 1170.2, AS/NZS 1170.3 and AS 1170.4.

2 *Delete* Note 1 and *replace* with the following:

1 For the design of bridges, loads specified in AS 5100.2 should be used.

AMDT No. 1 FEB 2012

Clause 3.2.3

1 *Delete* 'AS 1170.1' and *replace* with 'AS/NZS 1170.0'.

2 In the Note, *delete* 'SAA HB77.2 or SAA HB77.8, as applicable' and *replace* with 'AS 5100.2'.

AMDT No. 1 FEB 2012

Clause 3.2.4

Fourth line, *delete* 'AS 1170.1' and *replace* with AS/NZS 1170.1'.

AMDT No. 1 FEB 2012

Clause 3.2.5 (new)

Add new Clause as follows:

3.2.5 Structural robustness

All steel structures, including members and connection components, shall comply with the structural robustness requirements of AS/NZS 1170.0.

AMDT No. 1 FEB 2012

Clause 3.3

Item (b), *delete* 'AS 1170.1' and *replace* with 'AS/NZS 1170.0.'

AMDT No. 1 FEB 2012

Table 3.4

Delete fifth row and *replace* with the following:

Connection component other than a bolt, pin or weld	9.1.9(a), (b), (c), and (d)	0.90
	9.1.9(e)	0.75

AMDT No. 1 FEB 2012

Clause 3.11

Second paragraph, *delete* 'SAA HB77.2 or SAA HB77.8' and *replace* with 'AS 5100.2'.

AMDT No. 1 FEB 2012

Clause 4.1.1

Delete second paragraph and *replace* with the following:

The design action effects for earthquake loads shall be obtained using either the equivalent static analysis of Section 6 of AS 1170.4 or the dynamic analysis of Section 7 of AS 1170.4.

AMDT No. 1 FEB 2012

Figure 4.6.3.2

Delete the last row and *replace* with the following:

Symbols for end restraint conditions	[symbol] = Rotation fixed, translation fixed	[symbol] = Rotation fixed, translation free
	[symbol] = Rotation free, translation fixed	[symbol] = Rotation free, translation free

AMDT No. 1 FEB 2012

Table 5.6.1

In the third column, last entry, *delete* '3.50' and *replace* with '2.50'.

AMDT No. 1 FEB 2012

Clause 5.13.1

Fourth line, *delete* 'AS 1163' and *replace* with 'AS/NZS 1163'.

AMDT No. 1 FEB 2012

Clause 5.13.3

1 Fourth line, *delete* 'AS 1163' and *replace* with 'AS/NZS 1163'.

2 Tenth line, *delete* 'AS 1163' and *replace* with 'AS/NZS 1163'

AMDT No. 1 FEB 2012

Clause 5.13.4

Delete text (excluding the 'NOTE') and *replace* with the following:

The nominal bearing buckling capacity (R_{bb}) of an I-section or C-section web without transverse stiffeners shall be taken as the axial compression capacity determined in accordance with Section 6, using the following parameters:

(a) $\alpha_b = 0.5$.

(b) $k_f = 1.0$.

(c) area of web = $t_w b_b$.

(d) geometrical slenderness ratio taken as $2.5d_1/t_w$ when the top and bottom flanges are effectively restrained against lateral movement out of the plane of the web or $5.0d_1/t_w$ when only one flange is effectively restrained against lateral movement.

(e) b_b is the total bearing width obtained by dispersions at a slope of 1:1 from b_{bf} to the neutral axis (if available), as shown in Figure 5.13.1.1.

The nominal bearing buckling capacity (R_{bb}) of a square or rectangular hollow section web to AS/NZS 1163 without transverse stiffeners shall be taken as the axial compression capacity determined in accordance with Section 6 using the following parameters:

(i) $\alpha_b = 0.5$.

(ii) $k_f = 1.0$.

(iii) area of web = $t_w b_b$.

(iv) geometrical slenderness ratio taken as $3.5d_5/t_w$ for interior bearing ($b_d \geq 1.5d_5$) or $3.8d_5/t_w$ for end bearing ($b_d < 1.5d_5$).

(v) b_b is the total bearing width as shown in Figure 5.13.1.3.

AMDT No. 1 FEB 2012

Clause 5.13.5

First line, *delete* 'AS 1163' and *replace* with 'AS/NZS 1163'.

AMDT No. 1 FEB 2012

Clause 6.3.3

1 *Delete* title and first sentence of Clause 6.3.3 and *replace* with the following:

6.3.3 Nominal capacity of a member of constant cross-section subject to flexural buckling

The nominal member capacity (N_c) of a member of constant cross-section subject to flexural buckling shall be determined as follows:

2 *Insert* the following new paragraph at the end of the Clause:

Fabricated monosymmetric and non-symmetric sections other than unlipped angles, tees and cruciform sections, and hot-rolled channels braced about the minor principal axis, shall be designed for flexural torsional buckling according to AS/NZS 4600 with a reduction factor of 0.85 applied to the nominal member capacity (N_c). A capacity factor of 0.90 shall also be used.

AMDT No. 1 FEB 2012

Table 6.3.3(1)

For compression member section constant value '–0.5', *add* a second section description, as follows:

	—	Welded H, I and box section fabricated from Grade 690 high strength quenched and tempered plate

AMDT No. 1 FEB 2012

Table 6.3.3(3)

In the following row , under fifth column, *delete* '1.161' and *replace* with '0.161'.

205	0.184	0.176	0.168	0.161	0.154

AMDT No. 1 FEB 2012

Clause 8.3.2

Second paragraph, second line, *delete* 'AS 1163' and *replace* with 'AS/NZS 1163'.

AMDT No. 1 FEB 2012

Clause 8.3.3

Item (b), *delete* 'AS 1163; and *replace* with 'AS/NZS 1163'.

AMDT No. 1 FEB 2012

Clause 8.3.4

Second paragraph, second line, *delete* 'AS 1163' and *replace* with 'AS/NZS 1163'.

AMDT No. 1 FEB 2012

Clause 8.4.2.2

Paragraph commencing 'Alternatively, for doubly.......' *delete* 'AS 1163' and *replace* with 'AS/NZS 1163'.

AMDT No. 1 FEB 2012

Clause 8.4.4.2

Delete definition of M_{ox} and *replace* with the following:

M_{ox} = the nominal out-of-plane member moment capacity

$$= M_{bx}\left(1+\frac{N^*}{\phi N_t}\right) \leq M_{rx}$$

AMDT No. 1 FEB 2012

Clause 9.1.4

Item (b), second line, *delete* the words 'for the minimum size of member'.

AMDT No. 1 FEB 2012

Clause 9.1.9

Delete text and *replace* with the following:

Connection components (cleats, gusset plates, brackets and the like) other than connectors shall have their design capacities assessed as follows:

(a) Connection components subject to shear—using Clause 5.11.3.

(b) Connection components subject to tension—using Clause 7.2.

(c) Connection components subject to compression—using Clauses 6.2.1 and 6.3.3.

(d) Connection components subject to bending—using Clause 5.2.1.

(e) A connection component, including a member framing onto the connection component, subject to a design shear force or design tension force (R^*_{bs}) shall satisfy the following equation:

$$R^*_{bs} \leq \phi R_{bs}$$

where

ϕ = capacity factor

= 0.75

R_{bs} = nominal design capacity in block shear

= $0.6 f_{uc} A_{nv} + k_{bs} f_{uc} A_{nt}$

$\leq 0.6 f_{yc} A_{gv} + k_{bs} f_{uc} A_{nt}$

f_{uc} = minimum tensile strength of connection element

f_{yc} = yield stress of connection element

A_{nv} = net area subject to shear at rupture

A_{nt} = net area subject to tension at rupture

A_{gv} = gross area subject to shear at rupture

k_{bs} = a factor to account for the effect of eccentricity on the block shear capacity

= 1.0 when tension stress is uniform

= 0.5 when tension is non-uniform

AMDT No. 1 FEB 2012

Clause 9.3.1

First line, second paragraph, *delete* 'AS 1110' and 'AS 1111' and *replace* with 'AS 1110 series' and AS 1111 series', respectively.

AMDT No. 1 FEB 2012

Table 9.3.1

Delete Table 9.3.1 and *replace* with the following:

TABLE 9.3.1

BOLTS AND BOLTING CATEGORY

Bolting category	Bolt Standard	Bolt grade	Method of tensioning	Minimum tensile strength (f_{uf}) (see Note 2) MPa
4.6/S	AS 1111 (series), AS 1110 (series)	4.6	Snug tight	400
8.8/S	AS/NZS 1252, AS 1110 (series)	8.8	Snug tight	830
8.8/TB	AS/NZS 1252	8.8	Full tensioning	830
8.8/TF (see Note 1)	AS/NZS 1252	8.8	Full tensioning	830

NOTES:

1 Special category used in connections where slip at the serviceability limit state is to be restricted (see Clauses 3.5.5 and 9.1.6).

2 f_{uf} is the minimum tensile strength of the bolt as specified in AS 4291.1—2000, except for grade 8.8 bolts less than 16 mm diameter where the minimum tensile strength is 800 MPa.

3 Bolts to AS 1110 (series) and AS 1111 (series) are not suitable for full tensioning.

AMDT No. 1 FEB 2012

Clause 9.3.2.5

In the first sentence, *delete* 'shall be reduced by 15%' and *replace* with the following:

'shall be reduced by multiplying by [1 – 0.0154(t – 6)], where t is the total thickness of the filler, including any paint film, up to 20 mm. Any filler plate shall extend beyond the connection and the extension of the filler plate shall be secured with enough bolts to distribute the calculated design force in the connected element over the combined cross-section of the connected element and filler plate.'

AMDT No. 1 FEB 2012

Clause 9.3.3.3

Second last line, *delete* 'Clause 15.2.5.1' and *replace* with Table 15.2.5.1'.

AMDT No. 1 FEB 2012

Table 9.6.2

Second column, heading row, *delete* the word 'flame'.

AMDT No. 1 FEB 2012

Clause 9.7.1.1

Delete text of Clause and *replace* with the following:

Welding shall comply with AS/NZS 1554.1, AS/NZS 1554.2, AS/NZS 1554.4 or AS/NZS 1554.5, as appropriate.

AMDT No. 1 FEB 2012

Clause 9.7.1.3

In the first sentence, following 'AS/NZS 1554.1' *add* 'or AS/NZS 1554.4, as appropriate'.

AMDT No. 1 FEB 2012

Clause 9.7.2.1

In the last paragraph, following 'AS/NZS 1554.1' *add* 'or AS/NZS 1554.4, as appropriate'.

AMDT No. 1 FEB 2012

Clause 9.7.2.3

Item (b)(i), following 'AS/NZS 1554.1' *add* 'or AS/NZS 1554.4, as appropriate'.

AMDT No. 1 FEB 2012

Clause 9.7.2.7

Item (a), fifth line, following 'AS/NZS 1554.1' *add* ', AS/NZS 1554.4'.

AMDT No. 1 FEB 2012

Figure 9.7.3.1

Add the following new diagrams and to the figure:

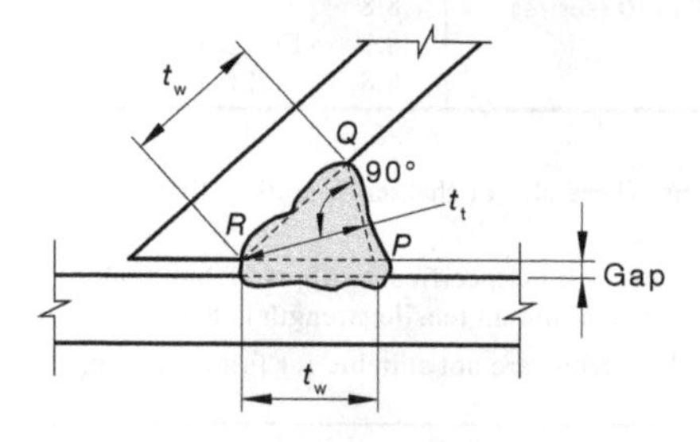

(d) Fillet weld at angled connection - Acute angle side

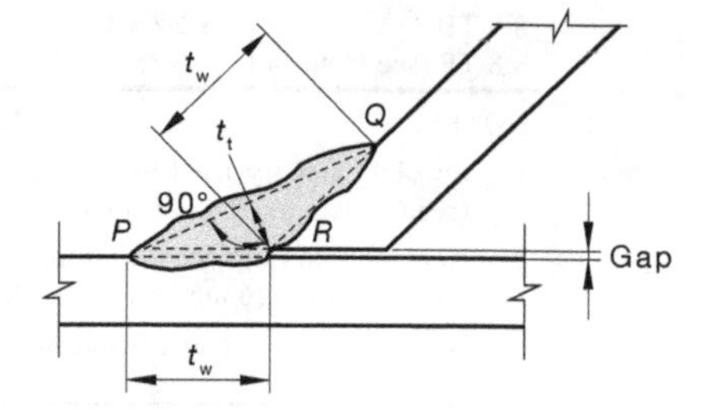

(e) Fillet weld at angled connection - Obtuse angle side

AMDT No. 1 FEB 2012

Table 9.7.3.10(1)

Delete table, including note, and *replace* with the following:

TABLE 9.7.3.10(1)

NOMINAL TENSILE STRENGTH OF WELD METAL (f_{uw})

(see Notes and Table 10.4.4)

Structural steel welding to AS/NZS 1554.1 and AS/NZS 1554.5—Steel Types 1–8C					
Manual metal arc (AS/NZS 4855)	**Submerged arc (AS 1858.1)**	**Flux cored arc (AS/NZS ISO 17632)**	**Gas metal arc (AS/NZS 2717.1) (ISO 14341)**	**Gas tungsten arc (ISO 636)**	**Nominal tensile strength of weld metal, f_{uw}**
A-E35, A-38 B-E43XX	W40X	A-T35, A-T38 B-T43	A-G35, A-G38 B-G43	A-W35, A-W38 B-W43	430
A-E42,A-E46 B-E49XX	W50X	A-T42, A-T46 B-T49	A-G42, A-G46 B-G49, W500	A-W42, A-W46 B-W49	490
A-E50 B-E55XX	W55X	A-T50 B-T55, B-T57	A-G50 B-G55, B-G57 W55X, W62X	A-W50 B-W55, B-W57	550

Structural steel welding to ASNZS 1554.4—Steel Types 8Q–10Q					
Manual metal arc (AS/NZS 4855, AS/NZS 4857)	**Submerged arc (AS 1858.1 AS 1858.2)**	**Flux cored arc (AS/NZS ISO 17632 AS/NZS ISO 18276)**	**Gas metal arc (AS/NZS 2717.1) (ISO 14341, ISO 16834)**	**Gas tungsten arc (ISO 636, ISO 16834)**	**Nominal tensile strength of weld metal, f_{uw}**
A-E35, A-38 B-E43XX	W40X	A-T35, A-T38 B-T43	A-G35, A-G38 B-G43	A-W35, A-W38 B-W43	430
A-E42,A-E46 B-E49XX	W50X	A-T42, A-T46 B-T49	A-G42, A-G46 B-G49 W50X	A-W42, A-W46 B-W49	490
A-E50 B-E55XX B-E57XX B-E59XX	W55X	A-T50 B-T55, B-T57, B-T59	A-G50 B-G55, B-G57, B-G59 W55X	A-W50 B-W55 B-W57, B-W59	550
A-E55 B-E62XX	W62X	A-T55 B-T62	A-G55 B-G62 W62X	A-W55 B-W62	620
A-E62 B-E69XX	W69X	A-T62 B-T69	A-G62 B-G69 W69X	A-W62 B-W69	690
A-E69 B-E76XX B-E78XX	W76X	A-T69 B-T76, B-78	A-G69 B-G76, B-G78 W76X	A-W69 B-W76, B-W78	760
A-E79 B-E83XX	W83X	A-T79 B-T83	A-G79 B-G83, W83X	A-W79 B-W83	830

NOTES:

1 The minimum tensile strength of the European type A classification series consumables is slightly higher than that shown in this Table.

2 The B–E57XX, B–E59XX, B-E78XX and equivalent strength consumables for other welding processes, may be difficult to source commercially.

3 The letter 'X' represents any flux type (manual metal arc welding process) or impact energy value (submerged arc and gas metal arc welding processes).

AMDT No. 1 FEB 2012

Clause 10.2

Delete the third paragraph.

AMDT No. 1 FEB 2012

Clause 10.3.1

Delete clause title and text and *replace* with the following:

10.3.1 General

The design service temperature shall be the estimated lowest metal temperature to be encountered in service or during erection or testing and taken as the basic design temperature as defined in Clause 10.3.2, except as modified in Clause 10.3.3.

AMDT No. 1 FEB 2012

Clause 10.3.2

Delete clause title and text except for the 'NOTE', and *replace* with the following:

10.3.2 Basic design temperature

Lowest one-day mean ambient temperature (LODMAT) isotherms for Australia are given in Figure 10.3. The basic design temperature shall be the LODMAT temperature, except that—

(a) structures that may be subject to especially low local ambient temperatures shall have a basic service temperature of 5°C cooler than the LODMAT temperature; and

(b) critical structures, located where the Bureau of Meteorology records indicate the occurrence of abnormally low local ambient temperatures for a significant time to cause the temperature of the critical structure to be lowered below the LODMAT temperature, shall have a basic design service temperature equal to such a lowered temperature of the critical structure.

AMDT No. 1 FEB 2012

Clause 10.3.3 (new)

Add new clause as follows:

10.3.3 Modifications to the basic design temperature

The design service temperature shall be the basic design temperature, except that for parts that are subject to artificial cooling below the basic design service temperature (for example, in refrigerated buildings), the design service temperature shall be the minimum expected temperature for the part.

AMDT No. 1 FEB 2012

Table 10.4.1

Delete table, including notes, and *replace* with the following:

TABLE 10.4.1

PERMISSIBLE SERVICE TEMPERATURES ACCORDING TO STEEL TYPE AND THICKNESS

Steel type (see Table 10.4.4)	Permissible service temperature, °C (see Note 1)					
	Thickness, mm					
	≤6	>6 ≤12	>12 ≤20	>20 ≤32	>32 ≤70	>70
1	−20	−10	0	0	0	5
2	−30	−20	−10	−10	0	0
2S	0	0	0	0	0	0
3	−40	−30	−20	−15	−15	−10
4	−10	0	0	0	0	5
5	−30	−20	−10	0	0	0
5S	0	0	0	0	0	0
6	−40	−30	−20	−15	−15	−10
7A	−10	0	0	0	0	—
7B	−30	−20	−10	0	0	—
7C	−40	−30	−20	−15	−15	—
8C	−40	−30	—	—	—	—
8Q	−20	−20	−20	−20	−20	−20
9Q	−20	−20	−20	−20	−20	−20
10Q	−20	−20	−20	−20	−20	−20

NOTES:

1 The permissible service temperature for steels with a L20, L40, L50, Y20 or Y40 designation shall be the colder of the temperature shown in Table 10.4.1, and the specified impact test temperature.

2 This Table is based on available statistical data on notch toughness characteristics of steels currently made in Australia or New Zealand. Care should be taken in applying this Table to imported steels as verification tests may be required. For a further explanation, see WTIA Technical Note 11.

3 (—) indicates that material is not available in these thicknesses.

AMDT No. 1 FEB 2012

Clause 10.4.2

At the end of the first paragraph, *add* 'or AS/NZS 1554.4, as appropriate'.

AMDT No. 1 FEB 2012

Clause 10.4.3.1

Delete clause title and text, including the Note, and *replace* with the following:

10.4.3.1 *Steel subject to strain between 1.0% and 10.0%*

Where a member or component is subjected to an outer bend fibre strain during fabrication of between 1.0% and 10.0%, the permissible service temperature for each steel type shall be increased by at least 20°C above the value given in Table 10.4.1.

NOTE: Local strain due to weld distortion should be disregarded.

AMDT No. 1 FEB 2012

Clause 10.4.3.2

Delete clause title and text, including note, and *replace* with the following:

10.4.3.2 *Steel subject to a strain of not less than 10.0%*

Where a member or component is subjected to an outer bend fibre strain during fabrication of not less than 10.0%, the permissible service temperature for each steel type shall be increased by at least 20°C above the value given in Table 10.4.1 plus 1°C for every 1.0% increase in outer bend fibre strain above 10.0%.

NOTE: Local strain due to weld distortion should be disregarded.

AMDT No. 1 FEB 2012

Clause 10.4.3.3

Delete clause title and text, including note, and *replace* with the following:

10.4.3.3 *Post-weld heat-treated members*

Where a member or component has been welded or strained and has been subjected to a post-weld heat-treatment temperature of more than 500°C, but not more than 620°C, the permissible service temperature given in Table 10.4.1 shall not be modified.

NOTE: Guidance on appropriate post-weld heat-treatment may be found in AS 4458.

AMDT No. 1 FEB 2012

Clause 10.4.3.4 (new)

Add a new clause as follows:

10.4.3.4 *Non-complying conditions*

Steels, for which the permissible service temperature (as modified where applicable) is not known or is warmer than the design service temperature specified by the designer, shall not be used, unless compliance with each of the following requirements is demonstrated:

(a) A mock-up of the joint or member shall be fabricated from the desired grade of steel, having similar dimensions and strains of not less than that of the service component.

(b) Three Charpy test specimens shall be taken from the area of maximum strain and tested at the design service temperature.

(c) The impact properties as determined from the Charpy tests shall be not less than the minimum specified impact properties for the grade of steel under test.

(d) Where the Standard to which the steel complies does not specify minimum impact properties, the average absorbed energy for three 10 mm × 10 mm test specimens shall be not less than 27 J, provided none of the test results is less than 20 J.

(e) Where a plate thickness prevents a 10 mm × 10 mm test piece from being used, the standard test thickness closest to the plate thickness shall be used and the minimum value energy absorption requirements shall be reduced proportionally.

AMDT No. 1 FEB 2012

Table 10.4.4

Delete table and *replace* with the following:

TABLE 10.4.4

STEEL TYPE RELATIONSHIP TO STEEL GRADE

Steel type (see Note)	Specification and grade of parent steel				
	AS/NZS 1163	AS/NZS 1594	AS/NZS 3678 AS/NZS 3679.2	AS/NZS 3679.1	AS 3597
1	C250	HA1 HA3 HA4N HA200 HA250 HA250/1 HU250 HA300 HA300/1 HU300 HU300/1	200 250 300	300	—
2	C250L0	—	—	300L0	—
2S	—	—	250S0 300S0	300S0	—
3		XF300	250L15 250L20 250Y20 250L40 250Y40 300L15 300L20 300Y20 300L40 300Y40	300L15	—
4	C350	HA350 HA400 HW350	350 WR350 400	350	—
5	C350L0	—	WR350L0	350L0	—
5S	—	—	350S0	350S0	—
6		XF400	350L15 350L20 350Y20 350L40 350Y40 400L15 400L20 400Y20 400L40 400Y40		—
7A	C450	—	450	—	—
7B	C450L0	—	—	—	—

(continued)

TABLE 10.4.4 *(continued)*

Steel type (see Note)	Specification and grade of parent steel				
	AS/NZS 1163	**AS/NZS 1594**	**AS/NZS 3678 AS/NZS 3679.2**	**AS/NZS 3679.1**	**AS 3597**
7C	—	—	450L15 450L20 450Y20 450L40 450Y40	—	—
8C	—	XF500	—	—	—
8Q		—	—	—	500
9Q		—	—	—	600
10Q	—	—	—	—	700

NOTE: Steel types 8Q, 9Q and 10Q are quenched and tempered steels currently designated as steel types 8, 9 and 10 respectively in AS/NZS 1554.4.

AMDT No. 1 FEB 2012

Clause 10.5

Delete clause text and note and *replace* with the following:

A fracture assessment shall be made, using a fracture mechanics analysis coupled with fracture toughness measurements of the steel selected, weld metal and heat-affected zones and non-destructive examination of the welds and their heat-affected zones.

NOTE: For methods of fracture assessment, see BS 7910 and WTIA Technical Note 10.

AMDT No. 1 FEB 2012

Clause 11.1.5

In the first paragraph, following 'AS/NZS 1554.1' *add* 'or AS/NZS 1554.4, as appropriate.'

AMDT No. 1 FEB 2012

Clause 11.2

Delete the list of referenced documents and *replace* with the following:

AS

1418 Cranes, hoists and winches
1418.1 Part 1: General requirements
1418.3 Part 3: Bridge, gantry, portal (including container cranes) and jib cranes
1418.5 Part 5: Mobile cranes
1418.18 Part 18: Crane runways and monorails

5100 Bridge design
5100.1 Part 1: Scope and general principles
5100.2 Part 2: Design loads

AMDT No. 1 FEB 2012

Table 11.5.1(4)

Delete the 'Detail category' relating to illustration (49), and *replace* with the following:

45 ($t \geq 8$ mm)
40 ($t < 8$ mm)

AMDT No. 1 FEB 2012

Clause 11.8.2

Delete Item (b) (ii) and *replace* with the following:

(ii) the summation $\sum_j$ is for *j* design stress ranges (f_j^*) for which $\phi f_{5c} \leq f_j^* < \phi f_{3c}$; and

AMDT No. 1 FEB 2012

Clause 12.5

Fourth line, *delete* 'AS 1170.1' and *replace* with 'Section 4 of AS/NZS 1170.0'.

AMDT No. 1 FEB 2012

Clause 12.6.1

Add a new fourth paragraph as follows:

Alternatively, recognized methods of assessment in accordance with ENV 13381-4 and EN 13381-8 may be used.

AMDT No. 1 FEB 2012

Clause 12.6.2.2

In Item (a), *delete* the existing note and *replace* with the following:

NOTE: Experience has shown that the above regression method can also be used for materials such as intumescent and ablative coatings subject to the coefficient of correlation exceeding 0.9.

AMDT No. 1 FEB 2012

Section 13

Delete text of Section 13, including Table 13.3.4.2, and *replace* with the following:

13.1 GENERAL

This Section sets out the additional minimum design and detailing requirements for steel structures, structural members, and connections which form the whole or parts of a building or structure subject to the earthquake forces specified in AS 1170.4.

13.2 DEFINITIONS

For the purposes of this Section, the definitions given in Clause 1.3 of AS 1170.4 shall apply for the following terms:

Bearing wall system

Braced frame

Braced frame, concentric

Braced frame, eccentric

Ductility (of a structure)

Moment-resisting frame

Moment-resisting frame, intermediate

Moment-resisting frame, ordinary

Moment-resisting frame, special

Seismic-force-resisting system

Space frame

Structural ductility factor

Structural performance factor

13.3 DESIGN AND DETAILING REQUIREMENTS

13.3.1 General

Design and detailing requirements for a structure shall be based on the earthquake design category and structural system assigned to the structure in accordance with AS 1170.4.

Limited ductile steel structures shall comply with Clause 13.3.5.

Moderately ductile steel structures shall comply with Clause 13.3.6.

Fully ductile steel structures shall comply with Clause 13.3.7.

13.3.2 Stiff elements

A stiff element that is deemed not to be part of the seismic-force-resisting system may be incorporated into a steel structure, provided its effects on the behaviour of the seismic-force-resisting system are considered and provided for in the analysis and design.

13.3.3 Non-structural elements

A non-structural element which is attached to or encloses the exterior of a steel structure shall be capable of accommodating the movements resulting from earthquake forces as follows:

(a) All connections and panel joints shall permit relative movement between storeys equal to the design storey deflection calculated in accordance with AS 1170.4, or 6 mm, whichever is the greater.

(b) Connections shall be ductile and shall have a rotation capacity to preclude brittle failure.

(c) Connections which permit movements in the plane of a panel shall include sliding connections using slotted or oversize holes, or connection details which permit movement by bending, or other connection details which have been demonstrated by test to be adequate.

13.3.4 Structural ductility factor and structural performance factor

The structural ductility factor (μ) and structural performance factor (S_p) for steel structures and members shall be as given in Table 13.3.4.

TABLE 13.3.4

STRUCTURAL DUCTILITY FACTOR (μ) AND STRUCTURAL PERFORMANCE FACTOR (S_p)—STEEL STRUCTURES

Description of structural system	μ	S_p
Special moment-resisting frames (fully ductile) (see Note)	4	0.67
Intermediate moment-resisting frames (moderately ductile)	3	0.67
Ordinary moment-resisting frames (limited ductile)	2	0.77
Moderately ductile concentrically braced frames	3	0.67
Limited ductile concentrically braced frames	2	0.77
Fully ductile eccentrically braced frames (see Note)	4	0.67
Other steel structures not defined above	2	0.77

NOTE: The design of structures with $\mu > 3$ is outside the scope of this Standard (see Clause 13.3.7).

13.3.5 Requirements for 'limited ductile' steel structures ($\mu = 2$)

Limited ductile steel structures shall comply with the following requirements:

(a) The minimum yield stress specified for the grade of steel shall not exceed 350 MPa.

(b) Concentrically braced frames—connections of diagonal brace members that are expected to yield shall be designed for the full member design capacity.

(c) Ordinary moment resisting frames—no additional requirements.

13.3.6 Requirements for 'moderately ductile' structures ($\mu = 3$)

13.3.6.1 *General*

The minimum yield stress specified for the grade of steel shall not exceed 350 MPa.

13.3.6.2 *Bearing wall and building frame systems*

Concentrically braced frames in bearing wall and building frame systems shall comply with the following:

(a) The design axial force for each diagonal tension brace member shall be limited to 0.85 times the design tensile capacity. Connections of each diagonal brace member shall be designed for the full member design capacity.

(b) Any web stiffeners in beam-to-column connections shall extend over the full depth between flanges and shall be butt welded to both flanges.

(c) All welds shall be weld category SP in accordance with AS/NZS 1554.1. Welds shall be subjected to non-destructive examination as given in Table 13.3.6.2 and all such non-destructive examination shall comply with AS/NZS 1554.1.

TABLE 13.3.6.2

MINIMUM REQUIREMENTS FOR NON-DESTRUCTIVE EXAMINATION

Weld type	Visual scanning %	Visual examination %	Magnetic particle or dye penetrant %	Ultrasonics or radiography %
Butt welds in members or connections in tension	100	100	100	10
Butt welds in members or connections others than those in tension	100	50	10	2
All other welds in members or connections	100	20	5	2

13.3.6.3 *Moment-resisting frames, intermediate*

Intermediate moment-resisting frames shall comply with the following additional requirements:

(a) The minimum yield stress specified for the grade of steel shall not exceed 350 MPa.

(b) Web stiffeners in beam to column connections shall extend over the full depth between flanges and shall be butt welded to both flanges.

(c) Members in which plastic hinges will form during inelastic displacement of the frame shall comply with the requirements for plastic analysis specified in Clause 4.5.

13.3.6.4 *Fabrication in areas of plastic deformation*

All areas of plastic deformation shall satisfy the following:

(a) *Edge* In parts of a member or connection subject to plastic deformation, a sheared edge shall not be permitted unless the edge is sheared oversize and machined to remove all signs of the sheared edge. A gas cut edge shall have a maximum surface roughness of 12 μm (Centre Line Average Method).

(b) *Punching* In parts subject to plastic deformation, fastener holes shall not be punched full size. If punched, holes shall be punched undersize and reamed or drilled to remove the entire sheared surface.

13.3.7 Requirements for 'fully ductile' structures (μ >3)

A steel structure which is fully ductile has a structural ductility factor >3 and is required by AS 1170.4 to be designed in accordance with NZS 1170.5. Steel members and connections for such structures shall be designed and detailed in accordance with NZS 3404.

AMDT No. 1 FEB 2012

Clause 14.3.3

1 *Delete* first paragraph and *replace* with the following:

Cutting may be by sawing, shearing, cropping, machining, thermal cutting (including laser cutting and plasma cutting) or water cutting processes, as appropriate.

2 In the third and fifth paragraphs, following 'AS/NZS 1554.1' *add* ', AS/NZS 1554.4 or AS/NZS 1554.5, as appropriate'.

AMDT No. 1 FEB 2012

Clause 14.3.4

1 First line, *add* ', AS/NZS 1554.4' after 'AS/NZS 1554.1'.

2 Second line, *delete* AS 1554.2' and *replace* with 'AS/NZS 1554.2'.

AMDT No. 1 FEB 2012

Clause 14.3.5.1

Delete the word 'flame' in the first, second and third paragraphs, in each instance.

AMDT No. 1 FEB 2012

Clause 14.3.5.2

1 *Add* the following text to the second paragraph:

'The plate washer shall completely cover the hole such that the minimum distance from the edge of the hole to the edge of the plate washer shall be 0.5 times the hole diameter.'

2 *Add* the following to Item (a)(iii):

', where the length of the slotted hole is taken as the total length from one hole edge to another along the longest dimension.'

3 *Add* the following to Item (b)(i):

'The plate washer shall completely cover the hole such that the minimum distance from the edge of the hole to the edge of the plate washer shall be 0.5 times the hole diameter.'

4 *Add* the following to the first paragraph of Item (b)(ii):

'The plate washer shall completely cover the hole such that the minimum distance from the edge of the hole to the edge of the plate washer shall be 0.5 times the hole diameter.'

5 *Add* the following to the first paragraph of Item (b)(iii):

'The plate washer shall completely cover the hole such that the minimum distance from the edge of the hole to the edge of the plate washer shall be 0.5 times the hole diameter.'

AMDT No. 1 FEB 2012

Clause 14.4.3

1 First paragraph, first line, *delete* 'plate' and *replace* with 'welded I-section'.

2 First paragraph, second line, *delete* 'AS/NZS 3678 or AS/NZS 3679.1' and *replace* with 'AS/NZS 3679.1 or AS/NZS 3679.2'.

3 Item (j), *delete* $\left(\frac{b_r}{150}\right)$mm and *replace* with $\left(\frac{b_f}{150}\right)$mm.

AMDT No. 1 FEB 2012

Clause 15.2.1

Last line, *delete* the reference to 'AS/NZS 1554.1' and *replace* with 'AS/NZS 1554 (series)'.

AMDT No. 1 FEB 2012

Table 15.2.5.1

Delete 'NOTE' and *replace* with the following:

NOTE: The minimum bolt tensions given in this Table are approximately equivalent to the minimum proof loads derived from a proof load stress of 600 MPa, as specified in AS 4291.1.

AMDT No. 1 FEB 2012

Clause 15.2.5.3

Item (c), *delete* reference to 'Clause 15.2.5.1' and *replace* with 'Table 15.2.5.1'.

AMDT No. 1 FEB 2012

Appendix A

Delete list of referenced documents and *replace* with following updated list:

AS
1101 Graphical symbols for general engineering
1101.3 Part 3: Welding and non-destructive examination

1110 ISO metric hexagon bolts and screws
1110.1 Part 1: Product grades A and B—Bolts
1110.2 Part 2: Product grades A and B—Screws

1111 ISO metric hexagon bolts and screws
1111.1 Part 1: Product grade C—Bolts
1111.2 Part 2: Product grade C—Screws

1112 ISO metric hexagon nuts
1112.1 Part 1: Style 1—Product grades A and B
1112.2 Part 2: Style 2—Product grades A and B
1112.3 Part 3: Product grade C
1112.4 Part 4: Chamfered thin nuts—Product grades A and B

1170 Structural design actions
1170.4 Part 4: Earthquake actions in Australia

1210 Pressure vessels

1275 Metric screw threads for fasteners

AS

1391	Metallic materials—Tensile testing at ambient temperature
1418	Cranes, hoists and winches
1418.1	Part 1: General requirements
1418.3	Part 3: Bridge, gantry and portal (including container cranes) and jib cranes
1418.5	Part 5: Mobile cranes
1418.18	Part 18: Crane runways and monorails
1530	Methods for fire tests on building materials, components and structures
1530.4	Part 4: Fire-resistance test of elements of construction
1657	Fixed platforms, walkways, stairways and ladders—Design, construction and installation
1735	Lifts, escalators and moving walks
1735.1	Part 1: General requirements
1858	Electrodes and fluxes for submerged-arc welding
1858.1	Part 1: Carbon steels and carbon manganese steels
2074	Cast steels
2205	Methods of destructive testing of welds in metal
2205.2.1	Part 2.1: Transverse butt tensile test
2327	Composite structures
2327.1	Part 1: Simply supported beams
2670	Evaluation of human exposure to whole-body vibration
2670.1	Part 1: General requirements
2670.2	Part 2: Continuous and shock-induced vibration in buildings (1 to 80 Hz)
3597	Structural and pressure vessel steel—Quenched and tempered plate
3600	Concrete structures
4291	Mechanical properties of fasteners made of carbon steel and alloy steel
4291.1	Part 1: Bolts, screws and studs
4291.2	Part 2: Nuts with specified proof load values—Coarse thread
4458	Pressure equipment—Manufacture
5100	Bridge design
5100.1	Part 1: Scope and general principles
5100.2	Part 2: Design loads
5100.6	Part 6: Steel and composite construction

AS/NZS

1163	Cold-formed structural steel hollow sections
1170	Structural design actions
1170.0	Part 0: General principles
1170.1	Part 1: Permanent, imposed and other actions
1170.2	Part 2: Wind actions
1170.3	Part 3: Snow and ice actions
1252	High strength steel bolts with associated nuts and washers for structural engineering

AS/NZS

1554	Structural steel welding
1554.1	Part 1: Welding of steel structures
1554.2	Part 2: Stud welding (steel studs to steel)
1554.4	Part 4 Welding of high strength quenched and tempered steels
1554.5	Part 5: Welding of steel structures subject to high levels of fatigue loading
1559	Hot-dip galvanized steel bolts with associated nuts and washers for tower construction
1594	Hot-rolled steel flat products
1873	Powder-actuated (PA) hand-held fastening tools (series)
2717	Welding—Electrodes—Gas metal arc
2717.1	Part 1: Ferritic steel electrodes
3678	Structural steel—Hot-rolled plates, floorplates and slabs
3679	Structural steel
3679.1	Part 1: Hot-rolled bars and sections
3679.2	Part 2: Welded I sections
4600	Cold-formed steel structures
4855	Welding consumables—Covered electrodes for manual metal arc welding of non-alloy and fine grain steels—Classification
4857	Welding consumables—Covered electrodes for manual metal arc welding of high-strength steels—Classification

NZS

1170	Structural design actions
1170.5	Part 5: Earthquake actions—New Zealand
3404	Steel structures Standard

ISO

636	Welding consumables—Rods, wires and deposits for tungsten inert gas welding of non-alloy and fine-grain steels—Classification
14341	Welding consumables—Wire electrodes and weld deposits for gas shielded metal arc welding of non alloy and fine grain steels—Classification
16834	Welding consumables—Wire electrodes, wires, rods and deposits for gas-shielded arc welding of high strength steels—Classification
17632	Welding consumables—Tubular cored electrodes for gas shielded and non-gas shielded metal arc welding of non-alloy and fine grain steels—Classification
18276	Welding consumables—Tubular cored electrodes for gas-shielded and non-gas-shielded metal arc welding of high-strength steels—Classification

EN

13381	Test methods for determining the contribution to the fire resistance of structural members
13381-4	Part 4: Applied passive protection products to steel members
13381-8	Part 8: Applied reactive protection to steel members

BS

7910	Guide to methods for assessing the acceptability of flaws in metallic structures

AMDT No. 1 FEB 2012

Appendix B, Paragraph B1

Add a new sentence after the existing sentence, as follows:

'Alternatively, the guidance given in Appendix C of AS/NZS 1170.0 may be used, where appropriate.'

AMDT No. 1 FEB 2012

Appendix B, Paragraph B2

Delete the text and *replace* with the following:

The relative horizontal deflection between adjacent frames at eaves level of industrial portal frame buildings under the serviceability wind load specified in AS/NZS 1170.0 and AS/NZS 1170.2 may be limited to the following:

(a) Building clad with steel or aluminium sheeting, with no ceilings, with no internal partitions against external walls and no gantry cranes operating in the building—frame spacing/200.

(b) As in (a) but with gantry cranes operating—frame spacing/250.

(c) As in (a) but with external masonry walls supported by steelwork in lieu of steel or aluminium sheeting—frame spacing/200.

The absolute horizontal deflection of a frame in an industrial portal frame building under the serviceability wind load specified in AS/NZS 1170.0 and AS/NZS 1170.2 may be limited to the following:

(i) Building clad with steel or aluminium sheeting, with no ceilings, with no internal partitions against external walls and no gantry cranes operating in the building—eaves height/150.

(ii) As in (i) but with gantry cranes operating—crane rail height/250.

(iii) As in (i) but with external masonry walls supported by steelwork in lieu of steel or aluminium sheeting—eaves height/250.

Alternatively, the guidance given in Appendix C of AS/NZS 1170.0 may be used where appropriate.

AMDT No. 1 FEB 2012

Appendix C, Paragraph C2

Second line, *delete* 'AS 2311' and *replace* with 'AS/NZS 2311'.

AMDT No. 1 FEB 2012

Appendix C, Paragraph C4

Second paragraph, *delete* 'AS 1163' and *replace* with 'AS/NZS 1163'.

AMDT No. 1 FEB 2012

Appendix C, Paragraph C7

Delete list of relevant Standards and *replace* with the following updated list:

AS

1192 Electroplated coatings—Nickel and chromium

1214 Hot-dip galvanized coatings and threaded fasteners (ISO metric coarse thread series)

1627 Metal finishing—Preparation and pretreatment of surfaces
1627.0 Part 0: Method selection guide
1627.1 Part 1: Removal of oil, grease and related contamination
1627.2 Part 2: Power tool cleaning

AS
1627.4 Part 4: Abrasive blast cleaning of steel
1627.5 Part 5: Pickling
1627.6 Part 6: Chemical conversion treatment of metals
1627.7 Part 7: Metal finishing—Preparation and pretreatment of surfaces—Hand tool cleaning of metal surfaces
1627.9 Part 9: Pictorial surface preparation standards for painting steel surfaces

1789 Electroplated zinc (electrogalvanized) coatings on ferrous articles (batch process)

1856 Electroplated coatings—Silver

1897 Electroplated coatings on threaded components (metric coarse series)

1901 Electroplated coatings—Gold and gold alloys

2239 Galvanic (sacrificial) anodes for cathodic protection

2832 Cathodic protection of metals
2832.3 Part 3: Fixed immersed structures
2832.4 Part 4: Internal surfaces
2832.5 Part 5: Steel in concrete structures

3730 Guide to the properties of paints for buildings (series)
4169 Electroplated coatings—Tin and tin alloys

AS/NZS
1580 Paints and related materials—Methods of test (series)

2311 Guide to the painting of buildings

2312 Guide to the protection of structural steel against atmospheric corrosion by the use of protective coatings

3750 Paints for steel structures
3750.6 Part 6: Full gloss polyurethane (two-pack)
3750.10 Part 10: Full gloss epoxy (two-pack)
3750.11 Part 11: Chlorinated rubber—High-build and gloss
3750.12 Part 12: Alkyd/micaceous iron oxide
3750.13 Part 13: Epoxy primer (two-pack)
3750.15 Part 15: Inorganic zinc silicate paint
3750.17 Part 17: Etch primers (single pack and two-pack)
3750.22 Part 22: Full gloss enamel—Solvent-borne

4534 Zinc and zinc/aluminium alloy coatings on steel wire

4680 Hot-dip galvanized (zinc) coatings on fabricated ferrous articles

4792 Hot-dip galvanized (zinc) coatings on ferrous hollow sections, applied by a continuous or a specialized process

---

AMDT No. 1 FEB 2012

INDEX

Delete Index.

Appendix E

The new AS 4100 Block Shear Failure Provisions

E.1 General

To understand the new Block Shear provisions in AS 4100 AMD 1 (see Clause 9.1.9 in Appendix D), we need to consider the nature of failure(s) associated with that overall failure mode. The best way to view this is to consider the failure paths (or planes) on a connection component subject to tension or shear. In these instances, the connection component may fail when a "block" of material pulls out (or "ruptures") from the balance of the component. Examples of such failures and their failure paths are shown in Figures E.2 to E.4 below. It is evident from these Figures that Block Shear Failure involves the development of a combination of individual failure planes subject to tension (denoted by subscript '*t*') and shear (subscript '*v*'). These failure modes are dependent on either the gross (subscript '*g*') or net section (e.g. deduction for bolt holes – subscript '*n*') of the failure plane. The Design Capacity for Block Shear is then developed from the design capacities from the contributing individual failure planes. Hogan & Munter [2007a] provide further useful background on the phenomena being described.

To further illustrate the use of the new Block Shear provisions in AS 4100 AMD 1, we will look at its application in Example 8.1 (Section 8.11.1).

E.2 EXAMPLE 8.1 (Section 8.11.1) TENSION MEMBER

(1) Connection Configuration

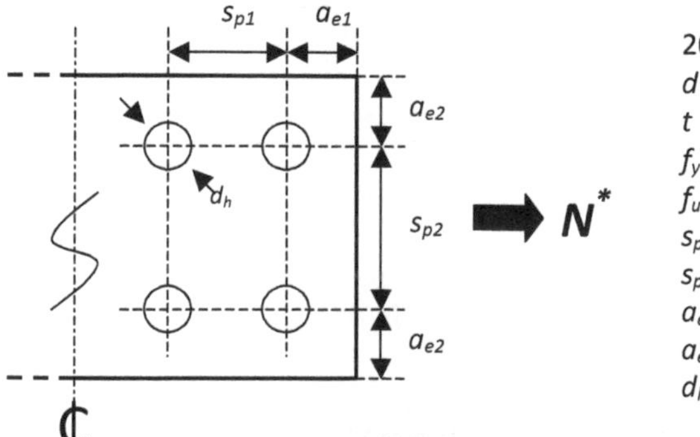

200 x 20 mm Flat:
d = 200 mm
t = 20 mm
$f_y = f_{ym}$ = 280 MPa
$f_u = f_{um}$ = 440 MPa
s_{p1} = 70 mm
s_{p2} = 90 mm
a_{e1} = 35 mm
a_{e2} = 55 mm
d_h = bolt hole diameter = 22 mm

Figure E.1 *Connection configuration for the Tension Member in Example 8.1 (Section 8.1.1).*

(2) Failure Mode A

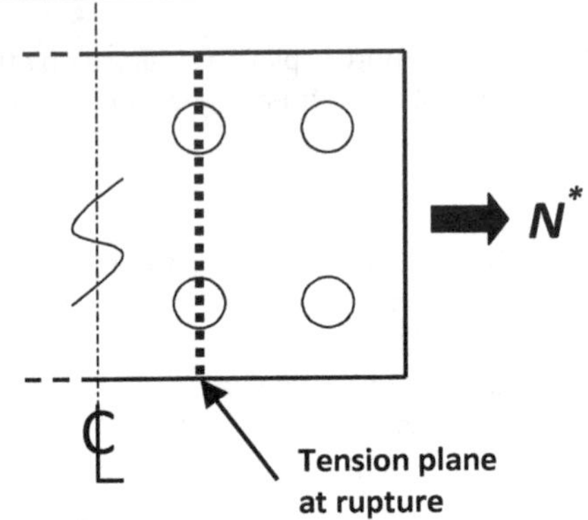

This failure mode is considered superfluous as it is covered by the member tension design capacity provisions in Clause 7.2 of AS 4100. That is, strictly speaking, the new AS 4100 Block Shear provisions should be applied when both tension and shear failure planes are present.

Figure E.2 *Failure paths for Mode A Block Shear rupture.*

Gross yielding (AS 4100 Clause 9.7.2):

ϕN_{t1} $= \phi A_g f_{ym} = 0.9 A_g f_{ym} = 0.9 \times 4000 \times 280 / 10^3$
$= 1010$ kN

Net fracture (AS 4100 Clause 9.7.2):

ϕN_{t2} $= \phi 0.85 k_t A_n f_u = 0.9 \times 0.85 \times k_t A_n f_u = 0.765 \times 1.0 \times 3120 \times 440 / 10^3$
$= 1050$ kN (assume uniform stress distribution $\therefore k_t = 1.0$)

$\therefore$ the Mode A controlling failure mode (min. ϕN_{t1}, ϕN_{t2}) is Gross Yielding with $\phi N_{t1} = 1010$ kN.

(3) Failure Mode B (Block Shear)

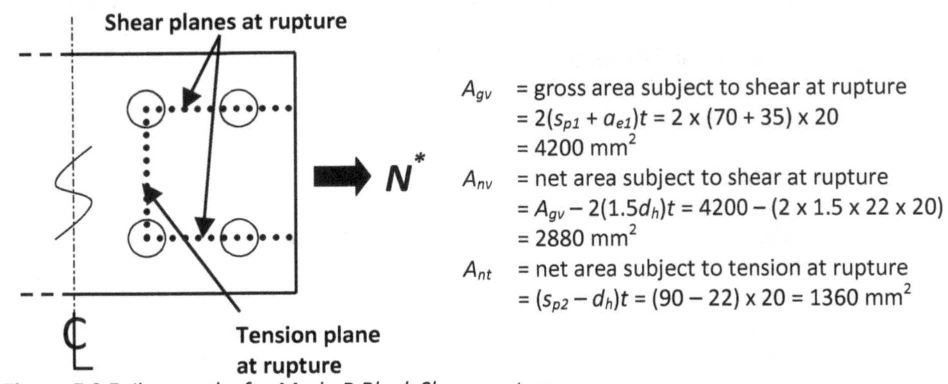

A_{gv} = gross area subject to shear at rupture
$= 2(s_{p1} + a_{e1})t = 2 \times (70 + 35) \times 20$
$= 4200$ mm^2

A_{nv} = net area subject to shear at rupture
$= A_{gv} - 2(1.5 d_h)t = 4200 - (2 \times 1.5 \times 22 \times 20)$
$= 2880$ mm^2

A_{nt} = net area subject to tension at rupture
$= (s_{p2} - d_h)t = (90 - 22) \times 20 = 1360$ mm^2

Figure E.3 *Failure paths for Mode B Block Shear rupture.*

Block Shear Fracture provisions from Clause 9.1.9 of AS 4100 AMD 1 (see Appendix D) with $k_{bs} = 1.0$ (due to symmetry):

ϕR_{bs1} $= \phi(0.6 f_u A_{nv} + k_{bs} f_u A_{nt}) = 0.75 \times (0.6 \times 440 \times 2880 + 1.0 \times 440 \times 1360) / 10^3$
$= 1020$ kN

ϕR_{bs2} $= \phi(0.6 f_y A_{gv} + k_{bs} f_u A_{nt}) = 0.75 \times (0.6 \times 280 \times 4200 + 1.0 \times 440 \times 1360) / 10^3$
$= 978$ kN

ϕR_{bs} $= \text{min.}[\ \phi R_{bs1};\ \phi R_{bs2}\ ] = 978$ kN

(4) Failure Mode C (Block Shear)

Note the "mirror" image failure mode with the shear rupture plane running along the bottom bolt holes and the tension rupture plane from the bottom left bolt hole running vertically to the top edge would yield the same result.

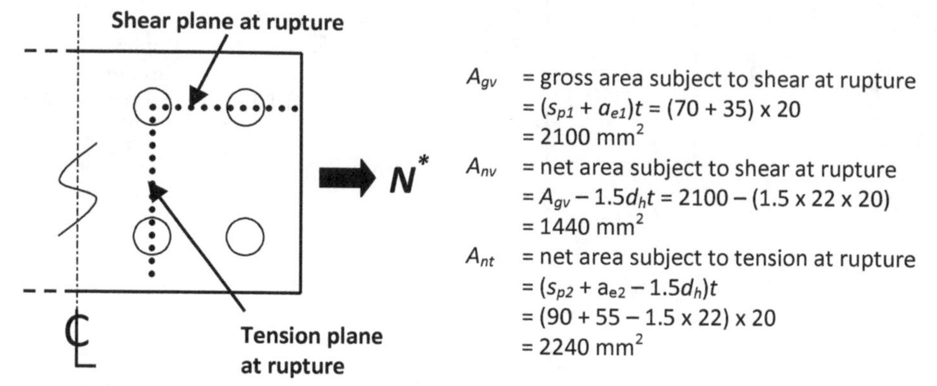

A_{gv} = gross area subject to shear at rupture
= $(s_{p1} + a_{e1})t$ = (70 + 35) x 20
= 2100 mm^2

A_{nv} = net area subject to shear at rupture
= $A_{gv} - 1.5d_h t$ = 2100 – (1.5 x 22 x 20)
= 1440 mm^2

A_{nt} = net area subject to tension at rupture
= $(s_{p2} + a_{e2} - 1.5d_h)t$
= (90 + 55 – 1.5 x 22) x 20
= 2240 mm^2

Figure E.4 *Failure paths for Mode C Block Shear rupture.*

Using the Block Shear Fracture provisions from Clause 9.1.9 of AS 4100 AMD 1 (see Appendix D) with k_{bs} = 1.0 (as before – assuming tension stress on the Block Shear is basically uniform):

ϕR_{bs1} = $\phi(0.6 f_u A_{nv} + k_{bs} f_u A_{nt})$ = 0.75 x (0.6 x 440 x 1440 + 1.0 x 440 x 2240) / 10^3
= 1020 kN

ϕR_{bs2} = $\phi(0.6 f_y A_{gv} + k_{bs} f_u A_{nt})$ = 0.75 x (0.6 x 280 x 2100 + 1.0 x 440 x 2240) / 10^3
= 1000 kN

ϕR_{bs} = min.[ϕR_{bs1}; ϕR_{bs2}] = 1000 kN

(5) Summary

Failure Mode A:

ϕN_{t1} = 1010 kN (Gross Yielding from Clause 7.2 of AS 4100)

ϕN_{t2} = 1050 kN (Net fracture from Clause 7.2 of AS 4100)

Failure Mode B :

ϕR_{bs} = 978 kN (Block Shear failure from Clause 9.1.9 of AS 4100 AMD 1 – Appendix D)

Failure Mode C :

ϕR_{bs} = 1000 kN (Block Shear failure from Clause 9.1.9 of AS 4100 AMD 1 – Appendix D)

Failure Mode B (Block Shear) governs for the Tension Member with a 978 kN design tension capacity. Calculations in these instances reveal that the (new) AS 4100 Block Shear provisions may govern over the (previous) Clause 7.2 provisions whenever deductions in gross area are required (e.g. whenever single point fasteners are required, etc).

E.3 EXAMPLE 8.1 (Section 8.11.1) SPLICE PLATES

The same failure modes considered above for the Tension Member can also be rationally applied to the Splice Plates.

(1) Connection Configuration

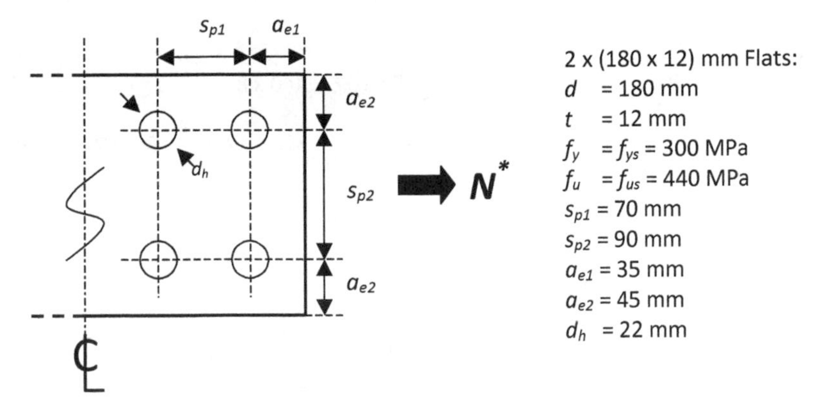

2 x (180 x 12) mm Flats:
d = 180 mm
t = 12 mm
f_y = f_{ys} = 300 MPa
f_u = f_{us} = 440 MPa
s_{p1} = 70 mm
s_{p2} = 90 mm
a_{e1} = 35 mm
a_{e2} = 45 mm
d_h = 22 mm

Figure E.5 *Connection configuration for the Splice Plates in Example 8.1 (Section 8.1.1).*

(2) Failure Mode A

Using the same methodology as for the Tension Member and initially calculate for one (1) splice plate:

A_g = d x t = 180 x 12 = 2160 mm^2
A_n = A_{nt} = $A_g - 2d_h t$ = 2160 – (2 x 22 x 12) = 1630 mm^2

Hence, for two (2) splice plates:

ϕN_{t1} = $2\phi A_g f_{ys}$ = 2 x (0.9 x 2160 x 300) / 10^3 = 1170 kN
ϕN_{t2} = $2\phi 0.85 k_t A_n f_{us}$ = 2 x (0.765 x 1.0 x 1630 x 440) / 10^3 = 1100 kN

∴ the Mode A controlling failure mode is net fracture (not Gross Yielding as for the Tension Member) with ϕN_{t2} = 1100 kN.

(3) Failure Mode B (Block Shear)

Using the same methodology as for the Tension Member and initially calculate for one (1) splice plate:

A_{gv} = $2(s_{p1} + a_{e1})t$ = 2 x (70 + 35) x 12 = 2520 mm^2
A_{nv} = $A_{gv} - 2(1.5d_h)t$ = 2520 – (2 x 1.5 x 22 x 12) = 1730 mm^2
A_{nt} = $(s_{p2} - d_h)t$ = (90 – 22) x 12 = 816 mm^2

Hence, for two (2) splice plates:

ϕR_{bs1} = $2\phi(0.6 f_u A_{nv} + k_{bs} f_u A_{nt})$ = 2 x [0.75 x (0.6 x 440 x 1730 + 1.0 x 440 x 816)] / 10^3
= 1220 kN
ϕR_{bs2} = $2\phi(0.6 f_y A_{gv} + k_{bs} f_u A_{nt})$ = 2 x [0.75 x (0.6 x 300 x 2520 + 1.0 x 440 x 816)] / 10^3
= 1220 kN
ϕR_{bs} = min.[ϕR_{bs1}; ϕR_{bs2}] = 1220 kN

(4) Failure Mode C (Block Shear)

Using the same methodology as for the Tension Member and initially calculate for one (1) splice plate:

A_{gv} $= (s_{p1} + a_{e1})t = (70 + 35) \times 12 = 1260$ mm^2

A_{nv} $= A_{gv} - 1.5d_h t = 1260 - (1.5 \times 22 \times 12) = 864$ mm^2

A_{nt} $= (s_{p2} + a_{e2} - 1.5d_h)t = (90 + 45 - 1.5 \times 22) \times 12$
$= 1220$ mm^2

Hence, for two (2) splice plates:

ϕR_{bs1} $= 2\phi(0.6f_u A_{nv} + k_{bs} f_u A_{nt}) = 2 \times [0.75 \times (0.6 \times 440 \times 864 + 1.0 \times 440 \times 1220)] / 10^3$
$= 1150$ kN

ϕR_{bs2} $= 2\phi(0.6f_y A_{gv} + k_{bs} f_u A_{nt}) = 2 \times [0.75 \times (0.6 \times 300 \times 1260 + 1.0 \times 440 \times 1220)] / 10^3$
$= 1150$ kN

ϕR_{bs} $= \min.[\ \phi R_{bs1};\ \phi R_{bs2}\ ] = 1150$ kN

(5) Summary

Failure Mode A:

ϕN_{t1} = 1170 kN (Gross Yielding from Clause 7.2 of AS 4100)

ϕN_{t2} = 1100 kN (Net fracture from Clause 7.2 of AS 4100)

Failure Mode B :

ϕR_{bs} = 1220 kN (Block Shear failure from Clause 9.1.9 of AS 4100 AMD 1 – Appendix D)

Failure Mode C :

ϕR_{bs} = 1150 kN (Block Shear failure from Clause 9.1.9 of AS 4100 AMD 1 – Appendix D)

Failure Mode A due to member net fracture governs in this instance for the Splice Plates with a 1100 kN design tension capacity. This is different for the Tension Member (above) where Failure Mode B governs.

OVERALL SUMMARY

This simple example illustrates the point that various failure path assessments are required when using the new AS 4100 Block Shear provisions (Clause 9.1.9 as noted in AS 4100 AMD 1 – see Appendix D). Additionally, the new Block Shear provisions may control over the AS 4100 Clause 7.2 provisions for tension member design at the connection.

It should be noted that AS 4100 requires experimental justification for failure modes in connection design models. Experimental testing has noted that Block Shear type failures in coped I-section beam ends subject to shear force, in angle members subject to tension, gusset plates subject to tension and in Tee sections. In these instances there is some eccentricity effect on the connection, however, it begs the question should the new AS 4100 Block Shear provisions be applied to all connections. There is a strong case for a thorough literature review on Block Shear and connection details it actually applies to. Hogan & Munter [2007a] provides some useful interim information in this area. It may be reasonable to argue that the Block Shear checks may not be applicable for the connection considered in this Appendix. However, apart from illustrating the use of the new AS 4100 provisions, the above checks were done to be "on the safe side" despite other references not containing evidence (to the knowledge of the authors) of such failures in the connection type considered above.

Notation

The symbols used in this Handbook conform to the notation in AS 4100 and, where applicable, AS/NZS 1170. When not noted in these Standards, the symbols are based on their particular definitions.

A	=	area of cross-section
	=	reaction at point A
A_a	=	area allowance for staggered bolt holes
A_c	=	minor diameter area of a bolt – i.e. core area (at the root of the threads)
A_{cw}	=	critical area of column web in bearing (crushing)
A_d	=	sum of area deductions for holes and penetrations
A_e	=	effective area of a cross-section
A_{ei}	=	effective area of element i
A_{ep}	=	area of end plate
A_f	=	flange area
A_g	=	gross area of a cross-section
A_{gm}	=	gross section area of tie member section
A_{gs}	=	gross section area of splice plate section
A_h, A_{h1}, A_{h2}	=	one-half of A (area definition)
A_i	=	area of element/member i
A_n	=	net area of a cross-section
A_{nm}	=	net section area of tie member section
A_{ns}	=	net section area of splice plate section
A_o	=	plain shank area of a bolt
	=	enclosed area of a hollow section (torsion)
A_p	=	cross-sectional area of a pin
A_{ref}	=	reference area, at height upon which the wind pressure acts
A_s	=	tensile stress area of a bolt
	=	web stiffener area plus an effective length of the web
	=	area of a stiffener or stiffeners in contact with a flange
	=	area of an intermediate web stiffener
A_w	=	gross sectional area of a web
	=	effective area of a web
A_{ws}	=	effective cross-section area of web-stiffener
a	=	distance, dimension
	=	intermediate value for minimum area of an intermediate stiffener
	=	torsion bending constant
	=	acceleration coefficient (earthquake actions)
	=	centroid spacing of longitudinal elements in laced/battened compression members
a, b, c	=	dimensions
a_e	=	minimum distance from the edge of a hole to the edge of a ply measured in the direction of the component of a force plus half the bolt diameter
a_{e2}	=	distance from the centre of a hole to the edge of the ply
ABCB	=	Australian Building Codes Board
AISC	=	Australian Institute of Steel Construction (now ASI)
AISC(US)	=	American Institute of Steel Construction
AS	=	Australian Standard
ASI	=	Australian Steel Institute (previously AISC)
AS/NZS	=	joint Australian & New Zealand Standard
B	=	width of section
	=	reaction at point B
b	=	width
	=	lesser dimension of a web panel
	=	long side of a plate element (i.e. $b \times t$)
	=	intermediate value for minimum area of an intermediate stiffener
	=	distance, dimension
b_b, b_{bf}, b_{bw}	=	bearing widths as defined in Chapter 5 (Section 5.8)

b_{cw} = clear element width of an element for compression section capacity calculations
= clear width of an element outstand from the face of a supporting plate element
= clear width of a supported element between faces of supporting plate elements

b_d = bearing width as defined in Chapter 5 (Section 5.8), equals b_o

b_e = effective width of a plate element

b_{eff} = effective width of a plate element

b_{ei} = effective width of the i-th plate element of a section

b_{es} = stiffener outstand from the face of a web

b_f = width of a flange

b_{fo} = flange restraint factor for buckling capacity of a web with intermediate stiffeners

b_i = element width of the i-th plate element of a section

b_o = bearing widths as defined in Chapter 5 (Section 5.8), equals b_d

b_{rc} = dimension of web thickness plus flange-web fillet radius

b_s = stiff bearing length

b_1, b_2 = batten width for laced and battened members
= widths

b_5 = flat width of web (RHS/SHS)

BCA = Building Code of Australia

C = earthquake design coefficient
= total bearing force in the compression region of a bolt group loaded out-of-plane
= reaction at point C
= length from neutral axis to outer fibre
= vector sum of non-permanent actions (e.g. imposed, wind, etc.)

C_{dyn} = wind dynamic response factor

C_{fig} = aerodynamic shape factor – internal and external wind pressures

C_1, C_2 = corner fillet weld lengths for battens

C_2 = lateral distance between centroids of the welds or fasteners on battens

c = distance, dimension
= shear force dimension for laced compression members
= distance from centroid to edge

c_1, c_2 = distances from centroid to edge

c_k = intermediate value used to calculate minimum area of an intermediate stiffener

c_m = factor for unequal moments

c_{mx}, c_{my} = c_m for bending/buckling about x- and y-axis

c_z = element slenderness ratio used to calculate Z_e for Non-compact sections

c_2, c_3 = intermediate values for evaluating the higher tier of M_{ix}

c_3 = intermediate variable for shear-storey calculation method of δ_s

CF = cold-formed (for hollow sections)

CFW = Continuous Fillet Weld

CG = centre of gravity

CHS = Circular Hollow Section (to AS/NZS 1163)

CIDECT = Comité International pour le Développement et l'Étude de la Construction Tubulaire (or International Committee for the Development and Study of Tubular Structures)

CPBW = Complete Penetration Butt Weld

D = depth of section
= reaction at point D
= depth (outside)

D_0 = depth (inside)

D_1 = width of 'dog-bone' form pin connection

D_2, D_3, D_4 = dimensions of 'dog-bone' form pin connection

D_3, D_5, D_{5r} = dimensions of Flush form pin connection

D_5 = width of Flush form pin connection

d = depth of a section
= diameter of circle
= maximum cross-sectional dimension of a member
= shear force dimension for laced compression members

d_b = lateral distance between centroids of the welds or fasteners on battens

d_c = depth of columns

d_d = height of any holes in a web

d_e, d_{e1}, d_{e2} = effective outside diameter of a CHS

d_f = diameter of a fastener (bolt or pin)
= distance between flange centroids

d_i = inside diameter of a CHS

d_n = web slenderness ratio for plastic design
d_o = outside diameter of a CHS
d_p = clear transverse dimension of a web panel
= depth of deepest web panel in a length
= pin diameter
d_r = reaction eccentricity from simply supported beams
d_s = length of a weld segment in a weld group
d_1 = clear depth between flanges ignoring fillets or welds
d_2 = twice the clear distance from the neutral axis to the compression flange
= distance from column centre line to column face adjacent to beam connection
= distance between flange centroids
d_5 = flat width of web (for RHS/SHS bearing loads)
DC = Detail Category (for fatigue design Section 12 of AS 4100)
DCT = Design Capacity Tables
DTI = Direct Tension Indication device
DTT = Design Throat Thickness (for welds)
dia = diameter
E = Young's modulus of elasticity, (200×10^3 MPa in AS 4100 and 205×10^3 MPa in NZS 3404)
= nominal action effect
= reaction at point E
E_d = design action effect (e.g. axial force, shear force, bending moment)
e = eccentricity
= distance between end plate and a load-bearing stiffener
= eccentricity of F_p^* from the web
e_c = bearing force eccentricity on column cap plate
e_d = edge distance
e_i = distance between element centroid and effective section centroid
e_{min} = minimum eccentricity
e_x, e_y, e_z = eccentricity
EA = Equal Angle (AS/NZS 3679.1)
ERW = electric resistance welding
E41XX/W40X = lower strength weld consumable type
E48XX/W50X = higher strength weld consumable type
F = full restraint at section for flexural-torsional buckling
F = action in general, force or load
= design wind force
F_x = force in the x-axis direction
F_y = force in the y-axis direction
= yield stress (commonly described as f_y)
F^* = total design load on a member between supports
F_h^* = design horizontal force acting on each flange from warping
F_n^* = design force normal to a web panel
F_p^* = design eccentric force parallel to a web panel
f = stress range (for fatigue)
f_{at} = tensile stress in a rod under catenary action
f_c = corrected fatigue strength for thickness of material
f_f = uncorrected fatigue strength
f_{max} = maximum stress in a stress range (fatigue)
f_{min} = minimum stress in a stress range (fatigue)
f_{rn} = detail category reference fatigue strength at n_r cycles – normal stress
f_{rnc} = corrected detail category reference fatigue strength – normal stress
f_{rs} = detail category reference fatigue strength at n_r cycles – shear stress
f_{rsc} = corrected detail category reference fatigue strength – shear stress
f_u = tensile strength used in design
f_{uf} = minimum tensile strength of a bolt
f_{ui} = design tensile strength of a pin member/connection eye plate/gusset element i
f_{um} = design tensile strength of a tie member section
f_{up} = design tensile strength of a ply
f_{us} = design tensile strength of a splice plate section
f_{uw} = nominal tensile strength of weld metal
f_v = shear stress, tangential shear stress (torsion)
f_y = yield stress used in design
f_{yc} = design yield stress of a column

	=	design yield stress for the compression region of a plastic hinge
f_{yf}	=	design yield stress of a flange element
f_{yi}	=	design yield stress of the *i*-th plate element of a section
	=	design yield stress of a pin member/connection eye plate/gusset element *i*
f_{ym}	=	design yield stress of tie member section
f_{yp}	=	yield stress of a pin used in design
f_{ys}	=	yield stress of a stiffener used in design
	=	design yield stress of splice plate section
f_{yt}	=	design yield stress for the tension region of a plastic hinge
f_{yw}	=	design yield stress of a web element
f_3	=	detail category fatigue strength at constant amplitude fatigue limit
f_{3c}	=	corrected detail category fatigue strength at constant amplitude fatigue limit
f_5	=	detail category fatigue strength at cut-off limit
f_{5c}	=	corrected detail category fatigue strength at cut-off limit
f^*	=	design stress range (for fatigue)
f_i^*, f_j^*, f_k^*	=	design stress range for loading event *i, j, k*
f_n^*	=	design normal stress
f_s^*	=	design shear stress
f_{va}^*	=	average design shear stress in a web
f_{vm}^*	=	maximum design shear stress in a web
FCAW	=	Flux Cored Arc Welding
FLT	=	Flats or Flat section (to AS/NZS 3679.1)
FW	=	fillet weld
G	=	shear modulus of elasticity, 80×10^3 MPa
	=	nominal dead load (AS 1170.1) or nominal permanent action (AS/NZS 1170)
G_g	=	gravity load on the structure (earthquake actions)
G^*	=	design dead load/permanent action, G
g	=	gravity constant (9.81 m/s^2)
	=	gap between plies required for pin connections
GAA	=	Galvanizers Association of Australia
GANZ	=	Galvanizing Association of NZ
GMAW	=	Gas Metal Arc Welding
GP	=	General Purpose weld category/quality
H	=	floor-to-floor/ceiling/roof height
	=	column height
H_Q	=	imposed/live horizontal/transverse load
H_T	=	factor for higher tier combined tension and bending check on M_{ry}
H^*	=	design horizontal load/action
h	=	height of column/building
	=	distance between flange centroids
	=	height
h_s	=	storey height
HAZ	=	Heat-Affected Zone (for weldments)
HERA	=	Heavy Engineering Research Association (NZ)
HR	=	Hot-Rolled or Hot-Finished
HS	=	High Strength (as in high strength bolts)
HW	=	Heavily welded longitudinally
I	=	second moment of area of a cross-section
	=	structure importance factor
I_b	=	I for a beam
I_c	=	I for a column
I_{cy}	=	I of a compression flange about the section minor principal *y*-axis
I_{min}	=	I about the minor principal *y*-axis
I_n	=	I about rectangular *n*-axis (e.g. for angles)
I_{np}	=	product second moment of area about the *n*- and *p*-axis
I_p	=	polar second moment of area of a bolt/weld group
	=	I about rectangular *p*-axis (e.g. for angles)
I_s	=	I of a pair of stiffeners or a single stiffener
I_w	=	warping constant for a cross-section
I_{wp}	=	polar second moment of area of a weld group
I_{ws}	=	I of web-stiffener taken about axis parallel to the web
I_{wx}	=	second moment of area about the *x*-axis of a weld group
I_x	=	I about the cross-section major principal *x*-axis
I_{xi}	=	I_x for element/member *i*
I_{xy}	=	product second moment of area about the *x*- and *y*-axis (=0)
I_y	=	I about the cross-section minor principal *y*-axis

I_{yi} = I_y for element/member i
i = integer, number
ICR = Instantaneous Centre of Rotation (for bolt and weld groups)
IISI = International Iron & Steel Institute (now Worldsteel Association
IPBW = Incomplete Penetration Butt Weld
ISO = International Organization for Standardisation
J = torsion constant for a cross-section
j = integer, number
= multiple of concentrated load to UDL
K_p = plastic moment coefficient for portal frames
K_1 = deflection correction coefficient
k = integer, number
k_c = dimension of flange thickness plus flange-web fillet radius
k_e = member effective length factor
k_{ex} = member effective length factor between restraints for column buckling about the x-axis
k_{ey} = member effective length factor between restraints for column buckling about the y-axis
k_f = form factor for members subject to axial compression
k_h = factor for different hole types
k_l = effective length factor for load height
k_p = bearing factor for pin rotation
k_r = effective length factor for restraint against lateral rotation
= reduction factor for the length of a bolted or welded lap splice connection
k_s = ratio used to calculate α_p and α_{pm} for RHS/SHS under bearing loads
= ratio used to calculate minimum area of an intermediate stiffener
k_t = effective length factor for twist restraint
= correction factor for distribution of forces in a tension member
k_{ti} = k_t for a pin member/connection eye plate/gusset element i
k_v = ratio of RHS/SHS flat width of web (d_5) to thickness (t)
k_y = yield stress normalisation factor
k_1, k_2 = factors for warping torsion
L = lateral restraint at section for flexural-torsional buckling
L = see relevant definitions for l (AS 4100 uses l instead of L though NZS 3404 use the latter symbol)
L_{eff} = effective buckling length
L_1 = load dispersion width in column web at the k_c dimension from the flange face
= the equivalent definition of l in permissible design Standards
l = span
= member length
= length of beam segment or sub-segment
l_b = length of a beam member
l_c = length of a column member
l_e = effective length of a compression member
= effective length of a laterally and/or torsionally restrained flexural member
l_{ex} = member effective length between restraints for column buckling about the x-axis
l_{ey} = member effective length between restraints for column buckling about the y-axis
l_f = spacing of F, P or L restraints along a beam for continuously lateral restraint
l_s = segment length
l_w = welded lap connection length
= total length of weld in a weld group
l_{wf} = perimeter length of each flange fillet weld
l_{ws} = effective length of the cross-section area on each side of a stiffener for column action
l_{ww} = perimeter length of web fillet welds
l_x = member actual length between restraints for column buckling about the x-axis
l_y = member actual length between restraints for column buckling about the y-axis
l_z = distance between partial or full torsional restraints
l_1, l_2 etc. = lengths
l/r = member slenderness ratio based on actual length
l_e/r = member slenderness ratio based on effective length
l_e/r_s = slenderness ratio of web-stiffener combination
l_e/r_y = load-bearing stiffener slenderness ratio

l_{ws}/r_{ws} = web-stiffener compression member slenderness ratio

LODMAT = Lowest One-Day Mean Ambient Temperature

LR = lateral restraint

LW = Lightly welded longitudinally

L0 = guaranteed impact properties of the steel from a low temperature impact test at 0°C

L15 = guaranteed impact properties of the steel from a low temperature impact test at −15°C

M = nominal bending moment

M_B = moment at point B

M_b = nominal member moment capacity

= moment on a batten in a battened compression member

M_{br} = M_b from DCT (ASI [2009a], OneSteel [2012b]) with α_m = 1.0

M_{bx} = M_b about major principal x-axis

M_{bxo} = M_{bx} for a uniform distribution of moment

M_{b1} = M_b from DCT (ASI [2009a], OneSteel [2012b]) with α_m = 1.0

M_C = moment at point C

M_{cx} = lesser of M_{ix} and M_{ox}

M_D = moment at point D

M_d = wind direction multiplier

M_{fy} = maximum flange moment (from twisting)

M_i = nominal in-plane member moment capacity

M_{ix} = M_i about major principal x-axis

M_{iy} = M_i about minor principal y-axis

M_m = same as M_{max}

M_{max} = maximum moment

M_{m1}, M_{m2} = same as M_{max}

M_o = reference elastic buckling moment for a member subject to bending

= nominal out-of-plane member moment capacity

M_{ox} = nominal out-of-plane member moment capacity about major principal x-axis

= maximum depth of overall envelope in a bending moment diagram

M_p = fully plastic moment

= nominal moment capacity of a pin

M_{pr} = nominal plastic moment capacity reduced for axial force

M_{prx} = M_{pr} about x-axis

M_{pry} = M_{pr} about y-axis

M_r = corrected $(\phi M_b)_{DCT}$ with α_m

M_{rx} = M_s about major principal x-axis reduced by axial force

M_{ry} = M_s about minor principal y-axis reduced by axial force

M_s = nominal section moment capacity

= wind shielding multiplier – upwind building effect

M_{sx} = M_s about x-axis

M_{sy} = M_s about y-axis

M_{syL} = M_{sy} causing compression in left-most outer-fibre (for bending about vertical y-axis)

M_{syR} = M_{sy} causing compression in right-most outer-fibre (for bending about vertical y-axis)

M_t = twisting moment

= wind topographical multiplier

M_{tx} = lesser of M_{rx} and M_{ox}

M_x = moment about x-axis

M_y = bending moment for outer-fibre yielding

M_z = twisting moment about z-axis

= nominal torsion capacity

$M_{z,cat}$ = wind multiplier for building height and terrain category

M_1, M_2 etc. = moment at point 1, 2 etc.

M_{1x}, M_{2x} = the smaller and larger end bending moments to evaluate β_m for the higher tier of M_{ix}

M^* = design bending moment

= design bending/twisting moment on a bolt/weld group

M_i^* = (resolved) design in-plane bending/twisting moment on a bolt/weld group

M_m^* = maximum calculated M^* along the length of a member or in a segment

= mid-span bending moment of a pin-ended tensile member due to self-weight w_s

M_{mx}^* = M^* about x-axis for the member/segment being considered

M_{mxt}^* = maximum M^* about x-axis at top of column

M_{my}^* = M^* about y-axis for the member/segment being considered

M_{myt}^* = maximum M^* about y-axis at top of column

M_o^* = design bending/twisting moment on a bolt/weld group
M_r^* = design moment reduced by α_m
M_s^* = serviceability design bending moment
M_x^* = M^* about x-axis
M_{xb}^* = M^* about x-axis at base
M_{xbr}^* = M^* about x-axis at brace
M_{xi}^* = M^* about x-axis for member i = 1, 2, 3 etc.
M_{xt}^* = maximum M^* about x-axis at top of column
$M_{x\_mid}^*$ = M^* about x-axis at mid-height beam
$M_{x\_top}^*$ = M^* about x-axis at top of column
M_y^* = M^* about y-axis
M_{yb}^* = M^* about y-axis at base
M_{ybr}^* = M^* about y-axis at brace
M_{yi}^* = M^* about y-axis for member i = 1, 2, 3 etc.
M_{yt}^* = maximum M^* about y-axis at top of column
$M_{y\_mid}^*$ = M^* about y-axis at mid-height beam
$M_{y\_top}^*$ = M^* about y-axis at top of column
M_z^* = M^* about z-axis
= design twisting moment
M_1^* = design in-plane bending/twisting moment on a bolt/weld group
M_1^*, M_2^* = M^* at points/region 1, 2
M_2^*, M_3^*, M_4^* = M^* at quarter-, mid- and three-quarter points of a beam segment/sub-segment
MMAW = Manual Metal Arc Welding
N = nominal axial force (tension or compression)
= loaded beam overhang dimension
N_G = permanent/dead axial load
N_Q = imposed/live axial load
N_c = nominal member capacity in compression
N_{cx} = N_c for member buckling about major principal x-axis
N_{cy} = N_c for member buckling about minor principal y-axis
N_{ol} = (classic) Euler buckling equation
N_{om} = elastic flexural buckling load of a member
N_{omb} = N_{om} for a braced member
N_{ombx} = N_{omb} buckling about the x-axis
N_{omby} = N_{omb} buckling about the y-axis
N_{oms} = N_{om} for a sway member
N_{oz} = nominal elastic torsional buckling capacity of a member
N_s = nominal section capacity of a compression member
= nominal section capacity for axial load
N_t = nominal section capacity in tension
N_{tf} = nominal tension capacity of a bolt
N_{ti} = minimum bolt tension at installation
= tension induced in a bolt during installation
N^* = design axial force, tensile or compressive
= member design axial load transmitted by the pin connection
N_f^* = segment flange force
N_r^* = design axial force in a restraining member
N_s^* = pin serviceability limit state bearing load
N_t^* = design axial tension force on a member
N_{tf}^* = design tensile force on a bolt
N_w^* = flange weld forces
n = number of (plate, etc.) elements in a section
= index for higher tier combined tension and bending check on M_{ry}
= number of bolts in a bolt group
= rectangular n-axis (e.g. for angles)
n_b = number of battens in a battened compression member
n_{ei} = number of effective interfaces
n_i, n_j, n_k = number of cycles of nominal loading event i, j, k
n_n = number of shear planes *with* threads in the shear plane – bolted connections
n_r = reference number of stress cycles
n_s = number of shear planes on a pin
n_{sc} = number of stress cycles
n_x = number of shear planes *without* threads in the shear plane – bolted connections
n_1 = number of bolts in the furthest bolt row in a bolt group loaded out-of-plane
n_5 = 5×10^6 stress cycles
NA = Not Applicable
= Neutral Axis

NCC = National Construction Code series (by the ABCB – the BCA is part of this)
NZ = New Zealand
NZS = NZ Standard
P = partial restraint at section for flexural-torsional buckling
P = concentrated load
P_G = concentrated permanent/dead load
P_Q = concentrated imposed/live load
P_{st} = diagonal stiffener load in column web
P_1, P_2 = concentrated load
= axial load
P^* = force/load
P_u^* = ultimate design concentrated load
p = rectangular p-axis (e.g. for angles)
PFC = Parallel Flange Channel (to AS/NZS 3679.1)
PLT = Plate (to AS/NZS 3678)
PSA = Period of structural adequacy (for fire resistance)
Q = nominal live load (AS 1170.1) or nominal imposed action (AS/NZS 1170)
= first moment of area
= unloaded beam overhang dimension
Q_1, Q_2 = prying reaction at end of bent plate
Q^* = design live load/imposed action
QA = Quality Assurance
R = nominal resistance
= return period
R_A = reaction at support A
R_b = nominal bearing capacity of a web
= lesser of R_{bb} and R_{by} for unstiffened webs
R_{bb} = nominal bearing buckling capacity
R_{bc} = nominal capacity of a column web from flange bearing
R_{by} = nominal bearing yield capacity
R_f = structural response factor (earthquake actions)
R_{G1}, R_{G2} = permanent/dead load reaction 1, 2 etc.
R_{Q1}, R_{Q2} = imposed/live load reaction 1, 2 etc.
R_{sb} = nominal buckling capacity of a stiffened web
R_{sy} = nominal yield capacity of a stiffened web
R_t = nominal column flange capacity at beam tension flange
R_u = nominal capacity
R^* = design bearing force
= design reaction
R_{bf}^* = flange load on column web for bearing (crushing)
R_x^*, R_y^* = design reaction about x- and y-axis
r = radius of gyration
= fillet radius
= corner radius
= radius (i.e. half-diameter)
r_{ext} = outside radius of a section
r_i = inside radius
r_{min} = radius of gyration about the minor principal y-axis
r_o = outside radius
r_s = radius of gyration of web-stiffener combination about the axis parallel to the web
r_{ws} = same as r_s
r_x = radius of gyration about major principal x-axis
r_y = radius of gyration about minor principal y-axis
RC = Reinforced Concrete
RHS = Rectangular Hollow Section (to AS/NZS 1163)
RND = Round Bar (to AS/NZS 3679.1)
S = plastic section modulus
= site factor (earthquake actions)
= spacing between lacing in laced compression members
S_b = flexural stiffness of beams at a joint
S_c = flexural stiffness of columns at a joint
S_{red} = reduced S due to presence of relatively large holes
S_x = plastic section
S_y = plastic section modulus about the y-axis
S^* = design action effect
s = spacing between stiffeners
= width of a web panel
= side dimension
= half perimeter of a triangle
s_b = batten spacing in battened and back-to-back compression members
s_{end} = web depth of end panel
s_p = staggered pitch, the distance measured parallel to the direction of force

s_g = gauge, perpendicular to the force, between consecutive centre-to-centre of holes
SAW = Submerged Arc Welding
SC = shear centre
SCNZ = Steel Construction New Zealand
SHS = Square Hollow Section (to AS/NZS 1163)
SP = Structural (or Special) Purpose weld category/quality
SQ = Square Bar (to AS/NZS 3679.1)
SR = stress relieved
SR = axial compression load ratio for plastic design of webs
SW = Stud Welding
S–N = fatigue strength versus number of cycles curve(s)
T = the structure period (earthquake actions)
= tension force
T_i = design tension force on bolt i loaded out-of-plane in a bolt group
= tension capacity of a pin member/connection eye plate/gusset element i
T_1, T_2 etc. = design tension forces on bolts loaded out-of-plane in a bolt group
t = thickness
t_e = external ply thickness of a pin connection
t_f = thickness of a flange
t_{fb} = thickness of beam flange
t_{fc} = thickness of a column flange
t_g = gusset thickness for back-to-back tension members
t_i = element thickness of the i-th plate element of a section
= thickness of a pin member/connection eye plate/gusset element i
= internal ply thickness of a pin connection
t_p = thickness of ply/plate
t_s = thickness of a stiffener
t_t = design throat thickness (DTT) of a weld
t_w = thickness of a web
= size of a fillet weld (leg length)
t_{wc} = thickness of column web
TFB = Taper Flange Beam (to AS/NZS 3679.1)
TFC = Taper Flange Channel (to AS/NZS 3679.1)
U = unrestrained at section for flexural-torsional buckling
UA = Unequal Angle (to AS/NZS 3679.1)
UB = Universal Beam (to AS/NZS 3679.1)
UC = Universal Column (to AS/NZS 3679.1)
UDL = Uniformly Distributed Load
UNO = Unless Noted Otherwise
UTS = Ultimate Tensile Strength
V = shear force, nominal shear force
= base shear force (earthquake actions)
V_R = gust speed applicable to the region for an annual probability of exceedance of $1/R$
V_b = nominal shear buckling capacity of a unstiffened/stiffened web
= nominal column web capacity in shear yielding and shear buckling
= nominal bearing capacity of a ply or a pin
= shear force on a batten in a battened compression member
V_{bi} = design bearing capacity *on* a pin from ply i
$V_{des,\theta}$ = design wind speed = maximum value of $V_{sit,\beta}$
V_f = nominal shear capacity of a bolt or pin – strength limit state
V_p = nominal ply tearout capacity from bolt bearing
V_{sf} = nominal shear capacity of a bolt – serviceability limit state
$V_{sit,\beta}$ = site wind speed
V_u = nominal shear capacity of a web with uniform shear stress distribution
V_v = nominal shear capacity of a web/member
V_{vm} = nominal web shear capacity in the presence of bending moment
V_{vx} = nominal shear capacity of a web/member (shear along the y-axis)
V_{vy} = nominal shear capacity of a web/member (shear along the x-axis)
V_w = nominal shear yield capacity of a web
V^* = design shear force
= design horizontal storey shear force at lower column end
= design transverse shear force

V_b^* = design bearing force on a ply at a bolt or pin connection
V_{bi}^* = design bearing force *on* a pin from ply *i*
V_{bs}^* = serviceability limit state bearing force on a pin
V_c^* = design shear force present in the column of a beam-column joint
V_f^* = design shear force on a bolt or pin – strength limit state
V_o^* = eccentric out-of-plane design shear force on a bolt/weld group
V_{res}^* = vector resultant of forces acting on a bolt/weld in a bolt/weld group
V_{sf}^* = design shear force on a bolt – serviceability limit state
V_x^* = design shear force on a bolt/weld group in the *x*-direction
V_y^* = design shear force on a bolt/weld group in the *y*-direction
V_z^* = design shear force on a bolt/weld group in the *z*-direction
V_1^*, V_2^* = design shear force
v_{vf} = nominal capacity of flange fillet welds per unit length
v_{vw} = nominal capacity of web fillet welds per unit length
v_w = nominal capacity of a fillet weld per unit length
v^* = design shear force on a fillet weld per unit length
v_{res}^* = resultant shear force on a bolt/weld in a bolt group
v_w^* = design shear force on a fillet weld per unit length
= minimum shear force per unit length on the connection between stiffener to web
v_x^* = resultant shear force in the *x*-axis direction on a bolt/weld in a bolt/weld group
v_y^* = resultant shear force in the *y*-axis direction on a bolt/weld in a bolt/weld group
v_z^* = resultant shear force in the *z*-axis direction on a bolt/weld in a bolt/weld group
W = total load from nominal UDL
= wind action/load
W_G = total load from permanent/dead UDL
W_Q = total load from imposed/live UDL
W^* = total load from design UDL

w = nominal UDL
w_G = permanent/dead UDL
w_Q = imposed/live UDL
w_s = self-weight UDL
w^* = design UDL
w_{hu}^* = w_u^* in the horizontal direction
w_u^* = ultimate design UDL
WB = Welded Beam (to AS/NZS 3679.2)
WC = Welded Column (to AS/NZS 3679.2)
WTIA = Welding Technology Institute of Australia
x = major principal axis coordinate
x_L = coordinate of centroid
x_c = *x*-axis coordinate to the centroid from a datum point/axis
x_e = distance from neutral axis to extreme fibre in the *x*-direction
x_i = dimension *i* in the *x*-direction
x_m = distance along *x*-axis to maximum bending moment
x_n = distance along *x*-axis to bolt *n*
x_o = coordinate of shear centre
x_s = *x*-axis coordinate of a weld segment with length d_s in a weld group
Y_1, Y_2 = coordinates
y = minor principal axis coordinate
= beam deflection
y_c = *y*-axis coordinate to the centroid from a datum point/axis
= mid-length deflection (sag) of a tie rod under catenary action
= distance from NA to compression block centroid in a bolt group loaded out-of-plane
y_e = distance from neutral axis to extreme fibre in *y*-direction
y_{eB} = distance from neutral axis to bottom extreme fibre in *y*-direction
y_{eT} = distance from neutral axis to top extreme fibre in *y*-direction
y_h = distance from overall section centroid to centroid of A_h
y_i = dimension *i* in the *y*-direction
= lever arm to design tension force on bolt *i* loaded out-of-plane in a bolt group
y_m = maximum beam deflection
y_n = distance along *y*-axis to bolt *n*

y_s = y-axis coordinate of a weld segment with length d_s in a weld group
y_1 = distance from NA to furthest bolt row in a bolt group loaded out-of-plane
y_1, y_2 = distance from NA to flange centroids 1, 2
Z = elastic section modulus
Z_c = Z_e for a compact section
Z_e = effective section modulus
Z_{ex} = Z_e about the x-axis
Z_{ey} = Z_e about the y-axis
Z_{eyL} = Z_y to left-most outer-fibre (for bending about vertical axis)
Z_{eyR} = Z_y to right-most outer-fibre (for bending about vertical axis)
Z_r = reduced Z calculated by neglecting ineffective widths for bending
Z_{red} = reduced Z due to presence of relatively large holes
Z_x = Z about the x-axis
Z_{xB} = Z about the x-axis to bottom outer fibre
Z_{xT} = Z about the x-axis to top outer fibre
Z_y = Z about the y-axis
Z_{yL} = Z_y to left-most outer-fibre
Z_{yR} = Z_y to right-most outer-fibre
z = principal axis coordinate
$(\phi M_b)_{DCT}$ = ϕM_b from the DCT (ASI [2009a], OneSteel [2012b])
/S = snug-tight bolt installation
/TB = fully tensioned (tension controlled) - bearing bolt installation
/TF = fully tensioned (tension controlled) - friction bolt installation
α = lacing angle for laced compression members
= angle
α_a = compression member factor, as defined in Chapter 6 (Section 6.2.10)
α_b = compression member section constant
α_{bc} = moment modification factor for bending and compression
α_c = compression member slenderness reduction factor
α_{cx} = α_c for x-axis column buckling
α_{cy} = α_c for y-axis column buckling
α_d = tension field coefficient for web shear buckling
α_f = flange restraint factor for web shear buckling
α_m = moment modification factor (flexural members)
α_{mr} = reduced α_m
α_p = coefficient used to calculate R_{by} (RHS/SHS)
α_{pm} = coefficient used to calculate α_p
α_s = slenderness reduction factor (flexural members)
= inverse of the slope of the S–N curve for fatigue design
α_T = coefficient of thermal expansion for steel, 11.7×10^{-6} per degree Celsius
α_t = factor for torsional end restraint
α_v = shear buckling coefficient for a web
α_{vm} = shear-bending interaction reduction factor
α_1, α_2 = slenderness factors for torsion
β = angle
β_e = modifying factor to account for conditions at the far ends of beam members
β_m = ratio of smaller to larger bending moment at the ends of a member
= ratio of end moment to fixed end moment
β_{mx}, β_{my} = β_m about x- and y-axis
β_t = measure of elastic stiffness of torsional end restraint
β_{tf} = thickness correctness factor (fatigue design)
β_x = monosymmetry section constant
γ = load factor
= factor used in combined action checks
= factor for transverse stiffener arrangement
= angle
$\gamma, \gamma_1, \gamma_2$ = ratios of compression member stiffness to end restraint stiffness for beam-columns
γ_i = load factor for condition i
Δ = deflection (or sway deflection)
Δ_{ct} = mid-span deflection of a member loaded by transverse loads with end moments
Δ_{cw} = same as Δ_{ct} but without end moments
Δ_s = translational relative displacement between top-to-bottom storey height h_s
Δ_x = deflection along the x-axis from M_y^* for combined actions

Δ_y	=	deflection along the y-axis from M_x^* for combined actions
Δ_1	=	deflection at point 1
Δ_2	=	deflection at point 2
Δ_4	=	deflection at point 4
ΔA_f	=	area reduction in flange from holes
δ	=	moment amplification factor
	=	deflection (or deflection from member curvature)
δ_b	=	moment amplification factor for a braced member
δ_{bx}, δ_{by}	=	δ_b for bending/buckling about the x- and y-axis
δ_m	=	moment amplification factor, taken as the greater of δ_b and δ_s
δ_p	=	moment amplification factor for plastic design
δ_s	=	moment amplification factor for a sway member
ξ	=	compression member factor
η	=	compression member imperfection factor
θ	=	angle, twist angle
	=	angle between principal axis and axis under consideration
θ_1, θ_2	=	angle dimension for 'dog-bone' form pin connection
π	=	pi ($\approx$ 3.14159)
ψ_a	=	reduction factor of imposed loads due to area
ψ_c	=	combination factor for imposed actions
ψ_l	=	factor for determining quasi-permanent values (long-term) of actions
ψ_s	=	factor for determining frequent values (short-term) of actions
λ	=	member slenderness ratio
	=	elastic buckling load factor
λ_c	=	elastic buckling load factor
λ_e	=	plate element slenderness
λ_{ed}	=	plate element deformation slenderness limit
λ_{ef}	=	flange element slenderness
λ_{ei}	=	plate slenderness of the i-th plate element of a section
λ_{ep}	=	plate element plasticity slenderness limit
λ_{ey}	=	plate element yield slenderness limit
λ_{eyi}	=	plate yield slenderness limit of the i-th plate element of a section
λ_{ew}	=	web element slenderness
λ_{ms}	=	modified compression member slenderness
λ_n	=	modified compression member slenderness
λ_{nx}	=	λ_n for column buckling about the x-axis
λ_{ny}	=	λ_n for column buckling about the y-axis
λ_s	=	section slenderness
λ_{sp}	=	section plasticity slenderness limit
λ_{sy}	=	section yield slenderness limit
λ_1, λ_2	=	plate element slenderness values for evaluating the higher tier of M_{rx}
μ	=	slip factor
ν	=	Poisson's ratio, 0.25
Ω	=	safety/permissible stress factor
ρ	=	density
ρ_{air}	=	density of air for wind actions (1.2 kg/m^3)
σ_a	=	tension/stocky compression member axial stress
σ_{ac}	=	compression member axial stress
σ_b	=	bending stress
σ_{eq}	=	equivalent (limiting) stress
σ_x	=	normal stress in the x-axis direction
σ_y	=	normal stress in the y-axis direction
σ_z	=	normal stress in the z-axis direction
σ_1	=	normal stress in the principal 1-axis direction
σ_2	=	normal stress in the principal 2-axis direction
σ_3	=	normal stress in the principal 3-axis direction
τ	=	shear stress, torsional shear stress
τ_{av}	=	average shear stress
τ_{max}	=	maximum shear stress
τ_{vm}	=	maximum permissible shear stress
τ_{xy}	=	shear stress in the x-y plane
τ_{yz}	=	shear stress in the y-z plane
τ_{zx}	=	shear stress in the z-x plane
ϕ	=	capacity factor
	=	angle of twist over length L (torsion)
ϕ_t	=	angle of twist (torsion)
ϕ'	=	$d\theta/dz$
ϕ''	=	$d^2\theta/dz^2$
ϕ'''	=	$d^3\theta/dz^3$

Index

Picture Credits

Building Name
One Shelley Street
Location
Sydney, NSW, Australia

Credits
Architect: Fitzpatrick and Partners
Structural and Fire Engineering: Arup
Builder: Brookfield Multiplex

Structural System
The floor plate utilises composite steel construction (slab and beams) supported by one row of internal columns and the perimeter is a steel diagrid. The steel diagrid supports the floor plate, maximises the flexibility of internal floor space, assists in the building lateral stability and acts as a deep transfer truss distributing load to a few hard supports at the building perimeter.

Cover photographs
Di Quick